Praise for *Darwin's Doubt*

"*Darwin's Doubt* represents an opportunity for bridge-building, rather than dismissive polarization—bridges across cultural divides in great need of professional, respectful dialog—and bridges to span evolutionary gaps."

—Dr. George Church, professor of genetics at Harvard Medical School and author of *Regenesis*

"*Darwin's Doubt* is an intriguing exploration of one of the most remarkable periods in the evolutionary history of life—the rapid efflorescence of complex body plans written in the fossils of the Burgess Shale . . . No matter what convictions one holds about evolution, Darwinism, or intelligent design, *Darwin's Doubt* is a book that should be read, engaged, and discussed."

—Dr. Scott Turner, professor of biology at the State University of New York and author of *The Tinkerer's Accomplice: How Design Emerges from Life Itself*

"Meyer elegantly explains why the sudden appearance of animal forms in the Cambrian period gave Darwin pause. He also demonstrates, based on cutting-edge molecular biology, why explaining the origin of animals is now not just a problem of missing fossils, but an even greater engineering problem at the molecular level. With mathematical precision, he shows why the neo-Darwinian mechanism cannot produce the genetic information and novel proteins—or systems for regulating their expression—that are required to build new animals. An excellent book and a must read for anyone who wants to gain understanding of the very real—though often unreported—scientific challenges facing neo-Darwinism."

—Dr. Russell Carlson, professor of biochemistry and molecular biology and director of the Complex Carbohydrate Research Center at the University of Georgia

"*Darwin's Doubt* is by far the most up-to-date, accurate, and comprehensive review of the evidence from all relevant scientific fields that I have encountered in more than forty years of studying the Cambrian explosion. An engaging investigation of the origin of animal life and a compelling case for intelligent design."

—Dr. Wolf-Ekkehard Lönnig, senior scientist emeritus (biologist) at the Max Planck Institute for Plant Breeding Research, Cologne, Germany

"It is hard for us paleontologists, steeped as we are in a tradition of Darwinian analysis, to admit that neo-Darwinian explanations for the Cambrian explosion have failed miserably. New data acquired in recent years, instead of solving

Darwin's dilemma, have rather made it worse. Meyer describes the dimensions of the problem with clarity and precision. His book is a game changer for the study of evolution and points us in the right direction as we seek a new theory for the origin of animals."

—Dr. Mark McMenamin, paleontologist at Mt. Holyoke College and author of *The Emergence of Animals*

"With the publication of *On the Origin of Species* in 1859, Darwin acknowledged that there wasn't an adequate explanation for the pattern in the fossil record in which a wide diversity of animal life suddenly appeared in the Cambrian geological period. His doubt about the 'Cambrian explosion' centered on the wide range of body forms, the missing fossil intermediates and the lack of evidence for antecedents. Meyer's book examines the implications of the 'Cambrian explosion.' It is a fascinating story and analysis of Darwin's doubt about the fossil record and the debate that has ensued. It is a tour de force . . . This book is well informed, carefully researched, up-to-date, and powerfully argued. Its value is that it confronts Darwin's doubt and deals with the assumptions of Neo-Darwinism. This book is much needed and I recommend it to students of all levels, to professionals, and to laypeople."

—Dr. Norman C. Nevin, OBE, BSc, MD, FRCPath, FFPH, FRCPE, FRCP; professor emeritus in medical genetics, Queen's University, Belfast

"Another excellent book by Stephen Meyer. I particularly like his refutation of the concept of self-assembly of biological systems. The book explains the difference between specified complexity and order and shows that natural forces cannot generate the kind of complexity we see in living systems. I know from my personal work in the Systems Centre at Bristol University that complex systems do not create themselves but require an intelligent designer. Stephen Meyer has clearly listened to the arguments of those who are skeptical about intelligent design and has addressed them thoroughly. It is really important that Darwinists read this book carefully and give a response."

—Dr. Stuart Burgess, professor of design and nature, head of mechanical engineering at Bristol University

"I spend my life reading science books. I've ready many hundreds of them over the years, and in my judgment *Darwin's Doubt* is the best science book ever written. It is a magnificent work, a true masterpiece that will be read for hundreds of years."

—George Gilder, technologist, economist, and *New York Times* bestselling author of *Wealth and Poverty*

"Meyer writes beautifully. He marshals complex information as well as any writer I've read . . . This book—and his body of work—challenges scientism with real science and excites in me the hope that the origins-of-life debate will soon be largely free of the ideology that has long colored it . . . a wonderful, most compelling read."

—Dean Koontz, *New York Times* bestselling author

"Dr. Meyer makes it clear that these well-documented facts of paleontology pose a serious challenge to Darwin's theory, the view that has held sway in biology (and well beyond) for nearly 150 years. The issue on the table is not now, nor has it ever been, the fact of evolution (change over time); the issue has always been the mechanism of evolution—is it blind and undirected or is it under the control of an intelligence that had a goal in mind? That's the nub of the question, and in *Darwin's Doubt,* Stephen Meyer has masterfully laid out one of the most compelling lines of evidence for the latter."

—Dr. William S. Harris, professor at Sanford School of Medicine, University of South Dakota

"Dr. Meyer has written a comprehensive and up-to-date analysis on the massive scientific evidence revealing the total failure of the neo-Darwinian explanation for life's history. *Darwin's Doubt* is important, clearly written with sound arguments, excellent illustrations, and examples that make the topic easily understandable even for non-specialists . . . Randomness as a source for biological innovations is a present day paradigm and its supporters have too much at stake to give it up easily. The vague claims of Darwinian evolution refer to historical developments and as such are hard to prove. However, molecular biology has given us tools to experimentally test its claims. As so convincingly shown by Dr. Meyer the experimental evidence refers to intelligence—not randomness."

—Dr. Matti Leisola, professor of bioprocess engineering, Aalto University, Finland (emeritus); editor in chief of *Bio-Complexity*

"It is no secret among professionals that recent findings by developmental and molecular biologists are challenging current Darwinian theories of evolution. Meyer has condensed the research, made it accessible to the non-specialist and put it in the context of the debate over the origins of biological novelty. He makes a case for intelligent design as the only currently viable scientific theory for the origin of biological novelty, as found in the explosion of new species during the Cambrian geologic era. Meyer's challenge to the dominant paradigm of naturalism will no doubt be strongly resisted by those committed to a materialist world view, but provide food for refection for those who are searching for truth."

—Dr. Donald L. Ewert, molecular biologist, associate member (retired), Wistar Institute

"A truly remarkable book. [Meyer] presents evidence associated with the serious weaknesses of materialistic theories of biological evolution, and positive evidence for the theory of intelligent design . . . Meyer's attack is really against what is called 'macroevolution' (large scale population change). Michael Behe (in his *Edge of Evolution*) points out that there is abundant evidence for 'microevolution' (smaller population change), but there is a boundary at which the evidence for microevolution stops and evidence for macroevolution either doesn't exist, or any clues that do exist are beset with problems so serious that explanatory attempts boil down to 'just-so-stories.' This leaves macroevolution sitting atop a boundary (or wall) with an outlook no better than that of Humpty Dumpty."

—Dr. Mark C. Biedebach, professor emeritus, department of biological sciences, California State University, Long Beach

"A great book on the origin of animal life and crises of Darwin evolution; very clear, factual, comprehensive, logical, and informative. An enjoyable reading for both non-expert and expert."

—Dr. Change Tan, molecular biologist/developmental biologist, associate professor, University of Missouri-Columbia

"Darwinists keep two sets of books. The first set is the real record within the peer-reviewed literature that discusses why the mechanism of the origin of life and the mode and tempo speciation are more baffling today than they were two centuries ago. The second set of books is the popular literature that promotes to the public a soothing, fanciful narrative claiming that the grand history of life is fully explained with only minor but exciting details left to be filled in. Steven Meyer gives an insightful and thoughtful treatment to this state of affairs, auditing the second set of books using the data found in the first. Justice Louis Brandies taught us that 'Sunlight is the best disinfectant,' and Dr. Meyer lets the sun shine in."

—Dr. Stephen A. Batzer P.E., forensic engineer

"Buckle your seatbelts and brace yourself for tremors from the world of science. The evolution debate is about to undergo a major shakeup, and the world is beginning to listen in. Steve Meyer's book is a much anticipated bombshell that details the swarm of problems of Darwinian evolution that come from Cambrian fossils. It also clearly presents the case for intelligent design. Ask yourself: how often does a book of this kind receive a warm welcome from leading geneticists and paleontologists? Never, until now! *Darwin's Doubt* has been praised by Dr. George Church, a geneticist at Harvard University; by Dr. Mark McMenamin, a Cambrian fossil specialist at Mt. Holyoke College, and by Dr. Scott Turner, an evolutionary theorist at the State University of New York. And that's just the tip of the iceberg. Charles Darwin's own *Origin of Species* launched a revolution in

1859 whose scientific, cultural and spiritual effects are still with us. Now a new revolution is on the horizon."

—Dr. Tom Woodward, research professor, Trinity College, and author of *Darwin Strikes Back: Defending the Science of Intelligent Design*

"Stephen C. Meyer is brilliant and his latest book, *Darwin's Doubt,* is a must read."

—David Limbaugh, syndicated columnist and author

"Stephen Meyer's new book, *Darwin's Doubt,* is a fascinating and rigorous study demonstrating not only that biologists and paleontologists do not have an adequate explanation for the Cambrian Explosion, but that there is an alternative view that makes more sense. Those who are open to the possibility of intelligent design will find a treasure trove of supporting evidence for their view in this book. Those who oppose intelligent design owe it to themselves to read this book to understand Meyer's position and to grapple with his arguments."

—Dr. Richard Weikart, professor of history at California State University, Stanislaus and author of *From Darwin to Hitler*

"Meyer is a talented writer with an easygoing voice who has blended interesting history with clear explanations in what may come to be seen as a classic presentation of this most fundamental of all debates."

—Terry Scambray, *New Oxford Review*

DARWIN'S DOUBT

THE EXPLOSIVE ORIGIN OF ANIMAL LIFE AND THE CASE FOR INTELLIGENT DESIGN

STEPHEN C. MEYER

HarperOne
An Imprint of HarperCollins*Publishers*

Interior figures herein have been used by permission. For a listing of credits, please see pages 525–29, which constitute a continuation of this copyright page.

HarperCollins books may be purchased for educational, business, or sales promotional use. For information, please e-mail the Special Markets Department at SPsales@harpercollins.com.

HarperCollins website: http://www.harpercollins.com

HarperCollins®, ®, and HarperOne™ are trademarks of HarperCollins Publishers

FIRST HARPERCOLLINS PAPERBACK EDITION PUBLISHED IN 2014

Library of Congress Cataloging-in-Publication Data
Meyer, Stephen C.
Darwin's doubt : the explosive origin of animal life and the case for intelligent design / by Stephen C. Meyer. — First edition.
pages cm
Includes bibliographical references.
ISBN 978-0-06-207148-4
1. Life-Origin. 2. Evolution (Biology) 3. Darwin, Charles, 1809–1882. 4. Intelligent design (Teleology) I. Title.
QH325.M47 2013
576.8'2—dc23 2013004594

COVER PHOTOGRAPH:
Courtesy Chen, J. Y., G. Q. Zhou, M. Y. Zhu, and K. Y. Yeh. *The Chengjiang Biota—A Unique Window of the Cambrian Explosion*. Taichung, Taiwan: National Museum of Natural Science (1996), 149, Figure 186. *Used by permission.*

17 18 RRD(H) 10 9 8 7 6

CONTENTS

PROLOGUE

When people today hear the term "information revolution," they typically think of silicon chips and software code, cellular phones and supercomputers. They rarely think of tiny one-celled organisms or the rise of animal life. But, while writing these words in the summer of 2012, I am sitting at the end of a narrow medieval street in Cambridge, England, where more than half a century ago a far-reaching information revolution began in biology. This revolution was launched by an unlikely but now immortalized pair of scientists, Francis Crick and James Watson. Since my time as a Ph.D. student at Cambridge during the late 1980s, I have been fascinated by the way their discovery transformed our understanding of the nature of life. Indeed, since the 1950s, when Watson and Crick first illuminated the chemical structure and information-bearing properties of DNA, biologists have come to understand that living things, as much as high-tech devices, depend upon digital information—information that, in the case of life, is stored in a four-character chemical code embedded within the twisting figure of a double helix.

Because of the importance of information to living things, it has now become apparent that many distinct "information revolutions" have occurred in the history of life—not revolutions of human discovery or invention, but revolutions involving dramatic increases in the information present within the living world itself. Scientists now know that building a living organism requires information, and building a fundamentally new form of life from a simpler form of life requires an immense amount of *new* information. Thus, wherever the fossil record testifies to the origin of a completely new form of animal life—a pulse of biological innovation—it also testifies to a significant increase in the information content of the biosphere.

In 2009, I wrote a book called *Signature in the Cell* about the first "information revolution" in the history of life—the one that occurred with the origin of the first life on earth. My book described how discoveries in molecular biology during the 1950s and 1960s established that DNA contains information in digital form, with its four chemical subunits (called nucleotide bases) functioning like letters in a written language or symbols in a computer code. And molecular biology also revealed that cells employ a complex information-processing system to access and express the information stored in DNA as they use that information to build the proteins and protein machines that they need to stay alive. Scientists attempting to explain the origin of life must explain how both information-rich molecules and the cell's information-processing system arose.

The type of information present in living cells—that is, "specified" information in which the sequence of characters matters to the function of the sequence as a whole—has generated an acute mystery. No undirected physical or chemical process has demonstrated the capacity to produce specified information starting "from purely physical or chemical" precursors. For this reason, chemical evolutionary theories have failed to solve the mystery of the origin of first life—a claim that few mainstream evolutionary theorists now dispute.

In *Signature in the Cell,* I not only reported the well-known impasse in origin-of-life studies; I also made an affirmative case for the theory of intelligent design. Although we don't know of a *material* cause that generates functioning digital code from physical or chemical precursors, we do know—based upon our uniform and repeated experience—of one type of cause that has demonstrated the power to produce this type of information. That cause is *intelligence* or *mind.* As information theorist Henry Quastler observed, "The creation of information is habitually associated with conscious activity."[1] Whenever we find functional information—whether embedded in a radio signal, carved in a stone monument, etched on a magnetic disc, or produced by an origin-of-life scientist attempting to engineer a self-replicating molecule—and we trace that information back to its ultimate source, invariably we come to a mind, not merely a material process. For this reason, the discovery of digital information in even the simplest living cells indicates the prior activity of a designing intelligence at work in the origin of the first life.

My book proved controversial, but in an unexpected way. Though I clearly stated that I was writing about the origin of the *first* life and about theories of chemical evolution that attempt to explain it from simpler preexisting chemicals, many critics responded as if I had written another book altogether. Indeed, few attempted to refute my book's actual thesis that intelligent design provides the best explanation for the origin of the information necessary to produce the first life. Instead, most criticized the book as if it had presented a critique of the standard neo-Darwinian theories of *biological* evolution—theories that attempt to account for the origin of *new* forms of life from simpler *preexisting* forms of life. Thus, to refute my claim that no chemical evolutionary processes had demonstrated the power to explain the *ultimate* origin of information in the DNA (or RNA) necessary to produce life from simpler preexisting chemicals in the first place, many critics cited processes at work in *already living* organisms—in particular, the process of natural selection acting on random mutations in *already existing sections of information-rich DNA*. In other words, these critics cited an undirected process that acts on preexistent information-rich DNA to refute my argument about the failure of undirected material processes to produce information in DNA in the first place.[2]

For example, the eminent evolutionary biologist Francisco Ayala attempted to refute *Signature* by arguing that evidence from the DNA of humans and lower primates showed that the genomes of these organisms had arisen as the result of an unguided, rather than intelligently designed, process—even though my book did not address the question of human evolution or attempt to explain the origin of the human genome, and even though the process to which Ayala alluded clearly presupposed the existence of another information-rich genome in some hypothetical lower primate.[3]

Other discussions of the book cited the mammalian immune system as an example of the power of natural selection and mutation to generate new biological information, even though the mammalian immune system can only perform the marvels it does because its mammalian hosts are already alive, and even though the mammalian immune system depends upon an elaborately *preprogrammed* form of adaptive capacity rich in genetic information—one that arose long after the origin of the first life. Another critic steadfastly maintained that "Meyer's main argument" concerns "the inability of random mutation and selection *to add* information to [preexisting] DNA"[4] and attempted to refute the book's presumed critique of the neo-Darwinian mechanism of biological evolution accordingly.

I found this all a bit surreal, as if I had wandered into a lost chapter from a Kafka novel. *Signature in the Cell* simply did not critique the theory of biological evolution, nor did it ask whether mutation and selection can *add* new information to preexisting information-rich DNA. To imply otherwise, as many of my critics did, was simply to erect a straw man.

To those unfamiliar with the particular problems faced by scientists trying to explain the origin of life, it might not seem obvious why invoking natural selection does not help to explain the origin of the first life. After all, if natural selection and random mutations can generate new information in living organisms, why can it also not do so in a prebiotic environment? But the distinction between a biological and prebiotic context was crucially important to my argument. Natural selection assumes the existence of living organisms with a capacity to reproduce. Yet self-replication in all extant cells depends upon information-rich proteins and nucleic acids (DNA and RNA), and the origin of such information-rich molecules is precisely what origin-of-life research needs to explain. That's why Theodosius Dobzhansky, one of the founders of the modern neo-Darwinian synthesis, can state flatly, "Pre-biological natural selection is a contradiction in terms."[5] Or, as Nobel Prize–winning molecular biologist and origin-of-life researcher Christian de Duve explains, theories of *prebiotic* natural selection fail because they "need information which implies they have to presuppose what is to be explained in the first place."[6] Clearly, it is not sufficient to invoke a process that commences only once life has begun, or once biological information has arisen, to explain the origin of life or the origin of the information necessary to produce it.

All this notwithstanding, I have long been aware of strong reasons for doubting that mutation and selection can add *enough* new information of the right kind to account for large-scale, or "macroevolutionary," innovations—the various information revolutions that have occurred after the origin of life. For this reason, I have found it increasingly tedious to have to concede, if only for the sake of argument, the substance of claims I think likely to be false.

And so the repeated prodding of my critics has paid off. Even though I did not write the book or make the argument that many of my critics critiqued in responding to *Signature in the Cell,* I have decided to write that book. And this is that book.

Of course, it might have seemed a safer course to leave well enough alone. Many evolutionary biologists now grudgingly acknowledge that no chemical evolutionary theory has offered an adequate explanation of the origin of life or the ultimate origin of the information necessary to produce it. Why press a point you never made in the first place?

Because despite the widespread impression to the contrary—conveyed by textbooks, the popular media, and spokespersons for official science—the orthodox neo-Darwinian theory of biological evolution has reached an impasse nearly as acute as the one faced by chemical evolutionary theory. Leading figures in several subdisciplines of biology—cell biology, developmental biology, molecular biology, paleontology, and even evolutionary biology—now openly criticize key tenets of the modern version of Darwinian theory in the peer-reviewed technical literature. Since 1980, when Harvard paleontologist Stephen Jay Gould declared that neo-Darwinism "is effectively dead, despite its persistence as textbook orthodoxy,"[7] the weight of critical opinion in biology has grown steadily with each passing year.

A steady stream of technical articles and books have cast new doubt on the creative power of the mutation and selection mechanism.[8] So well established are these doubts that prominent evolutionary theorists must now periodically assure the public, as biologist Douglas Futuyma has done, that "just because we don't know *how* evolution occurred, does not justify doubt about *whether* it occurred."[9] Some leading evolutionary biologists, particularly those associated with a group of scientists known as the "Altenberg 16," are openly calling for a new theory of evolution because they doubt the creative power of the mutation and natural selection mechanism.[10]

The fundamental problem confronting neo-Darwinism, as with chemical evolutionary theory, is the problem of the origin of new biological information. Though neo-Darwinists often dismiss the problem of the origin of life as an isolated anomaly, leading theoreticians acknowledge that neo-Darwinism has also failed to explain the source of novel variation without which natural selection can do nothing—a problem equivalent to the problem of the origin of biological information. Indeed, the problem of the origin of information lies at the root of a host of other acknowledged problems in contemporary Darwinian theory—from the origin of new body plans to the origin of complex structures and systems such as wings, feathers, eyes, echolocation, blood clotting, molecular machines, the amniotic egg, skin, nervous systems, and multicellularity, to name just a few.

At the same time, classical examples illustrating the prowess of natural selection and random mutations do not involve the creation of novel genetic information. Many biology texts tell, for example, about the famous finches in the Galápagos Islands, whose beaks have varied in shape and length over time. They also recall how moth populations in England darkened and then lightened in response to varying levels of industrial pollution. Such episodes are often presented as conclusive evidence for the power of evolution. And indeed they are, depending on how one defines "evolution." That term has many meanings, and few biology textbooks distinguish between them. "Evolution" can refer to anything from trivial cyclical change within the limits of a preexisting gene pool to the creation of entirely novel genetic information and structure as the result of natural selection acting on random mutations. As a host of distinguished biologists have explained in recent technical papers, small-scale, or "microevolutionary," change cannot be extrapolated to explain large-scale, or "macroevolutionary," innovation.[11] For the most part, microevolutionary changes (such as variation in color or shape) merely utilize or express existing genetic information, while the macroevolutionary change necessary to assemble new organs or whole body plans requires the creation of entirely new information. As an increasing number of evolutionary biologists have noted, natural selection explains "only the survival of the fittest, not the arrival of the fittest."[12] The technical literature in biology is now replete with world-class biologists[13] routinely expressing doubts about various aspects of neo-Darwinian theory, and especially about its central tenet, namely, the alleged creative power of the natural selection and mutation mechanism.

Nevertheless, popular defenses of the theory continue apace, rarely if ever acknowledging the growing body of critical scientific opinion about the standing of the theory. Rarely has there been such a great disparity between the popular perception of a theory and its actual standing in the relevant peer-reviewed scientific literature. Today modern neo-Darwinism seems to enjoy almost universal acclaim among science journalists and bloggers, biology textbook writers, and other popular spokespersons for science as the great unifying theory of all biology. High-school and college textbooks present its tenets without qualification and do not acknowledge the existence of any significant scientific criticism of it. At the same time, official scientific organizations—such as the National Academy of Sciences (NAS), the American Association for the Advancement of Sciences (AAAS), and the National Association of Biology Teachers (NABT)—routinely assure the public that the contemporary version of Darwinian theory enjoys unequivocal support among qualified scientists and that the evidence of biology overwhelmingly supports the theory. For example, in 2006 the AAAS declared, "There is no significant controversy within the scientific community about the validity of the theory of evolution."[14] The media dutifully echo these pronouncements. As *New York Times* science writer Cornelia Dean asserted in 2007, "There is no credible scientific challenge to the theory of evolution as an explanation for the complexity and diversity of life on earth."[15]

The extent of the disparity between popular representations of the status of the theory and its actual status, as indicated in the peer-reviewed technical journals, came home to me with particular poignancy as I was preparing to testify before the Texas State Board of Education in 2009. At the time the board was considering the adoption of a provision in its science education standards that would encourage teachers to inform students of both the strengths and weaknesses of scientific theories. This provision had become a political hot potato after several groups asserted that "teaching strengths and weaknesses" were code words for biblical creationism or for removing the teaching of the theory of evolution from the curriculum. Nevertheless, after defenders of the provision insisted that it neither sanctioned teaching creationism nor censored evolutionary theory, opponents of the provision shifted their ground. They attacked the provision by insisting that there was no need to consider weaknesses in modern evolutionary theory because, as Eugenie Scott, spokeswoman for the National Center for Science Education, insisted in *The Dallas Morning News*, "There are no weaknesses in the theory of evolution."[16]

At the same time, I was preparing a binder of one hundred peer-reviewed scientific articles in which biologists described significant problems with the theory—a binder later presented to the board during my testimony. So I knew—unequivocally—that Dr. Scott was misrepresenting the status of scientific opinion about the theory in the relevant scientific literature. I also knew that her attempts to prevent students from hearing about significant problems with evolutionary theory would have likely made Charles Darwin himself uncomfortable. In *On the Origin of Species*, Darwin openly acknowledged important weaknesses in his theory and professed his own doubts about key aspects of it. Yet today's public defenders of a Darwin-only science curriculum apparently do not want these, or any other scientific doubts about contemporary Darwinian theory, reported to students.

This book addresses Darwin's most significant doubt and what has become of it. It examines an event during a remote period of geological history in which numerous animal forms appear to have arisen suddenly and without evolutionary precursors in the fossil record, a mysterious event commonly referred to as the "Cambrian explosion." As he acknowledged in the *Origin,* Darwin viewed this event as a troubling anomaly—one that he hoped future fossil discoveries would eventually eliminate.

The book is divided into three main parts. Part One, "The Mystery of the Missing Fossils," describes the problem that first generated Darwin's doubt—the missing ancestors of the Cambrian animals in the earlier Precambrian fossil record—and then tells the story of the successive, but unsuccessful, attempts that biologists and paleontologists have made to resolve that mystery.

Part Two, "How to Build an Animal," explains why the discovery of the importance of information to living systems has made the mystery of the Cambrian explosion more acute. Biologists now know that the Cambrian explosion not only represents an explosion of new animal form and structure but also an explosion of information—that it was, indeed, one of the most significant "information revolutions" in the history of life. Part Two examines the problem of explaining how the unguided mechanism of natural selection and random mutations could have produced the biological *information* necessary to build the Cambrian animal forms. This group of chapters explains why so many leading biologists now doubt the creative power of the neo-Darwinian mechanism and it presents four rigorous critiques of the mechanism based on recent biological research.

Part Three, "After Darwin, What?" evaluates more current evolutionary theories to see if any of them explain the origin of form and information more satisfactorily than standard neo-Darwinism does. Part Three also presents and assesses the theory of intelligent design as a possible solution to the Cambrian mystery. A concluding chapter discusses the implications of the debate about design in biology for the larger philosophical questions that animate human existence. As the story of the book unfolds, it will become apparent that a seemingly isolated anomaly that Darwin acknowledged almost in passing has grown to become illustrative of a fundamental problem for all of evolutionary biology: the problem of the origin of biological form and information.

To see where that problem came from and why it has generated a crisis in evolutionary biology, we need to begin at the beginning: with Darwin's own doubt, with the fossil evidence that elicited it, and with a clash between a pair of celebrated Victorian naturalists—the famed Harvard paleontologist Louis Agassiz and Charles Darwin himself.

PART ONE

THE MYSTERY OF THE MISSING FOSSILS

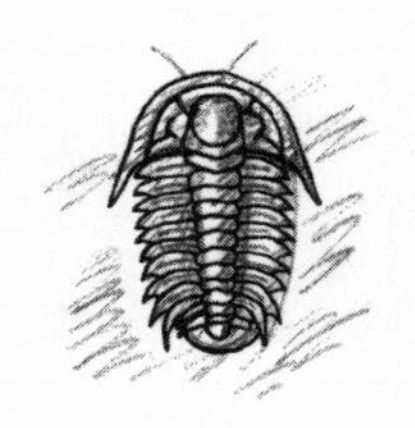

1

DARWIN'S NEMESIS

When Charles Darwin finished his famous book, he thought that he had explained every clue but one.

By anyone's measure, *On the Origin of Species* was a singular achievement. Like a great Gothic cathedral, the ambitious work integrated many disparate elements into a grand synthesis, explaining phenomena in fields as diverse as comparative anatomy, paleontology, embryology, and biogeography. At the same time, it was impressive for its simplicity. Darwin's *Origin* explained many classes of biological evidence with just two central organizing ideas. The twin pillars of his theory were the ideas of *universal common ancestry* and *natural selection.*

The first of these pillars, universal common ancestry, represented Darwin's theory of the history of life. It asserted that all forms of life have ultimately descended from a *single common ancestor* somewhere in the distant past. In a famous passage at the end of the *Origin,* Darwin argued that "all the organic beings which have ever lived on this earth have descended from some one primordial form."[1] Darwin thought that this primordial form gradually developed into new forms of life, which in turn gradually developed into other forms of life, eventually producing, after many millions of generations, all the complex life we see in the present.

Biology textbooks today usually depict this idea just as Darwin did, with a great branching tree. The trunk of Darwin's tree of life represents the first primordial organism. The limbs and branches of the tree represent the many new forms of life that developed from it (see Fig. 1.1). The

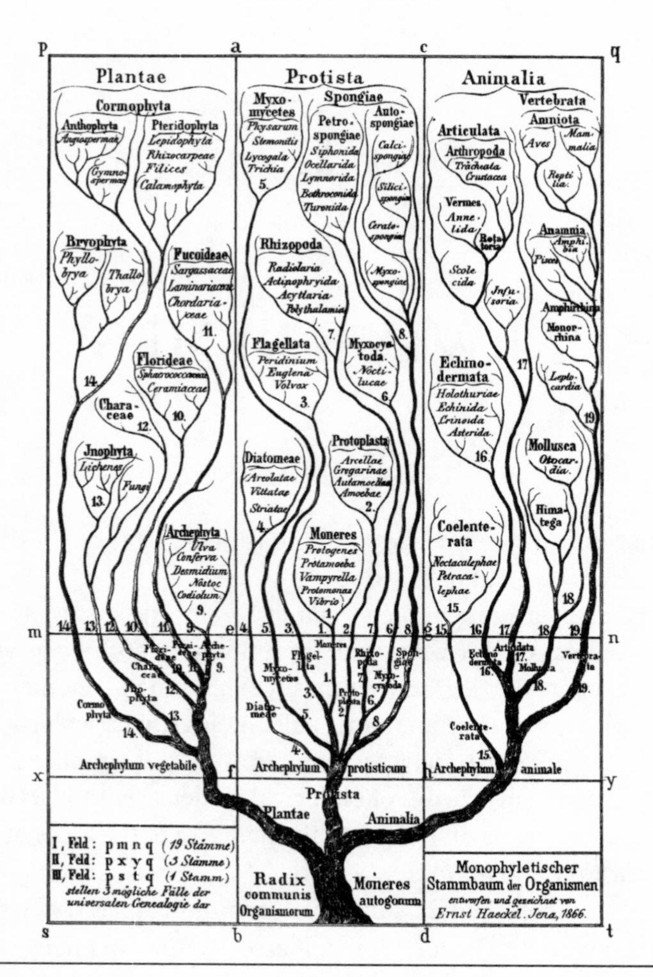

FIGURE 1.1
Darwin's evolutionary tree of life, as depicted by the nineteenth-century German evolutionary biologist Ernst Haeckel.

vertical axis on which the tree is plotted represents the arrow of time. The horizontal axis represents changes in biological form, or what biologists call "morphological distance."

Biologists often call Darwin's theory of the history of life "universal common descent" to indicate that *every* organism on earth arose from a single common ancestor by a process of "descent with modification." Darwin argued that this idea best explained a variety of biological evidences: the succession of fossil forms, the geographical distribution of various species (such as Galápagos finches), and the anatomical and embryological similarities among otherwise highly distinct organisms.

The second pillar of Darwin's theory affirmed the creative power of a process he called *natural selection,* a process that acted on random variations in the traits or features of organisms and their offspring.[2] Whereas the theory of universal common descent postulated *a pattern* (the branching tree) to represent the history of life, Darwin's idea of natural selection referred to a *process* that he said could generate the change implied by his branching tree of life.

Darwin formulated the idea of natural selection by analogy to a well-known process, that of "artificial selection" or "selective breeding." Anyone in the nineteenth century familiar with the breeding of domestic animals—dogs, horses, sheep, or pigeons, for example—knew that human breeders could alter the features of domestic stock by allowing only animals with certain traits to breed. A sheepherder from the north of Scotland might breed for a woollier sheep to enhance its chances of survival in a cold northern climate (or to harvest more wool). To do so, he would choose only the woolliest males and woolliest ewes to breed. If generation after generation he continued to select and breed only the woolliest sheep among the resulting offspring, he would eventually produce a woollier breed of sheep. In such cases, "the key is man's power of accumulative selection," wrote Darwin. "Nature gives successive variations; man adds them up in certain directions useful to him."[3]

Darwin noted that pigeons have been coaxed into a dizzying variety of breeds: the carrier, with its elongated eyelids and a "wide gape of mouth"; the "short-faced tumbler," with its "beak in outline almost like that of a finch"; the common tumbler, with its penchant for flying in close formation and "tumbling in the air head over heels"; and, perhaps strangest of all, the pouter, with its elongated legs, wings, and body overshadowed by its "enormously developed crop, which it glories in inflating" for its astonished patrons.[4]

Of course, pigeon breeders achieved these startling metamorphoses by carefully sifting and selecting. But, as Darwin pointed out, nature also has a means of sifting: defective creatures are less likely to survive and reproduce, while those offspring with beneficial variations are more likely to survive, reproduce, and pass on their advantages to future generations. In the *Origin,* Darwin argued that this process, natural selection acting on random variations, could alter the features of organisms just as intelligent selection by human breeders can. Nature itself could play the role of the breeder.

Consider once more our flock of sheep. Imagine that instead of a human selecting the woolliest males and ewes to breed, a series of very cold winters ensures that all but the very woolliest sheep in a population die. Now again only very woolly sheep will remain to breed. If the cold winters continue over several generations, will the result not be the same as before? Won't the population of sheep eventually become discernibly woollier?

This was Darwin's great insight. Nature—in the form of environmental changes or other factors—could have the same effect on a population of organisms as the intentional decisions of an intelligent agent. Nature would favor the preservation of certain features over others—specifically, those that conferred a functional or survival advantage upon the organisms possessing them—causing the features of the population to change. And the resulting change will have been produced not by an intelligent breeder choosing a desirable trait or variation—not by "artificial selection"—but by a wholly natural process. What's more, Darwin concluded that this process of natural selection acting on randomly arising variations had been "the chief agent of change" in generating the great branching tree of life in all its variety.

On the Origin of Species seized the attention of the scientific community like a thunderclap. Darwin's analogy to artificial selection was powerful, his proposed mechanism of natural selection and random variation easily grasped, and his skill in dispensing with potential objections unrivalled. Moreover, the explanatory scope of his argument for universal common descent constituted something of a *tour de force.* By the close of the *Origin,* it seemed to many that Darwin had dispensed with every conceivable objection to his theory but one.

THE ANOMALY: DARWIN'S DOUBT

Despite the scope of his synthesis, there was one set of facts that troubled Darwin—something he conceded his theory couldn't adequately explain, at least at present. Darwin was puzzled by a pattern in the fossil record that seemed to document the geologically sudden appearance of animal life in a remote period of geologic history, a period that at first was commonly called the Silurian, but later came to be known as the Cambrian.

During this geological period, many new and anatomically sophisticated creatures appeared suddenly in the sedimentary layers of the geologic column without any evidence of simpler ancestral forms in the earlier layers below, in an event that paleontologists today call the Cambrian explosion. Darwin frankly described his concerns about this conundrum in the *Origin:* "The difficulty of understanding the absence of vast piles of fossiliferous strata, which on my theory were no doubt somewhere accumulated before the Silurian [i.e., Cambrian] epoch, is very great," he wrote. "I allude to the manner in which numbers of species of the same group suddenly appear in the lowest known fossiliferous rocks."[5] The sudden appearance of animals so early in the fossil record did not easily accord with Darwin's new theory of gradual evolutionary change, and there was one scientist who would not let him forget it.

THE ANTAGONIST

Swiss-born paleontologist Louis Agassiz, of Harvard University, was one of the best-trained scientists of his age, and he knew the fossil record better than any man alive. Hoping to enlist Agassiz as an ally, Darwin sent him a copy of *On the Origin of Species* and asked him to consider the argument with an open mind (see Fig. 1.2). One can almost see the great, aging naturalist receiving the unremarkable package from the postman,

FIGURE 1.2
Figure 1.2a (left): Louis Agassiz. *Figure 1.2b (right):* Charles Darwin.

unwrapping the small green volume that had stirred such a tempest on both sides of the Atlantic. Perhaps he retired to his study the better to concentrate, scrutinizing the book's prepossessing title, recalling what he had already heard about the work. He read the book with deep interest, making notes in the margin as he moved through it, but in the end his verdict would disappoint its author. Agassiz concluded that the fossil record, particularly the record of the explosion of Cambrian animal life, posed an insuperable difficulty for Darwin's theory.

THE TWO-PRONGED CHALLENGE

To see why, consider brachiopods and trilobites, two of the best-documented creatures in the Cambrian fossil record by 1859. The brachiopod (see Fig. 1.3), with its two shells, looks like a clam or an oyster, but is very different

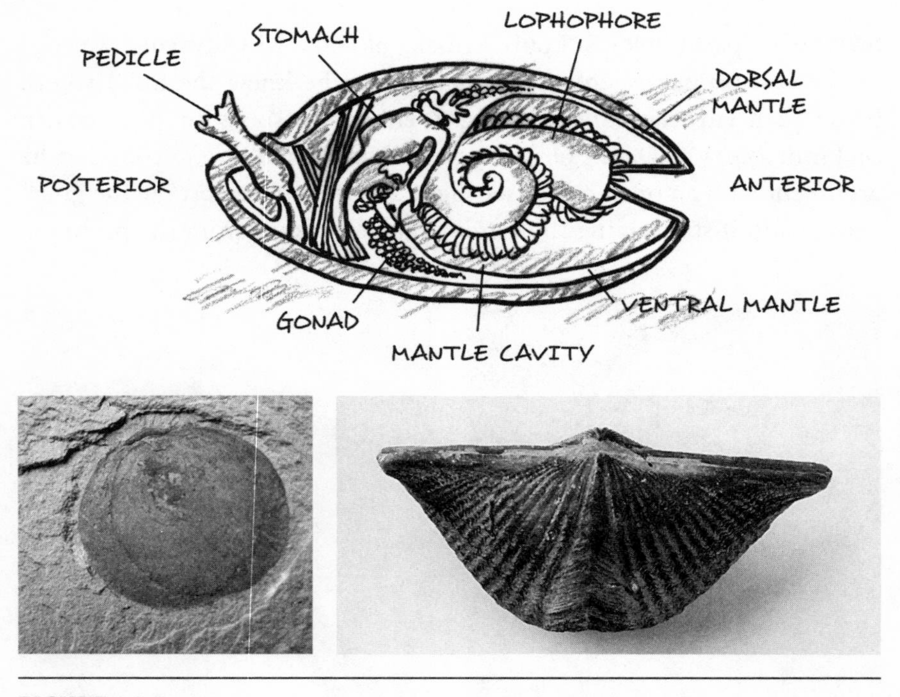

FIGURE 1.3
Figure 1.3a (top): Brachiopod internal anatomy. *Figure 1.3b (bottom, left):* Brachiopod fossil showing remains of internal structure. *Courtesy Paul Chien. Figure 1.3c (bottom, right):* Fossil showing exterior structure of brachiopod shell. *Courtesy Corbis.*

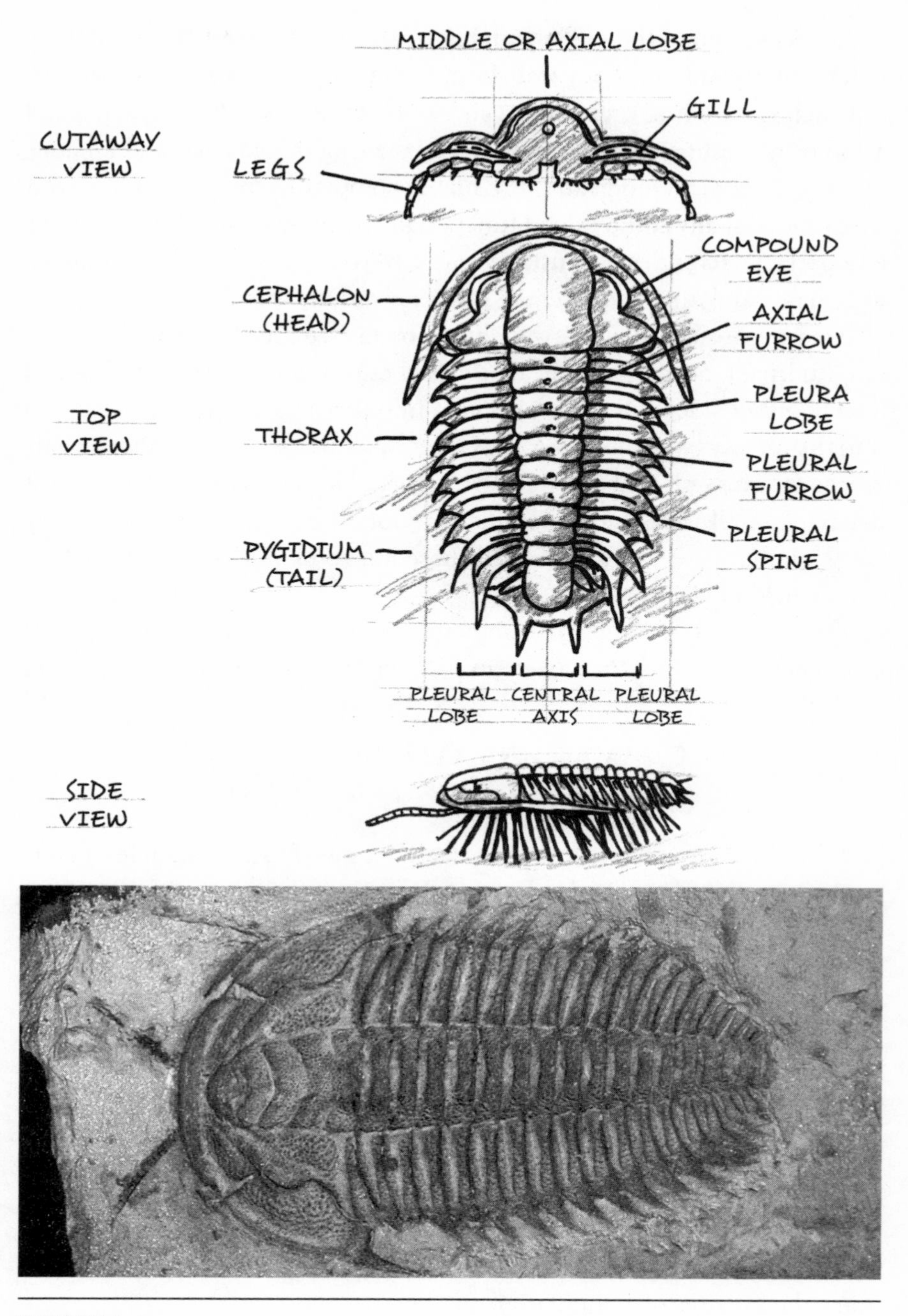

FIGURE 1.4

Figure 1.4a (top): Trilobite anatomy. *Figure 1.4b (bottom):* Trilobite fossil of the species *Kuanyangia pustulosa. Courtesy Illustra Media.*

inside. As shown in the accompanying figure, it possesses a gonad, mantle, mantle cavity, anterior body wall, body cavity, gut, and lophophore, the last of which is a feeding organ like a ring of tentacles, usually in the shape of a coil or horseshoe, with a mouth inside the ring of tentacles, and an anus outside. The brachiopod exhibits a highly complex overall body plan, with many individually complex and functionally integrated anatomical systems and parts. Its tentacles, for instance, are covered by cilia precisely arranged to generate and direct a current of water toward the mouth.[6]

Even more sophisticated was the trilobite (see Fig. 1.4), with its three longitudinal lobes across its head (a raised middle lobe and a flatter pleural lobe to either side) and a body divided into three parts—head, chest, and tail, the former two consisting of as many as thirty segments. It had a pair of legs for every pleural groove and another three pairs for the head. Most dramatic of all were the compound eyes found on even some of the very early trilobites—eyes that afforded these not so primitive animals a 360-degree field of vision.[7]

The abrupt appearance of such complex anatomical designs presented a challenge to each of the two main parts of Darwin's theory of evolution.

THE CAMBRIAN EXPLOSION AND THE ACTION OF NATURAL SELECTION

The Cambrian fossil evidence represented a significant challenge to Darwin's claim that natural selection had the capacity to produce novel forms of life. As Darwin described it, the ability of natural selection to produce significant biological change depends upon the presence of three distinct elements: (1) randomly arising variations, (2) the heritability of those variations, and (3) a competition for survival, resulting in differences in reproductive success among competing organisms.

According to Darwin, variations in traits arise *randomly.* Some variations (such as thicker fleece) might confer advantages in the *competition for survival* in particular environmental conditions. Those variations that are heritable and that impart functional or survival advantage will be preserved in the next generation. As nature "selects" these successful variations, the features of a population change.

Darwin conceded that the beneficial variations responsible for permanent change in species are both rare and necessarily modest. Major variations in forms, what later evolutionary biologists would term "macromutations," inevitably produce deformity and death. Only minor variations meet the test of viability and heritability.

It followed that, over human timescales, the benefits of this evolutionary mechanism would be difficult or impossible to spot. But given enough time, favorable variations would *gradually* accumulate and give rise to new species and, given more time, even fundamentally new groups of organisms and body designs. If artificial selection could conjure so many strange breeds from a wild strain in a few centuries, Darwin argued, imagine what natural selection could achieve over many millions of years. Even the origin of complex structures such as the mammalian eye—which seemed at first to present a significant challenge to his theory—could be explained if one postulated the existence of an initially simpler structure (such as a light sensitive spot) that could be gradually modified over long periods of time.

And that was the rub. Darwin's mechanism of natural selection and random variation necessarily required a lot of time to generate wholly novel organisms, creating a dilemma that Agassiz was keen to expose.

In an 1874 *Atlantic Monthly* essay titled "Evolution and the Permanence of Type," Agassiz explained his reasons for doubting the creative power of natural selection. Small-scale variations, he argued, had never produced a "specific difference" (i.e., a difference in species). Meanwhile, large-scale variations, whether achieved gradually or suddenly, inevitably resulted in sterility or death. As he put it, "It is a matter of fact that extreme variations finally degenerate or become sterile; like monstrosities they die out."[8]

Darwin himself insisted that the process of evolutionary change he envisioned must occur very gradually for the same reason. Thus, Darwin realized that building, for instance, a trilobite from single-celled organisms by natural selection operating on small, step-by-step variations would require countless transitional forms and failed biological experiments over vast stretches of geologic time. As University of Washington paleontologist Peter Ward would later explain, Darwin had very specific expectations for what paleontologists would find below the lowest known strata of animal fossils—in particular, "intervening strata showing fossils of increasing complexity until finally trilobites appeared."[9] As Darwin noted, "If my theory be true, it is indisputable that before the lowest Silurian [Cambrian] stratum was deposited, long periods elapsed, as long as, or probably far longer than, the whole interval from the Silurian age to the present day; and that during these vast, yet quite unknown, periods of time, the world swarmed with living creatures."[10]

The mechanism of natural selection necessarily had to work gradually on small incremental variations. And, indeed, the kinds of variations that Darwin actually observed and described in developing his analogy between natural and artificial selection were in every case minor. Only by selecting and accumulating minor variations over many generations were breeders able to produce the striking changes in the features of a breed, changes that were, nevertheless, extraordinarily modest compared to the radical differences in form between, say, Precambrian and Cambrian forms of life. At the end of the day, as Agassiz hastened to note, the pigeons Darwin cited in support of the creative power of artificial and, by analogy, natural selection were still pigeons. More significant changes to the form and anatomical structure of organisms would, by the logic of Darwin's mechanism, require untold millions of years, precisely what seemed unavailable in the case of the Cambrian explosion.

THE CAMBRIAN EXPLOSION AND THE TREE OF LIFE

The abrupt appearance of the Cambrian fauna also posed a separate but related difficulty for Darwin's picture of a continuously branching tree of life. To produce truly novel animal forms, the Darwinian mechanism would—by its own internal logic—require not only millions of years, but untold generations of ancestors. Thus, even the discovery of a handful of plausible intermediates allegedly linking a Precambrian ancestor to a Cambrian descendant wouldn't come close to fully documenting Darwin's picture of the history of life. If Darwin is right, Agassiz argued, then we should find not just one or a few missing links, but innumerable links shading almost imperceptibly from alleged ancestors to presumed descendants. Geologists, however, had found no such myriad of transitional forms leading to the Cambrian fauna. Instead, the stratigraphic column seemed to document the abrupt appearance of the earliest animals.

Agassiz thought the evidence of abrupt appearance, and the absence of ancestral forms in the Precambrian, refuted Darwin's theory.[11] Of these earlier forms, Agassiz asked, "Where are their fossilized remains?" He insisted that Darwin's picture of the history of life "contradict[ed] what the animal forms buried in the rocky strata of our earth tell us of their own introduction and succession upon the surface of the globe. Let us therefore hear them;—for, after all, their testimony is that of the eye-witness and the actor in the scene."[12]

MURCHISON, SEDGWICK, AND THE CAMBRIAN FOSSILS OF WALES

Darwin, for his part, responded with more than civility. Far from dismissing Agassiz, he conceded that his objection carried considerable force. Nor was Agassiz alone in pressing these concerns. Other leading naturalists thought the fossil evidence presented a significant obstacle to Darwin's theory. At the time, perhaps the best place to investigate the lowest known strata of fossils was Wales, and one of its leading experts was Roderick Impey Murchison, who named the earliest geologic period the Silurian after an ancient Welsh tribe. Five years before *On the Origin of Species,* he called attention to the sudden appearance of complex designs like the compound eyes of the first trilobites, creatures already thriving at the apparent dawn of animal life. For him, this discovery ruled out the idea that these creatures had evolved gradually from some primitive and relatively simple form: "The earliest signs of living things, announcing as they do a high complexity of organization, entirely exclude the hypothesis of a transmutation from lower to higher grades of being."[13]

The other pioneering explorer of Wales's rich fossil record, Adam Sedgwick, also thought that Darwin had leaped beyond the evidence, as he told him in a letter in the fall of 1859: "You have deserted—after a start in that tram-road of all solid physical truth—the true method of induction."[14] Sedgwick might have had in mind the same evidence the two men had studied together some twenty-eight years before when the Cambridge professor had brought Darwin along as his field assistant to explore, in the Upper Swansea Valley in northwestern Wales, the very strata that seemed to testify so powerfully to the sudden appearance of animal life. It was these strata that Sedgwick named after a Latinized English term for the country of Wales—"Cambria," a designation that eventually replaced "Silurian" as the name for the earliest strata of animal fossils.

Sedgwick emphasized that these Cambrian animal fossils appeared to pop out of nowhere into the geological column. But he also stressed what

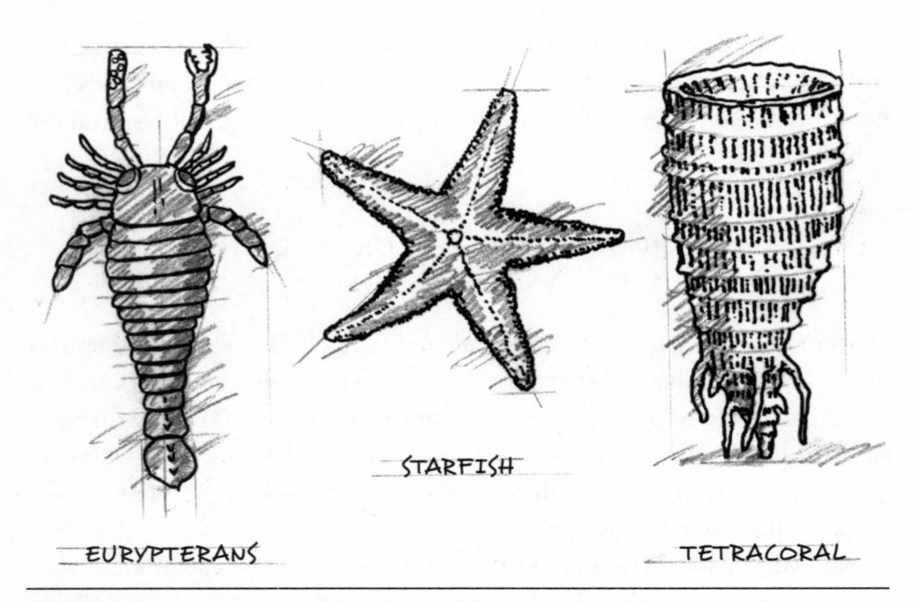

FIGURE 1.5
Three organisms that first appear in the Ordovician period: eurypterans (sea scorpions), starfish, and tetracoral.

he viewed as a broader reason to doubt Darwin's evolutionary model: the sudden appearance of the Cambrian animals was merely the most outstanding instance of a pattern of discontinuity that extends throughout the geologic column. Where in the Ordovician strata, for instance, are many of the families of the trilobites and brachiopods present in the Cambrian just below it?[15] These creatures along with numerous other types suddenly *disappear*. But just as suddenly one finds newcomers in the Ordovician strata like the eurypterans (sea scorpions), starfish, and tetracorals (see Fig. 1.5).[16] In a later Paleozoic period called the Devonian, the first amphibians (e.g., *Ichthyostega*) arise. Much later, many staples of the Paleozoic era (which encompasses the Cambrian, Ordovician, and four subsequent periods) suddenly go extinct in a period called the Permian.[17] Then, in the Triassic period that follows, completely novel animals such as turtles and dinosaurs emerge.[18] Such discontinuity, Sedgwick argued, is not the exception, but the rule.

DATING BY DISCONTINUITY

Already by Sedgwick's time, the various strata of fossils had proved so distinct one from another that geologists had come to use the sharp discontinuities between them as a key means for dating rocks. Originally, the best tool for determining the relative age of various strata was based on the notion of superposition. Put simply, unless there is a reason to believe otherwise, a geologist provisionally assumes that lower rocks were put down before the rocks above them. Now, contrary to a widespread caricature, no respected geologist, then or now, adopts this method uncritically. The most basic training in geology teaches that rock formations can be twisted, upended, even mixed pell-mell by a variety of phenomena. This is why geologists have always looked for other means to estimate the relative age of different strata.

In 1815, Englishman William Smith had hit upon just such an alternative means.[19] While studying the distinct fossil strata exposed during canal construction, Smith noted that so dissimilar are the fossil types among different major periods and so sharp and sudden the break between them, that geologists could use this as one method for determining the relative age of strata. Even when layers of geological strata are twisted and turned, the clear discontinuities between the various strata often allow geologists to discern the order in which they were deposited, particularly when there is a broad enough sampling of rich geological sites from the period under investigation to study and cross-reference. Although not without its pitfalls, this approach has become a standard dating technique, used in conjunction with superposition and other more recent radiometric dating methods.[20]

Indeed, it's difficult to overemphasize how central the approach is to modern historical geology. As Harvard paleontologist Stephen Jay Gould explains, it is the phenomenon of fossil succession that dictates the names of the major periods in the geological column (see Fig. 1.6). "We might take the history of modern multi-cellular life, about 600 million years, and divide this time into even and arbitrary units easily remembered as 1–12 or A-L, at 50 million years per unit," Gould writes. "But the earth scorns our simplifications, and becomes much more interesting in its derision. The history of life is not a continuum of development, but a record

ERAS	PERIODS	ALTERNATE PERIODS	EPOCHS (NORTH AMERICA)	DURATION (IN MILLIONS OF YEARS)	
CENOZOIC ERA	NEOGENE PERIOD	QUARTERNARY PERIOD	HOLOCENE EPOCH	23 M.Y.	2.6 M.Y.
			PLEISTOCENE EPOCH		
		TERTIARY PERIOD	PLIOCENE EPOCH		63.4 M.Y.
			MIOCENE EPOCH		
	PALEOGENE PERIOD		OLIGOCENE EPOCH	43 M.Y.	
			EOCENE EPOCH		
			PALEOCENE EPOCH		
			66 M.Y. AGO		
MESOZOIC ERA	CRETACEOUS PERIOD			79 M.Y.	
			145 M.Y. AGO		
	JURASSIC PERIOD			56.3 M.Y.	
			201.3 M.Y. AGO		
	TRIASSIC PERIOD			52.9 M.Y.	
			254.2 M.Y. AGO		
PALEOZOIC ERA	PERMIAN PERIOD			44.7 M.Y.	
			298.9 M.Y. AGO		
	CARBONIFEROUS PERIOD			60 M.Y.	
			358.9 M.Y. AGO		
	DEVONIAN PERIOD			60.3 M.Y.	
			419.2 M.Y. AGO		
	SILURIAN PERIOD			24.6 M.Y.	
			443.8 M.Y. AGO		
	ORDOVICIAN PERIOD			41.6 M.Y.	
			485.4 M.Y. AGO		
	CAMBRIAN PERIOD 530 MILLION YEARS AGO			55.6 M.Y.	
			541 M.Y. AGO		
NEOPROTEROZOIC ERA	EDIACARAN PERIOD			94 M.Y.	
			635 M.Y. AGO		
(PRECAMBRIAN TIME)	OTHER PRECAMBRIAN PERIODS			APPROX. 4,000 M.Y.	

FIGURE 1.6
The geological timescale.

punctuated by brief, sometimes geologically instantaneous, episodes of mass extinction and subsequent diversification."[21] The question that Darwin's early critics posed was this: How could he reconcile his theory of gradual evolution with a fossil record so discontinuous that it had given rise to the names of the major distinct periods of geological time, particularly when the first animal forms seemed to spring into existence during the Cambrian as if from nowhere?

A SOLUTION UNSEEN

Of course, Darwin was well aware of these problems. As he noted in the *Origin,* "The abrupt manner in which whole groups of species suddenly appear in certain formations has been urged by several paleontologists—for instance, by Agassiz, Pictet, and Sedgwick—as a fatal objection to the belief in the transmutation of species. If numerous species, belonging to the same genera or families, have really started into life all at once, the fact would be fatal to the theory of descent with slow modification through natural selection."[22] Darwin, however, proposed a possible solution. He suggested that the fossil record may be significantly incomplete: either the ancestral forms of the Cambrian animals were not fossilized or they hadn't been found yet. "I look at the natural geological record, as a history of the world imperfectly kept, and written in a changing dialect," Darwin wrote. "Of this history we possess the last volume alone, relating only to two or three countries. Of this volume, only here and there a short chapter has been preserved; and of each page, only here and there a few lines. . . . On this view, the difficulties above discussed are greatly diminished, or even disappear."[23]

Darwin himself was less than satisfied with this explanation.[24] Agassiz, for his part, would have none of it. "Both with Darwin and his followers, a great part of the argument is purely negative," he wrote. They "thus throw off the responsibility of proof. . . . However broken the geological record may be, there is a complete sequence in many parts of it, from which the character of the succession may be ascertained." On what basis did he make this claim? "Since the most exquisitely delicate structures, as well as embryonic phases of growth of the most perishable nature, have been preserved from very early deposits, we have no right to infer the disappearance of types because their absence disproves some favorite [i.e., Darwinian] theory."[25]

Though Darwin himself *was* less than enthusiastic about his response to Agassiz's objection, it seemed adequate to satisfy the needs of the moment. The overwhelming preponderance of evidence that Darwin had marshaled seemed to support his theory. In any case, many leading naturalists—Joseph Hooker, Thomas Huxley, Ernst Haeckel, and Asa Gray—all younger than Agassiz, quickly aligned themselves with his evolutionary line of thinking. True, some scientists, notably the Scottish engineering professor Fleeming Jenkin and (later) the English geneticist William Bateson, expressed persistent doubts about the efficacy of natural selection. But despite the views of some weighty scientific critics, Darwin's revolutionary theory won increasingly wide support and soon defined the terms of the debate about the history of life. Those who rejected it wholesale, as Agassiz did, consigned themselves to increasing irrelevance.

AGASSIZ UNDER THE MICROSCOPE

So did Agassiz identify a genuine problem for Darwin's theory, a mystery, at least, waiting to be solved? If so, whatever became of this problem? And if not, how could such a brilliant and knowledgeable scientist, someone so steeped in the evidence, fall so far outside the mainstream of scientific opinion?

Historians of science in the post-Darwinian era have typically attempted to answer this later question by portraying Agassiz as a brilliant and respected scientist who nevertheless was too ossified to catch the new wave, a figure past his prime and mired in philosophical prejudice.[26] Biographer Edward Lurie describes the Harvard naturalist as a "giant of the nineteenth century . . . a person deeply involved in his surroundings, a man who understood the possibilities of life with an uncommon awareness."[27] Similarly, historian Mabel Robinson says that she long awaited a biography of Agassiz that "would re-create this man of genius and his headlong splendid race through life." He was, she said, "a man to remember because genius is rare," "an immortal Pied Piper."[28] These scholars are merely echoing what Agassiz's contemporaries, even Darwin himself, said. "What a set of men you have at Harvard!" Darwin told the American poet Henry Wadsworth Longfellow. "Both our universities put together cannot furnish the like. Why, there is Agassiz—he counts for three."[29]

Even so, many historians argue that Agassiz was too infected by German idealism to properly assess the factual basis of Darwin's case. According to idealist philosophers of biology, living forms exemplified transcendent ideas and in their organization provided evidence of purposive design in nature. Comments historian A. Hunter Dupree, "Agassiz's idealism was of course the basis of his concepts of species and their distribution," of his insistence that a divine or intellectual cause must stand behind the origin of each type.[30] The ship of science was transitioning from idealism to modern empiricism. Agassiz had fallen overboard, since he had imbibed too deeply an outmoded idealism from his teacher, the French anatomist Georges Cuvier, and from philosophers like Friedrich Schelling, who "ran wild in trying to put all nature into a unified and absolute system of ideas."[31] Agassiz wasn't merely wrong, Dupree explains, but an annoying obscurantist, actively fighting "against the extension of empiricism into natural history."[32]

Edward Lurie offers a similar if somewhat more nuanced assessment: although "quite capable of making the most admirable scientific discoveries reflecting complete devotion to scientific method," Agassiz "would then interpret the data through the medium of what seemed to be the most absurd metaphysics."[33] The very man who made "the most careful, exact, and precise descriptions" of the natural world would, in his generalizations from those observations, "indulge in flights of idealistic fancy."[34] In short, Lurie thought that "Agassiz's cosmic philosophy shaped his entire reaction to the evolution idea."[35]

As science advanced in the late nineteenth century, it increasingly excluded appeals to divine action or divine ideas as a way of explaining phenomena in the natural world. This practice came to be codified in a principle known as methodological naturalism. According to this principle, scientists should accept as a working assumption that all features of the natural world can be explained by material causes without recourse to purposive intelligence, mind, or conscious agency.

Proponents of methodological naturalism argue that science has been so successful precisely because it has assiduously avoided invoking creative intelligence and, instead, searched out strictly material causes for previously mysterious features of the natural world. In the 1840s, the French philosopher August Comte argued that science progresses through three distinct phases. In its theological phase, it invokes the mysterious action of the gods to explain natural phenomena, whether thunderbolts or the spread of disease. In a second, more advanced metaphysical stage, scientific explanations refer to abstract concepts like Plato's forms or Aristotle's final causes. Comte taught that science only reaches maturity when it casts aside such abstractions and explains natural phenomena by reference to natural laws or strictly material causes or processes. Only in this third and final stage, he argued, can science achieve "positive" knowledge.

During the late nineteenth century, scientists increasingly embraced this "positivistic" vision.[36] Agassiz, by insisting that the Cambrian fossils pointed to "acts of mind"[37] and an "intervention of an intellectual power," stood firmly against this new vision. For many, his reference to the work of a transcendent mind merely demonstrated that he was unable to abandon an outmoded idealistic approach. The train of scientific progress had left Agassiz behind.

AN OLD FOSSIL RECOVERED

Though clearly Agassiz did reject the principle of methodological naturalism, as it is now named, there are problems with portraying him as a fossil of another age. First, Agassiz was unsurpassed in his commitment to the empirical method. It is Agassiz about whom the story is told of the professor instructing one of his students to observe a fish for three arduous days, a story iconic enough that it is reprinted in freshman composition textbooks. In the story, the student, Samuel Scudder, pulls out his hair trying to see anything new about the slimy creature, wondering why Professor Agassiz is torturing him with this "hideous fish." But in the end Scudder breaks through to new levels of observational depth and precision. Mabel Robinson notes that if such teaching methods seem less revolutionary to contemporary readers than they did to Scudder, that's because Agassiz trained an army of able young naturalists who took his method to other universities, and they in turn passed them on to their students, themselves future professors.[38]

William James, the founder of American pragmatism, extolled Agassiz's commitment to empirical rigor in a letter he wrote to his father while on an expedition with Agassiz in 1865 to South America. In the letter the young man commented that he felt a "greater feeling of weight and solidity about the presence of this great background of special facts than about the mind of any other man I know,"[39] a storehouse of precise data made possible by "a rapidity of observation, and a capacity to recognize them again and remember everything about them."[40] James would eventually enter the field of psychology, but he took with him the empirical approach to problem solving that Agassiz had modeled so impressively.[41]

As Lurie concedes, Agassiz's stature among American scientists grew out of his unrivaled knowledge of geology, paleontology, ichthyology, comparative anatomy, and taxonomy. So passionate was Agassiz for the particulars of the natural world that he began organizing a system of information-sharing among naturalists, sailors, and missionaries around the world. He collected more than 435 barrels of specimens, among them an extremely rare group of fossil plants.[42] In a single year, Agassiz amassed more than 91,000 specimens and identified close to 11,000 new species,[43] making Harvard's natural history museum preeminent among such museums in the world.

He also appears to have gone to great lengths, literally and figuratively, to assess *On the Origin of Species* empirically, going so far as to make a research voyage retracing Darwin's trip to the Galápagos Islands. As he explained to German zoologist Carl Gegenbauer, he "wanted to study the Darwin theory free from all external influences and former prejudices."[44]

The idea that religious or philosophical prejudice compromised Agassiz's scientific judgment raises other questions. As historian Neal Gillespie explains, Agassiz was "second to no man in his opposition to sectarian religious interference with science."[45] Moreover, Agassiz showed himself perfectly willing to accept natural mechanisms where before supernatural intervention had been the preferred explanation. Since he regarded material forces, and the laws of nature that described them, as the products of an underlying design plan, he saw any creative work they did as deriving ultimately from a creator. For instance, he assumed this was the case with the development of embryos: he attributed their natural evolution from zygote to adult as a natural phenomenon and considered this no threat to his belief in a creator.[46] He also readily accepted the notion of a naturally evolving solar system.[47] He thought a skillful cosmic architect could work through secondary natural causes every bit as effectively as through direct acts of agency. The marginalia in his copy of *On the Origin of Species* suggest that he had this same attitude concerning biological evolution. "What is the great difference," he wrote, "between supposing that God makes variable species or that he makes laws by which species vary?"

A third problem with the official portrait of Darwin's chief rival concerns Lurie's suggestion that Agassiz was a master of particulars, but not of generalizing from those particulars. The historical record suggests otherwise. For example, Agassiz was the man who ably generalized from a wide array of particular clues in his work on the ice age, winning over the geological establishment by demonstrating how a range of facts were best explained by the action of retreating glaciers.

Here a direct comparison between Darwin and Agassiz is possible. Each searched for an explanation of a curious geological phenomenon in the Scottish Highlands, the parallel roads of Glen Roy. Glen Roy is the valley of the River Roy and, although it's a place of breathtaking beauty, what visitors found most intriguing about it over the years were its three parallel roads that wind along the canyon wall on either side of the river (see Fig. 1.7).[48] Scottish legend held that they were hunting paths built for use by early Scottish kings or perhaps even for the mythical warrior Fingal. Scientists later argued that the roads were natural rather than artificial. Darwin and Agassiz were both convinced that natural processes were the cause, but they nevertheless arrived at different explanations. What was the end of the matter? In his autobiography, Darwin explained, "Having been deeply impressed with what I had seen of the elevation of the land in S. America, I attributed the parallel lines to the action of the sea; but I had to give up this view when Agassiz propounded his glacier-lake theory."[49] Subsequent investigations in the late nineteenth and early twentieth centuries confirmed that Agassiz's interpretation was the correct one.[50]

Agassiz, then, was far more than just a walking encyclopedia or an insatiable gatherer of fossils who couldn't see the proverbial forest for the trees. Those who insist otherwise can point to but one example to support their position, namely, his rejection of Darwin's theory; but they cannot use that example to establish his general inability to interpret evidence and then turn around and use that supposed inability to explain his failure to accept Darwin's theory. That is to argue in a circle.

There is a far more obvious solution to the historical puzzle posed by the great Agassiz's objection to Darwin's theory: the fossils of the Cambrian

FIGURE 1.7
Parallel roads of Glen Roy.

strata do, in fact, arise abruptly in the geological record, in clear defiance of what Darwin's theory would lead us to expect. In short, a genuine mystery is at hand.

Two final considerations lend support to this view. First, as already noted, Darwin himself accepted the validity of Agassiz's objection.[51] As he acknowledged elsewhere in the *Origin,* "To the question why we do not find rich fossiliferous deposits belonging to these assumed earliest periods prior to the Cambrian system, I can give no satisfactory answer. . . . The case at present must remain inexplicable; and may be truly urged as a valid argument against the views here entertained."[52]

Second, Darwin's attempt to account for the absence of the expected fossil ancestors of the Cambrian forms failed to address the full strength and subtlety of Agassiz's objection. As Agassiz explained, the problem with Darwin's theory was not just the general incompleteness of the fossil record or even a pervasive absence of ancestral forms of life in the fossil record. Rather, the problem, according to Agassiz, was the *selective* incompleteness of the fossil record.

Why, he asked, does the fossil record always happen to be incomplete at the nodes connecting major branches of Darwin's tree of life, but rarely—in the parlance of modern paleontology—at the "terminal branches" representing the major already known groups of organisms? These terminal branches were well represented (see Fig. 1.8), often stretching over many generations and millions of years, while the "internal branches" at the connecting nodes on Darwin's tree of life were nearly always—and selectively—absent. As Agassiz explained, Darwin's theory "rests partly upon the assumption that, in the succession of ages, *just those transition types* have dropped out from the geological record which would have proved the Darwinian conclusions had these types been preserved."[53] To Agassiz, it sounded like a just-so story, one that *explains away* the absence of evidence rather than genuinely explaining the evidence we have.

Was there any easy answer to Agassiz's argument? If so, beyond his stated willingness to wait for future fossil discoveries, Darwin didn't offer one.

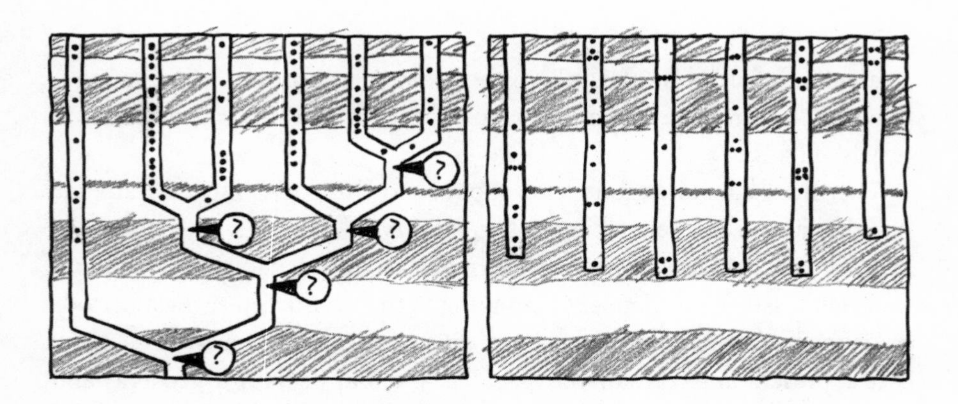

FIGURE 1.8
The vertical lines in these diagrams represent known animal phyla. The dots within the vertical lines represent animals from those phyla that have been found fossilized in different strata. The diagram on the left shows the animal tree of life as expected based upon Darwinian theory. The diagram on the right shows a simplified representation of the actual pattern of the Precambrian–Cambrian fossil record. Notice that fossils representing the internal branches and nodes, but not the terminal branches, are missing.

AN ENDURING MYSTERY

In the years immediately following the publication of *On the Origin of Species,* many of Agassiz's concerns were temporarily swept aside as public and scientific fascination with Darwin's ideas grew. Even so, a persistent mystery lay at the feet of biologists, one that subsequent generations of scientists would revisit and repeatedly seek to resolve. As Darwin noted, in his time, the fossils of the Cambrian were relatively few, and the period of the explosion only vaguely understood. But perhaps future scientists would come to his rescue with fresh discoveries.

The story of the successive attempts to solve the Cambrian mystery stretches from Darwin's time to the present, and from the Swansea Valley in southern Wales to remote fossil sites in southern China. In the next chapter, the work of detection moves from the late nineteenth century to the early twentieth, from the British Isles to British Columbia, and to a fossil site above the Kicking Horse River so astonishing that, even today, paleontologists and some of the most skeptical and hardened of scientific rationalists speak its name with childlike reverence.

2

THE BURGESS BESTIARY

Only in fiction can we expect such fine orchestration of setting and dramatic action. Gothic tales haunted by demons of the past have their thunderstorms and crumbling mansions, the existentialist novel its disorienting cityscapes, the romance its unattainable balconies laced with jasmine. In ordinary life, the staging is usually less precise. Intricate family tragedies unravel in tidy suburban ranch-style homes, while enchanted romances blossom over cubicle walls. But the twentieth century's most revolutionary fossil discovery was more like fiction: the setting was commensurate with the moment.

Photographs taken during the summer's expedition show a lean and balding man with pleasant crinkles at the corners of his eyes and a deep thought line slashing down between his brows; he stands precariously on rocky ascents, pick and hoe at hand, gazing far into the distance from a stony peak, at ease among the forbidding slopes and treacherous ridges. Working his way over one ridge and then above the tree line of the next, Charles Doolittle Walcott reached a place where he could see for miles. To the northwest, the crude arrowhead of Mount Wapta jutted skyward. Below lay Emerald Lake, its waters green from the mineral-rich glacial till. To the east and west, snowy peaks stretched to the horizon (see Figs. 2.1 and 2.2). Only the view to the northeast lacked a vista. Here was the homely shale of a barren ridge. Of course, as in any fairy tale, there lay the real prize, a hidden vista measured not in miles, but in ages.

Walcott, already the director of the Smithsonian Institution, was about to enter the most significant phase of his professional life. More than this,

FIGURE 2.1
The scenery of the Burgess Shale and surrounding area. *Courtesy Corbis.*

FIGURE 2.2
Charles Doolittle Walcott in the field (c. 1911). *Courtesy Smithsonian Institution Archives.*

he was about to make perhaps the most dramatic discovery in the history of paleontology, a rich trove of middle Cambrian–era fossils, including many previously unknown animal forms, preserved in exquisite detail, suggesting an event of greater suddenness than had been known even in Darwin's time and detailing a greater diversity of biological form and architecture than had hitherto been imagined.

Where did this wealth of biological form come from, and why, again, did it seem to arise so suddenly during the Cambrian period? Walcott was the first to explore the Burgess Shale, and he would be the first to suggest an answer to the questions it raised.

THE BESTIARY

Among paleontologists, the fateful clue that led to the Burgess Shale's discovery is the stuff of legend. Paleontologist Stephen Jay Gould considered it to have been rendered best in an obituary of Charles Walcott written by Walcott's former research assistant, Charles Schuchert:

> *One of the most striking of Walcott's faunal discoveries came at the end of the field season of 1909, when Mrs. Walcott's horse slid on going down the trail and turned up a slab that at once attracted her husband's attention. Here was a great treasure—wholly strange Crustacea of Middle Cambrian time—but where in the mountain was the mother rock from which the slab had come? Snow was even then falling, and the solving of the riddle had to be left to another season, but next year the Walcotts were back again on Mount Wapta, and eventually the slab was traced to a layer of shale—later called the Burgess Shale—3000 feet above the town of Field.*[1]

Gould quotes the legend to celebrate its archetypal appeal even as he debunks it: "Consider the primal character of this tale—the lucky break provided by the slipping horse, . . . the greatest discovery at the very last minute of a field season (with falling snow and darkness heightening the drama of finality), the anxious wait through a winter of discontent, the triumphant return and careful, methodical tracing of errant block to mother lode."[2] A compelling story, Gould concludes, but pure fiction. Walcott's own diaries reveal that his team had plenty of time to begin excavating the site that very summer amid cooperative weather and even warm nights.

As for their return the following summer, locating the mother lode was apparently the work of a single day rather than a full week, a conclusion Gould drew from both Walcott's diaries and his knowledge of Walcott's expertise as a geologist.[3]

The motifs of the lucky break, the frustrating delay, and the final and fortuitous triumph will resurface later (see Chapter 7) as a tall tale of Gould's own, but for now consider only the scientific community's weakness for staging the Burgess discovery with various fictional props, as if the stunning scenery around it were not setting enough. This weakness for theater is understandable, considering what Walcott and later investigators found there. Over the next several years, Walcott's team alone collected more than 65,000 specimens, many of them astonishingly well preserved, some so bizarre that paleontologists would cast about for more than half a century for the proper categories in which to contain them.

Consider just one odd couple from Walcott's quarry, *Marrella* and *Hallucigenia*. *Marrella*, also called a lace crab, is an unusual form. Walcott described it as a type of trilobite, but later studies by Cambridge paleontologist Harry Whittington classified it not as a trilobite, nor a chelicerate (the subgroup of arthropods that includes spiders), and not even as a crustacean, but rather as a fundamentally distinct form of arthropod.[4] The creature is divided into twenty-six segments, each with a jointed leg for walking and a feathery gill branch for swimming. Its head shield has two long pairs of spikes directed backwards, and the underside of the head features two pairs of antennae. One is short and stout, the other long and sweeping (see Fig. 2.3).

Hallucigenia belongs to a genus and family of one. It has a rounded mass at one end (possibly the head) connected to a cylinder-shaped trunk sporting seven pairs of spines projecting upward and to either side, each of them almost as long as the trunk itself (see Fig. 2.4). On the underside of the creature are seven pairs of limbs, each corresponding in position to one of the pairs of spines on the back, though with the tentacle farthest back offset. The underbelly also features three pairs of shorter tentacles before the trunk tapers and curves upward in what was probably a flexible extension from the body. Each of the larger tentacles appears to have a hollow tube connected to the gut and a pincer at the tip. This ancient creature was so peculiar that paleontologists feigned disbelief at what they saw, giving it its memorable name.

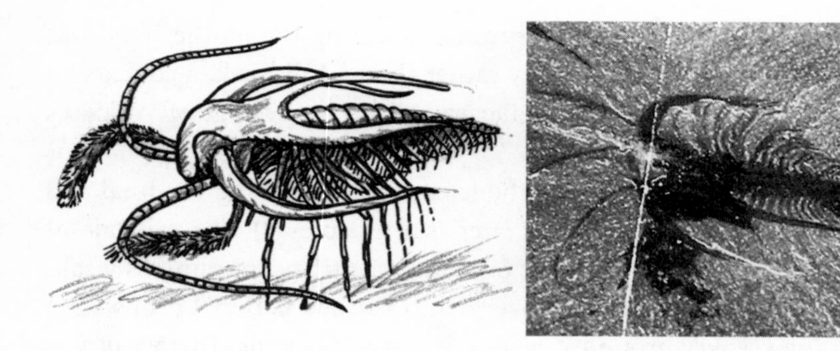

FIGURE 2.3
Figure 2.3a (left): Artist rendering of *Marrella splendens. Figure 2.3b (right):* Photograph of *Marrella splendens* fossil. *Courtesy Wikimedia Commons, user Smith609.*

FIGURE 2.4
Figure 2.4a (top): Artist rendering of *Hallucigenia sparsa. Figure 2.4b (bottom):* Photograph of *Hallucigenia sparsa* fossil. *Courtesy Smithsonian Institution.*

The term "Cambrian explosion" was to become common coin, because Walcott's site suggested the geologically abrupt appearance of a menagerie of animals as various as any found in the gaudiest science fiction. During this explosion of fauna, representatives of about twenty of the roughly twenty-seven total phyla present in the known fossil record made their first appearance on earth (see Fig. 2.5).[5]

The term "phyla" (singular: "phylum") refers to divisions in the biological classification system. The phyla constitute the highest (or widest) categories of biological classification in the animal kingdom, with each exhibiting a unique architecture, organizational blueprint, or structural body plan. Familiar examples of phyla are cnidarians (corals and jellyfish), mollusks (squid and clams), echinoderms (sea stars and sea urchins), arthropods (trilobites and insects), and the chordates, to which all vertebrates including humans belong.

The animals within each phylum exhibit distinguishing features that enable taxonomists to divide and group them further into other, progressively smaller divisions, beginning with classes and orders, and eventually coming to families, genera, and individual species. The broadest and highest categories within the animal kingdom—such as phyla and classes—designate the major categories of animal life, typically designating unique body plans. Lower taxonomic categories—like genus and species—designate smaller degrees of difference among organisms that typically exemplify similar overall ways of organizing their body parts and structures.

Throughout the book I will use these conventional categories of classification, as do most Cambrian paleontologists. Nevertheless, I am aware that some paleontologists and systematists (experts in classification) today prefer "phylogenetic classification," a method that often uses a "rank-free" classification scheme.[6] Advocates of modern phylogenetic classification argue that the traditional classification system lacks objective criteria by which to decide whether a certain group of organisms should be assigned a particular rank of, for example, phylum or class or order.[7] Proponents of rank-free classification attempt to eliminate subjectivity in classification (and ranking) by grouping together animals that are thought, based upon studies of similar molecules in different groups, to share a common ancestor. This method of classification treats groups that emerge at roughly the same time on the tree of life as equivalent. Nevertheless, even proponents

GEOLOGICAL TIME PERIOD	ESTIMATED NUMBER OF ANIMAL PHYLA FIRST APPEARING	CUMULATIVE NUMBER OF PHYLA	NAMES OF PHYLA
PRECAMBRIAN	3	3	CNIDARIA(?) MOLLUSCA(?) PORIFERA
CAMBRIAN	20	23	ANNELIDA BRACHIOPODA BRYOZOA CHAETOGNATHA CHORDATA COELOSCLERITOPHORA CTENOPHORA ECHINODERMATA ENTOPROCTA EUARTHROPODA HEMICHORDATA HYOLITHA LOBOPODIA LORICIFERA NEMATOMORPHA PHORONIDA PRIAPULIDA SIPUNCULA TARDIGRADA VETULICOLIA
LATER GEOLOGICAL PERIODS	4	27	NEMATODA (CRETACEOUS) NEMERTEA (CARBONIFEROUS) PLATYHELMINTHES (EOCENE) ROTIFERA (EOCENE)
DO NOT APPEAR IN THE FOSSIL RECORD	9	36	ACANTHOCEPHALA CYCLIOPHORA DICYEMIDA GASTROTRICHA GNATHOSTOMULIDA KINORHYNCHA ORTHONECTIDA PENTASTOMA PLACOZOA

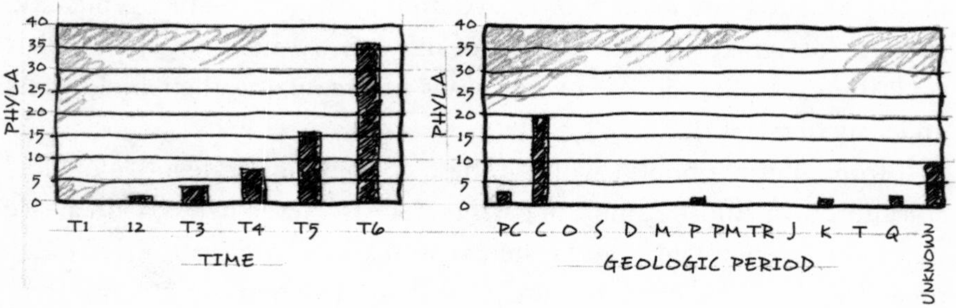

PHYLA
40 35 30 25 20 15 10 5 0
T1 T2 T3 T4 T5 T6
TIME

PHYLA
40 35 30 25 20 15 10 5 0
PC C O S D M P PM TR J K T Q UNKNOWN
GEOLOGIC PERIOD

FIGURE 2.5

Figure 2.5a (top): Chart showing when representatives of the different animal phyla first appeared in the fossil record. According to Darwinian theory, differences in biological form should increase gradually, steadily increasing the number of distinct body plans and phyla, over time. References for first appearances are found in note 5 of this chapter. *Figure 2.5b (bottom, left)* expresses that expectation graphically, showing the number of new phyla increasing steadily as members of one phylum diversify and give rise to new phyla. *Figure 2.5c (bottom, right)* shows the actual pattern of first appearance showing a spike in the number of phyla that first appear in the Cambrian, followed by either few or no new phyla arising in subsequent periods of geological history.

of phylogenetic classification often use the conventional taxonomic categories in their technical discussions of specific organisms because of their common scientific usage. So despite my own sympathy with some of the concerns of rank-free advocates (see below), I have chosen to do the same.

In any case, it's worth noting that using a rank-free classification system does not minimize the mystery of the Cambrian explosion. The Cambrian explosion presents a puzzle for evolutionary biologists, not just because of the number of phyla that arise, but rather because of the number of unique animal forms and structures that arise (as measured, perhaps, by the number of phyla)—however biologists decide to classify them. Thus, whether scientists decide to use newer rank-free classification schemes or older, more conventional, Linnaean categories, the "evolutionary novelties"—that is, the new anatomical structures and modes of organization—that arise suddenly with the Cambrian animals remain as facts of the fossil record, requiring explanation. (For an expanded technical discussion of these issues, go to this endnote.)[8]

One especially dramatic fact of the Cambrian explosion is the first appearance of many novel marine invertebrate animals (representatives of separate invertebrate[9] phyla, subphyla, and classes in the traditional

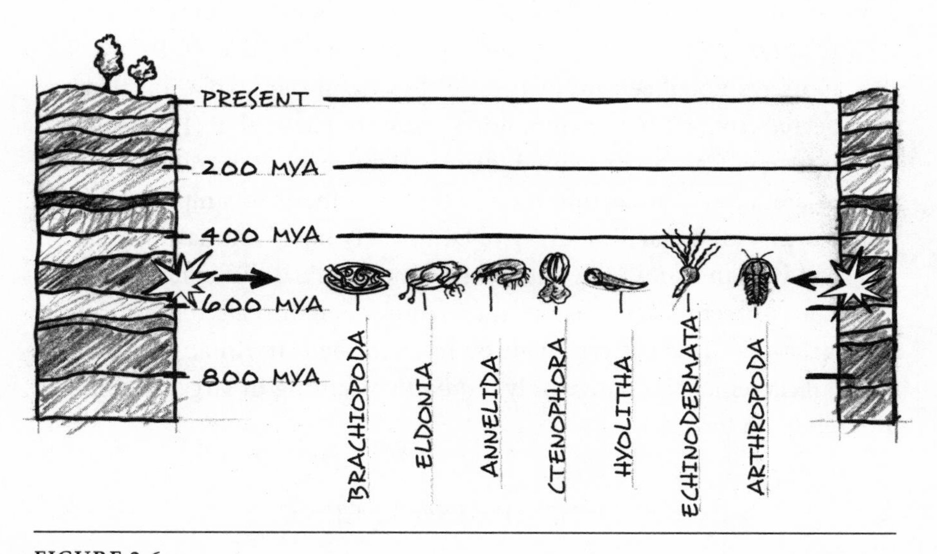

FIGURE 2.6
Representatives of some of the major animal groups that first appeared in the sedimentary rock record during the Cambrian period.

classification scheme). Some of these animals have mineralized exoskeletons, including those representing phyla, such as echinoderms, brachiopods, and arthropods, and each represent clearly distinct and novel body plans. Further, these are just three of dozens of novel body plans exemplified by the Burgess animals—animals in which both soft and hard parts are well preserved (see Fig. 2.6).

The variety in the Burgess Shale was so extreme it took several decades for paleontologists to grasp it fully. Walcott, for instance, attempted to fit all of the new forms into existing phyla. However, even in the midst of this attempt, he realized that this revolutionary quarry posed a problem more fundamental than a need to tidy up the existing taxonomy. He had met Louis Agassiz at a young age, having sold him some of his first fossils, and later described him as "a guide in whom I could trust and follow," one in whose work "I find this tribute to the Great Mind that created the objects of his study."[10] But in the great debate between Agassiz and Darwin, Walcott sided with the Englishman. Thus, the Burgess Shale struck Walcott as not only fascinating, but puzzling.

A PUZZLING PATTERN

Over the years, as paleontologists have reflected on the overall pattern of the Precambrian–Cambrian fossil record in light of Walcott's discoveries, they too have noted several features of the Cambrian explosion that are unexpected from a Darwinian point of view[11] in particular: (1) the sudden appearance of Cambrian animal forms; (2) an absence of transitional intermediate fossils connecting the Cambrian animals to simpler Precambrian forms; (3) a startling array of completely novel animal forms with novel body plans; and (4) a pattern in which radical differences in form in the fossil record arise before more minor, small-scale diversification and variations. This pattern turns on its head the Darwinian expectation of small incremental change only *gradually* resulting in larger and larger differences in form.

THE MISSING TREE

Figures 2.7 and 2.8 illustrate the difficulty posed by the first two of these features: sudden appearance and missing intermediates. These diagrams graph morphological change over time. The first shows the Darwinian expectation that changes in morphology should arise only as tiny changes accumulate. This Darwinian commitment to gradual change through microevolutionary variations produces the classic representation of evolutionary history as a branching tree.

Now compare this branching tree pattern with the pattern in the fossil record. The bottom part of Figure 2.7 and Figure 2.8 show that the Precambrian strata do not document the expected transitional intermediates between Cambrian and Precambrian fauna. Instead, the Precambrian–Cambrian fossil record, especially in light of the Burgess Shale after Walcott, points to the geologically sudden appearance of complex and novel body plans.

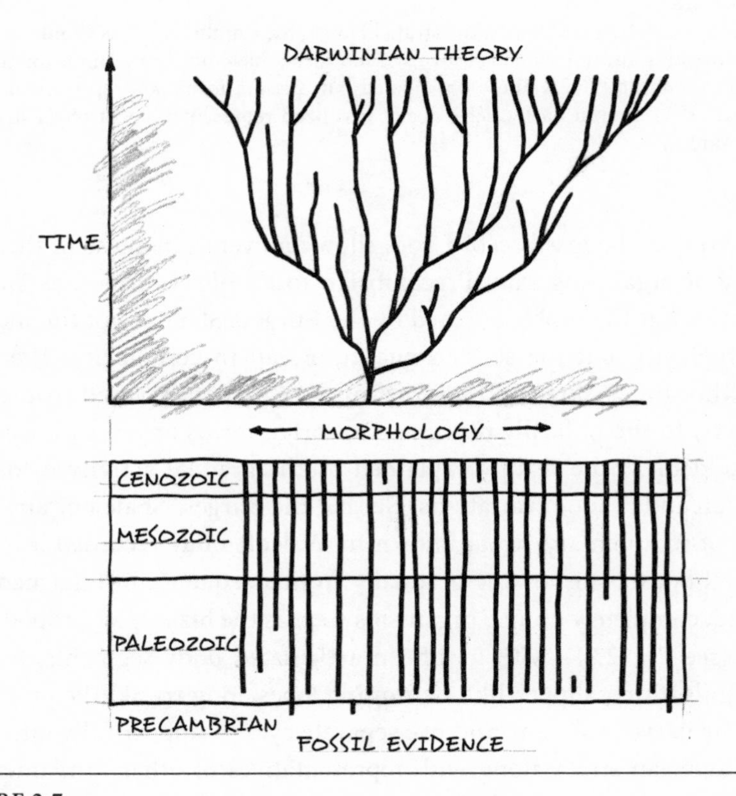

FIGURE 2.7
The origin of animals. Darwinian theory (*top*) predicts gradual evolutionary change in contrast to the fossil evidence (*bottom*), which shows the abrupt appearance of the major animal groups.

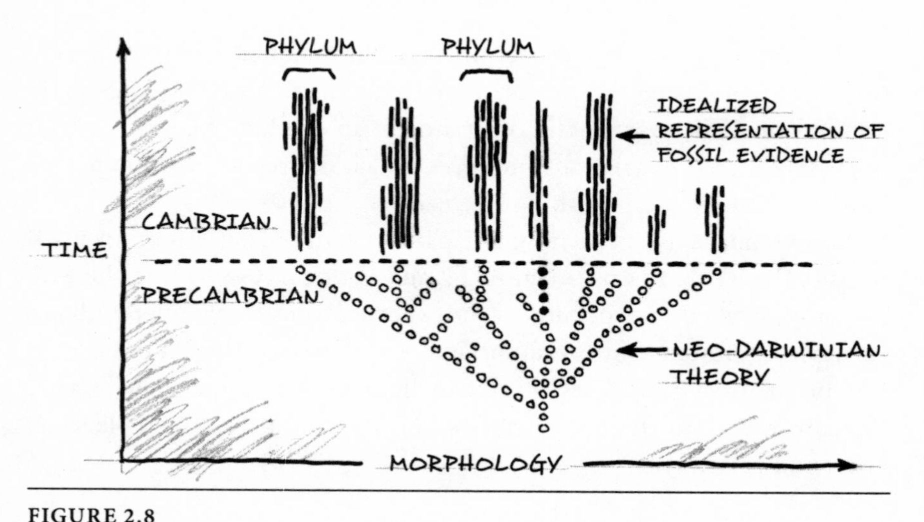

FIGURE 2.8
According to Darwinian theory, the strata beneath the Cambrian rocks should evidence many ancestral and intermediate forms. Such forms have not been found for the vast majority of animal phyla. These anticipated but missing forms are represented by the gray circles. Lines and dark circles depict fossilized representatives of phyla that have been found.

Of course, the fossil record does show an overall increase in the complexity of organisms from Precambrian to Cambrian times, as Darwin expected. But the problem posed by the Burgess Shale is not the increase in complexity, but the sudden quantum leap in complexity. The jump from the simpler Precambrian organisms (further explored in the next chapters) to the radically different Cambrian forms appears to occur far too suddenly to be readily explained by the gradual activity of natural selection and random variations. Neither the Burgess Shale nor any other series of sedimentary strata known in Walcott's day recorded a pattern of novel body plans arising gradually from a sequence of intermediates. Instead, completely unique organisms such as the bizarre arthropod *Opabinia* (see Fig. 2.9)—with its fifteen articulated body segments, twenty-eight gills, thirty flipper-like swimming lobes, long trunk-like proboscis, intricate nervous system, and five separate eyes[12]—appear fully formed in the Cambrian strata along with representatives of other fundamentally different body plans and designs of equal complexity.

Darwin, as we know, regarded the sudden appearance of the Cambrian

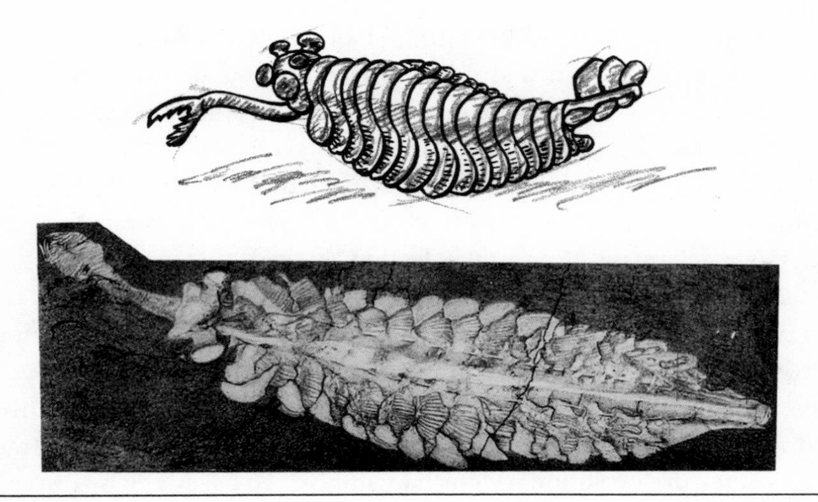

FIGURE 2.9
Figure 2.9a (top): Artist rendering of *Opabinia*. *Figure 2.9b (bottom):* Photograph of *Opabinia* fossil.

animals as a significant challenge to his theory.[13] Where natural selection had to bridge yawning chasms from relatively simple life-forms to exquisitely complex creatures, it would require great expanses of time.[14]

Darwin's recognition[15] of this constraint was prescient. Geologists in his day employed relative dating methods. They did not have modern radiometric methods for determining the "absolute" ages of rocks. For this reason, they did not yet fully understand how long it would have taken to accumulate the great columns of sedimentary rock and, thus, the great expanses of time that were available to the evolutionary process. Neither had scientists yet discovered the sophisticated inner workings of the cell, and the information-rich structures (DNA, RNA, and proteins) that had to be significantly altered to achieve even modest evolutionary changes. Nevertheless, Darwin was able—based upon what he knew of the complexity of organisms and his own understanding of how the mechanism of natural selection must operate—to deduce that descent with modification required time, and lots of it.

Recalling the context of Darwin's original argument reveals why. In the *Origin*, he sought to counter the famous watch-to-watchmaker design argument offered by theologian William Paley. Paley had argued that just as complex structures such as watches necessarily issue from intelligent watchmakers, the complex structures in living organisms must likewise owe their origin to a designing intelligence. With natural selection, Darwin proposed a purely natural mechanism for constructing the complex organs and structures (such as eyes) present in many forms of life. His mechanism of natural selection worked by constructing such systems one tiny step at a time, discarding the harmful variations and seizing upon the rare improvement. If evolution progressed by "whole watches"—that is, by entire anatomical systems like the trilobite's eye—then biology would have fallen back to the old absurdity of imagining that a watch could fall together purely at random and all at once. Thus, unless Darwin's evolutionary mechanism worked gradually by preserving the tiniest of random changes over many millions of years, it didn't work at all.

MORE MISSING LINKS

Two other features of the Cambrian explosion revealed in the Burgess Shale, features (3) and (4) described earlier, not only confirmed the reality of the Cambrian mystery, but broadened and deepened it—and at just the time when paleontologists were looking to resolve the mystery with new fossil discoveries.

First, the great profusion of completely novel forms of life in the Burgess assemblage (feature 3) demanded even more transitional forms than had previously been thought missing. Each new and exotic Cambrian creature—the anomalocarids (see Fig. 2.10), *Marrella, Opabinia,* and the bizarre and appropriately named *Hallucigenia*—for which there were again no obvious ancestral forms in the lower strata, required its own series of transitional ancestors. But where were they?

Darwin had hoped that later fossil discoveries would eventually eliminate what he regarded as the one outstanding anomaly associated with his theory. Walcott's discovery was not that discovery. Not only did the Burgess Shale fail to reveal the expected ancestral precursors of the known Cambrian animal forms, but it revealed a motley crew of previously unknown animal forms and body plans that now demanded their own

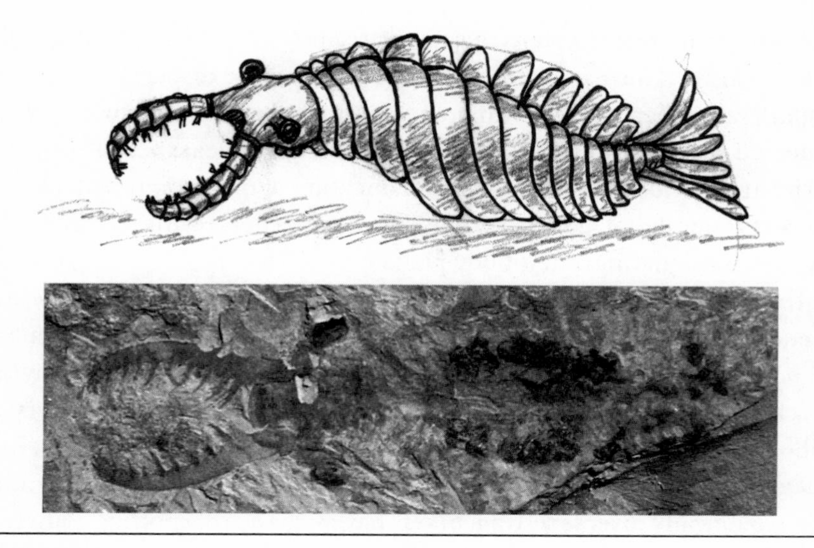

FIGURE 2.10
Figure 2.10a (top): Artist rendering of *Anomalocaris. Figure 2.10b (bottom):* Photograph of *Anomalocaris* fossil. *Courtesy J. Y. Chen.*

lengthy chain of evolutionary precursors, only complicating the task of explaining the Cambrian explosion in Darwinian terms.

ORDERS FROM THE TOP

The Burgess Shale raised an additional difficulty (feature 4, discussed earlier), though not one that Walcott recognized during his lifetime. Instead, its exposition would await a later generation of Cambrian experts, particularly Stephen Jay Gould. Darwin's theory implied that as new animal forms first began to emerge from a common ancestor, they would at first be quite similar to each other, and that large differences in the forms of life—what paleontologists call disparity—would only emerge much later as the result of the accumulation of many incremental changes. In its technical sense, *disparity* refers to the major differences in form that separate the higher-level taxonomic categories such as phyla, classes, and orders. In contrast, the term *diversity* refers to minor differences among organisms classified as different genera or species. Put another way, *disparity* refers to life's basic themes; *diversity* refers to the variations on those themes. The more body plans in a fossil assembly, the greater the disparity. And the animal forms preserved in the Burgess Shale display considerable disparity. Further, the large differences in form between the first animals appeared suddenly in the Burgess Shale, and the appearance of such disparity arose *before*, not after, the diversification of many representatives of lower taxonomic categories (such as species or genera) within each higher category, designating a new body plan.

The site of the Burgess Shale and its setting nicely illustrates the difference between diversity and disparity. Walcott's celebrated quarry is tucked away in the Canadian Rockies near the Continental Divide. Reaching it involves a six-mile hike through the picturesque scenery of Yoho National Park—Takakkaw Falls, Emerald Lake, and glaciers and glacier-cut mountain peaks thrusting into view at almost every turn. In this ecologically diverse setting, hikers have a chance of spotting squirrels, marmots, deer, moose, elk, wolves, and mountain goats. Rare sightings might include a grizzly bear or Canadian lynx, while alert birdwatchers might glimpse a horned lark, a white-tailed ptarmigan, the rare water pipit, or a gray-crowned rosy finch; an eagle, hawk, or grassland falcon; dippers, jays, migrating warblers, or harlequin ducks.[16]

As richly various as this array of animals is, all of them come from a single phylum, Chordata—and even from a single subphylum, Vertebrata. Imagine hiking to the quarry to excavate it and, on the hike, being lucky enough to spot every one of these animals along the way. After having feasted your eyes on such animal variety, when you arrive at Walcott's quarry, it yields not merely dozens of fossilized species from a single subphylum, but wildly disparate creatures from dozens of *phyla*.

According to Darwin's theory, the differences in form, or "morphological distance," between evolving organisms should increase gradually over time as small-scale variations accumulate by natural selection to produce increasingly complex forms and structures (including, eventually, new body plans). In other words, one would expect small-scale differences or *diversity* among species to precede large-scale morphological *disparity* among phyla. As the former Oxford University neo-Darwinian biologist Richard Dawkins puts it, "What had been distinct species within one genus become, in the fullness of time, distinct genera within one family. Later, families will be found to have diverged to the point where taxonomists (specialists in classification) prefer to call them orders, then classes, then phyla."[17]

Darwin himself made this point in *On the Origin of Species*. Explaining his famous tree diagram (see Fig. 2.11a), he noted that it illustrated more than just the theory of universal common descent. The tree diagram also illustrated how higher taxa should emerge from lower taxa by the accumulation of numerous slight variations. He said, "The diagram illustrates the steps by which small differences distinguishing varieties are increased into larger differences distinguishing species."[18] He went on to assert that the process of modification by natural selection would eventually move beyond the formation of species and genera to form "two distinct families, or orders, according to the amount of divergent modification supposed to be represented in the diagram."[19] In his view, this process would continue until it produced differences in form that were great enough that taxonomists would classify them as new classes or phyla. In short, diversity would precede disparity, and phyla-level differences in body plan would emerge only after species-, genus-, family-, order-, and class-level differences appeared.

The actual pattern in the fossil record, however, contradicts this expectation (compare Fig. 2.12 to Fig. 2.11b). Instead of more and more species eventually leading to more genera, leading to more families, orders, classes, and phyla, the fossil record shows representatives of separate phyla appearing first followed by lower-level diversification on those basic themes.

This is nowhere more dramatically apparent than in the Cambrian period explains Roger Lewin in the journal *Science:* "Several possible patterns exist for the establishment of higher taxa, the two most obvious of which are the bottom-up and the top-down approaches. In the first, evolutionary novelties emerge, bit by bit. The Cambrian explosion appears to conform to the second pattern, the top-down effect."[20] Or as paleontologists Douglas Erwin, James Valentine, and Jack Sepkoski note in their study of skeletonized marine invertebrates: "The fossil record suggests that the major pulse of diversification of phyla occurs before that of classes, classes before that of orders, orders before that of families. . . . The higher taxa do not seem to have diverged through an accumulation of lower taxa."[21] In other words, instead of a proliferation of species and

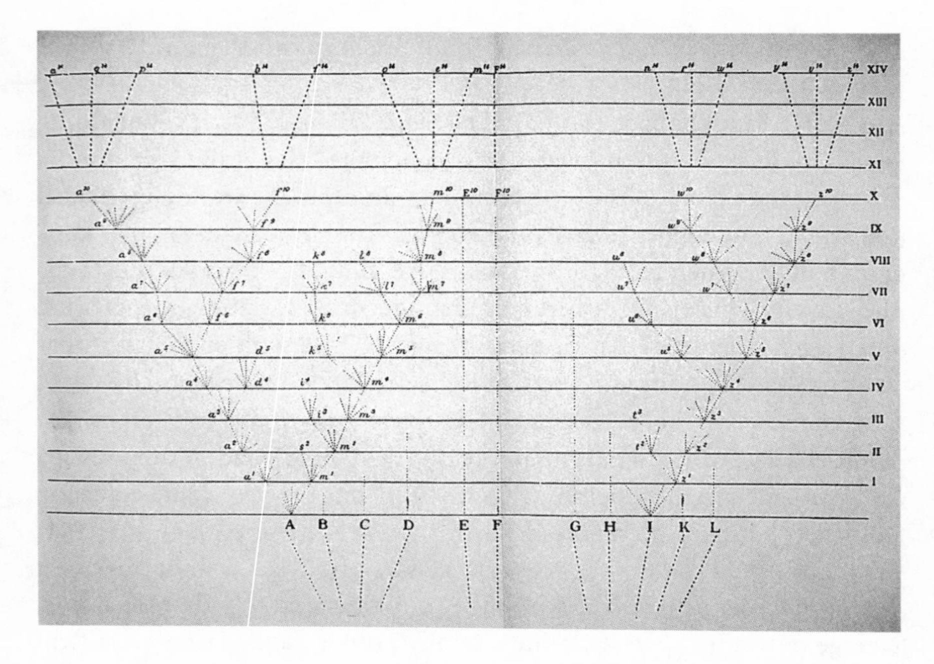

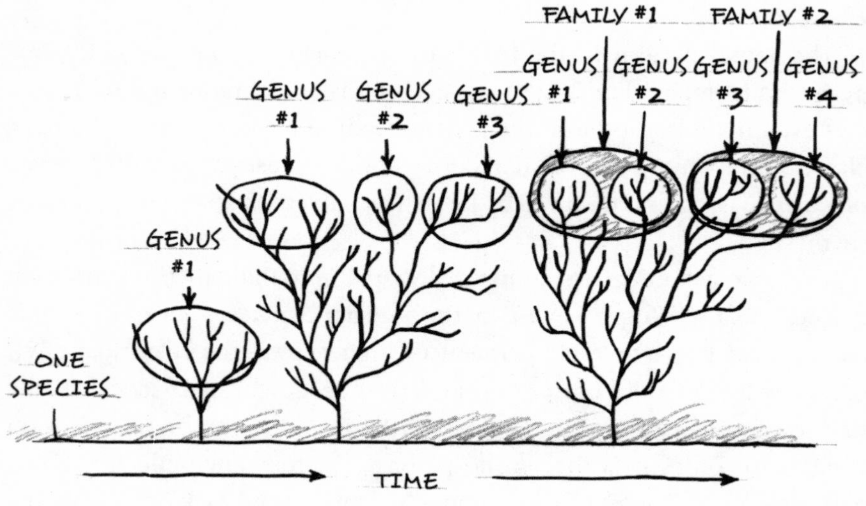

FIGURE 2.11

Figure 2.11a (top): Darwin's theory of common descent illustrated here with his famous branching tree of life diagram reproduced from the *Origin of Species,* 1859. *Figure 2.11b (bottom):* Growth of the tree of life over time in the manner envisioned by Darwin with new species giving rise to new genera and families, eventually giving rise to new orders, classes, and phyla (these higher taxonomic categories not depicted).

other representatives of lower-level taxa occurring first and then building to the disparity of higher taxa, the highest taxonomic differences such as those between phyla and classes appear first (instantiated by relatively few species-level representatives). Only later, in more recent strata, does the fossil record document a proliferation of representatives of lower taxa: different orders, families, genera, and so on. Yet we would not expect the neo-Darwinian mechanism of natural selection acting on random genetic mutations to produce the top-down pattern that we observe in the history of life following the Cambrian explosion.

Of course, advocates of modern phylogenetic classification, with their "rank-free" approach, don't describe this phenomenon as a "top-down" pattern, because their system of classification dispenses with taxonomic ranks and hierarchies. In their system, there is no "top" and no "down." Nevertheless, advocates of phylogenetic classification do acknowledge that different combinations of "character" states (characteristics or features of organisms) can mark either *bigger* or *smaller* morphological differences between clades (closely related groups of organisms that presumably share a common ancestor). And some leading advocates of phylogenetic classification have noted that the fossil record exhibits a pattern in which a few character traits marking *large* morphological differences between clades

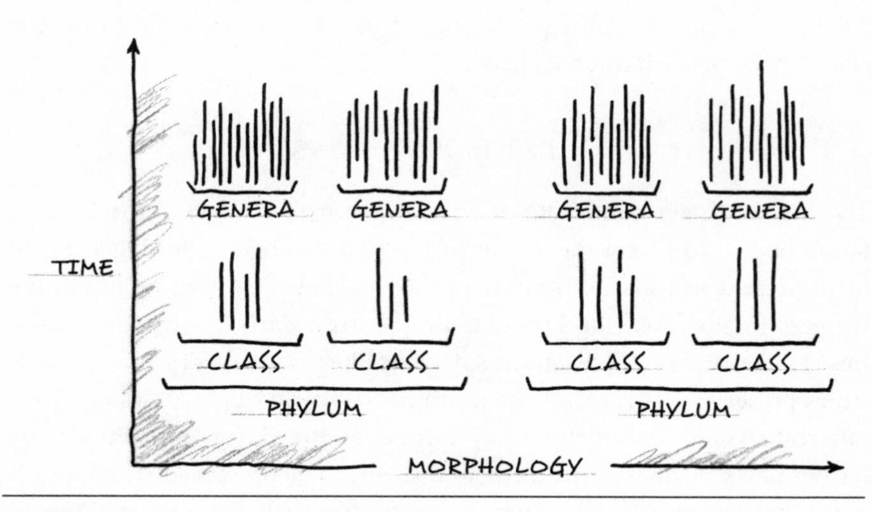

FIGURE 2.12
The top-down pattern of appearance in the fossil record: disparity precedes diversity.

arise *first,* followed *later* in each clade by the addition of other combinations of characters that mark *smaller* differences *within* those clades. Larger differences between clades arise first, smaller differences within them arose later—the themes precede the variations.

The founder of the modern phylogenetic classification, Willi Hennig, for example, noted that once particular groups arise, the range of allowable variability within those groups narrows. In his classic work *Phylogenetic Systematics,* Hennig quoted another paleontologist approvingly who observed: "The breadth of evolution of successive groups shows a distinct narrowing, since the basic divergences of organization became progressively smaller. The type of mammals is more uniform and closed than that of the reptiles, which in turn is unquestionably uniform compared to that of the Amphibia-Stegocephalia." Hennig goes on to explain that "the same phenomenon is repeated in every systematic unit of higher or lower order."[22]

Yet, on a Darwinian view, small-scale variations and differences should arise first, gradually giving rise to larger-scale differences in form—just the opposite of the pattern evident in the fossil record. Thus, the discovery, and later analysis, of the Burgess revealed another puzzling feature of the fossil record from a Darwinian point of view, regardless of which system of classification paleontologists prefer to use. Indeed, Walcott's discovery turned Darwin's anticipated bottom-up—or small changes first, big changes later—pattern on its head.

FIRST IMPRESSIONS

The extraordinary conditions at work in the preservation of the Burgess fauna helped to reveal the extent of the rich diversity (and disparity) of form present in the Cambrian period. On shale of a very fine grain, the Burgess fossils look like lithographic pictures, dark on light (see color insert plates 15 and 16). Even the soft parts like the gills and guts are sometimes preserved. This is not the norm in the world of paleontology. Usually soft tissues decay before they can be fossilized, leaving behind only harder parts, such as bone, teeth, and shells, to be preserved. The Burgess event that captured the Cambrian fauna for future discovery was different. Although it took the lives of untold Cambrian animals, it did so with an exquisite delicacy that preserved soft tissue.

Visualizing how this occurred will drive home why the conditions were so unusual. All of the fossilized animals of the Burgess Shale were sea creatures living near an enormous carbonate reef that later was thrust upward by plate tectonic activity to form what is now called the Cathedral Escarpment. Long after the sea creatures of the Burgess Shale were entombed, these tectonic forces drove the fossils upward from the seafloor carrying them many miles eastward along faults, at the same time building the mountains that Walcott would climb millions of years later.

Thanks to this tectonic movement of earth's major plates, the continents are now located in very different places than they were millions of years ago. At the time these Cambrian creatures were alive, the land masses that would later form North America lay on the equator.

Plate tectonic activity explains why a trove of sea creatures were found fossilized in the mountains of Yoho National Park rather than along a seafloor somewhere. But there's still the question of why so many different types of marine invertebrates, including soft-bodied ones, were so unusually well preserved. Paleontologists think they know the answer. They think the marine animals that were later fossilized in the Burgess Shale lived near the bottom of an ancient sea in front of an underwater cliff or escarpment. Due to tectonic activity, blocks at the edge of this underwater cliff began to break off. These blocks slumped, creating underwater mudflows in their wake. These slumps and flows transported the Burgess animals several kilometers into deeper waters where they were buried in such a way as to leave them not only undamaged, but also protected from scavengers and bacteria. Very probably, the mudflows were highly turbulent, for paleontologists found the creatures dumped and preserved in a variety of angles in relation to the bedding. The speed and pressure of these mudflows quickly produced a preservation-friendly, oxygen-free environment. Then the turbulent and muddy currents pressed fine silt and clay into the crevices of the bodies at just the right consistency and pressure to fossilize them without tearing their delicate appendages, an ideal set of circumstances for ensuring later observation by future paleontologists.[23]

Due in part to the unusual circumstances under which these fossils were preserved, there's now little doubt about the unparalleled disparity of the Cambrian fauna. Based on available evidence from the Burgess Shale and other sites around the world, the Cambrian period witnessed the appearance of, arguably, more disparate body plans than ever before or since. And this disparity arose at a most unexpected time, assuming Darwinian theory, namely, right at the dawn of animal life.

THE ARTIFACT HYPOTHESIS

Walcott grasped these difficulties, and had a deep enough professional commitment to Darwinism, to search for a solution. He realized that the Precambrian fossil record could, in principle, assist in explaining the pattern of the Cambrian fossil record. The discovery of a rich Precambrian fossil history detailing variations accumulating little by little would serve to cast the pattern of the Burgess in a different light. Yet the Precambrian strata of his day showed no signs of providing any obvious transitional forms, much less a well-articulated bottom-up pattern of animals representing lower taxa proliferating into forms exemplifying higher and higher taxonomic categories. Nevertheless, an idea occurred to Walcott that gave him fresh hope.

Perhaps his awareness of the dramatic ways that the surface of the earth had changed over geologic time, making preservation of the Burgess fauna itself possible, inspired him. Finding marine animals so high above sea level no doubt made Walcott acutely aware of the way in which continents and seas had changed locations over the course of geological time. And so, Walcott, ever the geologist, proposed an ingenious geological solution to the biological problem of the origin of animal life. He noted that the Precambrian period was a period of dramatic continental uplift. He then suggested that the ancestors of the trilobites first evolved at a time when the Precambrian seas had receded from the landmasses. Then, at the beginning of the Cambrian, the seas again rose, covering the continents and depositing recently evolved trilobites. Thus, according to Walcott, ancestral precursors to the trilobites and other distinctive Cambrian forms had existed, but they were not fossilized in sediments that would later be elevated above sea level until early in the Cambrian; instead, before the Cambrian, during a period when sea levels were lower, trilobites and their ancestral forms were being deposited offshore in what are now only deep-sea sediments.[24] Walcott named this cryptic period of time in which trilobites and other animals were rapidly evolving offshore as the "Lipalian interval." (The term "Lipalian" is derived from the Greek word for lost.)

In this view, the abrupt appearance of the Cambrian body plans in the geological column was merely an "artifact" of incomplete sampling of the fossil record and, indeed, the inability to access the undersea sedimentary layers where the ancestors of the Cambrian fauna presumably lay encased. In short, the transgression and regression of ancient seas made the ancestral precursors of the Cambrian fauna inaccessible to discovery.

His artifact hypothesis (also known as the "Lipalian interval" hypothesis) was a distinct advance over Darwin's unadorned claim that the fossil ancestors of the Cambrian animals had not yet been discovered. Walcott's hypothesis had the advantage of accounting for the sudden appearance of the trilobites and the absence of ancestral and transitional forms by reference to known geological processes. It also could be tested, at least once offshore drilling technology advanced to allow for the sampling of the buried offshore sedimentary rocks.

Although Walcott conceded that his hypothesis was essentially a negative argument that attempted to explain away the absence of evidence, he insisted that it was a sensible inference from his broad sampling of the paleontological data. "I fully realize that the conclusions above outlined are based primarily on the absence of a marine fauna in Algonkian [Precambrian] rocks," he wrote, "but until such is discovered I know of no more probable explanation of the abrupt appearance of the Cambrian fauna than that I have presented."[25]

LUMPING AND SPLITTING

Walcott used another strategy for squaring the Burgess Shale with Darwin's theory of evolution. Taxonomists, tasked with identifying and naming distinct groups of life-forms, have been divided into two types: "lumpers" and "splitters." "Lumpers" tend to group disparate organisms together in the same large classificatory categories and then make distinctions between them at lower taxonomic levels. "Splitters" tend to separate similar organisms into numerous higher taxonomic divisions. Walcott favored lumping, and his doing so with the Burgess fossils seemingly minimized the difficulties associated with the sudden proliferation of so many new Cambrian forms.

On his return to the Smithsonian, he placed all of the exotic forms of the Burgess into modern phyla. One of his efforts at lumping placed *Marrella splendens* not only in the same phylum, but also in the same class (Trilobita) as the trilobites, despite obvious morphological differences. He justified this classification by arguing that the organism foreshadowed the trilobite (compare Figs. 1.4 and 2.3). Gould later criticized Walcott's method of classification as "shoehorning." He noted that even one of Walcott's fellow lumpers, Yale paleontologist Charles Schuchert, called the classification of *Marrella* into question.[26] Gould also noted that Walcott used this strategy to minimize the challenge posed by the morphological disparity of the Burgess forms.[27]

Some paleontologists today reject Gould's criticism of Walcott's inclusion of so many Burgess animal forms into existing taxonomic categories. Nevertheless, few paleontologists think Walcott's use of lumping explained away the Cambrian explosion. Most, for example, classify *Marrella splendens* within an existing modern phylum, namely, Arthropoda, even if they also classify it within a new and separate class, Marrellomorpha. Yet, whether *Marrella,* for example, falls within a novel phylum or class, matters less than explaining why so many clearly novel forms, and the novel structures these forms exhibit, first arose with such apparent suddenness.

RESOLUTION—FOR A TIME

Walcott thought that he had solved the mystery of the Cambrian explosion, as did many other Darwinists who gratefully adopted both his taxonomy and his version of the artifact hypothesis. And since Walcott's approach held out hope of one day uncovering evidence of a Precambrian trunk for the animal phyla along with its primary limbs, adherents could not be accused of moving the paleontological case for Darwinism into the realm of untestable dogma. They had only to wait for the technologies of seafloor drilling to emerge and hope that nature had seen fit, unmolested under the ocean deeps, to leave concrete evidence of the gradual emergence of the major Cambrian body plans.

Walcott's theoretical accomplishment was no mean feat. His discovery of the Burgess Shale was like a defense attorney with absolute faith in his client stumbling upon a room stuffed with clues that would seem to discredit him. Through his grouping of disparate body types into existing phyla and his ingenious version of the artifact hypothesis, Walcott had found an elegant way to explain all this seemingly uncooperative evidence in a Darwinian way.

In defending Walcott for overlooking significant features of the Burgess fossils, Gould points out that Walcott's multiple and growing administrative demands hardly left time to revisit the foundational categories of animal taxonomy. How much less, then, was Walcott likely to revisit the most fundamental assumption of all—the assumption that animals originated gradually in a Darwinian way as the result of natural selection acting on small incremental variations? Consideration of that possibility would come decades later, only after Walcott's version of the artifact hypothesis had itself exploded.

3

SOFT BODIES AND HARD FACTS

In the spring of 2000, Discovery Institute, where I do my research, sponsored a lecture at the University of Washington geology department by renowned Chinese paleontologist J. Y. Chen (see Fig. 3.1). As the result of his role in excavating a new discovery of Cambrian-era fossils in southern China, Professor Chen's standing in the scientific world was on the rise. The discovery, near the town of Chengjiang in the Kunming Province, revealed a trove of early Cambrian animal forms. After *TIME* magazine mentioned the Chengjiang discovery in a 1995 cover story about the Cambrian explosion,[1] interest in the fossils surged. When he came to Seattle, Professor Chen had already published numerous scientific papers about this profusion of novel life-forms and had established himself as one of the foremost experts on the fossils in this unique geological setting.

Not surprisingly, Chen's visit generated considerable interest among the University of Washington faculty. He came bearing intriguing photographs and samples of the oldest and most exquisitely preserved Cambrian fossils in the world from an exotic site halfway around the globe, a site, moreover, that was now widely acknowledged to surpass even the legendary Burgess Shale as the most extensive and significant Cambrian-era locality.

The fossils from the Maotianshan Shale near Chengjiang (see Fig. 3.2) had established an even greater variety of Cambrian body plans from an even older layer of Cambrian rock than those of the Burgess, and they did so with an almost photographic fidelity. The Chinese fossils also helped to establish that the Cambrian animals appeared even more explosively than previously realized.

So there was little doubt about the significance of the discoveries that Chen came to report that day. What was soon in doubt, however, was Chen's scientific orthodoxy. In his presentation, he highlighted the

FIGURE 3.1
J. Y. Chen.

FIGURE 3.2
Figure 3.2a (left) shows the Moatianshan Shale outcrop. *Courtesy Illustra Media. Figures 3.2b and c (center and right)* show a Precambrian-Cambrian boundary marker at the Moatianshan site. *Courtesy Paul Chien.*

apparent contradiction between the Chinese fossil evidence and Darwinian orthodoxy. As a result, one professor in the audience asked Chen, almost as if in warning, if he wasn't nervous about expressing his doubts about Darwinism so freely—especially given China's reputation for suppressing dissenting opinion. I remember Chen's wry smile as he answered. "In China," he said, "we can criticize Darwin, but not the government. In America, you can criticize the government, but not Darwin."

Nevertheless, those in the audience that day soon learned that Professor Chen had good reasons for questioning Darwin's picture of the history of life. As Chen explained, the Chinese fossils turn Darwin's tree of life "upside down." They also cast doubt on a surviving version of Charles Walcott's artifact hypothesis, a crucial prop in the case for Darwinian gradualism.

BURGESS REVISITED

By the time Charles Walcott finished his last excavation of the Burgess Shale in 1917, he and his team had collected over 65,000 fossil specimens, all of which were shipped to the Natural History Museum at the Smithsonian Institution for cataloguing. In 1930, another American paleontologist, Harvard professor Percy Raymond, initiated another investigation of the Burgess. His specimens were also eventually stored in the United States.

As a result of these two prominent American-led excavations, there were initially no collections of Burgess fossils on public display in Canada. Many Canadian scientists regarded this as a national embarrassment, so in the 1960s the Canadian Geological Survey commissioned a British team to resume digging at the Walcott quarry, in order to "repatriate the Burgess Shale" by keeping most of the newly discovered fossils on permanent display in Canada.[2] The team was led by paleontologist Harry Whittington (see Fig. 3.3), of the University of Cambridge, who was assisted by two of his graduate students, Simon Conway Morris and Derek Briggs, both of whom would eventually distinguish themselves as internationally renowned experts on the Burgess Shale.

As Whittington analyzed the Cambrian fauna at the Burgess, he realized that Walcott had grossly underestimated the morphological disparity of this group of animals. Many of the creatures in the assemblage featured unique body designs, unique anatomical structures, or both.[3] *Opabinia*,

FIGURE 3.3
Harry Whittington. *Courtesy Archives of the Museum of Comparative Zoology, Ernst Mayr Library, Harvard University.*

with its five eyes, fifteen distinct body segments, and a claw at the end of a long proboscis, exemplified the unique forms on display in the Burgess. But so did *Hallucigenia, Wiwaxia, Nectocaris,* and many other Burgess animals. To this day, paleontologists describing *Nectocaris,* for example, can't decide whether it more closely resembles an arthropod, a chordate, or a cephalopod (a class of mollusk; see Fig. 3.4).

Whittington found that grouping such forms within well-established taxonomic categories, even *higher* taxonomic categories such as the class or phylum, strained the limits of these classifications. Even many of those animals that fell easily into existing phyla represented clearly unique subphyla or classes of organisms. *Anomalocaris* (literally, "abnormal shrimp") and *Marrella,* for example, both had hard exoskeletons or body parts and clearly represent either arthropods or creatures closely related to them. Yet each of these animals possessed many distinct anatomical parts and exemplified different ways of organizing these parts, thus clearly distinguishing themselves from better-known arthropods such as the previous staple of Cambrian paleontological studies, the trilobite.

Whittington, a trilobite expert, understood this as well as anyone. In 1971, he published the first comprehensive taxonomic review of the Burgess biota. In his review, he broke decisively with Walcott's previous attempt to lump all Cambrian forms into a few preexisting taxonomic categories.

In so doing, he reemphasized the morphological disparity present in the Burgess animal biota and, in the process, deprived evolutionary biologists of one part of Walcott's two-part strategy for minimizing the Cambrian

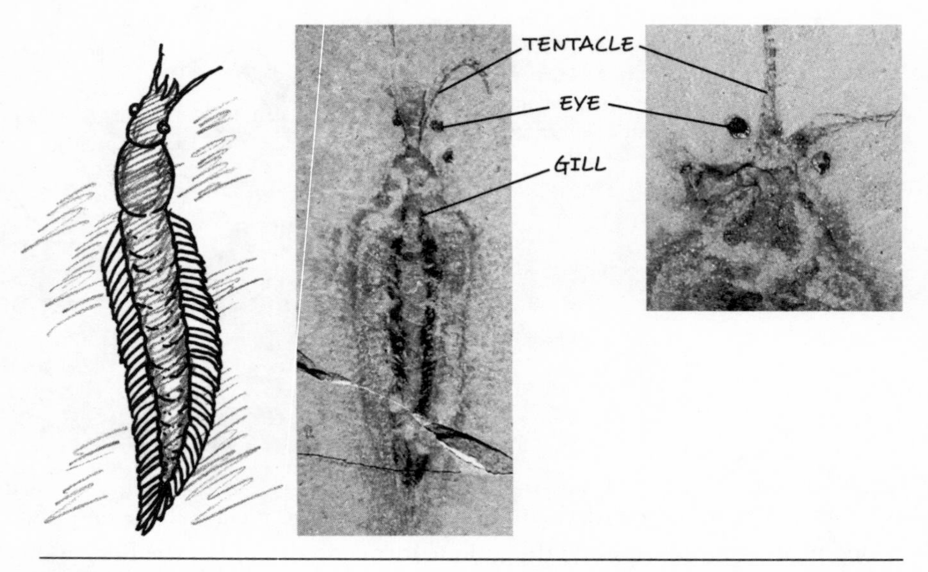

FIGURE 3.4
Figure 3.4a (left): Artist rendering of *Nectocaris. Figure 3.4b (middle, right):* Photograph of *Nectocaris* fossil. Reprinted by permission from Macmillan Publishers Ltd.: *Nature,* Martin R. Smith and Jean-Bernard Caron, "Primitive Soft-Bodied Cephalopods from the Cambrian," *Nature,* 465 (May 27, 2010): 469–72. Copyright 2010.

problem. By lumping all Burgess animals into existing phyla and classes, Walcott had seemingly diminished the problem of disparity by reducing the number of novel phyla for which connecting intermediates were required. By recognizing the disparity clearly on display in the Burgess, Whittington partially undercut Walcott's solution to the Cambrian mystery and highlighted what would become its central unsolved problem: the origin of novel biological form.

SOUNDING THE SEAFLOOR

Although Whittington, and later Gould, rejected Walcott's early attempt to "shoehorn" all the animals of the Burgess Shale into preexisting taxonomic categories, many paleontologists now also reject Stephen Jay Gould's characterization of many Burgess Shale creatures as being so exotic as to defy affinity in classification with any modern groups.[4] Many of these paleontologists would also recognize fewer total phyla first appearing in the Cambrian than Gould did, and perhaps even as few as Walcott did. As discussed in the previous chapter, still other paleontologists now favor "rank-free" approaches to classification.

Regardless, most paleontologists recognize that the Burgess Shale attests to an extraordinary profusion of new animal forms—including many manifesting unique anatomical structures and arrangements of body parts. Thus, whatever differences of opinion exist about *how to classify* these animals—and any five-year-old child can distinguish them from each other and from all previously known forms of life—their origin still requires explanation. Thus, as noted, Walcott's use of "lumping" did not solve the Cambrian mystery.[5]

But what of the second part of Walcott's proposal, the artifact hypothesis? To evaluate this hypothesis, Walcott devised a more clear-cut and less subjective test. Recall that Walcott argued that the ancestral precursors of the Cambrian animals were missing from the Precambrian fossil record because of the transgression and regression of seas. He posited an interval of geologic time in which the ancestors of the Cambrian fauna were evolving offshore in a Precambrian ocean and being deposited only in layers of marine sedimentary rock. In this hypothesis, only after the ancient ocean rose and covered the continent were the remains of Cambrian sea animals preserved in sediment that today is above sea level.

When Walcott proposed his ingenious geological scenario, it could not yet be tested. But with the development of offshore drilling technology in the 1940s, 1950s, and 1960s, oil companies began to drill through thousands of feet of marine sedimentary rock.[6] As geologists evaluated the contents of these drill cores, they did not find Walcott's predicted Precambrian fossils.

Instead, an even more fundamental problem for the hypothesis arose. At the time that Walcott proposed his version of the artifact hypothesis, geologists considered the oceanic and continental plates to be essentially stable and fixed with respect to one another. Mountain building, faulting, and other geologic processes were attributed to worldwide changes in sea level, accumulating troughs of sediment called geosynclines, rising mounds of igneous rocks from beneath the earth's crust, and even a shrinking earth.[7]

The idea that enormous, solid plates actually *move*, recycling themselves through the plate-tectonic processes of subduction and seafloor spreading, had not yet been proposed. Yet modern plate tectonic theory now affirms that oceanic crustal material eventually plunges back into the earth and melts in a process known as subduction. After surface rocks melt during subduction, they form a new supply of molten magma. Eventually, magma from other locations deep in the earth wells up at mid-oceanic ridges to form new igneous rocks, in a process known as seafloor spreading. It follows that any oceanic sediments deposited atop the oceanic igneous crust will have a limited "life span" on the surface of the earth. Eventually, these sedimentary rocks collide with the continental margin, plunge deep into the upper mantle, and melt to form magma.

As a consequence of this cycle, the maximum age of any marine sediment is strictly limited. And according to modern estimates, the oldest section of oceanic crust has existed only since the Jurassic (or about 180 million years ago[8])—far too young to contain fossil ancestors of the trilobites. As the evidence for plate tectonics mounted, scientists discarded Walcott's artifact hypothesis and Lipalian interval as nonstarters. Paleontologists today do not expect to find any Precambrian ancestors of the trilobites in oceanic sediments, since they realize that there are no Precambrian *sediments* in the ocean basins. If Precambrian strata are to be found anywhere, continents are the place.

OTHER VERSIONS OF THE ARTIFACT HYPOTHESIS

Although Walcott's proposals to explain away the absence of fossilized ancestors of the Cambrian animals came to naught, other versions of the artifact hypothesis continued to circulate. These proposals take two basic forms. Some scientists claimed, though for different reasons, that the expected Precambrian fossil ancestors had simply *not yet been found*—that missing fossils were an artifact of the incomplete *sampling* of the fossil record. Others suggested that Precambrian sedimentary rocks had not *preserved* the missing fossils—that the incomplete preservation of the Precambrian animals meant the missing fossils *were no longer there* to be found.

Walcott rejected the idea that paleontologists simply had not looked in, or sampled, enough places. He noted that geologists already had extensively investigated "the great series of Cambrian and Precambrian strata in eastern North America." Though they had looked "from Alabama to Labrador; in western North America [and] from Nevada and California far into Alberta and British Columbia, and also in China" their investigations had turned up nothing of significant interest.[9] In Walcott's view the continents simply had not preserved the fossilized remains of the Cambrian ancestors.

Before Walcott, some geologists had gone a step farther and suggested that *all* Precambrian sedimentary rocks had been destroyed via extreme heat and pressure, a process called "universal metamorphism." Walcott rejected this hypothesis, since he himself had encountered a "great series of pre-Cambrian sedimentary rocks on the North American continent" among other places. Other geologists suggested that major bursts of evolutionary innovation occurred only during periods when sedimentary deposition had ceased, thus again resulting in a lack of fossil preservation. But, as Gould remarked of Walcott's artifact hypothesis, this explanation also appeared to many scientists "forced and ad hoc . . . born of frustration, rather than the pleasure of discovery."[10]

CONTEMPORARY VERSIONS OF THE ARTIFACT HYPOTHESIS: TOO SOFT OR TOO SMALL

After the demise of the "universal metamorphism" idea, some paleontologists proposed simpler, more intuitively plausible versions of the artifact hypothesis. They claimed that the proposed intermediate forms leading to the Cambrian animals may have been either too small or too soft, or both, to have been preserved.

Developmental biologist Eric Davidson, of California Institute of Technology, has suggested that the transitional forms leading to the Cambrian animals were "microscopic forms similar to modern marine larvae" and were thus too small to have been reliably fossilized.[11] Other evolutionary scientists, such as Gregory Wray, Jeffrey Levinton, and Leo Shapiro, have suggested that the ancestors of the Cambrian animals were not preserved, because they lacked hard parts such as shells and exoskeletons.[12] They argue that since soft-bodied animals are difficult to fossilize, we shouldn't expect to find the remains of the supposedly soft-bodied ancestors of the Cambrian fauna in the Precambrian fossil record. University of California, Berkeley, paleontologist Charles R. Marshall summarizes these explanations:

[I]t is important to remember that we see the Cambrian "explosion" through the windows permitted by the fossil and geological records. So when talking about the Cambrian "explosion," we are typically referring to the appearance of large-body (can be seen by the naked eye) and preservable (and therefore largely skeletonized) forms. . . . If the stem lineages were both small and unskeletonized, then we would not expect to see them in the fossil record.[13]

Though intuitively plausible, several discoveries call into question both of these versions of the artifact hypothesis. As for the idea that the ancestors of the Cambrian animals were too *small* to be preserved, paleontologists have known for some time that the cells of filament-shaped microorganisms (probably cyanobacteria) have been preserved in ancient Precambrian rocks. Paleobiologist J. William Schopf, of the University of California, Los Angeles, has reported an extremely ancient example of these fossils in the Warrawoona Group strata of western Australia. These fossilized cyanobacteria are preserved in 3.465-billion-year-old bedded cherts (microcrystalline sedimentary rocks).[14] The same strata have also preserved stromatolite mats, an organic accretionary growth structure usually indicating the presence of bacteria, within slightly younger dolostone sediments of roughly 3.45 billion years in age (see Fig. 3.5).[15]

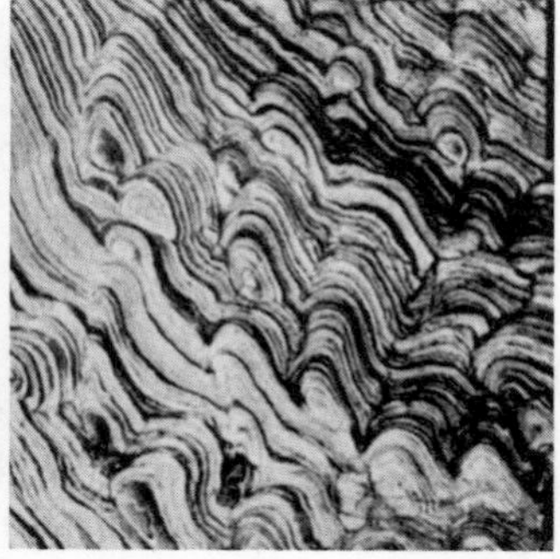

FIGURE 3.5

Figure 3.5a (left): Photographs of Cambrian-age fossil stromatolites. *Courtesy Wikimedia Commons, user Rygel, M. C. Figure 3.5b (right):* Alternating fine- and coarse-layered structure of Precambrian stromatolite fossils shown in cross section. *Courtesy American Association for the Advancement of Science,* Figure 2B, Hoffman, P., "Algal Stromatolites: Use in Stratigraphic Correlation and Paleocurrent Determination," *Science,* 157 (September 1, 1967): 1043–45. Reprinted with permission from AAAS.

These discoveries pose a problem for the idea that the Cambrian ancestors were too small to survive in the fossil record. The sedimentary rocks that preserve the fossilized cyanobacteria and single-celled algae are far older and, therefore, far more likely to have been destroyed by tectonic activity than those later sedimentary rocks that should have preserved the near-ancestors of the Cambrian animals. Yet these rocks, and the fossils they contain, have survived just fine. If paleontologists can find tiny fossilized cells in these far older and rarer formations, shouldn't they also be able to find some ancestral forms of the Cambrian animals in younger and more abundant sedimentary rocks? Yet few such precursors have been found.

There are also several reasons to question the second version of this hypothesis—the idea that the presumed Cambrian ancestors were too soft to be preserved. First, some paleontologists have questioned whether soft-bodied ancestral forms of the hard-bodied Cambrian animals would have even been anatomically viable.[16] They argue that many animals representing phyla such as brachiopods and arthropods could not have evolved their soft parts first and then added shells later, since their survival depends upon their ability to protect their soft parts from hostile environmental forces. Instead, they argue that soft and hard parts had to arise together.[17] As paleontologist James Valentine, of the University of California, Berkeley, has noted in the case of brachiopods, "The brachiopod *Bauplan* [body plan] cannot function without a durable skeleton."[18] Or as J. Y. Chen and his colleague Gui-Qing Zhou observe: "Animals such as brachiopods . . . cannot exist without a mineralized skeleton. Arthropods bear jointed appendages and likewise require a hard, organic or mineralized outer covering."[19]

Because these animals typically require hard parts, Chen and Zhou assume that the ancestral forms of these animals should have been preserved somewhere in the Precambrian fossil record if in fact they were ever present. Thus, the absence of hard-bodied ancestors of these Cambrian animals in the Precambrian strata shows that these animals first arose in the Cambrian period. As they rather emphatically insist: "The observation that such fossils are absent in Precambrian strata proves that these phyla arose in the Cambrian."[20]

It should be pointed out that this argument cannot be made for all Cambrian animal groups and, in my view, does not achieve the standing of a "proof" in any case. Many Cambrian phyla, including phyla characterized by mostly hard-shelled animals such as mollusks and echinoderms, do have soft-bodied representatives. The earliest known mollusk, *Kimberella*, for example, lacked a hard external shell (though it did have other hard parts).[21] So, clearly, some mainly hard-shelled Cambrian groups could have had soft-bodied ancestors.

It is also possible to postulate the existence of an arthropod or brachiopod ancestor—especially some extremely distant ancestor—lacking a hard shell. Soft-bodied onychophorans (velvet worms) were once proposed as ancestors of the arthropods, though more recent studies challenge this idea. Onychophorans themselves arise well *after* arthropods in the fossil record and cladistic analysis suggests that onychophorans may be a sister, rather than an ancestral, group to arthropods.[22] Even so, it's difficult to disprove a negative: in particular, to foreclose the possibility that arthropods or brachiopods *might* have had a soft-bodied ancestor deep in the Precambrian.

Nevertheless, it seems unlikely on a Darwinian view of the history of life that *all* Cambrian arthropod or brachiopod ancestors, especially the relatively recent ancestors of these animals, would have lacked hard parts entirely. There are many types of arthropods that arise suddenly in the Cambrian—trilobites, *Marrella, Fuxianhuia protensa, Waptia, Anomalocaris*—and all of these animals had hard exoskeletons or body parts. Moreover, the only known extant group of arthropods without a hard exoskeleton (the pentastomids) have a parasitic relationship with arthropods that do.[23] Thus, surely, it seems likely that *some* of the near ancestors of the many arthropod animals that arose in the Cambrian would have left at least *some* rudimentary remains of exoskeletons in the Precambrian fossil record—if, in fact, such ancestral arthropods existed in the Precambrian and if arthropods arose in a gradual Darwinian way.

Moreover, the arthropod exoskeleton is part of a tightly integrated anatomical system. Specific muscles, tissues, tendons, sensory organs—and a special mediating structure between the soft tissue of the animal and the exoskeleton called the endophragmal system—are all integrated to support the process of molting and exoskeletal growth and maintenance that is integral to the arthropod mode of existence. A best-case Darwinian scenario for the origin of such a system would, therefore, envision the "co-evolution" of these separate anatomical subsystems in a coordinated fashion, since some of these anatomical subsystems confer a functional advantage to the animal largely by supporting, and promoting, the growth and maintenance of the exoskeleton (and vice versa). Others would be vulnerable to damage without it. Thus, it seems unlikely that these interdependent subsystems would evolve independently first without an exoskeleton, only to have the exoskeleton arise suddenly as a kind of accretion atop an already integrated system of soft parts at the end of a long evolutionary process.

This, again, makes it reasonable to expect that at least *some* rudimentary arthropod hard parts would have been preserved in the Precambrian if arthropods were present then. That such parts are unknown for *all* Cambrian arthropods (and brachiopods) in a fossil record that presumably favors hard-part preservation, seems at least curious. And it appears, on its face, to support the assertions of those Cambrian paleontologists such as Chen and Zhou who take the absence of *any* hard parts in the Precambrian record as evidence of the absence of those groups that typically depend on hard parts for their existence.

In any case, advocates of the artifact hypothesis must at least explain a Cambrian explosion of hard body parts, if not whole Cambrian animals. As paleontologist George Gaylord Simpson noted in 1983, even if it's true that Precambrian ancestors were not preserved simply because they lacked hard parts, "there is still a mystery to speculate about: Why and how did many animals begin to have hard parts—skeletons of sorts—with apparent suddenness around the beginning of the Cambrian?"[24]

There is an additional, more formidable, difficulty for this version of the artifact hypothesis. Although the fossil record generally does not preserve soft body parts as frequently as hard parts, it has preserved many soft-bodied animals, organs, and anatomical structures from both the Cambrian and the Precambrian periods.

As we saw earlier, Precambrian sedimentary rocks in several places around the world have preserved fossilized colonial blue-green algae, single-celled algae, and cells with a nucleus (eukaryotes).[25] These microorganisms were not only small, but they also entirely lacked hard parts. Another class of late Precambrian organisms called the Vendian or Ediacaran biota included the fossilized remains of many soft-bodied organisms, including many that may well have been lichens, algae, or protists (microorganisms with cells containing nuclei). Cambrian-era strata themselves preserve many soft-bodied creatures and structures. The Burgess

Shale in particular preserved the soft parts of several types of hard-bodied Cambrian animals, such as *Marrella splendens,*[26] *Wiwaxia,*[27] and *Anomalocaris.* The Burgess Shale also documents *entirely* soft-bodied representatives[28] of several phyla, including:

- Cnidaria (represented by an animal called *Thaumaptilon,* a feather-shaped colonial organism formed from smaller soft sea anemone–like animals)[29]
- Annelida (represented by the polychaete worms *Burgessochaeta* and *Canadia*)[30]
- Priapulida (represented by *Ottoia, Louisella, Selkirkia*—all worms with a distinctive proboscis)[31]
- Ctenophora (represented by *Ctenorhabdotus,* a gelatinous animal with a translucent body similar to a modern comb jelly)[32]
- Lobopodia (represented by *Aysheaia* and *Hallucigenia,* segmented soft-bodied animals with many legs)[33]

The Burgess also preserves soft-bodied animals of unknown affinities, such as *Amiskwia,* a gelatinous air mattress–like animal;[34] *Eldonia,* a jellyfish-like animal with a much more complex anatomy than a modern jellyfish;[35] and the aforementioned, difficult to classify, *Nectocaris.*[36] As Simon Conway Morris notes, "The existing [Burgess] collections represent approximately 70,000 specimens. Of these, about 95 percent are either soft-bodied or have thin skeletons."[37]

THE CHENGJIANG EXPLOSION

Any doubts about the ability of sedimentary rocks to preserve soft and small body parts were permanently laid to rest by a series of dramatic fossil finds in southern China beginning in the 1980s.

In June 1984, paleontologist Xian-Guang Hou journeyed to Kunming, in southern China, to prospect for fossilized samples of a bivalved arthropod called a bradoriid.[38] The area around Kunming in the Yunnan Province was well known for its lower Cambrian strata and typical Cambrian-era fossils, such as bradoriids and trilobites, both of which were relatively easy to preserve because of their characteristically hard exoskeletons. In 1980,

Hou had found many bradoriid samples in a geological formation called the Qiongzhusi Section near Kunming.

In the summer of 1984 Hou traveled to the town of Chengjiang to look for bradoriids in another geological formation called the Heilinpu Formation. His efforts there yielded little success. As a result, he turned his attention to another outcrop, a sedimentary sequence now called the Maotianshan Shale. Hou's team set farmworkers to digging out and scouring the mudstone blocks. His book, *The Cambrian Fossils of Chengjiang, China,* describes what happened next:

> *At about three o'clock in the afternoon of Sunday July 1, a semicircular white film was discovered in a split slab, and was mistakenly thought to represent the valve of an unknown crustacean. With the realization that this . . . represented a previously unreported species, breaking of the rock in a search for additional fossils continued apace. With the find of another specimen, a 4–5 cm long animal with limbs preserved, it became apparent that here was nothing less than a soft-bodied biota.*[39]

Hou remembers the Cambrian specimen vividly, for it appeared "as if it was alive on the wet surface of the mudstone."[40] Redoubling their efforts, the researchers quickly uncovered the fossilized remains of one extraordinary soft-bodied animal after another. Most of the fossils were preserved as flattened two-dimensional imprints of three-dimensional organisms, although, as Hou observes, "some retain a low three-dimensional relief."[41] Most important, he notes, "The remains of hard tissues, such as the shells of brachiopods or the carapaces of trilobites, are well represented in the Chengjiang fauna, but less robust tissues, which are usually lost through decomposition, are also beautifully preserved."[42]

As the result of the very fine, small-grained sediments in which they were deposited, the Chengjiang fossils preserved anatomical details with a fidelity surpassing even that of the Burgess fauna.[43] The Maotianshan Shale also preserved an even greater variety of soft-bodied animals and anatomical parts than the Burgess Shale had done. In the ensuing years, Hou and his closest colleagues, J. Y. Chen and Gui-Qing Zhou, found many excellent examples of well-preserved animals lacking even a keratinized exoskeleton, including soft-bodied members of phyla such as Cnidaria (corals and jellyfish), Ctenophora (comb jellies), Annelida (a type of "ringed" segmented worm), Onychophora (segmented worms with legs), Phoronida (a tubular, filter-feeding marine invertebrate), and Priapulida (another distinctive type of worm).[44] (See Fig. 3.6.)

They found fossils preserving the anatomical details of numerous soft tissues and organs such as eyes, intestines, stomachs, digestive glands, sensory organs, epidermes, bristles, mouths, and nerves.[45] They also discovered jellyfish-like organisms called *Eldonia,* which exhibit delicate, soft body parts such as radiating water canals and nerve rings. Other fossils even revealed the contents of the guts of several animals.[46]

The discoveries near Chengjiang demonstrated beyond any reasonable doubt that sedimentary rocks can preserve soft-bodied fossils of great antiquity and in exquisite detail, thereby challenging the idea that the absence of Precambrian ancestors is a consequence of the fossil record's inability to preserve soft-bodied animals from that period. At the same time, the sedimentary rocks near Chengjiang had other surprises in store.

PRECAMBRIAN SECRETS

Paul Chien is a Chinese-American marine biologist who, as a boy, left mainland China with his family to escape the 1949 Communist takeover under Mao Tse-tung. Eventually, after completing Ph.D. studies in the United States, he became a biology professor at the University of San Francisco. He first learned about the discoveries in southern China after reading the *TIME* magazine cover story in 1995. Then he learned, ironically, from a story in *People's Daily,* the official newspaper of the Chinese Communist party, that some Chinese paleontologists thought that these discoveries challenged a Darwinian view of the history of life.

The great variety of the marine invertebrates present in Chengjiang fascinated Chien and convinced him to return to the country of his birth. After making his first trip in the summer of 1996, he met J. Y. Chen. As Paul Chien returned over several successive summers to do research of his own, he and J. Y. Chen continued to share their findings and compare notes.

Upon arriving in China in 1998 for his third summer of research, Paul Chien learned that Chen had discovered a fossil of an adult sponge in the late Precambrian rocks of a sedimentary formation called the Doushantuo Phosphorite—a formation that lies beneath the Maotianshan Shale. As the

FIGURE 3.6

Cambrian explosion fossils from Chengjiang fauna, artist depictions and fossil photos. *Photos in 3.6a, b, e, and f courtesy J. Y. Chen. Photos in 3.6c and d courtesy Paul Chien.*

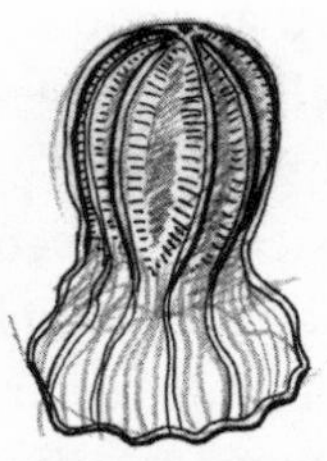

Figure 3.6a: A ciliated comb jelly, from the phylum Ctenophora.

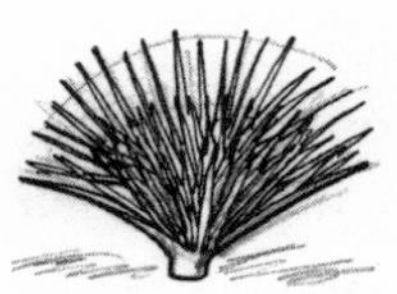

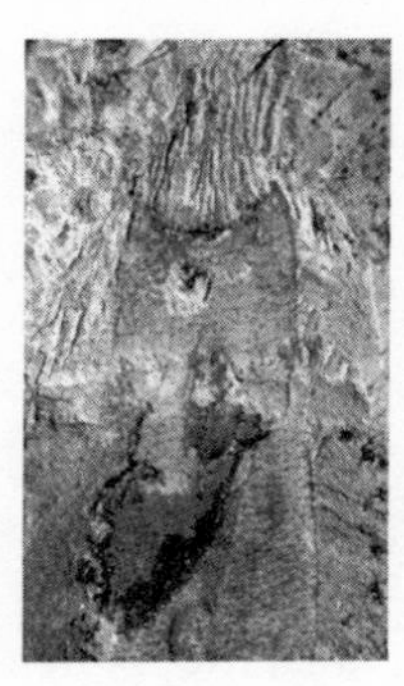

Figure 3.6b: A filter-feeding worm-like member of the phylum Phoronida.

Figure 3.6c: The shrimp-like arthropod, *Waptia.*

Figure 3.6d: A priapulid worm with its distinctive proboscis.

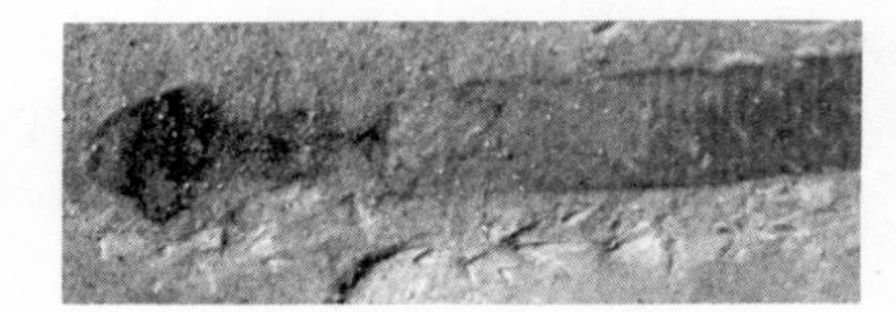

Figure 3.6e: The enigmatic jellyfish-like *Eldonia.*

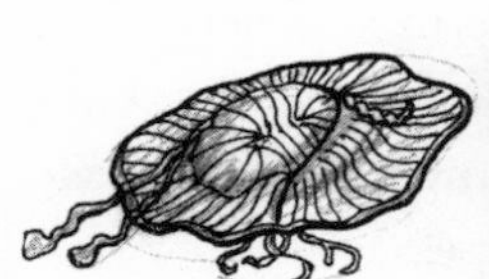

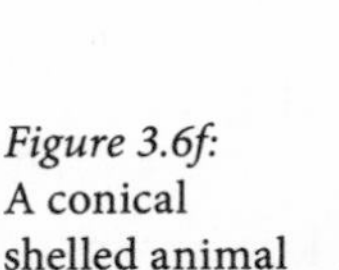

Figure 3.6f: A conical shelled animal from the phylum Hyolitha.

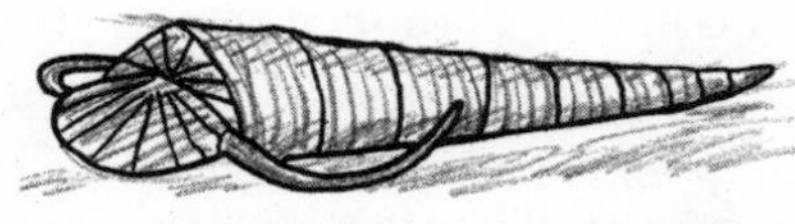

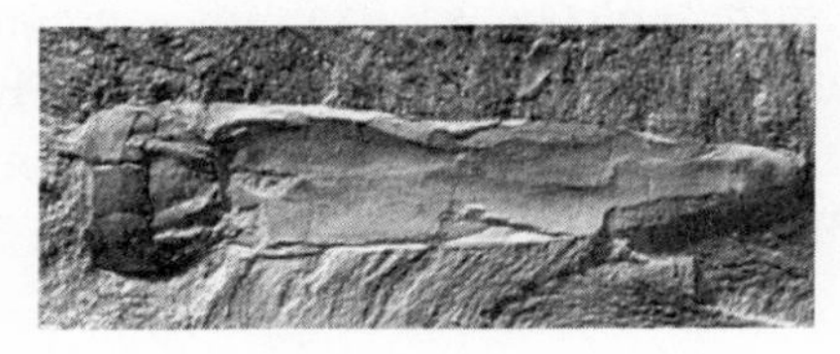

two scientists examined the sediments that encased Chen's fossil sponge, they made a discovery that would doom the most popular remaining version of the artifact hypothesis.

As J. Y. Chen began to examine the sedimentary rocks that enclosed his fossilized sponge, he decided to look at them in a so-called thin section under a light microscope. Chen wondered whether smaller embryonic forms of these Precambrian animals might also have been preserved in these phosphorite rocks. Sure enough, under magnification he found little round balls that he and Paul Chien identified as sponge embryos. In 1999, at a major international conference about the Cambrian explosion held near Chengjiang, J. Y. Chen, Paul Chien, and three other colleagues presented their findings.[47]

A number of Chinese paleontologists questioned them at first, suggesting that the little round balls were not sponge embryos at all, but instead the remains of brown and green algae.[48] Here Paul Chien's expertise came to the fore. Early in his career, Chien had perfected a technique for examining the embryos of living sponges under a scanning electron microscope. He now adapted his technique to examine these microscopic fossilized structures using a more powerful microscope. What he found startled him and amazed other scientists.

Sponges are nature's glassworks. They are made of a soft and flexible lattice of cells from which protrude silica-encrusted "spicules." Though sponges come in a variety of shapes and sizes, they are one of the simplest known forms of animal life, with between six and ten distinctive cell types.[49] In comparison, the typical arthropod has between thirty-five and ninety cell types.

As Chien examined the little balls of the Doushantuo Phosphorite under the scanning microscope, he noticed what looked like cells undergoing cell division. At first, he had no way to determine what type of cells they might be. But as he examined cross sections of these cells more carefully, he identified a distinctive structure that he knew from his prior research.

Sponges may be recognized by their characteristic spicules, and the fossilized cells he was examining preserved microscopic spicules in the early stages of their development.[50] Clearly, these were not algal balls; they were sponge embryos. Even more surprising, Chien was able to observe the internal structure of these embryonic cells, allowing him to identify the nuclei of some of these cells within the fossilized remains of the larger outer cell membrane (see Fig. 3.7).

The discovery of these sponge embryos has proven decisive in the case against the remaining versions of the artifact hypothesis, for several reasons.

First, though spicules in sponges are encased in a thin layer of glassy silica, sponges are generally considered to be a soft-bodied organism because of the predominantly soft tissues out of which the rest of their bodies are made. Moreover, the cells of all embryos during their earliest embryonic stages are soft. Even in animals that have internal or external skeletons, the nascent forms of these hard parts do not emerge until gastrulation, partway through the process of embryological development. Thus, discovery of an embryo in the earliest stages of cell division shows beyond a doubt that Precambrian sedimentary rocks can, under the right circumstances, preserve soft-bodied organisms.

It also established something else. J. Y. Chen found these sponge embryos beneath the Cambrian–Precambrian boundary in late Precambrian rock. Yet these Precambrian layers did not preserve remains of

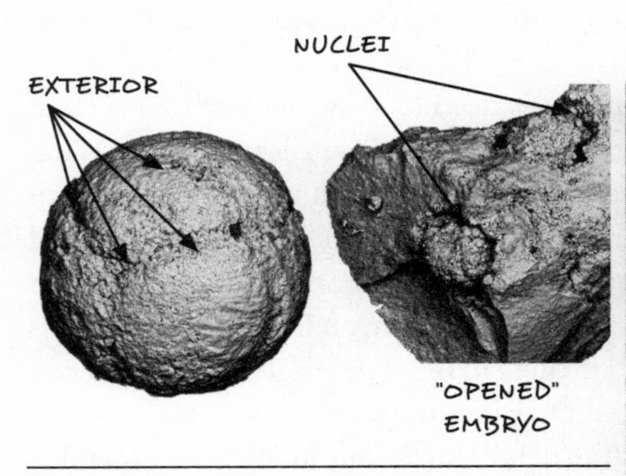

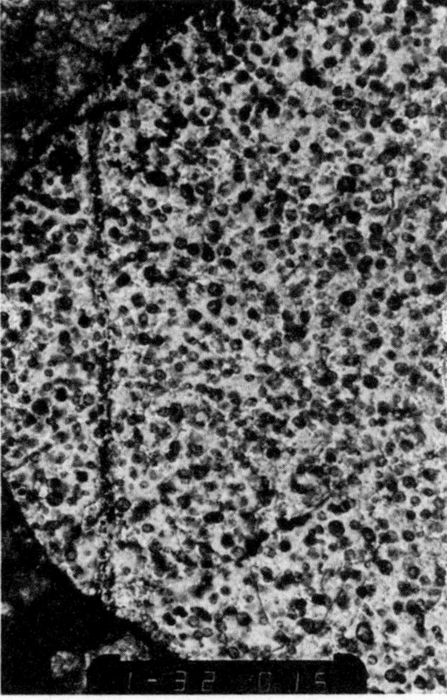

FIGURE 3.7
Figure 3.7a (above): Photographs of fossilized sponge embryos in the early stages of cell division, showing the sponge at the 8-cell stage of division, with four of its cells marked in the foreground.
Figure 3.7b (right): A close-up image of a fossil of a sponge embryo cell revealing numerous yolk granules inside. *Courtesy Paul Chien.*

any clearly ancestral or intermediate forms leading to the other main groups of Cambrian animals. This raised an obvious question. If the Precambrian sedimentary strata beneath the Maotianshan Shale preserved the soft tissues of tiny, microscopic sponge embryos, why didn't they also preserve the near ancestors of the *adult* animals that arose in the Cambrian, especially since some of those animals must have had at least some hard parts as a condition of their viability? If these strata could preserve embryos, then they should have preserved fully developed animals—at least, if such animals were present at the time. That well-developed, clearly ancestral animal forms were not preserved, when tiny sponge embryos[51] were, strongly indicates that such forms were simply not present in the Precambrian layers.

Of course, there are conditions under which fossils are unlikely to be preserved. We know, for example, that near-shore sands do not favor preservation of detail, let alone the fine detail of very small organisms a millimeter or less in length.[52] Even so, such considerations do little to bolster the artifact hypothesis. The sedimentary environments that produced the carbonates, phosphorites, and shales of the Precambrian strata beneath the Maotianshan Shale, for example, would have provided a congenial environment for fossilizing all sorts of creatures during Precambrian times.

Advocates of the artifact hypothesis need to show not just that certain factors discourage preservation in general. No one disputes that. What they need to show is that these factors were ubiquitous in Precambrian depositional environments worldwide. If near-shore sands characterized *all* Precambrian sedimentary deposits, then paleontologists would not expect to find any fossils there, at least not any tiny ones. Yet clearly this is not the case. Precambrian strata include many types of sediments that can preserve—and in the case of the Doushantuo formation in China, *have preserved*—animal remains in fine detail, including small and vulnerable sponge embryos.

Moreover, geologists Mark and Dianna McMenamin have noted that in many other Cambrian locales around the world, including one in Newfoundland that they have studied extensively, the pattern of sedimentation changes very little across the Cambrian–late Precambrian boundary, suggesting that many Precambrian environments would have provided equally good environments for the preservation of fossils.[53]

In their 2013 book, *The Cambrian Explosion*, paleontologists James Valentine and Douglas Erwin go further. They note that many late Precambrian depositional environments actually provide *more* favorable settings for the preservation of fossils than those present in the Cambrian period. As they note, "a revolutionary change in the sedimentary environment—from microbially stabilized sediments during the Ediacaran [late Precambrian] to biologically churned sediments as larger, more active animals appeared—occurred during the early Cambrian. Thus, the quality of fossil preservation in some settings may have actually declined from the Ediacaran to the Cambrian, the opposite of what has sometimes been claimed, yet we find a rich and widespread explosion of [Cambrian] fauna."[54]

STATISTICAL PALEONTOLOGY

Recent work in a field known as statistical paleontology casts further doubt on the artifact hypothesis. Since the discovery of the Burgess Shale, Precambrian- and Cambrian-era discoveries have repeatedly uncovered fossil forms that either establish radically disparate new forms of life or, increasingly, forms that fall into existing higher taxonomic groups (such as class, subphylum, or phylum).

As a result, the fossil record amply documents organisms corresponding to the terminal branches on the Darwinian tree of life (animal forms representing new phyla or classes, for example), but it fails to preserve those organisms representing the internal branches or nodes leading to these representatives of novel phyla and classes of Cambrian-era animals. Yet these intermediates are the very forms required to connect the terminal branches to form a coherent evolutionary tree and establish that the representatives of the Cambrian animals did arise by means of a gradual evolutionary process from simpler Precambrian ancestors.

Recall that Louis Agassiz thought that this pattern could not be explained by appealing to an incomplete fossil record, because the fossil record was strangely selective in its incompleteness, preserving abundant evidence of the terminal branches but consistently neglecting to preserve the representatives of the internal branches or nodes.

Contemporary paleontologists, such as Michael Foote at the University of Chicago, have come to a similar conclusion. Foote has shown, using statistical sampling analysis, that as more and more fossil discoveries fall within existing higher taxonomic groups (e.g., phyla, subphyla, and classes), and as they fail to document the rainbow of intermediate forms expected in the Darwinian view of the history of life, it grows ever more improbable that the absence of intermediate forms reflects a sampling bias—that is, an "artifact" of either incomplete sampling or preservation.

This kind of analysis merely quantifies what, in other circumstances, we would sense intuitively. Imagine that you reach into an enormous barrel full of marbles and randomly pull out a yellow, a red, and a blue marble. At this point your brief sampling should leave you undecided as to whether you have a representative sample of the barrel's contents. You might at first imagine that the barrel also contains marbles representing a rainbow of intermediate colors. But as you continue to sample from every place in the barrel and find that the barrel disgorges only those same three colors you begin to suspect that it may offer a much more limited selection of colors than, say, the rack of color samples at your local paint store.

Over the past 150 years or so, paleontologists have found many representatives of the phyla that were well known in Darwin's time (by analogy, the equivalent of the three primary colors) and a few completely new forms altogether (by analogy, some other distinct colors such as green and orange, perhaps). And, of course, within these phyla there is a great deal of variety. Nevertheless, the analogy holds at least insofar as the differences in form between any member of one phylum and any member of another phylum are vast, and paleontologists have utterly failed to find forms that would fill these yawning chasms in what biologists call "morphological space." In other words, they have failed to find the paleontological equivalent of the numerous finely graded intermediate colors (Pendleton blue, dusty rose, gun barrel gray, magenta, etc.) that interior designers covet. Instead, extensive sampling of the fossil record has confirmed a strikingly discontinuous pattern in which representatives of the major phyla stand in stark isolation from members of other phyla, without intermediate forms filling the intervening morphological space.

Foote's statistical analysis of this pattern, documented by an ever increasing number of paleontological investigations, demonstrates just how improbable it is that there ever existed a myriad of as yet undiscovered intermediate forms of animal life—forms that could close the morphological distance between the Cambrian phyla one tiny evolutionary step at a time.

In effect, Foote's analysis suggests that since paleontologists have reached repeatedly into the proverbial barrel, sampled it from one end to the other, and found only representatives of various radically distinct phyla but no rainbow of intermediates, we shouldn't hold our breath expecting such intermediates to eventually emerge. He asks "whether we have a representative sample of morphological diversity and therefore can rely on patterns documented in the fossil record." The answer, he says, is yes.[55]

By this affirmation, he doesn't mean that there are no biological forms left to discover. He means, rather, that we have good reason to conclude that such discoveries will not alter the largely discontinuous pattern that has emerged. "Although we have much to learn about the evolution of form," he writes, the statistical pattern created by our existing fossil data demonstrates that "in many respects our view of the history of biological diversity is mature."[56]

CHENGJIANG AND THE CAMBRIAN CONUNDRUM[57]

The Cambrian and Precambrian fossils from southern China have rendered the mystery associated with the Cambrian explosion more acute in other ways as well. First, the fossil finds in southern China coupled with advances in radiometric dating techniques applied to other Cambrian-era strata have allowed scientists to reassess the duration of the Cambrian explosion. As the name implies, the fossils documenting the Cambrian explosion appear within a relatively narrow slice of geologic time. Until the early 1990s, most paleontologists thought the Cambrian period began 570 million and ended 510 million years ago, with the Cambrian explosion of novel animal forms occurring within a 20- to 40-million-year window during the lower Cambrian period.

Two developments have led paleontologists and geochronologists to revise those estimates downward. First, in 1993, radiometric dating of zircon crystals from formations just above and below Cambrian strata in Siberia allowed for a precise redating of Cambrian strata. Radiometric analyses of these crystals fixed the start of the Cambrian period at 544 million years ago,[58] and the beginning of the Cambrian explosion itself to about 530 million years ago (see Fig. 3.8). These studies also suggested that the explosion of the novel Cambrian animal forms occurred within a window of geologic time much shorter than previously believed,

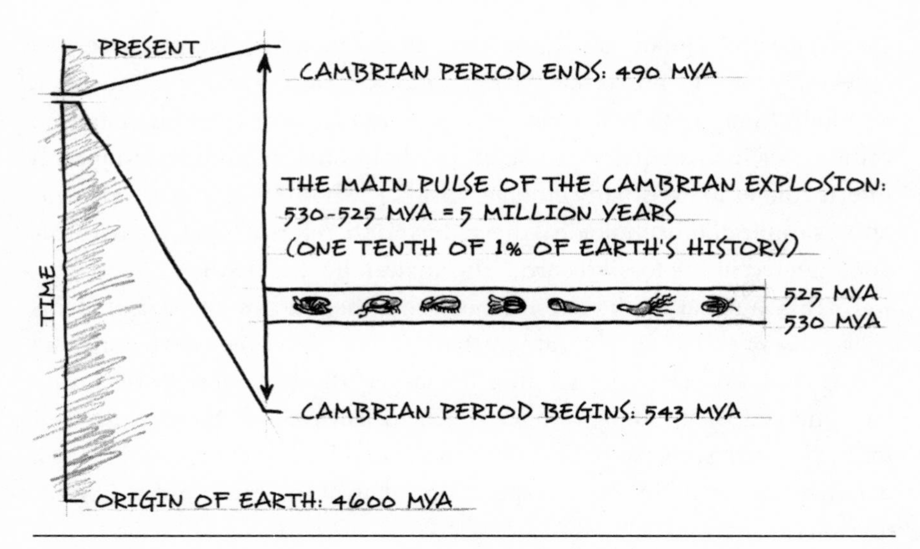

FIGURE 3.8
The Cambrian explosion occurred within a narrow window of geological time.

lasting no more than 10 million years, and the main "period of exponential increase of diversification" lasting only 5 to 6 million years.[59]

Geologically speaking, 5 million years represents a mere 1/10 of 1 percent (0.11 percent, to be precise) of earth's history. J. Y. Chen explains that "compared with the 3-plus-billion-year history of life on earth, the period [of the explosion] can be likened to one minute in 24 hours of one day."[60]

Some geologists or evolutionary biologists dispute these numbers, but they do so by redefining the Cambrian explosion as a series of separate events rather than using the term to refer to the main radiation of new body plans in the lower Cambrian. In 2009, I participated in a debate in which one of my opponents, paleontologist Donald Prothero, from Occidental College, used this common rhetorical strategy to minimize the severity of the Cambrian mystery. In his opening statement, he claimed that the Cambrian explosion actually took place over an 80-million-year period of time and that consequently those who cited the Cambrian as a challenge to the adequacy of the neo-Darwinian theory were mistaken. As I was listening to his opening statement, I consulted his textbook to see how he had derived his 80-million-year figure. Sure enough, he had included in the Cambrian explosion three separate pulses of new innovation or diversification, including the origin of a group of late *Pre*cambrian organisms called the Ediacaran or Vendian fauna. He also included not only the origin of the animal body plans in the lower Cambrian, but also the subsequent minor diversification (variations on those basic architectural themes) that occurred in the upper Cambrian. He included, for example, not just the appearance of the first trilobites, which occurred suddenly in the lower Cambrian, but also the origin of a variety of different trilobite species later from the upper Cambrian.

In my response to Prothero, I noted that he was, of course, free to redefine the term "Cambrian explosion" any way he liked, but that by using the term to describe several separate explosions (of different kinds), he had done nothing to diminish the difficulty of explaining the origin of the first explosive appearance of the Cambrian animals with their unique body plans and complex anatomical features. Beyond this, as we'll see in the next chapter, the Vendian organisms may not have been animals at all, and they bear little resemblance to any of the animals that arise in the Cambrian. We'll also see that most, if not all,[61] of these organisms actually went extinct well before the origin of the animals that first appear in the lower Cambrian and so they do little to minimize the problem of the explosive origin of animals.

In any case, expanding the definition of the Cambrian explosion only obscures the real challenge posed by the event, a challenge underscored by the discoveries at Chengjiang. An analysis by MIT geochronologist Samuel Bowring has shown that the main pulse of Cambrian morphological innovation occurred in a sedimentary sequence spanning no more than 6 million years.[62] Yet during this time representatives of at least sixteen completely novel phyla and about thirty classes first appeared in the rock record. In a more recent paper using a slightly different dating scheme, Douglas Erwin and colleagues similarly show that thirteen new phyla appear in a roughly 6-million-year window.[63] As we've seen, among these animal forms were the first trilobites, with their lens-focusing compound eyes among other complex anatomical features. The problem of explaining how so many new forms and structures arose so rapidly in the first explosive period of the Cambrian remains, whether or not one decides to include within the designation "Cambrian explosion" other distinct events (see Part Two).

THE TOP-DOWN PATTERN IN STARK RELIEF

The Chengjiang fauna makes the Cambrian explosion more difficult to reconcile with the Darwinian view for yet another reason. The Chengjiang discoveries intensify the top-down pattern of appearance in which individual representatives of the higher taxonomic categories (phyla, subphyla, and classes) appear and only later diversify into the lower taxonomic categories (families, genera, and species).

Discoveries at Chengjiang contradict the bottom-up pattern that neo-Darwinism expects. The site does not show the gradual emergence of unique species followed slowly but surely by the emergence of representatives of ever higher and more disparate taxa, leading to novel phyla. Instead, like the Burgess Shale, it shows body plan–level disparity arising first and suddenly, with no evidence of a gradual unfolding and ranging through the lower taxonomic groups.

Consider the case of the early chordates, a phylum consisting of creatures possessing a flexible, rod-shaped structure called a notochord. Mammals, fish, and birds are familiar members of this phylum. Prior to the discovery of the Chengjiang biota, chordates were unknown in the Cambrian period and were thought to have first appeared only much later, during the Ordovician period.[64] Now, following the discoveries at Chengjiang, the first appearance of chordates in the Cambrian period has been amply documented.

For example, J. Y. Chen and several other Chinese paleontologists have found a spindle-shaped eel-like animal called *Yunnanozoon lividum,* which many paleontologists have interpreted as an early chordate because it possesses, among other features, a digestive tract, branchial arches, and a large notochord much like a spinal cord.[65] In addition, J. Y. Chen and colleagues have reported the discovery of a sophisticated craniate-like chordate called *Haikouella lanceolata* from the lower Cambrian Maotianshan Shale. According to Chen and others, *Haikouella* has many of the same features of *Yunnanozoon lividum* as well as several additional anatomical features including a "heart, ventral and dorsal aorta, an anterior branchial arterial, gill filaments, a caudal (posterior) projection, a neural cord with a relatively large brain, a head with possible lateral eyes, and a ventrally situated buccal cavity with short tentacles."[66]

Simon Conway Morris, with D. G. Shu and several Chinese colleagues, has reported an even more dramatic find. They have discovered the fossilized remains of two small Cambrian fish, *Myllokunmingia fengjiaoa* and *Haikouichthys ercaicunensis* (see Fig. 3.9), suggesting a much earlier appearance for both fishes and vertebrates (a class of chordates), both of which were first thought to have originated in the Ordovician period, about 475 million years ago. Both of these taxa are jawless fish (agnathans) and are considered by Shu and others to be closely allied to modern lampreys.[67] Finally, a paper by Shu and others reports the first convincing specimen of another type of chordate from the Cambrian, a urochordate (tunicate).[68] This specimen, *Cheungkongella ancestralis,* is likewise found in the early Cambrian shales (Qiongzhusi Formation) near Chengjiang. These recent finds demonstrate that not only did the phylum Chordata first appear in

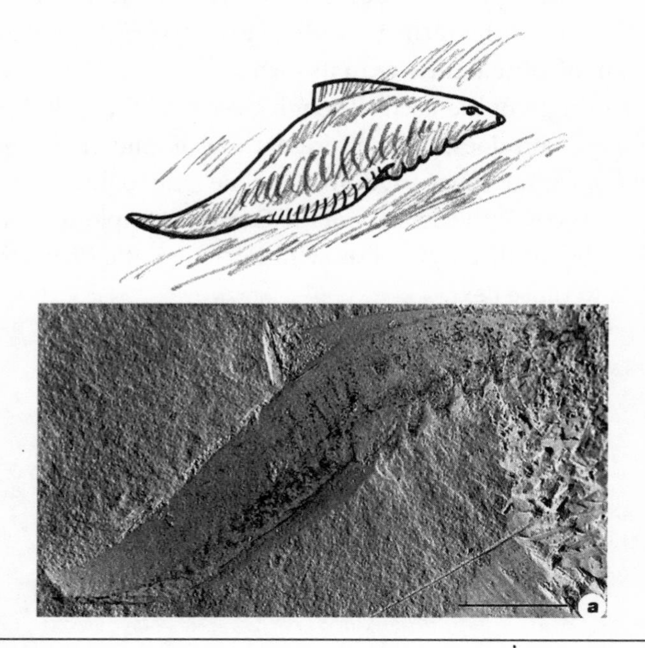

FIGURE 3.9

Figure 3.9a (top): Drawing of the Cambrian fish, *Myllokunmingia fengjiaoa. Figure 3.9b (bottom):* Photograph of *Myllokunmingia fengjiaoa* fossil. Reprinted by permission from Macmillan Publishers Ltd.: *Nature,* Shu et al., "Lower Cambrian Vertebrates from South China," *Nature,* 402 (November 4, 1999): 42–46. Copyright 1999.

the Cambrian, but also each one of the chordate subphyla (Cephalochordata, Craniata, and Urochordata) first emerged then as well. In any case, the discovery in China of chordates, and other previously undiscovered phyla in the Cambrian, only accentuates the puzzling top-down pattern of appearance that other Cambrian discoveries had previously established.[69]

MORE QUESTIONS THAN ANSWERS

Thus, despite the efforts to explain the Cambrian explosion using various versions of the artifact hypotheses, the mystery of the Cambrian explosion has only grown more acute as a result of the dramatic discoveries in southern China that have turned Darwin's tree of life upside down.

When I first heard J. Y. Chen describe these discoveries in 2000, I had been investigating another unsolved question about the history of life: What caused the first living cell, and the information it contains, to arise? As I heard Dr. Chen speak that day in Seattle, my interest in another puzzling question about the history of life began to germinate. Could it be that the origin of animal life was in its own way just as difficult a problem as that of the origin of life itself? Though I eventually concluded that the Cambrian explosion does, indeed, present a profound challenge to contemporary Darwinian theory, it didn't take me long to discover that some scientists believed that the mystery of the Cambrian explosion had already been resolved by the discovery of some rather enigmatic *Pre*cambrian fossils. We turn to those next.

4

THE *NOT* MISSING FOSSILS?

The atmosphere in the auditorium of the Sam Noble Science Museum at the University of Oklahoma was uncomfortably tense, with a security detail of local Norman, Oklahoma, police officers on hand to keep the peace—a conspicuous change from the campus security guard who might be present at a typical university event. The occasion? A new documentary, *Darwin's Dilemma,* that Jonathan Wells, a colleague of mine from the Discovery Institute, and I were scheduled to show. The film would explore the challenge to Darwin's theory posed by the Cambrian fossil record.

For weeks before our event in September of 2009, outspoken evolutionary biology students and an atheist student group, both egged on by militant off-campus bloggers, had threatened to disrupt the screening. Members of the biology faculty pledged to come, so that well before the official start time a large crowd had gathered.

The museum and the geology department, not wanting to complicate matters in their own minds by watching the film first, decided to launch a preemptive first strike by issuing a disclaimer and scheduling an official lecture designed to rebut the film. In the disclaimer the museum stated that, given its public funding, it had no choice but to rent the auditorium to groups irrespective of their "religious beliefs" or "scientific literacy." The disclaimer further noted that the Sam Noble Science Museum did "not support unscientific views masquerading as science, such as those of the Discovery Institute." The museum flyer also announced the lecture of one of their curators, a paleontologist from the university, and mocked the topic of the film—the Cambrian "explosion"—with carefully placed scare quotes. The lecture's start time of 5:00 P.M. also ensured a confluence afterward of the audience made upset by the lecture and the more friendly audience coming to see the film.

Jonathan Wells, a biologist well known for his skepticism about contemporary Darwinian theory, attended the pre-film lecture. Ignoring a few hostile glares, he listened as the paleontologist from the university argued that the Cambrian explosion presented no actual dilemma for Darwinian evolution, and as this same paleontologist speculated that if Darwin had only known what paleontologists today know of the Cambrian fossil record, he (Darwin) would have celebrated it as confirmation of his theory. This particular paleontologist also denied that novel animal forms emerged suddenly in the Cambrian. Instead, he argued that they arose in rudimentary form much earlier in the late Precambrian. He noted that paleontologists had discovered in late Precambrian sediments fossilized sponges, a type of primitive mollusk, and the burrows of worms.

He also laid particular stress on the significance of a group of enigmatic organisms first discovered in the Ediacaran Hills in southern Australia dating from about 565 million years ago, in a Precambrian period known as the Vendian or Ediacaran. Jonathan and I were well aware that most paleontologists do not regard these fossilized organisms as plausible ancestors to the Cambrian fauna. But on that evening the expert from the university claimed the opposite. He also claimed that some obscure Ediacaran organisms (with exotic names such as *Vernanimalcula, Parvancorina,* and *Arkarua*) represented early bilaterians (bilaterally symmetrical animals), arthropods, and echinoderms. He insisted that these organisms pushed back the explosion of animal life by some 40 million years, establishing a "fuse" for the Cambrian explosion in the form of primitive and presumably ancestral animal forms for several of the most significant Cambrian phyla and body designs.

An hour before I was to walk over to lead a discussion and answer questions about the Cambrian film from what proved to be an intensely hostile audience, Jonathan Wells called me with a report on the museum's attempt to refute the film preemptively. The presentation claimed to have resolved the Cambrian mystery—Darwin's dilemma—by showing that the ancestral precursors to the major groups of Cambrian animals had been found after all. But is this true?

THE EDIACARAN FAUNA AND VENDIAN RADIATION

In the previous chapter, we saw that many prominent paleontologists have sought to explain the Cambrian explosion as an artifact of our incomplete sampling of an incomplete fossil record. The lecture that my colleague heard that night in Oklahoma took a very different approach, giving the strong impression that the Precambrian fossil record actually *does* preserve the ancestral forms of the Cambrian animals and that the Ediacaran fauna, in particular, provide several striking examples of such forms.

In public presentations about the Cambrian explosion, I've often encountered this claim, though usually in the form of an unfocused question: "What about the Ediacaran?" Nevertheless, in writing about the Cambrian, I take care not to attribute the idea that the Ediacaran fauna represent Cambrian ancestors to leading Ediacaran or Cambrian experts, lest I critique a straw man. Most paleontologists doubt that well-known Ediacaran forms represent ancestors of the Cambrian animals and few think the late Precambrian fossil record as a whole makes the Cambrian explosion appreciably less explosive. The claim is important to address, however, since it persists as a kind of paleontological urban legend, one that even occasionally finds its way into the mouths of paleontologists.

The Ediacaran fauna derive their name from their most notable discovery site, the Ediacaran Hills in the outback of southeastern Australia. These fauna date from late Precambrian time, a period that the International Union of Geological Sciences has recently renamed the "Ediacaran period."[1] Since geologists used to call the last period of Precambrian time the "Vendian period," paleontologists also refer to the Ediacaran fauna as the Vendian fauna or biota (see Fig. 1.6). Paleontologists have made additional discoveries of Ediacaran- or Vendian-era creatures in England, Newfoundland, the White Sea in northwestern Russia, and the Namibian desert in southern Africa, suggesting a near worldwide distribution. Although these fossils were originally dated to between 700 and 640 million years old, volcanic ash beds both below and above the Namibian site have recently provided more accurate radiometric dates. These studies fix the date for the first appearance of the Ediacaran fauna at about 570–565 million years ago, and the last appearance at the Cambrian boundary about 543 million years ago, or about 13 million years before the onset of the Cambrian explosion itself.[2]

The late Precambrian-era sediments around the world have yielded four main types of fossils, all of which are dated between about 570 and

---

FIGURE 4.1

Examples of enigmatic Ediacaran fossils: *Dickinsonia, Spriggina,* and *Charnia.* Fossil photos in Figures 4.1a and 4.1b courtesy Peterson, K. J., Cotton, J. A., Gehling, J. G., and Pisani, D., "The Ediacaran Emergence of Bilaterians: Congruence Between the Genetic and the Geological Fossil Records," *Philosophical Transactions of the Royal Society B,* 2008, 363 (1496): 1435–43, Figure 2, by permission of the Royal Society.

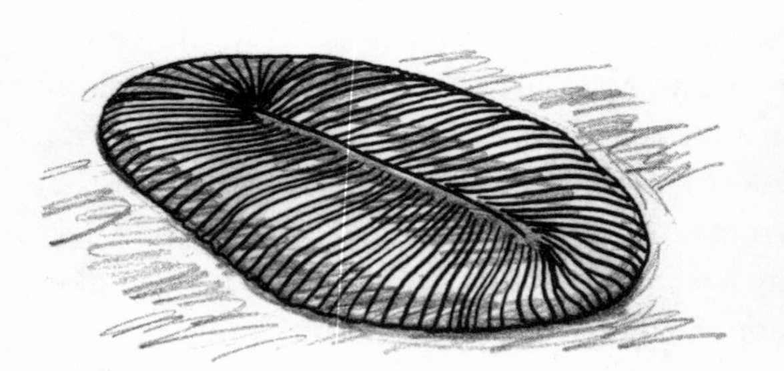

Figure 4.1a: Artist depiction of *Dickinsonia* (*left*) and photograph of *Dickinsonia* fossil (*right*).

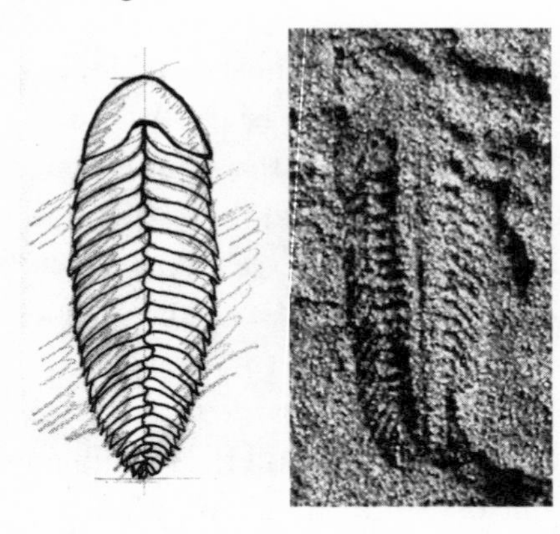

Figure 4.1b: Artist depiction of *Spriggina* (*left*) and photograph of *Spriggina* fossil (*right*).

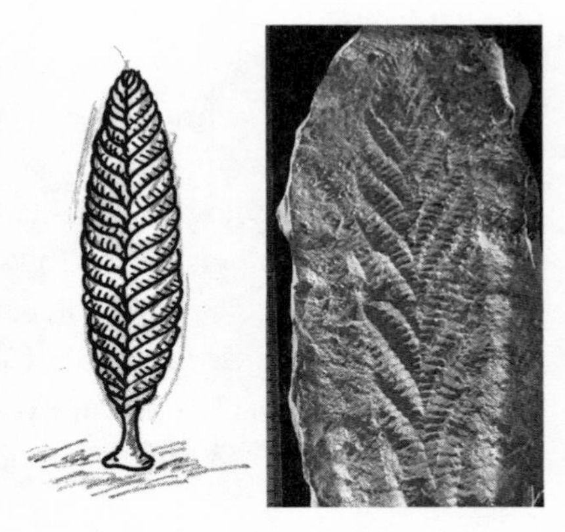

Figure 4.1c: Artist depiction of *Charnia* (*left*), and photograph of *Charnia* fossil (*right*). *Courtesy Wikimedia Commons, user Smith609.*

543 million years ago. The first group consists of the Precambrian sponges mentioned in the previous chapter. These animals first arose about 570 to 565 million years ago.

The second is the distinctive group of fossils from the Ediacaran Hills. The creatures fossilized there include such well-known forms as the flat, air mattress-like body of *Dickinsonia*; the enigmatic *Spriggina,* with its elongated and segmented body and possible head shield; and the frond-like *Charnia* (see Fig. 4.1). These organisms were at least mostly soft-bodied and are large enough to identify with the naked eye.

The third group includes what are called trace fossils, the possible remains of animal activity such as tracks, burrows, and fecal pellets. Some paleontologists have attributed these trace fossils to ancient worms.

The fourth group is the fossils of what may be primitive mollusks, a possibility that received support from a recent discovery in the cliffs along the White Sea in northwestern Russia. There, Russian scientists have discovered thirty-five distinctive specimens of a possible mollusk called *Kimberella,* probably a simple animal form. These new White Sea specimens, dated to 550 million years ago, suggest that *Kimberella* "had a strong [though not hard], limpet-like shell, crept along the sea floor, and resembled a mollusk."[3] Paleontologist Douglas Erwin, of the Smithsonian Institution, has commented: "It's the first animal that you can convincingly demonstrate is more complicated than a flatworm."[4] Additionally, seafloor tracks from Precambrian sediments in both Canada and Australia have been attributed to mollusks, since the tracks resemble what might have been left by a row of small teeth on the tongue-like ribbon of some mollusks as they scraped food particles off the seafloor. In this case, *Kimberella* may well have been the track maker.[5] The authors of the original descriptive paper in *Nature,* Mikhail Fedonkin, from the Russian Academy of Sciences, and Benjamin Waggoner, then at the University of California at Berkeley, conclude as much and suggest that such creatures "began to diversify before the beginning of the Cambrian."[6] Paleontologists, however, are still weighing the evidence.[7]

THE SIGNIFICANCE OF THE EDIACARAN

So do either the remains of the specific organisms from the Ediacaran Hills or the Ediacaran or Vendian biota as a whole solve the problem of the Cambrian explosion? Do these exotic forms represent a kind of fuse to the Cambrian explosion that eliminates the need to explain the rapid emergence of novel body plans and forms of animal life? There are many good reasons to doubt this idea.

First, with the exception of sponges and the possible exception of *Kimberella,* the body plans of visibly fossilized organisms (as opposed to trace fossils) bear no clear relationship to any of the organisms that appear in the Cambrian explosion (or thereafter).[8] The most noted Ediacaran organisms such as *Dickinsonia, Spriggina,* and *Charnia* do not have an obvious head, a mouth, bilateral symmetry (see below), a gut, or sense organs such as eyes. Some paleontologists question whether these organisms even belong in the animal kingdom.

Dickinsonia, for example, has been interpreted by University of Oregon paleontologist Gregory Retallack as having "fungal-lichen" affinities, since its mode of fossil preservation "is comparable not with that of soft-bodied jellyfish, worms, and cnidarians, but with the fossil record of fungi and lichens." *Dickinsonia*'s taxonomic position, Retallack notes, has long been an unsolved puzzle. "Biological affinities of *Dickinsonia* remain problematic," he writes, since it has been "variously considered a polychaete, turbellarian or annelid worm, jellyfish, polyp, xenophyophoran protist, lichen or mushroom."[9]

Similar disputes have characterized attempts to classify *Spriggina.* In 1976, Martin Glaessner, the first paleontologist to study the Ediacaran in detail, described *Spriggina* as a possible annelid polychaete worm based largely upon its segmented body. Nevertheless, Simon Conway Morris later rejected that hypothesis because *Spriggina* shows no evidence of the distinguishing "chaetes," leg-like bristled protrusions that polychaete worms possess. Glaessner himself later repudiated his original hypothesis that *Spriggina* was ancestral to polychaetes, noting that *Spriggina* "cannot be considered as a primitive polychaete, having none of the possible ancestral characters indicated . . . by specialists on the systematics and evolution of this group."[10]

In 1981, paleontologist Sven Jorgen Birket-Smith produced a reconstruction of a *Spriggina* fossil showing that it possessed a head and legs similar to those of trilobites, though examinations of subsequent *Spriggina* specimens have shown no evidence of it possessing limbs of any kind.[11] In 1984, Glaessner weighed in on this discussion as well.

He argued that "*Spriggina* shows no specific characters of the arthropods, particularly of the trilobites."[12] He also noted that the body segmentation of *Spriggina,* and "its known appendages are at the level of polychaete annelids"[13] (although, as noted, by this time he had rejected *Spriggina* as a possible polychaete ancestor). Instead, he proposed that *Spriggina* represented a side branch on the animal tree of life—one that resulted, "metaphorically" perhaps, in "an unsuccessful attempt to make an arthropod."

In a presentation to the Geological Society of America in 2003, geologist Mark McMenamin revived the idea that *Spriggina* might represent a trilobite ancestor. He argued that several features present in *Spriggina* fossils are comparable to those in trilobites such as "the presence of genal spines" and an effaced head or "cephalic region."[14] Nevertheless, many Ediacaran experts, including McMenamin, have also noted that *Spriggina* specimens show no evidence of eyes, limbs, mouths, or anuses, most of which are known from fossil trilobites.[15] Other paleontologists remain skeptical about whether *Spriggina* does in fact exhibit genal spines, noting that good specimens seem to show relatively smooth edges with no protruding spines.[16] In addition, analysis of the best specimens of *Spriggina* shows that it does not exhibit bilateral symmetry, undermining earlier attempts to classify it as a bilaterian animal, and by implication an arthropod.[17] Instead, *Spriggina* exhibits something called "glide symmetry" in which the body segments on either side of its midline are offset rather than aligned.[18] As geologist Loren Babcock of Ohio State University notes, "The zipper-like body plans of some Ediacaran (Proterozoic) animals such as *Dickinsonia* and *Spriggina* involve right and left halves that are not perfect mirror images of each other."[19] The lack of such symmetry, a distinctive feature of all bilaterian animals, and the absence in *Spriggina* specimens of many other distinguishing features of trilobites, has left the classification of this enigmatic organism uncertain.

Paleontologists James Valentine, Douglas Erwin, and David Jablonski distill the confusing welter of conflicting views about the Ediacaran fossils: "Although the soft-bodied fossils that appear about 565 million years ago are animal-like, their classifications are hotly debated. In just the past few years these fossils have been viewed as protozoans; as lichens; as close relatives of the cnidarians; as a sister group to cnidarians plus all other animals; as representatives of more advanced, extinct phyla; and as representatives of a new kingdom entirely separate from the animals."[20] What's more, Valentine, Erwin, and Jablonski note that those paleontologists who do regard the Ediacaran fauna as animals rarely classify them the same way, underscoring their lack of clear affinities to any known animal groups. As they note, "Still other specialists have parceled the fauna out among living phyla, with some assigned to the Cnidaria and others to the flatworms, annelids, arthropods and echinoderms."[21] The uncertain standing of these fossilized forms is partly due to their early extinction, but it also stems from an absence of defining characteristics shared with known groups. They conclude: "This confusing state of affairs arose because these body fossils do not tend to share definitive anatomical details with modern groups, and thus the assignments must be based on vague similarities of overall shape and form, a method that has frequently proved misleading in other cases."[22]

Other leading paleontologists also doubt that the Cambrian animals descended from these Ediacaran forms. In a phylogenetic diagram showing the evolutionary relationship of Precambrian and Cambrian fossils, Oxford biologists Alan Cooper and Richard Fortey depict the Ediacaran fauna as lying on a line of descent separate from the Cambrian animals rather than being ancestral to them.[23] In another paper, Fortey asserts that the beginning of the Cambrian "saw the sudden appearance in the fossil record of almost all the main types of animals (phyla) that still dominate the biota today." He concedes that there are a variety of fossils in older strata, but insist that "they are either very small (such as bacteria and algae) or their relationships to the living fauna are highly contentious, as is the case with the famous soft-bodied fossils from the late Precambrian Pound Quartzite, Ediacara, South Australia."[24]

Similarly, paleontologist Andrew Knoll and biologist Sean B. Carroll have argued: "It is genuinely difficult to map the characters of Ediacaran fossils onto the body plans of living invertebrates."[25] Although many paleontologists initially showed interest in the possibility that the Cambrian animal forms might have evolved from the Ediacaran organisms, paleontologist Peter Ward explains that "later study cast doubt on the affinity between these ancient remains preserved in sandstones [the Australian Ediacaran] and living creatures of today" (that is, animals representing phyla that first arose in the Cambrian).[26] As *Nature* recently noted, if the

Ediacaran fauna "were animals, they bore little or no resemblance to any other creatures, either fossil or extant."[27]

This absence of clear affinities has led an increasing number of paleontologists to reject ancestor-descendant relationships between all but (at most) a few of the Ediacaran and Cambrian fauna. Nevertheless, some have suggested that trace fossils may establish a link. In an authoritative 2011 paper in the journal *Science,* Douglas Erwin and colleagues described the discovery of Ediacaran trace fossils consisting of surface tracks, burrows, fecal pellets, and feeding trails, which, they argue, though small, could only have been made by animals such as worms with a relatively high degree of complexity.[28] On the basis of these findings, Erwin and other paleontologists have argued that these trace fossils suggest the existence of organisms with a head and tail, nervous systems, a muscular body wall allowing creeping or burrowing, and a gut with mouth and anus.[29] Other paleontologists suggest that these characteristics may indicate the presence of a Precambrian mollusk or a worm phylum.[30]

Graham Budd, a British paleontologist who works at Uppsala University in Sweden, and others, have disputed these associations. Budd and geologist colleague Sören Jensen argue that many alleged trace fossils actually show evidence of inorganic origin: "There are numerous reports of older trace fossils, but most can be immediately shown to represent either inorganic sedimentary structures or metaphytes [land plants], or alternatively they have been misdated."[31] Still others have suggested that surface tracks and trails could have been left by mobile single-celled organisms, including a known form of a giant deep-sea protist that leaves bilaterian-like impressions. As one paper explains, "Some such traces date back to 1.5 billion to 1.8 billion years ago, which outdates even the boldest claims of the time of origin of animal multi-cellularity and forces researchers to contemplate the possibility of an inorganic or bacterial origin."[32]

Even the most favorable interpretations of these trace fossils suggest that they indicate the presence of no more than two animal body plans (of largely unknown characteristics). Thus, the Ediacaran record falls far short of establishing the existence of the wide variety of transitional intermediates that a Darwinian view of life's history requires. The Cambrian explosion attests to the first appearance of organisms representing at least twenty phyla and many more subphyla and classes, each manifesting distinctive body plans. In a best case, the Ediacaran forms represent possible ancestors for, at most, four distinct Cambrian body plans, even counting those documented only by trace fossils. This leaves the vast majority of the Cambrian phyla with no apparent ancestors in the Precambrian rocks (i.e., at least nineteen of the twenty-three phyla present in the Cambrian have no representative in Precambrian strata).[33]

Third, even if representatives of four animal phyla were present in the Ediacaran period, it does not follow that these forms were necessarily *transitional* or *intermediate* to the Cambrian animals. The Precambrian sponges (phylum Porifera), for example, were quite similar to their Cambrian brethren, thus demonstrating, not a gradual transformation from a simpler precursor or the presence of an ancestor common to many forms, but quite possibly only an earlier first appearance of a known Cambrian form. The same may be true of whatever kind of worm may be attested by Precambrian tracks and burrows.

Moreover, even assuming, as some evolutionary biologists do,[34] that later Cambrian animals had a sponge-like Precambrian ancestor, the gap in complexity as measured by the number of cell types alone, to say nothing of the specific anatomical structures and distinct modes of body plan organization that are present in later animals but not in sponges, leaves a massive discontinuity in the fossil record that requires explanation (much like the morphological gap between *Spriggina* and actual arthropods).

AN EDIACARAN MINI-EXPLOSION

The Ediacaran fossils themselves provide evidence of a puzzling leap in biological complexity, though not one nearly great enough (or of the right kind) to account for the Cambrian explosion. Before organisms like *Kimberella, Dickinsonia,* and sponges appeared, the only living forms documented in the fossil record for over 3 billion years were single-celled organisms and colonial algae. Producing sponges, worms, and mollusks from single-celled organisms is a little like transforming a spinning top into a bicycle. The bicycle isn't remotely as complex as the automobile sitting beside it, but it represents an enormous leap in technological sophistication over the spinning top. Likewise, although the humble Ediacaran biota look simple beside most of the Cambrian animals, they represent an enormous leap in functional complexity over the single-celled organisms and colonial algae that preceded them.

Thus, the Ediacaran biota attest to a separate sudden increase in biological complexity within a short window of geological time (about 15 million years), following roughly 3 billion years in which only single-celled organisms inhabited the earth.[35] This leap in complexity, in a relatively short span of geological time, may well exceed the explanatory resources of natural selection working on random mutations. We will return to that question in Part Two.

The Ediacaran fossils therefore do not solve the problem of the sudden increase in biological form and complexity during the Cambrian. Instead, they represent an earlier, if less dramatic, manifestation of the same kind of problem. To biology's "big bang,"[36] the Ediacaran biota add a significant "pow." As paleobiologist Kevin Peterson, of Dartmouth College, and his colleagues note, these fauna represent "an apparent quantum leap in ecological complexity as compared with the 'boring billions' [of years] that characterize Earth before the Ediacaran," even if these organisms are "still relatively simple when compared with the Cambrian," which they characterize as another "quantum leap in organismal and ecological complexity."[37]

Many paleontologists now refer to the Ediacaran radiation as an explosion in its own right.[38] This Precambrian "pow" makes the problem of fossil discontinuity only more acute, since credible intermediates leading to the Ediacaran layers are completely nonexistent in the even more sparsely populated strata beneath them.

Finally, even if one regards the appearance of the Ediacaran fossils as evidence of a "fuse" leading to the Cambrian explosion as some have proposed,[39] the total time encompassed by the Ediacaran and Cambrian radiations still remains exceedingly brief relative to the expectations and requirements of a modern neo-Darwinian view of the history of life. As I will explain in more detail in Chapter 8, neo-Darwinism is the modern version of Darwin's theory that invokes random genetic changes called mutations as the source of much of the new variation upon which natural selection acts. Like classical Darwinism, the neo-Darwinian mechanism requires great stretches of time to produce novel biological form and structure. Yet, current studies in geochronology suggest that only 40 to 50 million years elapsed between the beginning of the Ediacaran radiation (570–565 million years ago) and the end of the Cambrian explosion (525–520 million years ago).[40] To anyone unfamiliar with the equations of population genetics by which neo-Darwinian evolutionary biologists estimate how much morphological change is likely to occur in a given period of time, 40 to 50 million years may seem like an eternity. But empirically derived estimates of the rate at which mutations accumulate imply that 40 to 50 million years does not constitute anything like enough time to build the necessary anatomical novelties that arise in the Cambrian and Ediacaran periods. I will describe this problem in more detail in Chapter 12.

Until recently, radiometric studies had estimated the duration of the Cambrian radiation itself at 40 million years, a period of time so brief, geologically speaking, that paleontologists had already dubbed it an "explosion." The relative suddenness of the Cambrian explosion, even on the earlier measure of its duration, had already raised serious questions about the adequacy of the neo-Darwinian mechanism; consequently, it had also raised questions about whether a Darwinian understanding of the history of life could be reconciled with the Cambrian and Precambrian fossil record. Thus, treating the Ediacaran and the Cambrian radiations as one continuous evolutionary event (itself an unrealistically generous assumption) only returns the problem to its earlier (pre-zircon redating) status.

For all these reasons, the late Precambrian fossils have not solved, but instead have deepened, the mystery of the origin of animal form. And few leading Cambrian paleontologists, of whom I was aware on that September evening in 2009 while preparing to answer questions at the University of Oklahoma, thought otherwise.

EDIACARAN EXOTICA

So what about the claim that certain exotic Ediacaran fossils are plausible ancestors to the Cambrian animal forms, even if better-known Ediacaran forms such as *Dickinsonia, Charnia,* and *Spriggina* are probably not? Did these exotic forms solve the mystery of the Cambrian explosion?

Only a few years before my visit to the University of Oklahoma, I had written a scientific review article with research help from several colleagues, including a paleontologist and a marine biologist.[41] (The latter was Paul Chien, who helped discover the Precambrian sponge embryos discussed in the previous chapter.) In our review article, I explained many of the problems with treating the Ediacaran as transitional intermediates discussed above. In the process of doing the research for that article, my colleagues and I encountered few paleontologists who thought that *Parvancorina, Arkarua* (see Fig. 4.2), or *Vernanimalcula* represented definitive ancestors of the Cambrian bilaterians, arthropods, or echinoderms. Could we have missed something?

In fact, leading Cambrian authorities have dismissed associations between these odd fossil forms and the Cambrian animals. Nevertheless, in his talk before the showing of our film, the local professor from the University of Oklahoma asserted that the rather indistinct fossil form found in the Ediacaran Hills called *Parvancorina* represented a plausible ancestor of the arthropods. Some have described *Parvancorina* as a shield-shaped fossil form with a raised anchor-shaped ridge impressed atop it, bearing a superficial resemblance in its shape to that of a trilobite—thus, the claim that it might have represented an early arthropod. Yet leading Cambrian paleontologists dispute this association. Cambrian expert James Valentine has argued that *Parvancorina* is not convincing as an arthropod ancestor, and for good reason. *Parvancorina* fossils lack a head, jointed limbs, and compound eyes, all distinctive features of arthropods. Thus, Valentine noted that *Parvancorina* fossils "have not been shown to share derived features" with arthropods.[42]

FIGURE 4.2

Figure 4.2a (left): Photograph of *Arkarua* fossil, courtesy Taylor & Francis, Ltd. *Figure 4.2b (right):* Photograph of *Parvancorina* fossil, courtesy Peterson, K. J., Cotton, J. A., Gehling, J. G., and Pisani, D., "The Ediacaran Emergence of Bilaterians: Congruence between the Genetic and the Geological Fossil Records," *Philosophical Transactions of the Royal Society B*, 2008, 363 (1496): 1435–43, Figure 2, by permission of the Royal Society.

Valentine makes much the same point about the small disc-like imprint called *Arkarua,* one of the other Ediacaran forms cited by the University of Oklahoma professor that night at the Sam Noble Museum. Valentine points out that it too lacks many distinctive features of the animal phylum to which it is typically assigned. Indeed, those who propose *Arkarua* as an ancestor of the Cambrian animals usually claim that it represents an early echinoderm (as the professor in Oklahoma did). Echinoderms include starfish, sand dollars, and other animals with five-fold symmetry extending from a central body cavity.[43] Some have perceived five tiny segmented divisions within the circular impressions left by *Arkarua,* making them seem roughly similar to some modern echinoderms. But that similarity has proven superficial at best. Other paleontologists observe that *Arkarua* lacks a stereom, or water vascular system, a definitive diagnostic feature of echinoderms; thus, its "echinoderm-specific features are not readily visible."[44] Valentine has argued that, absent such telltale features, the relationship of *Arkarua* to echinoderms "remains uncertain."[45]

In the case of *Vernanimalcula,* the story is more complicated but equally problematic. *Vernanimalcula* is the name that Chinese paleontologists gave to an imprint in phosphorite sediment found in the Doushantuo Formation in 2004. They found the structure in 580- to 600-million-year-old rocks, making the impression even older than the Ediacaran strata. The paleontologist David Bottjer of the University of Southern California, and some Chinese paleontologists (at least, initially), speculated that the *Vernanimalcula* imprint might be the remains of an early bilaterian.[46]

Recall that bilaterians are animals whose parts found on one side of the body midline are also found in mirror image on the other (as opposed to, say, a radially symmetric animal[47]). Figure 4.3 shows a picture of the structure of *Vernanimalcula* first found in the Doushantuo Phosphorite formation. Some paleontologists think that *Vernanimalcula* exhibits such bilateral symmetry and thus might be ancestral to the bilaterian animals that later first appeared in the Cambrian period.

But problems have emerged with this argument. First, the form of the *Vernanimalcula* does not resemble any specific bilaterian animal. In addition, recent scientific analyses of these remains have questioned whether this imprint preserves the remains of animals and, therefore, bilaterians at all. For example, in 2004, Stefan Bengtson and Graham Budd, two paleontologists and Cambrian experts, published a detailed chemical and

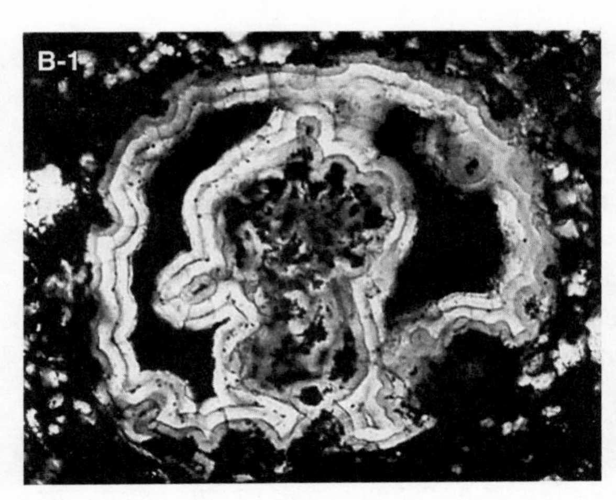

FIGURE 4.3
Photograph of *Vernanimalcula fossil. Courtesy American Association for the Advancement of Science,* from Chen, J.-Y., Bottjer, D. J., Oliveri, P., Dornbos, S. Q., Gao, F., Ruffins, S., "Small Bilaterian Fossils from 40 to 55 Million Years Before the Cambrian," *Science,* 305 (July 9, 2004): 218–22, Figure 1b. Reprinted with permission from AAAS.

microscopic analysis of these fossils in the journal *Science.*[48] They concluded that the structures preserved in phosphorite rocks had undergone significant alteration by so-called diagenesis and taphonomic processes. Diagenesis refers mainly to processes of chemical alteration that occur after sediments are deposited and before sedimentary rocks are fully hardened, or "lithified." Taphonomic processes are those that alter once living organisms after burial and preservation in sediments.

Based on their microscopic analysis, Bengtson and Budd rejected the hypothesis that these structures preserved the remains of an animal form. Instead, they argued that the phosphorite imprint exhibited distinctive features of the chemically altered remains of one-celled microfossils—microfossils that had been encrusted with layers of chemical residue from various diagenetic processes.[49]

More recently, in 2012, Bengtson and three other colleagues published another paper sharply critical of the view that *Vernanimalcula* represents an ancestor of the bilaterian animals—or even an animal. They show that the "structures key to animal identity are effects of mineralization that do not represent biological tissues." For this reason they conclude: "There is no evidential basis for interpreting *Vernanimalcula* as an animal, let alone a bilaterian."[50]

Though the paper was titled "A Merciful Death for the 'Earliest Bilaterian,' *Vernanimalcula,*" the authors were anything but merciful in wielding their arguments. Their article upbraided David J. Bottjer, the main paleontologist who has promoted the interpretation of *Vernanimalcula* as a bilaterian ancestor, for seeing what he wanted to see and disregarding the clear evidence of nonbiological mineralization. In a 2005 *Scientific American* article, Bottjer interpreted *Vernanimalcula* as the "oldest fossil animal with a bilaterian body plan yet discovered." In that article, Bottjer claimed that *Vernanimalcula* confirmed the "suspicion that complex animals have a much deeper root in time" and "that the Cambrian was less of an explosion and more of a flowering of animal life."[51] After unequivocally rejecting Bottjer's interpretation on the basis of their geochemical analysis, Bengtson and his coauthors rebuked Bottjer in rather personal terms:

> *It is likely that the fossils referred to [as] Vernanimalcula were interpreted as bilaterians because this was . . . the explicit quarry of its authors. If you know from the beginning not only what you are looking for, but what you are going to find, you will find it, whether or not it exists. As Richard Feynman (1974) famously remarked: "The first principle is that you must not fool yourself—and you are the easiest person to fool." . . . Once you have fooled yourself you will fool other scientists.*[52]

Bengtson and his colleagues insist that however much a few paleontologists such as Bottjer might have wanted to "heap" "evolutionary significance" on *Vernanimalcula* in order to relieve their cognitive dissonance about the Cambrian explosion, the evidence does not bear the weight of interpretation that had been placed upon it. Thus, they conclude *Vernanimalcula* should be "laid to a merciful rest," since the interpretation of it had "taken on a life of its own, a life it never had."[53]

A DEEPER PROBLEM

Though back in 2009, Jonathan Wells and I didn't know about the most recent critical analysis of *Vernanimalcula*'s pretentions, we did know that many leading paleontologists had rejected attempts to identify *Vernanimalcula* as an animal form. Thus, during the question-and-answer period following the film, Jonathan Wells explained why these and other obscure and enigmatic Precambrian fossils (or imprints) failed to qualify as convincing precursors to any of the Cambrian animals, citing the work of leading authorities in paleontology. In each case, he noted that similarities between the Ediacaran forms and later Cambrian animals had proven superficial, because the Ediacaran forms lacked many key diagnostic features of specific Cambrian phyla.

At the same time, as I later reflected on the lecture, I recognized a deeper problem with attempts to resolve the mystery of the Cambrian explosion by pointing to a few Precambrian fossils. Many defenders of the Darwinian picture of the history of life seemed to assume that the discovery of any alleged Precambrian animal forms, however implausible as ancestors of specific Cambrian animals or however sparsely distributed in the vast sequences of Precambrian strata, would solve the mystery of the Cambrian explosion, especially if these forms exemplified some abstractly perceived commonality such as bilateral symmetry.

To see what's wrong with this way of thinking, imagine an ambitious distance swimmer claiming that it would be possible to swim between California and Hawaii over a period of many months or years because of the small islands that provide way stations where he could eat, rest, and overnight at each stage along his marathon journey. But instead of showing that an archipelago dotting the route between California and Hawaii at reasonably accessible intervals actually exists, he points to a couple of barren atolls in the South Pacific far from the most likely course to Hawaii. Clearly, in that case, the claims of our intrepid hypothetical swimmer would not be credible. In a similar way, the claims of those who assert that a few isolated and anatomically enigmatic forms of life in the Precambrian solve the problem of the Cambrian explosion also lack credibility.

To appreciate another aspect of this problem, let's revisit the claims about *Vernanimalcula* as a possible ancestor of all the bilaterian phyla. On the one hand, for such a form to qualify as the ancestor common to a large number of specific phyla (such as the many bilaterian phyla that arise in the Cambrian), it must exhibit the basic bilaterian characteristics such as bilateral symmetry and what is called "triploblasty," the presence of three distinct tissue layers (endoderm, mesoderm, and ectoderm). At the same time, a viable candidate for the role of common ancestor cannot by definition manifest any of the *differentiating* characteristics that distinguish the individual Cambrian phyla and their respective body plans from each other. For example, any bilaterian that manifests the characteristic *exo*skeleton of, say, an arthropod cannot also qualify as a plausible ancestor of a chordate, because chordates have *internal* skeletons or notochords.

The logic of these distinct body designs precludes sharing *both* anatomical characteristics. For this reason, any hypothetical bilaterian common ancestor could only have existed as a kind of lowest anatomical common denominator, or what evolutionary biologists call a "ground plan," having only those few features that are common to all of the animal forms that allegedly evolved from it.

But this creates a dilemma. If a fossilized form is simple enough to qualify as the common ancestor of later highly differentiated bilaterian phyla, then it will necessarily lack most of the important distinguishing anatomical features of those specific phyla. That means that all the interesting anatomical novelties that differentiate one phylum from another must arise along the separate lineages branching out from the alleged common ancestor well after its origin in the fossil record. Heads, jointed limbs, compound eyes, guts, anuses, antennae, notochords, stereoms, lophophores (a tentacled feeding organ), and numerous other distinguishing characteristics of many different animals must come later on many distinct lines of descent. Yet the gradual evolutionary origin of these characteristics is not documented in the Precambrian fossil record. These characteristics do not appear until they arise suddenly in the Cambrian explosion.

For this reason, indistinct fossils such as *Vernanimalcula*—even if we take them as representing a common ancestor of many bilaterians—document little of the Darwinian story of the history of animal life. Hugely significant gaps in the fossil record would still remain, because the Precambrian fossil record simply does not document the gradual emergence of the crucial distinguishing characteristics of the Cambrian animals. The important anatomical novelties that define the individual Cambrian phyla as well as their first clear representatives arise as suddenly as ever.

To say that a form such as *Vernanimalcula,* or any of the other relatively indistinct Ediacaran forms, solves the problem of the missing Precambrian fossil record would be a bit like saying that a metal cylinder demonstrates all the steps involved in the construction of a toaster, automobile, submarine, or jet airplane simply because all these technological objects utilize "metal enclosures." It's true that each of these complex systems uses metal enclosures, but the presence of an enclosed metal surface is only a necessary, and not nearly a sufficient, condition of the origin of these various technological systems. Similarly, finding a simple but otherwise unadorned bilaterally symmetric form of life would hardly solve the problem of fossil discontinuity, because it would not by itself document the emergence of the unique characteristics of the individual bilaterian animals.

This paradox is well known to paleontologists who work on the Cambrian radiation. Charles Marshall and James Valentine, for instance, describe the difficulty of attempting to characterize an "undiagnostic" group, by which they mean a possible ancestral "stem" group that lacks the specialized characteristics of its presumptive evolutionary progeny. They write:

> *When trying to unravel the origins of the animal phyla . . . the hardest to examine is the phase between the actual cladogenic origin of a phylum and the time that it acquired its first phylum-specific characteristic(s). Even if we have fossils from this phase in a phylum's history, we will not be able to prove their kinships at the level of phyla.*[54]

Thus, even if *Vernanimalcula,* or some other fossil form, were simple enough and animal-like enough to qualify as a so-called ur (or original) form of animal life, it would paradoxically, for just that reason, be incapable of establishing itself as an unequivocal common ancestor of some *specific* Cambrian phylum.

And there's no relief in the other direction. If an alleged ancestral form manifests the distinguishing features of one of the specific Cambrian phyla—if, for example, *Vernanimalcula* or some other isolated form presented a convincing set of distinctively arthropod or chordate or echinoderm characteristics, then the very presence of those features would necessarily preclude the possibility of that specific animal form representing the *common* ancestor of all other Cambrian forms. The more an animal form manifests the characteristics of one phylum or group within the phylum, the less plausible it becomes as the ancestor of all the other animal phyla.

And that is the dilemma in a nutshell. Highly differentiated and complex Precambrian forms *by themselves* could not have been ancestors common to all the Cambrian phyla; whereas undifferentiated forms simple enough to have been ancestral to all the Cambrian phyla leave no evidence, by themselves, of the gradual emergence of the complex anatomical novelties that define the Cambrian animals. Either way—whether the few alleged Precambrian ancestors are viewed as simple and relatively undifferentiated or complex and highly differentiated—the fossil record, *given its otherwise pervasive pattern of discontinuity,* does not establish the gradual evolution of numerous anatomical and morphological novelties. Instead, only a true series of transitional intermediates in which the fossil record documents *both* the existence of an original animal form and the gradual appearance of the key distinguishing anatomical features and novelties (and the Cambrian animals themselves) would remedy this deficiency. And yet that is precisely what the Precambrian fossil record has failed to document.

As Graham Budd and Sören Jensen state, "The known [Precambrian/Cambrian] fossil record has not been misunderstood, and there are no convincing bilaterian candidates known from the fossil record until just before the beginning of the Cambrian (c. 543 Ma), even though there are plentiful sediments older than this that should reveal them."[55] Thus they conclude, "The expected Darwinian pattern of a deep fossil history of the bilaterians, potentially showing their gradual development, stretching hundreds of millions of years into the Precambrian, has singularly failed to materialize."[56]

DILEMMA ON DISPLAY

During the question-and-answer session that followed the screening of *Darwin's Dilemma,* none of the University of Oklahoma Ph.D. students or science faculty who attended the museum-sponsored lecture challenged my colleague Jonathan Wells when he explained why leading paleontologists do not think the exotic Precambrian forms cited in the lecture were ancestors of Cambrian forms. This nonreaction seemed a little odd given the stress the museum's own expert had laid on these claims, and given that he had made these claims rather emphatically in the same building to many of the same people just three hours earlier.

On our flight out of town the next day, Jonathan Wells told me something that cast our experience there in an even odder light. He'd had a chance to walk around the Sam Noble Science Museum after the lecture and before our event. He discovered that the museum has a display that vividly illustrates the severity of what we called Darwin's dilemma. Wells recorded some of his observations on his return to Seattle. One part of the account of what he saw while touring the museum's own display about the Cambrian explosion is worth quoting at length:

> *[The display] seemed factually accurate for the most part, emphasizing (among other things) that many of the Cambrian explosion fossils were soft-bodied—which puts the lie to the common explanation that their precursors are absent from the fossil record because they lacked hard parts. The exhibit also made it clear that the Ediacaran fossils went extinct at the end of the pre-Cambrian, so (with a few possible exceptions) they could not have been ancestral to the Cambrian phyla. One particular panel in the exhibit caught my attention. It showed over a dozen of the Cambrian phyla at the top of a branching tree with a single trunk, but none of the branch points corresponded to a real living thing. Instead, the branch points were artificial technical categories such as "Ecdysozoa," "Lophotrochozoa," "Deuterostomia," and "Bilateria." The artificiality of the branch points emphasized that the branching-tree pattern imposed on the fossil evidence was itself an artificial construct.*

So, after all the controversy, it turned out that the museum that had sponsored the lecture denying Darwin's Cambrian dilemma itself has an excellent display indicating that the expected ancestral forms of the Cambrian animals—the very ones that Darwin hoped to find a hundred and fifty years ago—are still missing from the Precambrian fossil record. But then why would the museum sponsor the presentation that it did? It's hard to say, I suppose, but I've seen this dynamic before in discussions about Darwinian evolution. Evolutionary biologists will acknowledge problems to each other in scientific settings that they will deny or minimize in public, lest they aid and abet the dread "creationists" and others they see as advancing the cause of unreason. Perhaps just our presence on campus raising questions about contemporary Darwinism made them feel defensive on behalf of "science." It's an understandable, if ironic, human reaction, of course, but one that in the end deprives the public of access to what scientists actually know. It also perpetuates the impression of evolutionary biology as a science that has settled all the important questions at just the time when many new and exciting questions—about the origin of animal form, for example—are coming to the fore.

5

THE GENES TELL THE STORY?

Reconstructing the history of life has a lot in common with detective work. Neither detectives nor evolutionary biologists can directly observe the events in the past that interest them most. Detectives typically did not see the crime occur. Evolutionary biologists did not witness the origin of animals or other groups of organisms. Yet this limitation doesn't mean either group of investigators lacks the evidence to determine with some confidence what happened. Detectives and evolutionary biologists as well as many other historical scientists—paleontologists, geologists, archeologists, cosmologists, and forensic scientists—do this regularly, based on careful inference from the clues or evidence left behind.

Many evolutionary biologists have commented on the forensic nature of their work. Here's how Richard Dawkins puts it: "I have used the metaphor of a detective, coming on the scene of the crime after it is all over and reconstructing from the surviving clues what must have happened."[1]

Perhaps the most obvious surviving traces of ancient life are fossils. But as evolutionary biologists and paleontologists have come to realize that the Precambrian fossil record has not furnished the confirmation that Darwin hoped it would, many have looked to other kinds of clues to establish the gradual emergence of Cambrian animal life from a common ancestor.

In this effort, contemporary evolutionary biologists have followed the example of Darwin himself. Though Darwin argued that a general progression of simpler to more complex forms of life in the fossil record meshed nicely with his theory, he was acutely aware that the discontinuity of the fossil record, particularly as evidenced in Precambrian and Cambrian strata, did not. This is why he emphasized other types of evidence to establish his theory of universal common descent.

In a famous chapter in *On the Origin of Species* titled "The Mutual Affinities of Organic Beings," Darwin made his case not on the basis of the fossil evidence, but on the basis of *similar* anatomical structures in many distinct organisms. He noted, for example, that the forelimbs of frogs, horses, bats, humans, and many other vertebrates exhibited a common five-digit ("pentadactyl") structure or organization (see Fig. 5.1). To explain such "homologies," as he called them, Darwin posited a vertebrate ancestor that possessed pentadactyl limbs in rudimentary form. As a menagerie of modern vertebrates evolved from this common ancestor, each retained in its own way the basic pentadactyl mode of organization. For Darwin, his theory of descent with modification from a common

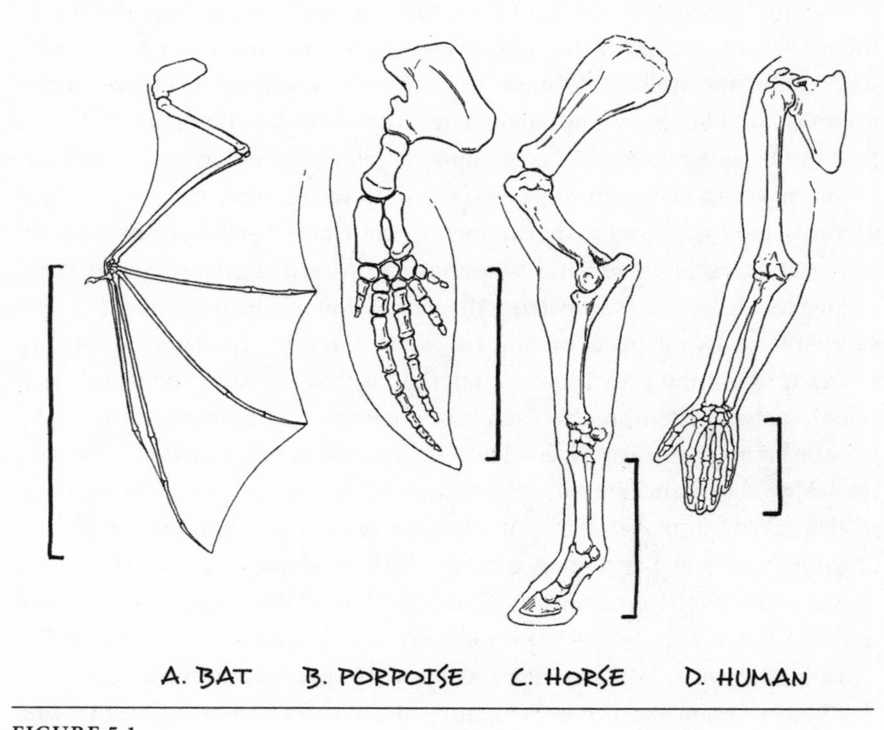

FIGURE 5.1
The common five-digit pattern of the pentadactyl limb as manifested in four modern animals. *Copyright Jody F. Sjogren. Used with permission.*

ancestor explained these similarities better than the received view of many older nineteenth-century biologists such as Louis Agassiz or Richard Owen, both of whom thought homologies reflected the common design plan of a creative intelligence.

In reconstructing the evolutionary history of life, most evolutionary biologists today emphasize the importance of homology. They assume that similarities in anatomy and in the sequences of information-bearing biomacromolecules such as DNA, RNA, and protein point strongly to a common ancestor.[2] They also assume that the *degree of difference* in such cases is on average proportional to the time elapsed since the divergence from a common ancestor. The greater the difference in the common feature or molecular sequence, the farther back the ancestor from which the feature or sequence arose.

Evolutionary biologists have used this approach to try to discern the evolutionary history of the Cambrian animals. If the Precambrian fossil record refuses to disclose the secrets of Precambrian evolution, so goes the thinking, perhaps the study of comparative anatomy and molecular homologies will. Given the well-established problem with the fossil evidence, many evolutionary biologists now particularly emphasize the importance of clues from molecular genetics. As evolutionary biologist Jerry Coyne, of the University of Chicago, notes, "Now we have a powerful, new, and independent way to establish ancestry: we can look directly at the genes themselves. By sequencing the DNA of various species and measuring how similar these sequences are, we can reconstruct their evolutionary relationships."[3]

There are two aspects of this endeavor. First, by analyzing the genes of existing animals representing phyla that first arose in the Cambrian, scientists have attempted to establish when the common ancestor of the Cambrian animal forms lived. This effort has generated what is known as the "deep-divergence hypothesis," which holds that the common ancestor of all animal life arose long before the Cambrian explosion. Second, by analyzing anatomical and molecular similarities, biologists have attempted to reconstruct the Precambrian–Cambrian tree of life, mapping the *course* of evolution during a cryptic period before the Cambrian.

Defenders of neo-Darwinism assert that these techniques have produced a coherent evolutionary picture of the early history of animal life. They assert that clues from the realm of genetics point unequivocally to

Precambrian ancestral forms and to an evolutionary history that fossils have failed to document.

This chapter will examine what genes tell us about the alleged universal common ancestor of all animals; the next chapter will consider whether the analysis of genes (and other features of organisms) yield a coherent treelike picture of the Precambrian prehistory of animal life. Genetic analyses have indeed revealed a trove of clues. The question is: Do those genetic clues establish the Precambrian ancestor and history that fossils have failed to document, or, as sometimes occurs in criminal investigations, has there been a rush to judgment?

DEEP DIVERGENCE

Many paleontologists and evolutionary biologists now concede that the long-sought-after Precambrian fossils, those necessary to document a Darwinian account of the origin of animal life, are missing.[4] Scientists are especially candid about this when addressing each other in the technical peer-reviewed literature. Often, however, defenders of evolutionary orthodoxy raise another possibility—that the common ancestor of the Cambrian animals has been documented after all, not by fossil evidence, but by molecular or genetic evidence of what they call a "deep divergence" of animal life. In making such claims, these biologists clearly privilege molecular evidence over evidence from the fossil record.

Proponents of deep divergence don't deny that the fossil evidence has come up short. Instead, they adopt one of the versions of the artifact hypothesis to account for that missing evidence. They then argue that there was no "explosion" of animal forms in the Cambrian, but rather a "long fuse" of animal evolution and diversification lasting many millions of years leading up to what only *looks* like an "explosion" of animal life in the Cambrian, but this evolutionary history was hidden from the fossil record. Indeed, they argue that molecular evidence establishes a long period of undetected or cryptic evolution in Precambrian times, beginning from a common ancestor some 600 million to 1.2 billion years ago, depending upon which study of the molecular genetic data they cite. If correct, the Cambrian phyla may have had many hundreds of millions of years to evolve from a common ancestor.[5]

THE MOLECULAR CLOCK

Advocates of deep divergence use a method of analysis known as the "molecular clock." Molecular clock studies also assume that the extent to which sequences differ in similar genes in two or more animals reflects the amount of time that has passed since those animals began to evolve from a common ancestor. A small difference means a short time; a big difference, a long time. To determine exactly how short or long, these studies estimate the mutation rate by analyzing genes in two species or taxa

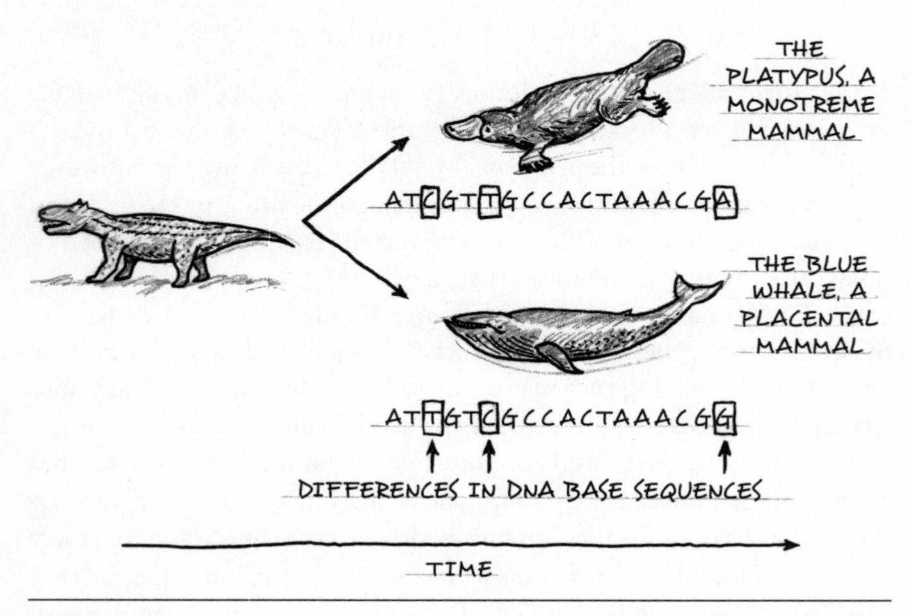

FIGURE 5.2
The idea behind the molecular clock. The two animals and their homologous gene sequences at the right of the figure show the molecular distance between two present-day animals, that is, how many mutational differences have accumulated over time since they diverged on the tree of life. The animal at the left of the figure (a mammal-like reptile) represents the common ancestor from which these animals presumably evolved. Knowing how long ago the common ancestor (the mammal-like reptile) lived, and how many mutational differences have accumulated in its descendants during that time, allows scientists to calculate a mutation rate. In theory, once the mutation rate has been determined, it can be used to calculate the divergence time of other present-day species, after their homologous genes have been compared for differences.

that are thought to have evolved from an ancestor whose presence in the fossil record can be discerned and dated accurately. For example, many molecular-clock studies of birds and mammals are calibrated based on the age of an early reptile thought to be the most recent common ancestor of both.

Genetic comparisons enable evolutionary biologists to estimate the number of mutational changes since divergence, and dating of the strata containing presumed fossil ancestors tells how long ago the divergence occurred. Assuming that different lineages evolve at the same rate,[6] together the two pieces of information enable evolutionary biologists to calculate a baseline mutation rate. They can then use that rate to determine how long ago some other pair of animals diverged from each other on the evolutionary tree (see Fig. 5.2).[7]

Advocates of the deep-divergence hypothesis have applied this method to analyzing similar genes, RNA molecules, or proteins in pairs of animals belonging to phyla that first arose in the Cambrian period. In this way they estimate how long it took for the different animal phyla to diverge from a common Precambrian ancestor.

DEEP AND DEEPER: EVIDENCE FOR DEEP DIVERGENCE

In the 1990s, evolutionary biologists Gregory A. Wray, Jeffrey S. Levinton, and Leo H. Shapiro performed a major study of Cambrian-relevant molecular sequence data. In 1996, they published their results in a paper entitled "Molecular Evidence for Deep Precambrian Divergences Among Metazoan Phyla."[8] Wray's team compared the degree of difference between the amino-acid sequences of seven proteins[9] derived from several different modern animals representing five Cambrian phyla (annelids, arthropods, mollusks, chordates, and echinoderms). They also compared the nucleotide base sequences of a ribosomal RNA molecule[10] from the same animal representatives of the same five phyla.

The Wray study concluded that the common ancestor of the animal forms lived 1.2 billion years ago, implying that the Cambrian animals took some 700 million years to evolve from this "deep-divergence" point before first appearing in the fossil record. Wray and his colleagues attempted to explain the absence of fossil ancestral forms during this period of time by postulating that Precambrian ancestors existed in exclusively soft-bodied forms, rendering their preservation unlikely.

More recently, Douglas Erwin and several colleagues performed a study comparing the degree of sequence difference between other genes—seven nuclear housekeeping genes[11] and three ribosomal RNA genes[12] across 113 different species of living Metazoa. (The term "Metazoa" refers to animals with differentiated tissue. The term "metazoan" refers to one such animal or can be used as an adjective, as in "the metazoan phyla.") They estimated that "the last common ancestor of all living animals arose nearly 800 million years ago."[13]

Many similar studies affirm a very ancient or "stratigraphically deep" divergence of the animal forms, in opposition to those who claim that the Cambrian animals appeared suddenly within just a few million years.[14] Each of these studies affirms the gradual emergence of animal life that most researchers expected to find on the basis of a Darwinian picture of the history of animal life. Indeed, a major aim of the Wray study was to challenge the view "that the animal phyla diverged in an 'explosion' near the beginning of the Cambrian period."[15] Wray and his colleagues argue instead that "all mean divergence time estimates between these four phyla and chordates, based on all seven genes, substantially predate the beginning of the Cambrian period."[16] They conclude: "Our results cast doubt on the prevailing notion that the animal phyla diverged explosively during the Cambrian or late Vendian, and instead suggest that there was an extended period of divergence during the mid-Proterozoic, commencing about a billion years ago."[17]

From an orthodox Darwinian point of view, the conclusions of these studies seem almost unavoidable since (1) the neo-Darwinian mechanism requires vast amounts of time to produce anatomical novelty and (2) such phylogenetic analyses assume that all the animal forms descended from a common ancestor. Many evolutionary biologists claim that clues long hidden in DNA now confirm these Darwinian axioms and, consequently, the existence of an extremely ancient, Precambrian ancestor of the Cambrian animals. As Andrew Knoll, a Harvard paleontologist, states, "The idea that animals should have originated much earlier than we see them in the fossil record is almost inescapable."[18]

REASONABLE DOUBT

Nevertheless, there is now good reason to doubt this allegedly overwhelming genetic evidence. In the idiom of our forensic metaphor, other material witnesses (fossils) have already come forward to testify, the testimony of the genes (and other key indicators of biological history) is grossly inconsistent, and that genetic testimony has come to us through a translator, who is shaping the way the jury perceives the evidence. Let's look at each of these problems in turn.

Fossil Testimony

Recall that the deep-divergence hypothesis has two components. One of them—a version of the artifact hypothesis—provides an explanation for why the Precambrian ancestral fossils are missing. And here the deep-divergence hypothesis first runs into trouble. As we saw in Chapter 3, there is no currently plausible version of the artifact hypothesis. The preservation of numerous soft-bodied Cambrian animals as well as Precambrian embryos and microorganisms undermines the idea of an extensive period of undetected soft-bodied evolution. In addition, the claim that exclusively soft-bodied ancestors preceded the hard-bodied Cambrian forms remains anatomically implausible. A brachiopod cannot survive without its shell. An arthropod cannot exist without its exoskeleton. Any plausible ancestor to such organisms would have likely left some hard body parts, yet none have been found in the Precambrian. Yet the deep-divergence hypothesis, whatever its other merits, requires a viable artifact hypothesis to explain the absence of fossilized Precambrian ancestors.

The Testimony of Genes: Conflicting Stories

There is a second, more telling reason to doubt the deep-divergence hypothesis: the results of different molecular studies have generated widely divergent results. Yet presumably there was only one common ancestor of all the Metazoa and only one ultimate divergence point.

For example, comparing the Wray-led study and the Erwin-led study generates a difference of 400 million years. In the case of other studies, even greater differences emerge. Many other studies have thrown their own widely varying numbers into the ring, placing the common ancestor of animals anywhere between 100 million and 1.5 billion years before the Cambrian explosion (some molecular clock studies, oddly, even place the common ancestor of the animals *after* the Cambrian explosion).[19] As Douglas Erwin, writing with fellow paleontologists James Valentine and David Jablonski, acknowledged in 1999: "Attempts to date those branching[s]" from a common Precambrian ancestor "by using molecular clocks have disagreed widely."[20] How can this be?

In the first place, different studies of *different molecules* generate widely divergent dates. In addition to the studies I have already cited, a 1997 paper by Japanese biologist Naruo Nikoh and colleagues examined two genes (aldolase and triose phosphate isomerase), and dated the split between eumetazoa and parazoa—animals with tissues (like cnidarians) from those without (like sponges)—at 940 million years ago.[21] Compare that to a 1999 paper by Daniel Wang, Sudhir Kumar, and S. Blair Hedges based on the study of 50 different genes, showing that "the basal animal phyla (Porifera, Cnidaria, Ctenophora) diverged between about 1200–1500 Ma."[22]

Sometimes contradictory divergence times are reported in the same article. For instance, a refreshingly forthright paper by evolutionary biologist Lindell Bromham of Australian National University and colleagues in *Proceedings of the National Academy of Sciences USA* analyzed two different molecules, mitochondrial DNA and 18S rRNA, to yield individual gene-based divergence dates that differed by as much as 1 billion years.[23] Another study investigating the divergence between arthropods and vertebrates found that depending on which gene was used, the divergence date might be anywhere between 274 million and 1.6 billion years ago, the former date falling almost 250 million years *after* the Cambrian explosion.[24] That paper in its conclusion chose to split the difference, confidently reporting an arithmetic average of about 830 million years ago. Likewise, bioinformaticians Stéphane Aris-Brosou, now at the University of Ottawa, and Ziheng Yang, at University College London, found that depending on which genes and which estimation methods were employed, the last common ancestor of protostomes or deuterostomes (two broadly different types of Cambrian animals) might have lived anywhere between 452 million and 2 billion years ago.[25]

A survey of recent deep-divergence studies, by molecular evolutionists Dan Graur and William Martin, notes one study in which the authors claim to be 95 percent certain that their divergence date for certain animal groups falls within a 14.2-billion-year range—more than three times the age of the earth and clearly a meaningless result.[26] Graur and Martin conclude that many molecular-clock estimates "look deceptively precise," but, given the nature of this field, their "advice to the reader is: whenever you see a time estimate in the evolutionary literature, demand uncertainty!"[27] The title of their paper, published in *Trends in Genetics,* made the point still more vividly: "Reading the Entrails of Chickens: Molecular Timescales of Evolution and the Illusion of Precision."

Sometimes even different studies of the *same or similar groups of molecules* have generated dramatically different divergence times. For example, Francisco Ayala and several colleagues have recalculated the divergence times of Metazoan phyla, using mostly the same protein-coding genes as Wray's team.[28] Correcting for "a host of statistical problems"[29] in the Wray study, Ayala and colleagues found that their own estimates "are consistent with paleontological estimates"—not with the deep-divergence hypothesis. "Extrapolating to distant times from molecular evolutionary rates estimated within confined data-sets," they conclude, is "fraught with danger."[30] Or as Valentine, Jablonski, and Erwin conclude, "The accuracy of the molecular clock is still problematical, at least for phylum divergences, for the estimates vary by some 800 million years depending upon the techniques or molecules used."[31] Reported Precambrian divergence times would vary even more dramatically were it not that evolutionary biologists and molecular taxonomists ignore certain molecules in their studies to avoid grossly contradictory results. Consider, for example, histones—proteins found in all eukaryotes involved in packing DNA into chromosomes. Histones exhibit little variation from one species to the next.[32] They are never used as molecular clocks. Why? Because the sequence differences between histones, assuming a mutation rate comparable to that of other proteins, would generate a divergence time at significant variance with those in studies of many other proteins.[33] Specifically, the small differences between histones yield an extremely recent divergence, contrary to other studies. Evolutionary biologists typically exclude histones from consideration, because those times do not confirm preconceived ideas about what the Precambrian tree of life ought to look like.

But that raises obvious questions. If we don't have fossils documenting a common animal ancestor, and if genetic studies produce such different and contradictory divergence times, how do we know what the tree of life *should* look like and when the first animals began to diverge from a common ancestor? If histones change too slowly to provide an accurate calibration of the molecular clock, then which molecules do change at the correct rate—and how do we know that they do? The answer to these questions for most evolutionary biologists usually runs something like this. We already know that the animal phyla evolved from a common ancestor and we also know roughly when they did; therefore, we must reject studies based on histone sequences because the conclusions of these studies would contradict that date.

But do we really know these things, and if so how? Assumptions about the window of time in which the first metazoan, the ancestor of all animals, must have lived are clearly not derived from the testimony of molecular genetics alone, since the results of sequence comparisons vary so greatly and include dates, depending upon the molecule studied, that fall outside that window. Instead, as one widely used textbook euphemistically puts it, evolutionary biologists must choose "phylogenetically informative" data.[34] By this, they mean sequences that exhibit neither too little nor too much variation—where *too much* and *too little* are determined by preconceived considerations of evolutionary plausibility, rather than by reference to independent criteria for determining the accuracy of molecular methods.

The subjective quality of these conclusions, where scientists "cherry-pick" evidence that conforms to favored notions and discard the rest, casts further doubt on the extent to which molecular comparisons yield any clear historical signal. Only one divergence point could represent the actual universal common ancestor of all animals. If, however, comparative sequence analyses generate divergence times that are consistent with nearly all possible evolutionary histories, with the divergence event ranging from a few million to a few *billion* years ago, then clearly most of these possible histories must be wrong. They tell us little about the actual time of the Precambrian divergence, if such an event really happened.

Questionable Assumptions

Other problems run even deeper, having to do with the assumptions that make comparative sequence analyses possible in the first place. These comparisons assume the accuracy of molecular clocks—that mutation rates of organisms have remained relatively constant throughout geological time. These studies also assume, rather than demonstrate, the theory of universal common descent. Both assumptions are problematic.

Even if we assume that mutation and natural selection (and other similarly undirected evolutionary processes) can account for the emergence of novel proteins and body plans, we cannot also assume that the protein molecular clock ticks at a constant rate. Unlike radiometric dating methods, molecular clocks depend on a host of contingent factors. As Valentine, Jablonski, and Erwin note, "Different genes in different clades evolve at different rates, different parts of genes evolve at different rates and, most importantly, rates within clades have changed over time."[35] So great is this variation that one paper in the journal *Molecular Biology and Evolution* cautions, "The rate of molecular evolution can vary considerably among different organisms, challenging the concept of the 'molecular clock.' "[36]

Keep in mind too that molecular clocks are calibrated based on the estimated age of presumably ancestral fossils. If, however, such estimates are incorrect by even a few million years, or if the fossil used to calibrate the mutation rate does not lie at the actual divergence point on the tree of life, the estimated mutation rate may be badly skewed. Calibration of molecular clocks depends on an accurate understanding of the ancestor-descendant relationships between fossils and their presumed descendant taxa. If the fossil used to calibrate the divergence time of two later groups was not actually a true ancestor, then the mutation-rate calculation based on that fossil's age may be grossly inaccurate. As Andrew Smith and Kevin Peterson note: "Molecular clocks are not error-free and come with their own suite of problems. . . . The accuracy of the technique depends upon having an accurate calibration point or points, and a reliable phylogeny with correct branching order and branch-length estimates."[37] Because these conditions are rarely met, "the idea that there is a universal molecular clock ticking away has long since been discredited."[38]

Applying the molecular clock to dating the alleged Precambrian ancestor of the animals further complicates matters. Because there are so few fossils in the Precambrian and no clear ancestor-descendant lineages, the calibration of the molecular clock must be done on the basis of very different fossil lineages arising hundreds of millions of years later. Indeed, without evidence from the fossil record (older than 550 million years ago) with which to calibrate the molecular clock, any attempt to date the origin of the Cambrian animal phyla becomes highly questionable.[39] Perhaps for this reason, Valentine, Jablonski, and Erwin have wondered whether "molecular clock dates can ever be applied reliably to such geologically remote events as Neoproterozoic branchings within the Metazoa."[40] (The Neoproterozic is the last Precambrian era.) These methodological problems may help account for the cacophony of conflicting results.

SMUGGLING IN DARWIN

A second crucial assumption behind the deep-divergence hypothesis is the idea of the common descent of all the animal forms—i.e., that all the Cambrian animals evolved from a common Precambrian ancestor. As the textbook *Understanding Bioinformatics* admits, "The key assumption made when constructing a phylogenetic tree from a set of sequences is that they are all derived from a single ancestral sequence, i.e., they are homologous."[41] Or as the Harvard University Press textbook *The Tree of Life* states, "We are obliged to assume at first that, for each character, similar states are homologous," whereby "homologous" the text means characters are similar *because* they share a common ancestor.[42]

This assumption (of universal common descent) raises the possibility that the ancestral entities represented by divergence points in these studies are artifacts of the assumptions by which molecular data are analyzed. Indeed, the computer programs that are used to compare molecular sequences have been written to produce trees showing common ancestors and branching relationships regardless of the extent to which the genes analyzed may or may not differ. Phylogenetic studies compare two or more gene sequences and then use degrees of difference to determine divergence points and nodes on a phylogenetic tree. Inherent in that procedure is the assumption that the nodes and divergence points existed in the past.

Thus, the deep-divergence studies do not, in any rigorous sense, *establish* any Precambrian ancestral forms. Did a single, original metazoan or bilaterian ancestor of the Cambrian animals actually exist? The Precambrian–Cambrian fossil record taken on its face certainly doesn't document such an entity. But neither do deep-divergence studies. Instead, these studies *assume* the existence of such ancestors, and then merely attempt, given that assumption, to determine how long ago such ancestors might have lived. One could argue that the conflicting divergence points do at least show that some common ancestor existed in the Precambrian, since, despite their conflicting results, all divergence studies indicate at least that. But, again, to invoke molecular studies that assume the existence of a common ancestor as evidence for such an entity only begs the question. Certainly it provides no reason for using molecular evidence to trump fossil evidence. Perhaps the Precambrian rocks do not record ancestors for the Cambrian animals because none existed. To foreclose that possibility, and to resolve the mystery of the missing Precambrian ancestral fossils, evolutionary biologists cannot use studies that assume the existence of the very entity their studies are thought to establish.

THE "SHMOO": A CATCH-22 REVISITED

The concept of deep divergence raises another issue related to my discussion at the end of the previous chapter about what would be required to document the missing ancestral forms of the Cambrian animals. Recall that I argued there that any plausible postulated common ancestor to all the animal phyla must have necessarily lacked most (or all) of the specific anatomical features that distinguish one phylum from another. The more arthropod-like a hypothetical animal form, the less plausible it would have been as an ancestor to the chordates, mollusks, echinoderms, annelid worms, and vice versa. In each case, the design logic and arrangement of parts necessary to provide the foundation for one mode of animal life preclude it from providing the foundation for other modes of animal life—just as a system of parts providing one mode of transportation (with a bicycle, for example) will typically preclude functioning as another (as with a submarine, for example).[43]

For this reason, biologists thinking about the characteristics of the earliest ancestor of all the metazoan phyla—the actual animal at the deep-divergence point—have typically postulated an extremely simple form of life—what one evolutionary biologist described to me as a "shmoo," after the blob-like cartoon character made famous by Al Capp in the 1940s and 1950s. Some have proposed that the ur-animal might have been something like a placozoan, a modern amorphous animal with only four cell types and no bilateral symmetry.[44] Other paleontologists have mainly characterized the hypothetical ur-metazoan negatively, by reference to the characteristics that it must *not* have had to be a plausible common ancestral form to all other metazoa. (This need to characterize the ur-metazoan negatively has led some leading paleontologists to question whether the ur-animal can be described affirmatively with any specificity.[45])

In any case, the need to characterize the ur-animal as an extremely simple "shmoo-like" form, lacking the numerous characteristics and anatomical novelties present in the Cambrian animals, highlights a deep dilemma for evolutionary theorists. On the one hand, to be plausible as a common ancestor of all the animal phyla, a hypothetical ur-metazoan must have few characteristics of later metazoan forms. Indeed, the more plausible the hypothetical ancestor, the simpler it must be, meaning it will lack more of the specific distinguishing features of the individual animal phyla. But that means any evolutionary scenario for the origin of the animals that postulates such a "stripped down" animal form as its starting point will need to envision those distinguishing characteristics arising later. And the fewer the number of characteristics in the hypothetical common ancestor, the more such characteristics will need to arise later. This logical requirement implies, in turn, the need for an even deeper divergence point in Precambrian history and the need for more time to produce these specific anatomical novelties—in turn, exacerbating the problem of fossil discontinuity. The more plausible the hypothetical common ancestor, the deeper the necessary divergence point and the greater the morphological discontinuity in the fossil record.

On the other hand, proposing a more complex (and more anatomically differentiated) common ancestor closer in its affinities to some Cambrian animal forms, would eliminate the need for so deep a divergence point. Nevertheless, it would also diminish the plausibility of such a hypothetical ancestor as an ur-metazoan common to all the other Cambrian animals. Again, the more a hypothetical form resembles one of the specific animal forms or phyla, the less plausible it will be as an ancestor of all the others. And that is the dilemma. Could there have been an animal form simple enough to serve as a viable ancestor common to all the animal phyla? Perhaps. But positing such a form only deepens the required depth of the divergence point and intensifies the already significant problem of Precambrian–Cambrian fossil discontinuity.

DEEP TROUBLE

Comparative genetic analyses do not establish a single deep-divergence point, and thus do not compensate for a lack of fossil evidence for key Cambrian ancestors—such as the ur-bilaterian or ur-metazoan ancestor. The results of different studies diverge too dramatically to be conclusive, or even meaningful; the methods of inferring divergence points are fraught with subjectivity; and the whole enterprise depends upon a question-begging logic. Many leading Cambrian paleontologists, and even some leading evolutionary biologists, now express skepticism about both the results and the significance of deep-divergence studies. For example, Simon Conway Morris has rejected the idea that such studies should trump fossil evidence of a more explosive, shallow and rapid Cambrian radiation. After assessing the inconsistent track record of deep-divergence studies, he concludes, "A deep history extending to an origination in excess of 1,000 Myr [million years] is very unlikely."[46] Conway Morris is one of several leading evolutionary biologists and Cambrian paleontologists who have expressed skepticism about these studies.[47] In any case, there is now little reason to regard the deep-divergence hypothesis as a genuine solution to the Cambrian conundrum.

6

THE ANIMAL TREE OF LIFE

In 2009, in honor of the bicentennial anniversary of Darwin's birth, a piece of artwork was created to adorn the ceiling of an exhibit room at the Natural History Museum in London. A paper in the journal *Archives of Natural History* noted that the inspiration for the artwork, titled "TREE," came from a diagram that Darwin had sketched in one of his notebooks—a diagram that later came to be known as the "tree of life" (see Fig. 2.11a). One BBC radio program called the TREE exhibit the "Darwinian Sistine Chapel."[1] Another article in *Archives of Natural History,* a journal published by the University of Edinburgh, remarked that, "TREE celebrates Darwinian evolutionism" and "secular science and reason."[2]

For many biologists, the iconic image of Darwin's tree of life represents perhaps the single best distillation of what the science of evolutionary biology has to teach, namely the "fact of evolution,"[3] apart from which "nothing in biology makes sense."[4] Though the fossil record does not directly attest to many of the expected intermediate forms represented on Darwin's tree, leading authorities assert that other lines of evidence, particularly from genetics, firmly establish Darwin's tree as the correct picture of the history of life.

In the previous chapter, we saw that there are many good reasons to doubt the deep-divergence hypothesis and its claim to have determined, based upon genetic evidence, the time at which the Cambrian animals began to evolve from specific Precambrian ancestors. Indeed, the idea that these studies can pinpoint when an ur-metazoan, or an ur-bilaterian, arose has engendered increasing skepticism among a growing number of evolutionary biologists and paleontologists.

The tree of life *as a whole,* however, is another matter. Many evolutionary biologists think the case for *universal* common descent is something close to unassailable because, they argue, analysis of both anatomical and genetic similarities converges on the same basic *pattern* of descent from a universal common ancestor. As Richard Dawkins asserts, "when we look comparatively at . . . genetic sequences in all these different creatures—we find the same kind of hierarchical tree of resemblance. We find the same *family tree*—albeit much more thoroughly and convincingly laid out—as we did with . . . the whole pattern of anatomical resemblances throughout all the living kingdoms."[5] Likewise, Jerry Coyne argues that gene sequences independently confirm the same set of evolutionary relationships—the same basic tree—established from the analysis of anatomy.[6] Oxford University chemist Peter Atkins is even more emphatic: "There is not a single instance of the molecular traces of change being inconsistent with our observations of whole organisms."[7]

As a result of this confidence, evolutionary biologists often dismiss the missing Precambrian fossil precursors and intermediates as a minor anomaly—one awaiting explanation by an otherwise completely adequate theory of the history of life. Because most evolutionary biologists are confident that a single continuous tree, with a single root, best represents the history of life—and explains so many other diverse facts of biology—they continue to think the same tree-like pattern also accurately describes the Cambrian explosion and the Precambrian history of animal life. Moreover, when evolutionary biologists reconstruct the phylogenetic history of a group (including animals), they typically do so in a time-independent manner. Their concern is usually to establish a relative order of branching along the tree of life, not to establish or "pinpoint" a series of absolute dates at which divergences occurred. Thus, although deep-divergence studies do not establish the existence of Precambrian animal ancestors for all the reasons argued in the previous chapter, the uncertainty surrounding the dates in these studies has not, for most evolutionary biologists, undermined their confidence in the overall tree-like pattern of animal life. Instead, many evolutionary biologists believe the strength of the case for the tree of life as a whole, based on other phylogenetic studies of similar genes and anatomical traits, indirectly establishes the existence of the missing evolutionary precursors of the Cambrian animals. As Coyne explains, "It stands to reason that if the history of life forms a tree, with all species originating from a single trunk, then one can find a common origin for every pair of twigs (existing species) by tracing each twig back through its branches until they intersect at the branch they have in common. This node, as we've seen, is their common ancestor."[8]

On the basis of similar logic, evolutionary biologists have typically assumed that what they think is true of all other forms of life is true of the Cambrian forms—that there *must be* a universal *animal* tree, the absence of fossil evidence, and the conflicting results of deep-divergence studies, notwithstanding.

To assess the other evidence from genetics that supports this conclusion, it's useful to review how the case for the universal animal tree is similar to the deep-divergence hypothesis and also how it is different. To establish both the fact and shape of the Darwinian *animal* tree of life, evolutionary biologists have long used methods that assume that both molecular sequences and anatomical similarities provide an accurate historical signal about the past. Like deep-divergence studies, these methods of "phylogenetic" reconstruction assume that the species or larger groupings (taxa) are related by descent from a common ancestor. (The term phylogeny, again, refers to the evolutionary history of a group of organisms. Thus, a "phylogenetic reconstruction" is an attempt to determine that history.) Such studies assume that the degree of difference between molecular or anatomical features in pairs of organisms indicates how long ago they diverged from a common ancestor. They also use independent calibrations of the molecular clock to calculate the exact divergence times.[9]

Unlike deep-divergence studies, however, which attempt to establish just a single divergence time—such as that of *the* common ancestor of all the animal phyla—these more detailed phylogenetic studies seek to establish the contours of the Precambrian tree of animal life. This involves assessing degrees of relatedness among representatives of *all* the Cambrian phyla to establish multiple divergence points and times (nodes on the tree of life) as well as the relationships of the major Cambrian groups.

Investigators employ these methods even in the absence of corroborating fossil evidence. In his textbook on fossils and evolution, following a full-page depiction of the discontinuous appearance of the Cambrian animals in the fossil record, Occidental College geologist Donald Prothero explains, "If the fossil record is poor in one particular group, we look to other sources of data." He concludes that two such sources of data, anatomical and molecular data, now "converge on a common answer"—one "that is almost certainly 'the truth' (as much as we can use that term in science)."[10]

But is all of this true? Does analysis of the genetic and anatomical similarity of the Cambrian animals really establish that the history of animal life is best depicted as a continuously branching tree? Does the pattern of a branching tree accurately depict the history of Precambrian and Cambrian animal life, and in so doing establish the existence of Precambrian forms that the fossil record fails to document?

THE PRECAMBRIAN AND CAMBRIAN TREE OF ANIMAL LIFE

History happened once. And if Richard Dawkins is correct that "there is, after all, one true tree of life, the unique pattern of evolutionary branchings that actually happened,"[11] then evolutionary history also happened once. Consequently, if we think of evolutionary trees describing the relationships of animal groups as *hypotheses* about an unobserved history (which is what they are), then having two or more *conflicting* hypotheses about only one history—the history that actually happened—means that we haven't figured out what did happen. A widely used textbook on phylogenetic methods explains this: "The fact that there is only one true tree . . . provides the basis for testing alternative hypotheses. If two hypotheses are generated for the same group of species, then we can conclude that at least one of these hypotheses is false. Of course, it is possible that both are false and some other tree is true."[12]

When a body of evidence supports multiple conflicting historical hypotheses, the evidence cannot be sending a definitive historical signal about what happened in the past. That raises the possibility that it may not be sending a signal at all. Conversely, when the evidence leads investigators to converge around a single historical hypothesis, when one hypothesis best explains a whole group of clues, it is much more likely that the evidence is telling us what actually happened.

Consider, by way of illustration, a case in which we know a true history of ancestor–descendant relationships to see how evidence can converge around a single (unequivocal) history. Between 1839 and 1856, Charles Darwin and his wife, Emma, produced ten children, listed below in alphabetical order:

Anne

Charles

Elizabeth

Francis

George

Henrietta

Horace

Leonard

Mary

William

This alphabetical listing, of course, is not their actual birth order. Instead, it is one of a large number of possible birth orders for Darwin's children, only one of which is the correct sequence. Indeed, only one of these arrangements *can* represent the actual Darwin family history.

Now, suppose I gave you and some friends a pile of historical evidence about Darwin's children, and asked you to "solve for their birth order." No one would consider the problem solved if you came back with more than one order. On the other hand, if you came back and presented a single coherent hypothesis of the birth order supported by evidence from birth records, family letters, and photographs from the Darwin family archives, that would provide persuasive evidence that you had obtained the correct solution. Since only one true history exists, once you find it the evidence will tend to fall naturally into place.

But does the evidence for a Precambrian animal tree of life fall similarly into place or does it generate multiple conflicting histories? We've already seen that fossil evidence does not point to a specific Precambrian tree of animal life, or perhaps to any tree at all. We've also seen that genetic evidence by itself does not establish a single divergence point for animal evolution. But what about the genetic and anatomical evidence taken together? Does *that* evidence converge on a single history of animal life? If so, then it could well make up for a lack of fossil evidence. Otherwise, it would seem to raise an obvious question: Are the observed genetic and anatomical "affinities" among the Cambrian phyla sending reliable historical signals at all?

CONFLICTING HISTORIES

There are several reasons to doubt that evidence of genetic and anatomical similarity is sending a reliable signal of the early history of animal life. First, comparisons of different molecules frequently generate divergent trees. Second, comparisons of anatomical characteristics and molecules frequently produce divergent trees. Third, trees based solely on different anatomical characteristics often contradict each other. Let's examine each problem.

Molecules vs. Molecules

Just as the molecular data do not point unequivocally to a single date for the last common ancestor of all the Cambrian animals (the point of deep divergence), they do not point unequivocally to a single coherent tree depicting the evolution of animals in the Precambrian. Numerous papers have noted the prevalence of contradictory trees based on evidence from molecular genetics. A 2009 paper in *Trends in Ecology and Evolution* notes that "evolutionary trees from different genes often have conflicting branching patterns."[13] Likewise, a 2012 paper in *Biological Reviews* acknowledges that "phylogenetic conflict is common, and frequently the norm rather than the exception."[14] Echoing these views, a January 2009 cover story and review article in *New Scientist* observed that today the tree-of-life project "lies in tatters, torn to pieces by an onslaught of negative evidence." As the article explains, "Many biologists now argue that the tree concept is obsolete and needs to be discarded," because the evidence suggests that "the evolution of animals and plants isn't exactly tree-like."

The *New Scientist* article cited a study by Michael Syvanen, a biologist at the University of California at Davis, who studied the relationships among several phyla that first arose in the Cambrian.[15] Syvanen's study compared two thousand genes in six animals spanning phyla as diverse as chordates, echinoderms, arthropods, and nematodes. His analysis yielded no consistent tree-like pattern. As the *New Scientist* reported, "In theory, he should have been able to use the gene sequences to construct an evolutionary tree showing the relationships between the six animals. He failed. The problem was that different genes told contradictory evolutionary stories." Syvanen himself summarized the results in the bluntest of terms: "We've just annihilated the tree of life. It's not a tree anymore, it's a different topology [pattern of history] entirely. What would Darwin have made of that?"[16]

Other studies trying to clarify the evolutionary history and phylogenetic relationships of the animal phyla have encountered similar difficulties. Vanderbilt University molecular systematist Antonis Rokas is a leader among biologists using molecular data to study animal phylogenetic relationships. Nevertheless, he concedes that a century and a half after *The Origin of Species,* "a complete and accurate tree of life remains an elusive goal."[17] In 2005, during the course of an authoritative study he eventually copublished in *Science,* Rokas was confronted with this stark reality. The study had sought to determine the evolutionary history of the animal phyla by analyzing fifty genes across seventeen taxa. He hoped that a single dominant phylogenetic tree would emerge. Rokas and his team reported that "a 50-gene data matrix does not resolve relationships among most metazoan phyla" because it generated numerous conflicting phylogenies and historical signals. Their conclusion was candid: "Despite the amount of data and breadth of taxa analyzed, relationships among most metazoan phyla remained unresolved."[18]

In a paper published the following year, Rokas and University of Wisconsin at Madison biologist Sean B. Carroll went so far as to assert that "certain critical parts of the TOL [tree of life] may be difficult to resolve, regardless of the quantity of conventional data available."[19] This problem applies specifically to the relationships of the animal phyla, where "[m]any recent studies have reported support for many alternative conflicting phylogenies."[20] Investigators studying the animal tree found that "a large fraction of single genes produce phylogenies of poor quality" such that in one case, a study "omitted 35% of single genes from their data matrix, because those genes produced phylogenies at odds with conventional wisdom."[21] Rokas and Carroll tried to explain the many contradictory trees by proposing that the animal phyla might have evolved too quickly for the genes to record some signal of phylogenetic relationships into the respective genomes. In their view, if the evolutionary process responsible for anatomical novelty works quickly enough, there would not be sufficient time for differences to accumulate in key molecular markers, in particular those used to infer evolutionary relationships in different animal phyla. Then, given enough time, whatever signal did exist might become lost. Thus, when groups of organisms branch rapidly and then evolve separately for long periods of time, this "can overwhelm the true historical signal"[22]—leading to the inability to determine evolutionary relationships.

Their article brings the discussion of the Cambrian explosion full circle from an attempt to use genes to compensate for the absence of fossil evidence to the acknowledgment that genes do not convey any clear signal about the evolutionary relationships of the phyla first preserved by fossils in the Cambrian. The logic of their analysis also leads them to a strangely familiar conclusion. Since the analysis of key genetic markers—like the genes tracked in molecular-clock studies that presumably accumulate mutations at a constant rate—shows a low number of mutational differences between the Cambrian animal phyla, Rokas and Carroll conclude from *specifically genetic evidence* that the phyla must have diverged rapidly. As they put it in another paper, "Inferences from these two independent lines of evidence (molecules and fossils) support a view of the origin of Metazoa as a radiation compressed in time."[23] Thus, the inability to reconstruct the evolutionary history of the animal phyla from the molecular data not only fails to establish a Precambrian pattern of descent; it ironically also reaffirms the extreme rapidity of the origin of the Cambrian animal forms.

Molecules vs. Anatomy

In 1965, chemist Linus Pauling and biologist Emile Zuckerkandl, often hailed as the fathers of the molecular-clock concept, proposed a rigorous way to confirm evolutionary phylogenies. They suggested that if studies of comparative anatomy and DNA sequences generated similar phylogenetic trees, then "the best available single proof of the reality of macroevolution would be furnished."[24] As they went on to explain, "only the theory of evolution . . . could reasonably account for such a congruence between lines of evidence obtained independently."[25] By focusing attention on these two independent lines of evidence and the possibility of their convergence (or conflict), Pauling and Zuckerkandl provided a clear and measurable way to test the neo-Darwinian thesis of universal common ancestry.

And according to some scientists, studies of molecular homologies have confirmed expectations about the history of the animal phyla derived from studies of comparative anatomy. After citing Pauling and Zuckerkandl's test, Douglas Theobald claims in his "29+ Evidences for Macroevolution" that "well-determined phylogenetic trees inferred from the independent evidence of morphology and molecular sequences match with an extremely high degree of statistical significance."[26]

In reality, however, the technical literature tells a different story. Studies of molecular homologies often *fail* to confirm evolutionary trees depicting the history of the animal phyla derived from studies of comparative anatomy. Instead, during the 1990s, early into the revolution in molecular genetics, many studies began to show that phylogenetic trees derived from anatomy and those derived from molecules often contradicted each other.

Probably the most protracted conflict of this type concerns a widely accepted phylogeny for the bilaterian animals. This classification scheme was originally the work of the influential American zoologist Libbie Hyman.[27] Hyman's view, generally known as the "Coelomata" hypothesis, was based on her analysis of anatomical characteristics, mainly germ (or primary tissue) layers, planes of body symmetry, and especially the presence or absence of a central body cavity called the "coelom," which gives the hypothesis its name. In the Coelomata hypothesis, the bilaterian animals were classified in three groups, the Acoelomata, the Pseudocoelomata, and the Coelomata, each encompassing several different bilaterian animal phyla.[28] (See Fig. 6.1a.)

Then, in the mid-1990s, a very different arrangement of these animal groups was proposed based on the analysis of a molecule present in each (the 18S ribosomal RNA; see Fig. 6.1b). The team of researchers who proposed this arrangement published a groundbreaking paper in *Nature* with a title that surprised many morphologists: "Evidence for a Clade of Nematodes, Arthropods and Other Moulting Animals."[29] The paper noted the conventional wisdom, based on Hyman's hypothesis, that arthropods and annelids

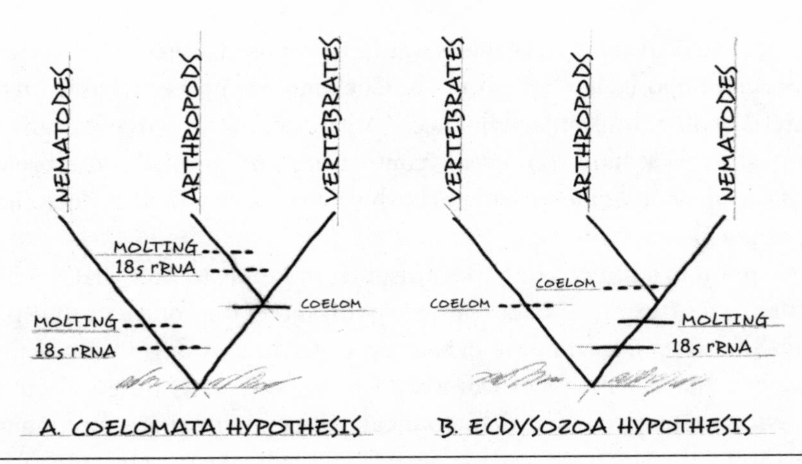

FIGURE 6.1
How scientists reconstruct evolutionary history depends on which similarities they regard as revealing the true history of descent (homology) and which similarities they regard as misleading (homoplasy). Advocates of the Coelomata hypothesis (Figure 6.1a) regard the coelom (body cavity) as a homologous feature. Thus, they think the presence of a coelom in both arthropods and vertebrates indicates a common ancestor that possessed a coelom (indicated by the solid horizontal line in 6.1a). But advocates of the Ecdysozoa hypothesis (6.1b) think the coelom evolved at least twice independently (indicated by the two dashed horizontal lines in Figure 6.1b). They regard the presence of the coelom as a historically misleading similarity—one that does not indicate the presence of that feature in the most recent common ancestor of the groups possessing it. These two hypotheses and their implied histories are not congruent, and cannot both be true.

were closely related because both phyla had segmented body plans.[30] But their study of the 18S ribosomal RNA suggested a different grouping, one that placed arthropods close to nematodes within a group of animals that molt, which they called "Ecdysozoa." This relationship surprised anatomists, since arthropods and nematodes don't exactly look like kissing cousins. Arthropods (such as trilobites and insects) have coeloms, whereas nematodes (such as the tiny worm *Caenorhabditis elegans*) do not, leading many evolutionary biologists to believe nematodes were early branching animals only distantly related to arthropods.[31] The *Nature* paper explained how unexpected this grouping of arthropods and nematodes was: "Considering the greatly differing morphologies, embryological features, and life histories of the molting animals, it was initially surprising that the ribosomal RNA tree should group them together."[32]

Since the Ecdysozoa hypothesis was first proposed, other scientists have vigorously opposed it, reaffirming the Coelomata hypothesis, based on the analysis of other molecular evidence.[33] Advocates of the Ecdysozoa grouping pushed back hard, however,[34] contending that, properly interpreted, available genetic evidence supports the Ecdysozoa, not the Coelomata hypothesis.[35]

My point in summarizing these disputes is simply to note that the molecular and anatomical data commonly disagree, that one can find partisans on every side, that the debate is persistent and ongoing, and that, therefore, the statements of Dawkins, Coyne, and many others about all the evidence (molecular and anatomical) supporting a single, unambiguous animal tree are manifestly false. As can readily be seen by comparing Figures 6.1a and 6.1b,[36] these hypotheses—Coelomata and Ecdysozoa—contradict each other. Although both might be false, both cannot be true.

Various papers analyzing other groups have found similar discrepancies between molecular and morphological versions of the animal tree. A paper by Laura Maley and Charles Marshall in the journal *Science* noted, "Animal relationships derived from these new molecular data sometimes are very different from those implied by older, classical evaluations of morphology."[37] For example, when tarantulas were used as the representative of arthropods, the arthropods were grouped more closely to mollusks than to deuterostomes (animals that develop anuses first and mouths later). This makes sense because both mollusks and arthropods are protostomes (animals that develop mouths first and anuses later). But when brine shrimp were used as the representative of the arthropods, the arthropods became the odd man out. Now mollusks were grouped most closely to deuterostomes, far away from arthropods—a result clearly at odds with the conventional phylogeny based upon anatomical characteristics.[38]

Examples of similar conflicts abound. The traditional phylogeny placed sponges at the bottom of the animal tree, with progressively more complex phyla (e.g., cnidarians, flatworms, nematodes) branching off. But Valentine, Jablonski, and Erwin note that molecules "indicate a very different configuration" of the tree, where some higher deuterostome phyla branch off very early and some comparatively less complex phyla branch very late.[39]

Likewise, morphological studies suggest phoronids (see Fig. 3.6) and brachiopods (see Fig. 1.3), both marine filter-feeding animals, are deuterostomes, but molecular studies classify them within protostomes.[40] Morphological studies typically imply that sponges are monophyletic (all part of an exclusive branch on the tree of life) because of their distinctive body architecture, but molecular studies suggest that sponges don't belong to a single unified group, with some sponges more closely related to jellyfish than they are to other sponges.[41] Cnidarians and ctenophores have similar body plans, leading many to expect they were closely related on the basis of morphology. But molecular data have distanced these phyla significantly.[42] As a major review article in *Nature* in 2000 observes, "Evolutionary trees constructed by studying biological molecules often don't resemble those drawn up from morphology."[43] And the problem isn't getting better over time. A 2012 paper admits that larger datasets are not solving this problem: "Incongruence between phylogenies derived from morphological versus molecular analyses and between trees based on different subsets of molecular sequences has become pervasive as datasets have expanded rapidly in both characters and species."[44]

Indeed, the widespread discrepancies between molecular data and morphological data and between various molecule-based trees have led some to conclude that Pauling and Zuckerkandl were wrong to assume that the degree of similarity indicates the degree of evolutionary relatedness.[45] As Jeffrey H. Schwartz and Bruno Maresca put it in the journal *Biological Theory:* "This assumption derives from interpreting molecular similarity (or dissimilarity) between taxa in the context of a Darwinian model of continual and gradual change. Review of the history of molecular systematics and its claims in the context of molecular biology reveals that there is no basis for the 'molecular assumption.'"[46]

Anatomy vs. Anatomy

Attempts to infer a consistent picture of the history of animal life based on analyzing the anatomical characteristics of different animals have also proven problematic. In the first place, there is a general and long-standing problem with attempts to infer the evolutionary history of the animal phyla from similar anatomical traits. At the level of the phyla—that is, when one compares the phyla to each other and tries to determine their branching order—the number of shared anatomical characteristics available for inferring evolutionary relationships drops off quite dramatically. There is an obvious reason for this. For example, an anatomical character such as the "leg," that is useful for diagnosing and comparing arthropods, which possess legs, proves useless for making comparisons between (for example) brachiopods or bryozoans, which do not. In the same way, basic structural features of human-designed systems, such as the distinctive submarine "trait" of an encapsulating watertight hull, might help to distinguish it from a cruise ship, which is only watertight on its underside. But this "trait" would be irrelevant for comparing and classifying, say, suspension bridges, motorcycles, or flat-screen televisions. In a similar manner, biologists find that there are only a handful of highly abstract characters, such as radial versus bilateral body symmetry, the number of fundamental tissue layers (triploblasty, three layers, versus diploblasty, two layers), or the type of body cavity present (true coelom, pseudocoelom, or no coelom), available for morphological comparisons of the many diverse animal forms. Yet evolutionary biologists have often disputed the validity of these rather abstract traits as guides to evolutionary history.[47] In addition, just as trees based upon the analysis of different sets of similar genes or proteins often conflict, trees constructed on the basis of different developmental and anatomical characteristics often conflict.

When biologists construct phylogenetic trees based upon anatomical characteristics, they typically group the animal phyla according to the presence or absence of several key characteristics. For example, the standard version of the animal tree, based upon anatomy, groups animals according to their style of body plan symmetry and by their mode of body plan development. As noted earlier, animals with mirror symmetry along their vertical head-to-tail axes all fall within the Bilaterian group. Animals with radial symmetry (or no symmetry) fall outside this group. Within the Bilateria, taxonomists distinguish other main groups—protostomes and deuterostomes—based upon their differing modes of body plan development—i.e., either "mouth first" or "anus first."

Yet a significant difficulty arises when evolutionary biologists consider how a particularly fundamental characteristic—the mode of germ-cell formation—is distributed among various groups on the canonical animal tree of life (see Fig. 6.2).[48] Germ cells produce eggs and sperm (in any sexually reproducing species) or gametes (in any asexually reproducing species), giving rise to the next generation.[49] Animals have two main ways of generating germ cells. In one mode of germ-cell formation, known as preformation, cells inherit *internal* signals from a region within their own cell structure to become germ cells (illustrated by solid black circles on Fig. 6.3a). In the other main way of generating germ cells, known as epigenesis, germ cells receive *external* signals from surrounding tissues to become primordial germ cells (PGCs, illustrated by solid white squares in Fig. 6.3b).

Germ-cell formation has indisputable evolutionary importance. To evolve, a population or a species must leave offspring; to leave offspring, species of animals must generate primordial germ cells. No PGCs, no reproduction; no reproduction, no evolution.

One might expect, therefore, that if a group of animals is all derived from a common ancestor (with a particular mode of gamete production), then the mode of germ-cell formation should also be essentially the same from one animal species to the next in that group. Further, assuming the common ancestry of all animals, our expectation of homologous modes of germ-cell formation among the animals ought to be higher than for any other tissue type, cell line, or mode of development. Why? Because mutations affecting the developmental mechanisms that govern PGC formation inevitably disrupt successful reproduction.[50] Again, if a species cannot reproduce, it cannot evolve.[51]

Thus, similar groups of animals—indeed, all animals, if they have descended from a common ancestor—ought to exhibit the same basic mode of germ-cell formation. Further, that the evolutionary tree derived from an analysis of the "mode of germ-cell formation" ought to be congruent with the trees derived from other such fundamental characteristics (such as body-plan symmetry, mode of development, number of primary tissues, and so forth).

But the mode of germ-cell formation is nearly randomly distributed among the different animal groups, making it impossible to generate a coherent tree based on this characteristic, let alone making any comparison between such a tree and the canonical tree. Note also the distribution of the two basic modes of germ-cell development within the animal phyla

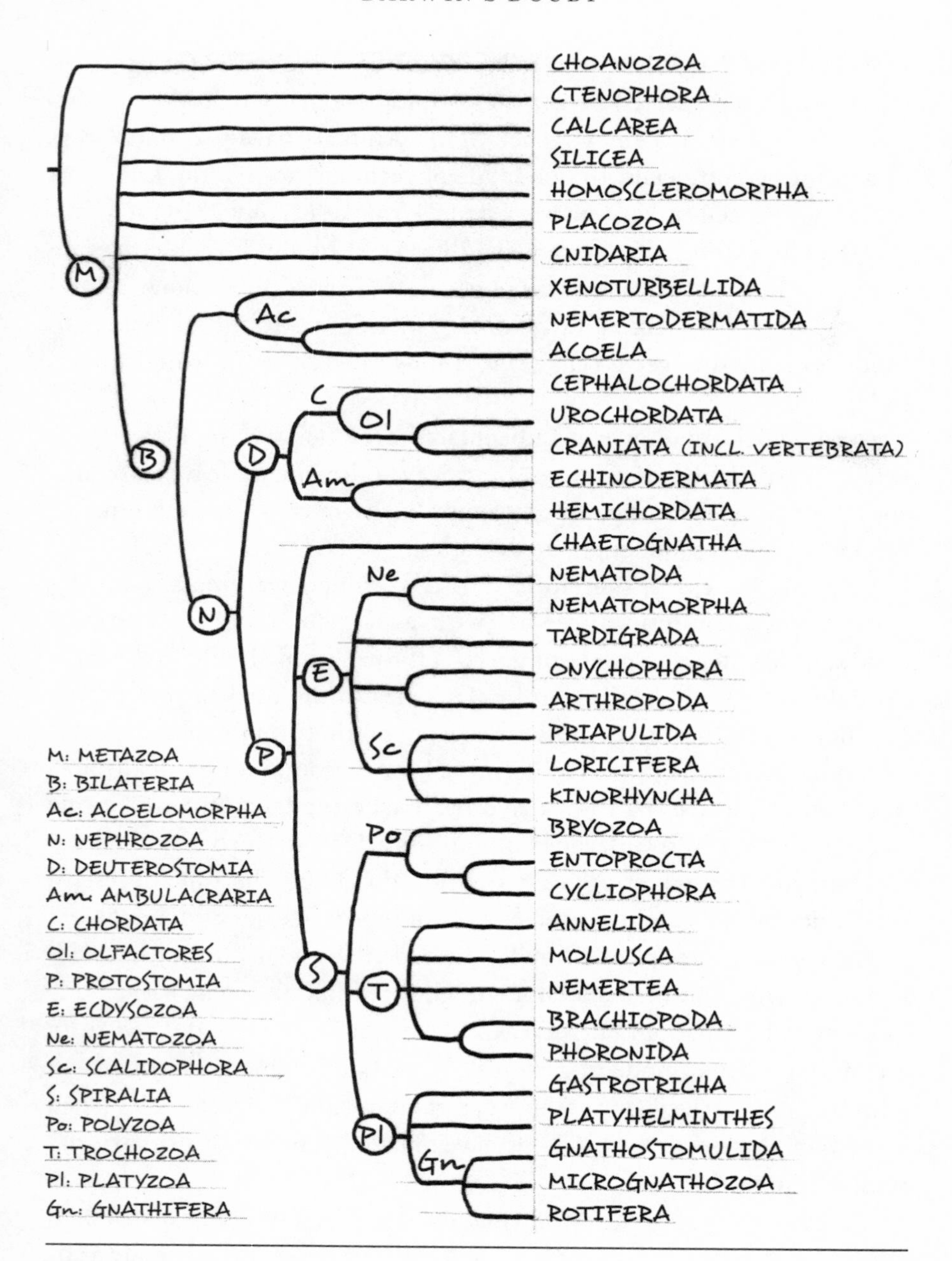

FIGURE 6.2
The canonical tree of the Metazoa as determined by the analysis of selected anatomical characters and genes.

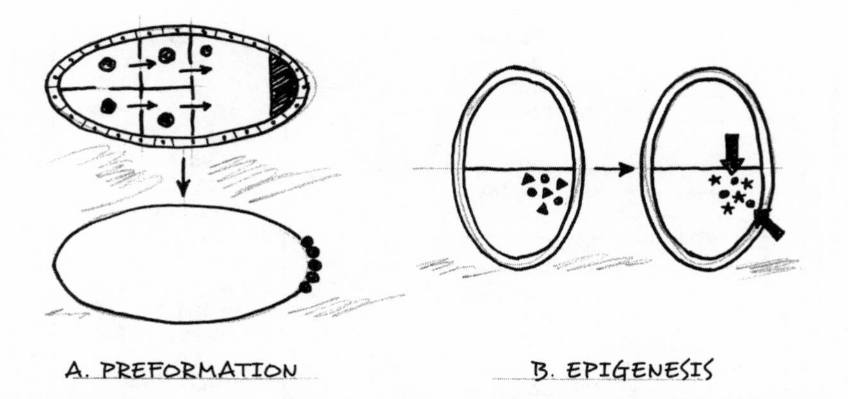

FIGURE 6.3
Two modes of germ-cell formation. *Figure 6.3a (left):* In fruit flies as the egg is being formed, the mother's nurse cells (the four cells on the left of the oval-shaped egg chamber, indicated by large black circles) deposit proteins and RNAs, which are transported to the posterior pole of the egg (indicated by the dark patch to the right of the large cell on the right). These maternally synthesized molecules then trigger the development of the germ cells and sex organs of the fly during embryogenesis. *Figure 6.3b (right):* In the eggs of mice there are no maternally deposited products that determine germ-cell formation. Rather, as the embryo develops a subpopulation of cells (represented by the triangles on the left) express "germline competence genes." These cells then "read" signals that arrive from other tissues (see arrows), causing the cells to differentiate into primordial germ cells (as indicated by the stars on the right).

as depicted on the canonical tree. Figure 6.4, derived from the work of Harvard developmental biologist Cassandra Extavour,[52] shows this distribution and provides another way of understanding the incongruence that arises when analyzing different anatomical characteristics.

Notice the two modes of germ-cell formation do not cluster together in separate parts of the canonical tree. Instead, they are distributed haphazardly among various phyla on different branches of the tree. In the protostomes, for instance, modes of germ-cell formation wink on and off between preformation and epigenesis. The same is the case within the deuterostomes: germ-cell formation varies almost randomly, and several groups exhibit both modes, rendering it difficult or impossible to determine which characteristic was present at different ancestral branching points. Noting this pattern of distribution, Cassandra Extavour concludes that "the data presently available cannot suggest homologies of the somatic components of metazoan gonads."[53]

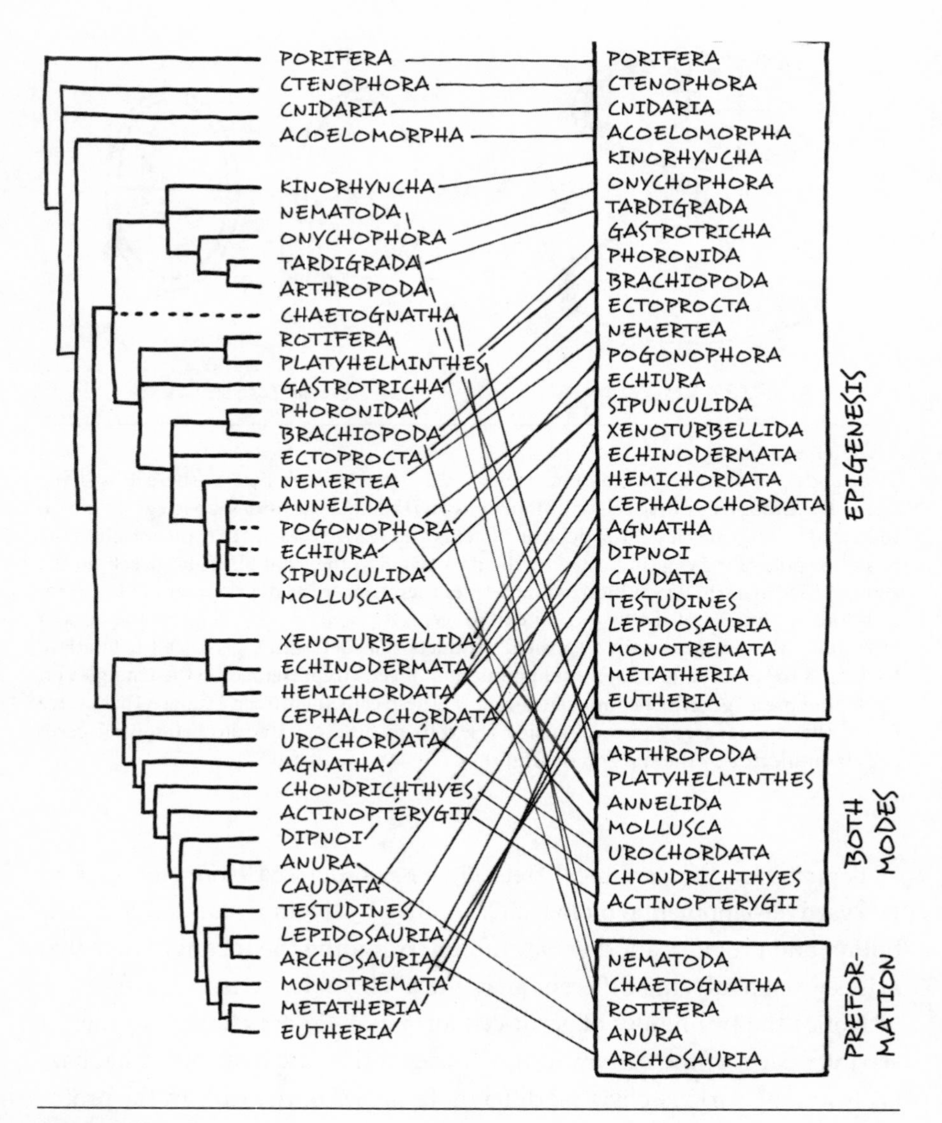

FIGURE 6.4

The distribution of modes of primordial germ-cell formation (epigenesis, preformation, or both) among various animal groups. The solid thin lines between the boxes on the right and the phyla names on the left show where the different modes of germ-cell formation are present in different phyla. The nearly random distribution of types of germ-cell formation among the various animal groups makes it impossible to generate an animal phylogeny (evolutionary history) based upon this character that will match the evolutionary history implied by the canonical animal tree of life.

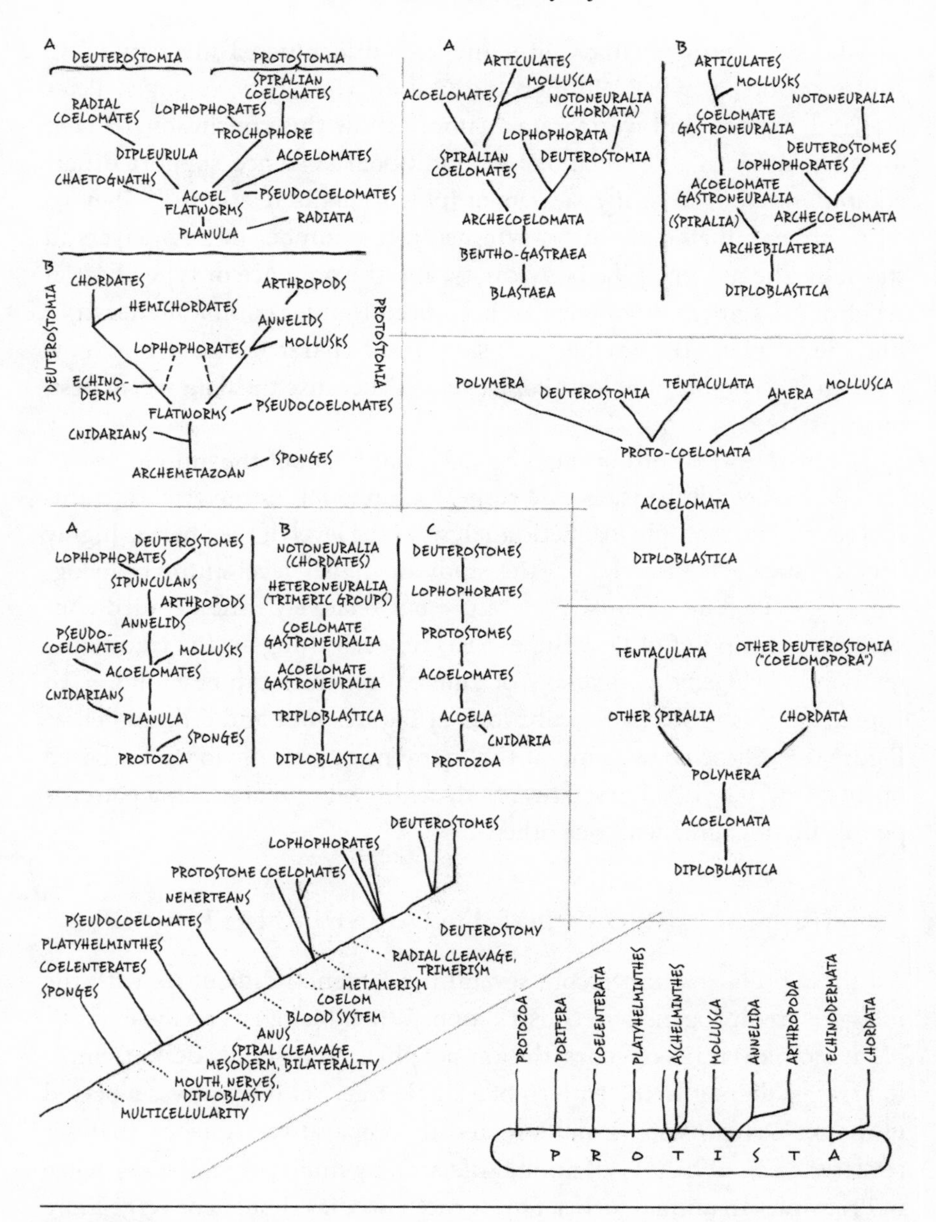

FIGURE 6.5
A selection of incompatible (mutually incongruent) phylogenetic trees representing the history of the major groups of animals, drawn from the zoological and evolutionary literature (1940–present). Note: The definitions of some taxonomic groups in some of these phylogenetic trees may have changed significantly since the time those phylogenies were originally constructed. Branch lengths may not always be drawn to scale.

After completing a survey of many such difficulties, University of St. Andrews zoologist Pat Willmer and Oxford University zoologist Peter Holland, experts on invertebrate anatomy, draw this conclusion: "Taken together, modern re-evaluations of traditional evidence support different *and mutually exclusive* subsets of [phylogenetic] relations."[54] They go on to observe that "patterns of symmetry, the number of germ layers in the body, the nature of the body cavity, and the presence or type of serial repetition [segmentation] have all been used to infer common ancestry." But, they explain, the phylogenetic story these characteristics tell is "now either unacceptable or at least controversial" because the data are, at best, inconsistent.[55]

The historical record of ongoing uncertainty about the animal tree of life since 1859 confirms, as one respected textbook on invertebrate animals explains, that "phylogenetic analysis at the level of the phyla is highly problematical."[56] As a result, "the study of higher level animal phylogeny has yielded an expansive literature but relatively little detailed consensus. . . . In point of fact, there exists no such thing as 'the traditional textbook phylogeny.' A diversity of different schemes can be found."[57] To appreciate this problem visually, look at Figures 6.1a and 6.1b as well as Figure 6.5. These show some of the many metazoan phylogenies, based on anatomy, published in the twentieth century. These branching patterns plainly do not agree with one other.

THE ASSUMPTIONS OF PHYLOGENETIC INFERENCE

All these problems underscore several fundamental difficulties with the methods of phylogenetic reconstruction. When biologists analyze multiple anatomical traits or genes, the animal phyla consistently defy attempts to arrange them into the pattern of a single tree. Yet if there was a period of hidden Precambrian evolution, and if comparative sequence analyses reveal the actual history of animal life and, by implication, the existence of Precambrian animal forms, phylogenetic studies should converge more and more around a single tree of animal life. Just as only one possible divergence point could represent the event at which animal forms began to evolve from a common animal ancestor, only one of the many trees produced by phylogenetic analysis can represent the actual Precambrian history of animal life. If, instead, phylogenetic analyses consistently generate different possible evolutionary histories, it's difficult to see how any one of them could be known to be sending a reliable historical signal. Again, the history of animal life only happened once.

One could argue that these conflicting trees do, at least, show that *some* tree-like evolutionary pattern of common ancestry preceded the Cambrian, since all conflicting trees do affirm that. But, again, they all "show" this because they all presuppose it, not because they demonstrate it.

CONVERGENT EVOLUTION

There is yet another reason to wonder whether studies of anatomical or molecular homology convey anything definitive about the history of life. Many animals have single traits or features in common with otherwise decidedly different animals. In such cases, it makes no evolutionary sense to classify these forms as closely related ancestors. For example, moles and mole crickets have remarkably similar forelimbs, though moles are mammals and mole crickets are insects. No evolutionary biologist regards these two animals as closely related, for understandable reasons.

The theory of universal common descent assumes that, generally, the more similar two organisms are, the more closely related they must be. Assuming common descent, animals with wildly differing body plans should not be closely related. The presence of nearly identical individual traits or structures within organisms exemplifying otherwise different body plans cannot, therefore, be attributed to evolution from a common ancestor. Instead, evolutionary biologists attribute similar traits or structures in such a context to so-called *convergent evolution,* the separate or independent origin of similar characters emerging on separate lines of descent after the point at which those lines diverged from their last common ancestor. Convergent evolution demonstrates that similarity does not always imply homology, or inheritance from a common ancestor.

For this reason, the repeated need to posit convergent evolution (and other related mechanisms)[58] casts further doubt on the method of phylogenetic reconstruction. Invoking convergent evolution negates the very logic of the argument from homology, which affirms that similarity implies common ancestry, except—we now learn—in those many, many cases when it does not. Repeatedly invoking convergence negates the assumption that justified the method of phylogenetic reconstruction in the first place, namely, that similarity is a reliable historical signal of common ancestry.

A FAMILY REUNION?

So what lesson should we draw from these many conflicting trees? Clearly, these contradictory results call into question the existence of a canonical tree of animal life. To see why, imagine being invited to an event billed as an extended-family reunion where you've been told you will encounter hundreds of your relatives, most of whom you have never met. Let's call the description in the invitation the "reunion hypothesis." The invitation says that a group photograph is planned, for which relatives will be grouped together according to their degree of relationship (first cousins with first cousins, and so on).

You show up and grab a coleslaw, eager to meet these many previously unknown relatives. You see the hundreds gathered and have every reason to believe that the "reunion hypothesis" is true. After all, the invitation in your mailbox described the event as a family reunion.

As the day goes on, however, something seems amiss. Here and there, you see familiar facial features—"Yes," you think, that person could be my cousin"—but the majority of attendees and all the strangers you engage in conversation exhibit no discernible family resemblances. Nor does anyone seem to share any personal relationships with anyone else at the reunion, no matter how long they chat and try to establish points of commonality. What's more, each person tells a different story of his or her family history. You try to group the strangers by physical characteristics (height, hair color, body type, and so on), but the characteristics that you find fail to yield evidence of family ties or genealogical connections. Whatever points of commonality you find refuse to assemble into a coherent, consistent story. The pedigrees are unclear. The reunion hypothesis is under major strain.

When the photographer arrives, she gamely tries to draw everyone together for the planned photograph. Chaos ensues. No one knows where to stand, because the family relationships are so unclear, if they exist at all. After milling around hopelessly for half an hour or so, people depart for their cars, wondering why they bothered to attend the picnic, and indeed why they were invited in the first place.

Do you now have good reasons to doubt the "reunion hypothesis"? You do. If the family reunion hypothesis were true, it would have been increasingly confirmed, because the evidence would have converged on a single consistent pattern. Had there been a true pattern of family relationships, the longer everyone talked, the more a single coherent pattern of relationship would have become apparent. But the people at the picnic were not your relatives, at least not in any way you could determine to be true. Instead of converging on a single pattern, the "evidence" was all over the map.

A FOREST OF TREES

Of course, my family reunion illustration breaks down as an analogy to the history of animal life, because if we could trace the history of all the people at the reunion back far enough we would find that they *are* all related by common ancestry. Though we can choose to assume that the same is true of the Cambrian animals, neither the fossil evidence nor the evidence of genetics and comparative anatomy actually establishes that. These three classes of evidence either provide no compelling evidence for Precambrian animal ancestors (in the case of fossils), or they provide question-begging and conflicting evidence (in the case of genes and anatomy).

And that is the point of my story. Since there can be only one true history of the Cambrian animals, the evidence should converge on a common family tree—if indeed we are looking at *evidence* of true history. The picture given by the evidence should be stable, not constantly changing. But the evidence from a variety of quarters has instead continually generated new, conflicting, and incoherent pictures of the history of animal life. As with the "cousins" in my illustration, there seems to be no consistent and coherent way to organize the animal groups into a family tree.

But if the genes don't tell the story of Precambrian ancestral forms, if they don't compensate for a dearth of fossil evidence and establish a unequivocal long cryptic history of animal life from an original animal, an ur-metazoan, then logically we are back to taking the fossil record at face value. In that case, the mystery of the missing ancestral fossils remains. If so, is there any way to explain the abrupt appearance of new forms of life in the fossil record within an evolutionary framework? During the 1970s, two young paleontologists thought that there just might be a way to do exactly that.

7

PUNK EEK!

Scientific discoveries are rarely made in Laundromats, but at least one great scientific breakthrough—an "aha" moment—occurred in one. The year was 1968, more than a decade before the discovery of the first Chengjiang fossils. The scientist overtaken by the muses was paleontologist Niles Eldredge. One day while standing in a Michigan Laundromat, following months of collecting trilobite fossils for his Ph.D. research, Eldredge happened to reach into his pocket. He removed one of the fossils he had been collecting, a specimen of a trilobite species called *Phacops rana*. Initially, as he examined the specimen, he felt "depressed." The fossil closely resembled many others that he had found across layers of strata during his fieldwork in the Midwest. His trilobites showed no evidence of gradual change, as classical neo-Darwinism had taught him to expect.[1]

As Eldredge explained in a lecture at the University of Pittsburgh in 1983, he then experienced a kind of scientific epiphany. He realized that the "absence of change itself" was "a very interesting pattern." Or as he later put it, "Stasis is data."[2] "Stasis" is the term that Eldredge and his scientific collaborator, Stephen Jay Gould (see Fig. 7.1), later gave to the pattern in which most species, "during their geological history, either do not change in any appreciable way or else they fluctuate mildly in morphology, with no apparent direction."[3] As Eldredge examined that solitary trilobite, he realized that he had been observing evidence of stasis for some time—however much he might have wanted it otherwise. As he explained, "Stasis . . . was by far the most important pattern to emerge from all my staring at *Phacops* specimens." He continued, "Traditionally seen as an artifact of a poor record, as the inability of paleontologists to find what evolutionary biologists going back to Darwin had told them must be there, stasis was, as Stephen Jay Gould put it, 'paleontology's trade secret'—an embarrassing one at that."[4]

This embarrassing realization proved pivotal, eventually leading Eldredge and Gould to reject both the gradualistic picture of evolutionary change articulated by Darwin and the neo-Darwinian understanding of the mechanism by which such change allegedly takes place. It also led them to formulate, in a series of scientific papers from 1972 to 1980, a new theory of evolution known as "punctuated equilibrium."[5]

As a consequence of this theory, neither Gould nor Eldredge expected to find a wealth of transitional intermediate forms in the fossil record. In their view, the main periods of biological innovation simply occurred too rapidly to leave many fossil intermediates behind.[6]

Gould and Eldredge sought to explain the occurrence of the rapid periods[7] of change (i.e., the punctuations) as the by-product of different kinds of evolutionary mechanisms or processes of change. They proposed, first, a mechanism called "allopatric speciation" to explain the rapid generation of new species. Gould, Eldredge, and another early advocate of punctuated equilibrium named Steven Stanley, a paleontologist at Johns Hopkins University, also proposed that natural selection operated on

FIGURE 7.1
Figure 7.1a (left): Stephen Jay Gould. *Courtesy Getty Images. Figure 7.1b (right):* Niles Eldredge. *Copyright © Julian Dufort 2011. Used with permission.*

higher levels. Rather than natural selection favoring the fittest individual organisms within a species—as it does in classical Darwinism and neo-Darwinism—these paleontologists proposed that it often selected the most fit *species* among a group of competing species. Because they thought that speciation occurred more rapidly, and because they thought that natural selection acted on whole species and not just individual organisms, the advocates of punctuated equilibrium theorized that morphological change typically occurs in larger, more discrete jumps than Darwin first envisioned.

Thus, in one sense, the theory of punctuated equilibrium, like the artifact hypothesis, sought to explain the absence of the transitional intermediate forms that were expected based on Darwin's theory. By repudiating Darwinian gradualism, the advocates of punctuated equilibrium sought to account for the absence of transitional forms in the fossil record apart from the artifact hypothesis or, at best, using what they imagined as a more modest version of it. But in repudiating Darwinian gradualism, punctuated equilibrium also represented a radically different view of the pace and mode of evolution—a new theory of evolution that purported to identify a new mechanism of evolutionary change. As historian of science David Sepkoski explains, "Gould and Eldredge proposed a radical revision of this standard [neo-Darwinian] narrative. They argued that the pattern of evolutionary history really was composed of fits and starts, consisting of long periods of evolutionary stasis (or 'equilibrium') 'punctuated' by shorter periods of rapid speciation."[8]

During the 1970s and 1980s, the theory of punctuated equilibrium, or "punk eek" as it is affectionately known, generated both intense scientific debate and extensive media coverage.[9] Critics called the model "evolution by jerks," leading Gould to reply that proponents of gradualism were offering "evolution by creeps."[10] Though initially Eldredge played more of a role in formulating the theory, Stephen Jay Gould emerged as the leading spokesman for it. As a result of his advocacy of the theory as well as his popular science writing, Gould achieved immense celebrity—celebrity that has in turn secured an enduring place for punctuated equilibrium in scientific awareness.

So what has become of this bold scientific proposal? Does punctuated equilibrium solve the problems that traditional neo-Darwinism does not?

Does it help explain the Cambrian explosion and the missing fossil intermediates that render it so mysterious?

WANTED: FAST ENGINE

Once they decided to take the fossil record at face value, the question for Gould and Eldredge was obvious: What could generate evolutionary change so rapidly? To explain the short bursts or punctuations, Gould and Eldredge proposed a mechanism of rapid speciation to which Stanley added (with their agreement) a new understanding of the mechanism of natural selection.[11]

Whereas the neo-Darwinian mechanism of natural selection acting on random mutations necessarily acts slowly and gradually, Gould and Eldredge invoked a process called "allopatric speciation" to explain how new species might arise quickly. The prefix *allo* means "other" or "different," and the suffix *patric* means "father." Thus, allopatric speciation refers to processes that generate new species from separate parent (or "father") populations. Allopatric speciation typically occurs when part of a population of organisms becomes geographically isolated—perhaps by the emergence of a mountain range or the shifting of a river's course—from a larger parent population and then a daughter population changes in response to differing environmental pressures.

Gould and Eldredge drew on insights from population genetics to explain why new genetic traits were more likely to spread and establish themselves within these smaller subpopulations. Population genetics, a subject to which I'll return in Chapter 12, describes the processes by which genetic traits change and become fixed in a population of organisms. It teaches that in typically *large* populations of organisms, it is difficult for a newly arising genetic trait to spread throughout an entire population. Yet for any evolutionary change to occur in a population, new genetic traits must become widespread, or "fixed," by a process called "fixation."

In smaller populations, however, the probability of a newly arising trait becoming fixed is much higher, since the new trait needs to spread to fewer organisms. By way of illustration, consider a bag containing fifty red and fifty blue marbles. Suppose that, by removing individual marbles at random, you seek to change the "population" of mixed-color marbles to one in which all the marbles are red. To produce a completely red "species," we must generate a population in which all the blue marbles have been eliminated. If someone randomly picks half of the one hundred marbles out of the bag, it is extremely improbable that all of the marbles thus eliminated will be of just one color. Indeed, there is less than 1 chance in 10^{30} that all of the marbles selected for removal will be blue.[12] Conversely, there is an extremely high probability that the remaining batch will still include both blue and red marbles.

In a much smaller group of, say, eight marbles, evenly divided between four red and four blue, the probability of selecting four blue marbles at random and leaving only red marbles, though unlikely, is not prohibitively small. There's now a much higher chance—1 in 70—that the remaining population of marbles will be all red.[13] By starting with a smaller number of marbles, the probability that random selection will result in a population of uniform color is much higher. In a similar way, the probability of fixing a genetic trait in a population of organisms *decreases* exponentially with the size of the population.

In formulating punctuated equilibrium, Gould realized that new species would inevitably have to arise in smaller populations, where random processes could have a greater chance of fixing traits. Prominent among those random processes is one called genetic drift. This occurs when genetic changes spread or disappear randomly through a population, without regard for their effect on survival and reproduction.

In Gould and Eldredge's view, allopatric speciation helped to explain how evolution could occur in larger, more discrete jumps than Darwinian gradualism predicts (see Fig. 7.2). As allopatric speciation occurs, it can generate what Gould and Eldredge conceived as sibling or offspring species. They thought that the processes that drive these speciation events occur relatively quickly in smaller populations, thus helping to explain the sudden jumps in the fossil record. As they put it: "Small numbers and rapid evolution virtually preclude the preservation of speciation events in the fossil record."[14] As they envisioned the evolutionary process working, the branches on the tree of life would split off so abruptly that they would appear as virtually "horizontal" lines, producing sudden discontinuities in the fossil record and therefore fewer fossilized intermediates. Eldredge and Gould explained it this way: "The theory of allopatric (or geographic) speciation suggests a different interpretation of paleontological data.

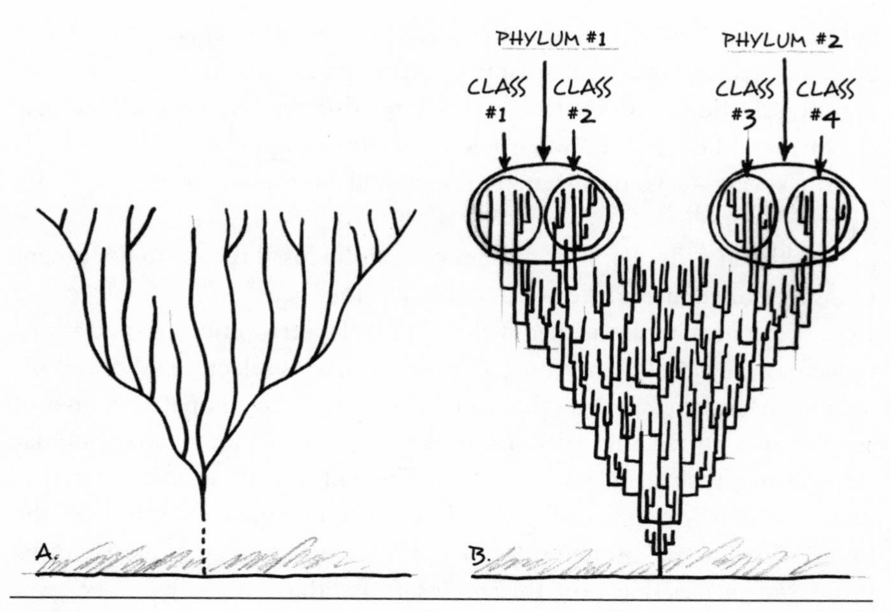

FIGURE 7.2
Two views of the history of life. *Figure 7.2a (left):* The traditional Darwinian picture showing slow, gradual change. *Figure 7.2b (right):* The history of life as depicted by the theory of punctuated equilibrium showing rapid speciation.

If new species arise very rapidly in small, peripherally isolated populations, then the expectation of insensibly graded fossils is a chimera. A new species does not evolve in the area of its ancestors; it does not arise from the slow transformation of all its forbears." Thus, they concluded, "Many breaks in the fossil record are real."[15]

Gould, Eldredge, and Stanley thought that members of these sibling or offspring species would, subsequent to their origin by allopatric speciation, compete against each other for resources and survival, just as, in neo-Darwinism, individual organisms or siblings may compete to survive and reproduce within a population. In their view, if members of one species succeed over another because of some selective advantages they possess, then that species will survive and predominate, passing on its traits. This process of interspecies or interpopulation competition (as opposed to intraspecies competition) Gould called "species selection."[16]

As Gould himself explained: "I propose, as the central proposition of macroevolution, that species play the same role of fundamental individual that organisms assume in microevolution. Species represent the basic units in theories and mechanisms of macroevolutionary change."[17] Since natural selection then would act upon large differences in overall biological form—differences between whole species as opposed to individuals within species—evolutionary change would take place in bigger, more discrete jumps.[18]

Gould and Eldredge thus did not expect the fossil record to document many intermediates. Instead, they thought the "gaps" in the fossil record were "the logical and expected result of the allopatric model of speciation" as well as the closely related mechanism of species selection.[19] Species selection made "the species" the unit of selection; the allopatric model of speciation asserted that new species quickly arise from smaller populations of organisms. Both mechanisms implied that fewer fossil intermediates would be preserved. According to punctuated equilibrium, the long "missing" transitional intermediates are not missing after all. In the process of species selection, the species, rather than the individual organism, competes for survival and thus becomes—in the jargon of evolutionary biology—the main "unit of selection" in macroevolution.

"PUNK EEK" AND THE FOSSIL RECORD

Eldredge and Gould devised the theory of punctuated equilibrium to eliminate the conflict between the fossil record and evolutionary theory. Nevertheless, punctuated equilibrium has its own problems accounting for the fossil record. In particular, the pattern of fossil appearance in the Cambrian period is inconsistent with both the way in which punctuated equilibrium depicts the history of life and with the idea that allopatric speciation and species selection are responsible for that pattern. There are several reasons for this.

First, the top-down pattern of appearance of Cambrian animal forms that we saw in Chapter 2 contradicts punctuated equilibrium's depiction of the history of life almost as much as it does the Darwinian picture (see Fig. 2.11). Recall that Darwin thought that the first representatives of the higher taxonomic categories emerged *after* the first appearance of representatives of each of the lower level taxa—that small differences distinguishing, for example, one species from another should gradually accumulate until they produced organisms different enough to be classified, first, as different genera, then, as different families, and eventually as different orders, classes, and so on. Instead, the first Cambrian animal forms are different enough from each other to justify classifying them as separate classes, subphyla, and phyla *from their first appearance in the fossil record* (see Fig. 7.3).

This pattern creates an acute difficulty for the theory of punctuated equilibrium. First, due to the action of allopatric speciation and species selection, advocates of punctuated equilibrium envision morphological change (represented as horizontal distance in Figure 7.2) arising in larger, more discontinuous increments of change. Nevertheless, like neo-Darwinists, they too see phyla-level differences arising from the "bottom up," starting with lower level taxonomic differences—albeit occurring in increments involving whole new species rather than individuals or varieties within species. Indeed, according to the theory of punctuated equilibrium, allopatric speciation first produces new *species* in smaller geographically isolated populations. For representatives of higher taxonomic categories to arise, these new species must accumulate new traits and evolve further. For this reason, punctuated equilibrium also expects small-scale diversity and differentiation of new species to precede the emergence of larger-scale morphological disparity and taxonomic differences. It also expects a "bottom-up" rather than a "top-down" pattern of appearance (see Fig. 7.3).

Second, for species selection to produce many new species, such as those that arise in the Cambrian explosion, a large pool of different species must first exist. The Precambrian fossil record does not document, however, the existence of such a large and diverse pool of competing Precambrian species upon which species selection (via allopatric speciation) might operate.

Paleontologists Douglas Erwin and James Valentine exposed this problem in 1987 in a seminal paper titled "Interpreting Great Developmental Experiments: The Fossil Record."[20] They questioned the ability of both of the main evolutionary theories of the time—punctuated equilibrium and neo-Darwinism—to explain the pattern of fossil appearance in the Precambrian–Cambrian fossil record.[21] Clearly, neo-Darwinism does not explain this pattern. But, as Valentine and Erwin argue, neither does punctuated equilibrium. As they conclude, the mechanism of species selection

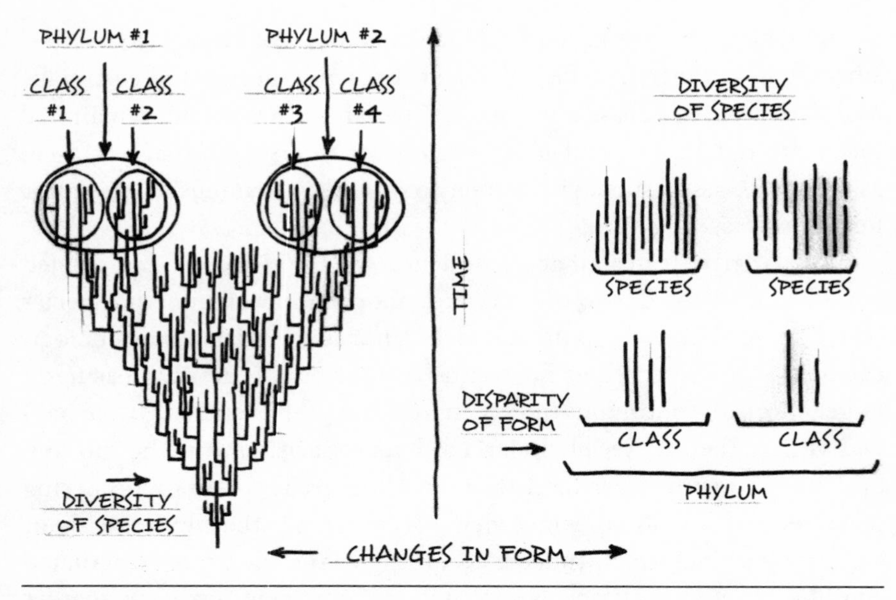

FIGURE 7.3
This figure shows that the theory of punctuated equilibrium (*on left*), like neo-Darwinism, anticipates small-scale diversity preceding large-scale disparity in form in contrast to the pattern in the fossil record (*shown on right*). Punctuated equilibrium also anticipates a "bottom-up," rather than a "top-down," pattern.

requires a large pool of species upon which to act. Thus, Valentine and Erwin conclude: "The probability that species selection is a general solution to the origin of higher taxa is not great."[22]

The late-Precambrian and Cambrian fossil records present another difficulty for punctuated equilibrium. Though Gould and Eldredge envisioned new traits becoming fixed in small isolated populations where speciation eventually occurs, they envisioned these traits first arising during periods of stasis in the large populations from which the smaller populations later separated. Gould realized that only stable large populations would afford enough opportunities for mutations to *generate* the new traits that macroevolution requires.[23] At the same time, he recognized that these new traits would have a far greater chance of *being fixed* into small, isolated populations where the random loss of some traits makes the fixation of others more likely (recall the marble example). By relying on large populations to generate new traits and small populations to fix them throughout a population, Gould wanted to provide both a plausible (if finely tuned) mechanism to explain both macroevolutionary change and the absence of fossil intermediates.[24] The late University of Chicago paleontologist Thomas J. M. Schopf described the balance this way, under punctuated equilibrium, evolution proceeds "in populations large enough to be reasonably variable, but small enough to permit large changes in gene frequencies due to random drift."[25]

But by relying on the accumulation of new traits within large parent populations, Gould undercut his own rationale for concluding that the fossil record should not preserve many intermediate forms. The reason for this is obvious: if novel genetic traits arise and spread within a large population of organisms, they are more likely to leave behind fossil evidence of their existence. Organisms with new and unique combinations or mosaics of traits represent nothing less than new forms of life. Thus, the process by which Gould envisions new genetic traits arising in large populations implies that new forms of life—some presumably transitional to other forms—should be preserved in the fossil record. Yet the Precambrian fossil record fails to preserve such a wealth of biological experiments during the long periods of relative stability in large populations that Gould's theory envisions.

THE TESTIMONY OF STATISTICAL PALEONTOLOGY

Studies in statistical paleontology have raised additional questions about whether the fossil record documents enough transitional intermediates to render punctuated equilibrium credible. In Chapter 3, I discussed the work of the statistical paleontologist Michael Foote. Foote used sampling theory to argue that the fossil record provides a reasonably complete picture of the forms of life that have existed on earth and to suggest that paleontologists are unlikely to find the many intermediate forms that neo-Darwinian theory requires.

Foote has also analyzed the question of whether the fossil record documents the number of intermediate forms that punctuated equilibrium requires. His answer: it depends.

Foote notes that whether a particular version of evolutionary theory can account for patterns in the fossil record depends upon the kind of mechanism of change it invokes. Neo-Darwinism relies upon a slow, gradually acting mechanism of change and, thus, has difficulty accounting for evidence of sudden appearance. Foote analyzes whether the number of proposed transitional species in the fossil record is consistent with punctuated equilibrium and concludes that it depends upon how fast the mechanisms upon which it relies can generate new forms of life. Though its proponents envision the evolution of new forms of life arising more abruptly than do the neo-Darwinists, they would still expect that the fossil record should have preserved some transitional fossils. Does the fossil record preserve even the relatively few intermediate forms that the theory of punctuated equilibrium implies that it should?

To answer that question, Foote developed a statistical method of testing the adequacy of different evolutionary models against several variables.[26] He observed that for punctuated equilibrium to succeed as an explanation for the data of the fossil record, it needs a mechanism capable of producing major evolutionary change quickly, because only such fast-acting change could account for the relative paucity of transitional forms in the fossil record. As Foote explained (writing with Gould in fact), the adequacy of punctuated equilibrium as an account of the fossil record depends upon the existence of a mechanism "of unusual speed and flexibility."[27]

But does the theory identify such a mechanism?

FROM WHENCE NEW TRAITS AND FORM?

Neither allopatric speciation nor species selection can generate the new genetic and anatomical traits necessary to produce animal forms, let alone in the relatively brief time of the Cambrian explosion. As conceived by Gould and the other advocates of punctuated equilibrium, allopatric speciation just allows for the possibility of the rapid *fixation* of preexisting traits, not the *generation* of new traits. When a parent population splits into two or more daughter populations, each of the daughter populations retains a part, but usually not the whole, of the gene pool of the original population. No new genetic traits are generated by the geographical isolation of one part of a population from another.

It could be argued, of course, that mutations might occur during the process of speciation, thus generating new genetic traits. But as Gould and Eldredge conceived of it, allopatric speciation occurs much too rapidly to have a reasonable chance of mutations generating anything fundamentally new. Darwin recognized in *On the Origin of Species* that evolution is a numbers game: larger population sizes and more generations offer more opportunities for favorable new variations to arise. As he explained: "Forms existing in larger numbers will always have a better chance . . . of presenting further favourable variations for natural selection to seize on, than will the rarer forms which exist in lesser numbers."[28] Yet for the mechanism of allopatric speciation to generate new traits, it would need to generate significant changes in form in small "peripherally isolated" populations over relatively few generations.[29] Because of these constraints, many biologists have concluded that allopatric speciation requires too much change too quickly to provide the theory of punctuated equilibrium with a biologically plausible mechanism for producing new traits or forms of animal life.

And that is why Gould and Eldredge, especially in their later formulations of the theory, envisioned new traits arising during long periods of stasis in larger populations rather than during short bursts of speciation. But a process in which traits arise "during long periods of stasis" does not constitute a "mechanism of unusual speed and flexibility," though that is precisely what, according to Gould and Foote, punctuated equilibrium requires in order to explain the abrupt appearance of new animal forms.

If allopatric speciation does not produce a fast-acting trait-generating mechanism, does species selection? Again, the answer is no. Species selection does not account for the origin of the different anatomical traits that distinguish one species from another. Species selection, as conceived by the proponents of punctuated equilibrium, acts on species and traits that *already* exist. Indeed, when Stanley, Gould, and Eldredge envisioned natural selection acting to favor the most fit species over another in a competition for survival, they presupposed the existence of a pool of different species and, therefore, also the existence of some mechanism for producing the traits that characterize those different species. That mechanism, however, would necessarily need to generate those differentiating traits before species could enter into competition with each other. Species selection *eliminates* less fit species in a competition for survival; it does not *generate* the traits that distinguish species and establish the basis for interspecies competition.

So where do these traits come from? When pressed, Gould eventually acknowledged that the origin of anatomical traits themselves result from good, old-fashioned natural selection acting on random mutations and variations—that is, from the neo-Darwinian mechanism acting over long periods of time on large relatively stable populations. But that meant that punctuated equilibrium, to the extent it relies on mutation and natural selection, is subject to the same evidential and theoretical problems as neo-Darwinism. And one of those problems is that the neo-Darwinian mechanism does not act quickly enough to account for the explosive appearance of new fossil forms in the Cambrian period. Like allopatric speciation, species selection does not qualify as the kind of rapid and flexible mechanism that Gould elsewhere insisted his theory must have in order to explain the abrupt appearance of animal forms in the fossil record.

NOVEL FORM AND MECHANISM

An even more profound difficulty with punctuated equilibrium as an explanation for the Cambrian explosion remains. Neither species selection nor allopatric speciation explains the origin of the representatives of the *higher* taxonomic categories—that is, the new animals representing new phyla and classes. Nor does it explain the structural and morphological features that distinguish animals from one another and earlier forms of life. Allopatric speciation explains how populations get separated from each other to form different *species*. Species selection describes how more fit *species* predominate over other species in a competition for survival. Neither mechanism gives any account of how the animals representing the specifically *higher* taxa or their distinctive anatomical novelties arose. Neither mechanism accounts, for example, for the origin of the compound eye of a trilobite, nor the gills of a Cambrian fish,[30] nor the echinoderm body plan.

Many critics of punctuated equilibrium have noted this problem. As Richard Dawkins wrote in 1986: "What I mainly want a theory of evolution to do is explain complex, well-designed mechanisms like hearts, hands, eyes and echolocation. Nobody, not even the most ardent species selectionist, thinks that species selection can do this."[31] Or as paleontologist Jeffrey Levinton argued in 1988, "It is inconceivable how selection among species can produce the evolution of detailed morphological structures. . . . Species selection did not form an eye."[32]

So where do these intricate structures come from? Again, when pressed, Gould resorted to the alleged power of the neo-Darwinian mechanism. As he wrote in his magisterial tome *The Structure of Evolutionary Theory,* published in 2002, the year of his death: "I do not deny either the wonder, or the powerful importance, of organized adaptive complexity." He went on to concede, "I recognize that we know no mechanism for the origin of such organismal features other than conventional natural selection at the organismic level."[33]

For this reason, few if any evolutionary biologists now regard punctuated equilibrium as a solution to the problem of the origin of biological form and novelty. As the evolutionary biologists Brian Charlesworth, Russell Lande, and Montgomery Slatkin have concluded, "genetic mechanisms that have been proposed [by proponents of punctuated equilibrium] to explain the abrupt appearance and prolonged stasis of many species are conspicuously lacking in empirical support."[34]

BURST OF INTEREST AND GRADUAL DECLINE

Still, it may not be entirely fair to criticize punctuated equilibrium for failing to account for the Cambrian explosion. Gould, in particular, equivocated about whether he meant punctuated equilibrium to serve as a comprehensive theory of macroevolutionary change or just an account of how new species emerged from a pool of preexisting species. Strictly speaking, the mechanisms of allopatric speciation and species selection sought to explain the pattern of stasis and discontinuity among different species and not among the higher taxa. Thus, near the end of his career, Gould complained about his critics "misunderstanding" his theory by asserting that he "proclaimed the total overthrow of Darwinism" and "intended punctuated equilibrium as both the agent of destruction and the replacement."[35]

Yet Gould and Eldredge, at least initially, advanced punctuated equilibrium as a bold new theory of evolutionary biology, giving the impression that it provided an ambitious solution to the problem of macroevolution and, by implication, events such as the Cambrian explosion. From 1972 to 1980, Eldredge and Gould presented a series of provocative scientific papers that portrayed punctuated equilibrium as a bold, and even revolutionary, alternative theory of macroevolution. Indeed, Gould himself referred to it explicitly as "a speciational theory of *macroevolution*."[36]

In their second main paper, published in 1977, Gould and Eldredge made explicit their intention to position their theory as a "radical"[37] challenge to neo-Darwinian gradualism and to replace it with a completely different understanding of the mode and mechanism of evolutionary change. Sepkoski notes that in this 1977 article "the authors were more explicit about the exact nature of the conceptual reconfiguration their theory brought to macroevolution."[38] In particular, he argues that Gould and Eldredge "extended their model to propose a new and 'general philosophy of change' in the natural world."[39] Gould was no less radical in a widely cited 1980 paper in the journal *Paleobiology* in which he offered punctuated equilibrium as "a new and general theory" of evolution. There he also famously declared the synthetic theory of neo-Darwinism "effectively dead, despite its persistence as textbook orthodoxy."[40]

Only after critics exposed punctuated equilibrium for lacking an adequate mechanism did Gould retreat to a more conservative formulation of the theory, making its reliance upon the neo-Darwinian mechanism explicit. From the early 1980s until his death in 2002, Gould made a series of concessions in particular about the inadequacy of speciation and species selection as mechanisms for generating complex adaptations. Thus, as Sepkoski notes, "Despite the brashness of many of his claims on behalf of punctuated equilibrium over the years, one is brought time and again back to the reconciliatory, even conservative justifications Gould made for his theory," particularly, he notes, in *The Structure of Evolutionary Theory*, written in the years just before Gould's death.[41]

In the end, Gould's concessions to neo-Darwinism brought his thinking back into conflict with the pattern of sudden appearance in the fossil record that the theory of punctuated equilibrium was designed to explain. If Gould and Eldredge were right about the abrupt appearance of new forms of life in the fossil record, and if the neo-Darwinian mechanism needs as much time as evolutionary biologists and population geneticists (see Chapters 8 and 12) calculate, then the mutation and selection mechanism does not have enough time to produce the new traits needed to build the forms of life that first appear in the Cambrian period. But punctuated equilibrium, as initially formulated to rely mainly on allopatric speciation and species selection, fared no better, since neither mechanism gives any explanation for the origin of new traits. And so, in the end, punctuated equilibrium highlighted rather than resolved a profound dilemma for evolutionary theory: neo-Darwinism allegedly has a mechanism capable of producing new genetic traits, but it appears to produce them too slowly to account for the abrupt appearance of new form in the fossil record; punctuated equilibrium attempts to address the pattern in the fossil record, but fails to provide a mechanism that can produce new traits whether abruptly or otherwise. No wonder, then, that leading Cambrian paleontologists such as James Valentine and Douglas Erwin concluded in 1987, that "neither of the contending theories of evolutionary change at the species level, phyletic gradualism or punctuated equilibrium, seem applicable to [explaining] the origin of new body plans."[42]

SPINNING IN CIRCLES

In a sudden flash of insight in the humdrum of a Laundromat, Niles Eldredge realized that stasis in the fossil record represented evidence rather than mere investigative failure. But like laundry spinning around in a washing machine, the theory of punctuated equilibrium itself became caught in a dreary cycle of contradiction. On the one hand, "punk eek" made a bold attempt to describe more accurately, and even explain, the decidedly discontinuous pattern of the fossil record. On the other, its advocates were forced to concede both the inadequacy of their proposed mechanisms and their need to rely upon the neo-Darwinian process of mutation and selection in order to account for the origin of new genetic traits and anatomical innovations. After Gould appeared to jettison both gradualism and a reliance on the neo-Darwinian mechanism in order to bring evolutionary theory into conformity with the fossil record, he eventually acknowledged that he could not explain the origin of the forms of life documented in the fossil record apart from that same slow and gradually acting mechanism. Thus, though the theory of punctuated equilibrium was initially presented as a solution to the mysterious and sudden origin of animal forms, upon closer inspection, it failed to offer such a solution.

Nevertheless, the failure of punctuated equilibrium to provide a sufficient mechanism has raised questions about the adequacy of the mechanism that Gould ultimately did reaffirm as the explanation for the origin of novel biological form. Can the neo-Darwinian mechanism of natural selection acting on random mutations build new forms of animal life with all their complex adaptations? If so, is it possible that it could do so in the brief time allowed by the fossil record? If not, is it reasonable to think that it could build new forms of animal life if only more time were available? If so, how much time would the Darwinian mechanism need to build complex adaptations and new forms of animal life? In the next several chapters, I will address these fundamental questions at the heart of the Cambrian mystery—questions, in brief, about how to build an animal.

PART TWO

HOW TO BUILD AN ANIMAL

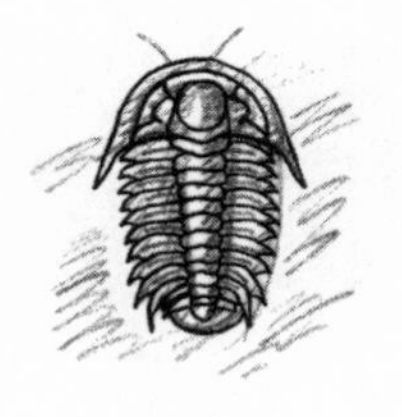

8

THE CAMBRIAN INFORMATION EXPLOSION

When I was a college professor, I used to ask my students a question: "If you want your computer to acquire a new function or capability, what do you have to give it?" Typically, I would hear a smattering of similar answers from the class: "code," "instructions," "software," "information." Of course, all these are correct. And thanks to discoveries in modern biology, we now know that something similar is true of life: to build a new form of life from a simpler preexisting form requires new information.

To this point I've examined one main aspect of the mystery surrounding the Cambrian explosion: the mystery of the missing Precambrian ancestral forms expected on the basis of Darwin's theory. The next group of chapters will examine a second, and perhaps more profound, aspect of the Cambrian mystery: that of the *cause* of the Cambrian explosion. By what means or process or mechanism could something as complex as a trilobite have arisen? Could natural selection have accomplished such a feat? To answer this question we will have to look more closely at what it takes to build a new form of animal life. And we'll see that an important part of the answer to that question will have something to do with the concept of *information*.

THE DARWINIAN ACCOUNT OF THE ORIGIN OF ANIMAL FORM

As Darwin envisioned the process, natural selection can accomplish nothing without a steady supply of variation as a source of new biological traits, forms, and structures. Only after useful new variations arise can natural selection sift them from the chaff of unhelpful variations. If, however, the amount of variation available to natural selection is limited, then natural selection will encounter limits on how much new biological form and structure it can build.

Even in the late nineteenth century, many leading scientists recognized this. For this reason, there has been a long history of scientific controversy about just how much novelty natural selection can produce and about whether natural selection is a truly creative process. In fact, between 1870 and 1920 classical Darwinism entered a period of eclipse, because many scientists thought that it could not explain the origin and transmission of new heritable variation.[1]

Darwin favored a theory of blended inheritance that seemed to imply limitations on the amount of genetic variability.[2] He thought that when parents with different traits combined germ cells during sexual reproduction, the resulting offspring would receive not one or the other set of their differing traits, but instead a compromise version. For example, if a male bird with red feathers on its wings mated with a female bird of the same species with white wing feathers, the theory implied that the two would likely produce offspring with pink wing feathers. As many of Darwin's contemporaries pointed out, such instances of blending inheritance involved strict limitations on the *range* of traits that could possibly arise, depriving natural selection of the wide-ranging supply of variation it would need to produce truly fundamental changes in the form of animals. The pink-feathered offspring might later reproduce with a white- or red-feathered bird of the same species, producing a slightly lighter or darker shade of pink feathers. Nevertheless, descendants of the original white- and red-feathered pair would never produce green, blue, or yellow feathers in subsequent generations. If correct, blending inheritance would eventually lead to a bland, homogenous, variationless state in a population.

In the 1860s, the Austrian monk Gregor Mendel, widely regarded as the founder of modern genetics, showed in his work on garden peas that Darwin's assumptions about blending inheritance were incorrect. The results of his studies created, at least initially, more problems for Darwinism. Mendel showed that the genetic traits of organisms typically have an integrity that resists blending. He showed this by cross-pollinating plants with yellow peas and plants with green peas. The plants in the subsequent generations produced *either* yellow or green peas, but nothing in between and nothing with an altogether different color.[3]

He also showed that the plants carried some kind of signal or instructions for building different traits even when the trait was not on display in a particular plant. He noticed, for example, that when he crossed pea plants with green and yellow seeds, the next generation had only yellow seeds, almost as if the capacity for generating green peas had been lost. But when he cross-pollinated the second-generation plants, the ones with only yellow peas, he found that both yellow *and* green peas emerged in the third generation, in a ratio of 3 to 1. From this Mendel hypothesized that the second-generation plants continued to carry signals, which he called "factors," and later scientists called genes, for generating green peas even when those plants themselves did not display that trait.

The classical Mendelian genetics that replaced Darwin's blending theory of inheritance also suggested limitations on the amount of genetic variability available to natural selection. If plant reproduction produced *either* green *or* yellow peas but never some intermediate form, and if the signals for producing the green traits and yellow traits persisted unchanged from generation to generation, it was difficult to see how sexual reproduction and genetic recombination could produce anything more than unique combinations of already existing traits.

In the decades after Mendel's work, geneticists came to understand genes as discrete units or packets of hereditary information that could be independently sorted and shuffled within the chromosome. This too suggested that a significant, but still strictly limited, amount of genetic variation could arise by genetic recombination during sexual reproduction. Thus, Mendelian genetics raised significant questions about whether the process of natural selection has access to enough variation (which, after Mendel, was conceived as *genetic* variation) to allow it to produce any significant morphological novelty. For a time, Darwin's theory was in retreat.

DARWINISM MUTATES

During the 1920s, 1930s, and 1940s, however, developments in genetics revived natural selection as the main engine of evolutionary change. Experiments performed by Hermann Muller in 1927 showed that X-rays could alter the genetic composition of fruit flies, resulting in unusual variations.[4] Muller called these X-ray-induced changes "mutations." Other scientists soon reported that they had produced mutations in the genes of other organisms, including humans. Whatever genes were made of—and biologists then still did not know—these developments suggested they could vary more than either Darwin or classical Mendelian genetics had assumed. Geneticists at the time also discovered that these small-scale changes in genes were potentially heritable.[5] But if variant versions of genes were heritable, then presumably natural selection could favor advantageous gene variants and eliminate the others. These mutations could then influence the future direction of evolution and, in theory at least, provide an unlimited supply of variation for natural selection's workshop.

The discovery of genetic mutations also suggested a way to reconcile Darwinian theory with insights from Mendelian genetics. During the 1930s and 1940s, a group of evolutionary biologists, including Sewall Wright, Ernst Mayr, Theodosius Dobzhansky, J. B. S. Haldane, and George Gaylord Simpson, attempted to demonstrate this possibility using mathematical models to show that small-scale variations and mutations could accumulate over time in whole populations, eventually producing large-scale morphological change.[6] These mathematical models formed the basis of a subdiscipline of genetics known as population genetics. The overall synthesis of Mendelian genetics with Darwinian theory came to be called "neo-Darwinism" or simply the "New Synthesis."

According to this new synthetic theory, the mechanism of natural selection acting upon genetic mutations suffices to account for the origin of novel biological forms. Small-scale "microevolutionary" changes can accumulate to produce large-scale "macroevolutionary" innovations. The neo-Darwinists argued that they had revived natural selection by discovering a specific mechanism of variation that could generate new forms of life from simpler preexisting ones. By the centennial celebration of Darwin's *Origin of Species* in 1959, it was widely assumed that natural selection and random mutations could indeed build new forms of life over the course of time with their distinctive body plans and novel anatomical structures. At the celebration, Julian Huxley, the grandson of T. H. Huxley, summarized this optimism in a grand proclamation:

> *Future historians will perhaps take this Centennial Week as epitomizing an important critical period in the history of this earth of ours—the period when the process of evolution, in the person of inquiring man, began to truly be conscious of itself. . . . This is one of the first public occasions on which it has been frankly faced that all aspects of reality are subject to evolution, from atoms and stars to fish and flowers, from fish and flowers to human societies and values—indeed, that all reality is a single process of evolution . . .*[7]

In a television broadcast leading up to the Centennial celebration, Huxley captured the optimistic mood more succinctly: "Darwinism has come of age, so to speak. We are no longer having to bother about establishing the fact of evolution."[8]

VARIATION AS INFORMATION

Initially, the elucidation of the structure of DNA by James Watson and Francis Crick in 1953 contributed to this euphoria.[9] Indeed, it seemed to lift the mechanism of genetic variation and mutation out of the mist and into the clear light of the emerging science of molecular biology. Watson and Crick's elucidation of the double helix structure of DNA suggested that DNA stored genetic information in the form of a four-character digital and chemical code (see Fig. 8.1). Later, following the formulation of Francis Crick's famed "sequence hypothesis," molecular biologists confirmed that the chemical subunits along the spine of the DNA molecule called nucleotide bases function just like alphabetic characters in a written language or digital characters in a machine code. Biologists established that the precise *arrangement* of these nucleotide bases conveyed instructions for building proteins.[10] (See Fig. 8.2.) Molecular biologists also determined that this store of genetic information in DNA is transmitted from one generation of cells and organisms to another. In short, it was established that DNA stores hereditary information for building proteins and thus, presumably, for building higher-order anatomical traits and structures as well.

The elucidation of the double helix seemed to resolve some longstanding issues in evolutionary biology. Darwinists had long maintained that natural selection produced new forms by separating the proverbial

FIGURE. 8.1 James Watson (*left*) and Francis Crick (*right*) presenting their model of the structure of the DNA molecule in 1953. *Courtesy A. Barrington Brown/Science Source.*

wheat from the chaff of genetic variation, but they didn't know where the raw material for all of the competing variations resided. Neither did they know how genes stored information for producing the traits associated with them. Further, even after geneticists discovered that stable genetic traits could be altered by mutations, they remained uncertain about what exactly it was that was being "mutated." Consequently, biologists were uncertain about exactly where variations and mutations occurred.

Watson and Crick's model suggested an answer to that question: genes correspond to long sequences of bases on a strand of DNA. Building on that insight, evolutionary biologists proposed that *new* variations arose, first, from the genetic recombination of different sections of DNA (different genes) during sexual reproduction and, second, from a special kind of variation called mutations that occur from random changes in the arrangement of nucleotide bases in DNA. Just as a few typographical errors in an English sentence might alter the meaning of a few words or even the whole sentence, so too might a change in the sequential arrangement of the bases in the genetic "text" in DNA produce new proteins or morphological traits.

Watson and Crick's discovery also raised new questions—in particular, questions about the information necessary to build completely new forms of life during the course of biological evolution. True, mutations

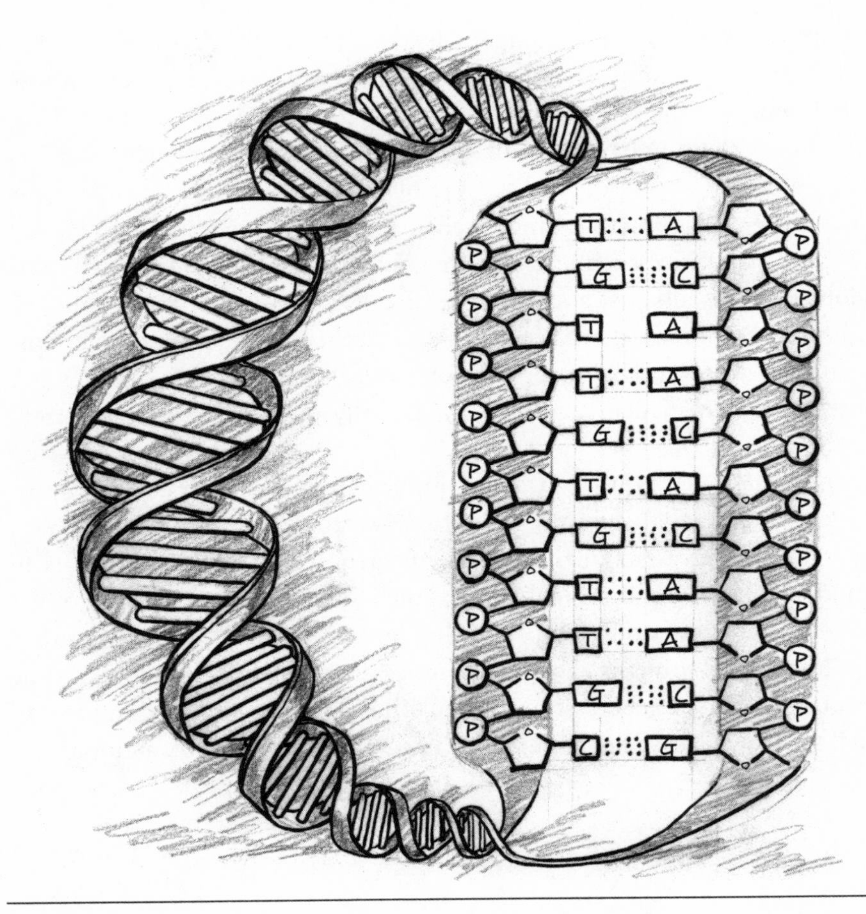

FIGURE 8.2
The model (or structural formula) of the DNA molecule showing the digital or alphabetic character of the nucleotide bases stored along the sugar-phosphate backbone of the molecule.

play a role in this process, but could they generate enough information to produce novel forms of animal life such as those that arose in the Cambrian period, an *explosion*—a vast proliferation—of new biological information?

THE CAMBRIAN INFORMATION EXPLOSION

Consider choanoflagellates, a group of single-celled eukaryotic organisms with a flagellum. What separates such organisms from a trilobite or a mollusk or even a lowly sponge? Clearly, all three of these higher forms of life are more complex than any one-celled organism. But exactly how much more complex?

James Valentine has noted that one useful way of comparing degrees of complexity is to assess the number of cell types in different organisms (see Fig. 8.3).[11] Though a single-celled eukaryote has many specialized internal structures such as a nucleus and various organelles, it still, obviously, represents just a single type of cell. Functionally more complex animals require more cell types to perform their more diverse functions. Arthropods and mollusks, for example, have dozens of specific tissues and organs, each of which requires "functionally dedicated," or specialized, cell types.

These new cell types, in turn, require many new and specialized proteins. An epithelial cell lining a gut or intestine, for example, secretes a specific digestive enzyme. This enzyme requires structural proteins to modify its shape and regulatory enzymes to control the secretion of the digestive enzyme itself. Thus, building novel cell types typically requires building novel proteins, which requires assembly instructions for building proteins—that is, genetic information. Thus, an increase in the number of cell types implies an increase in the amount of genetic information.

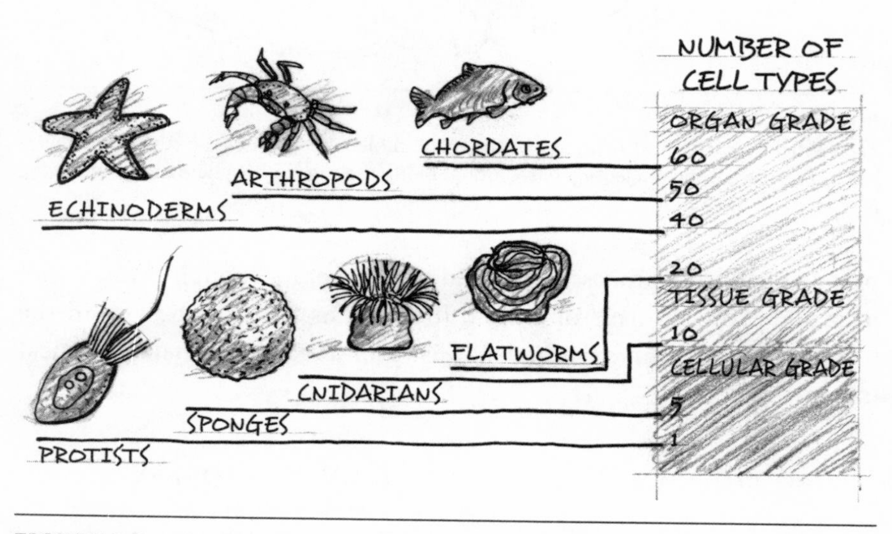

FIGURE 8.3
Biological complexity scale as measured in number of cell types of different organisms.

Applying this insight to ancient life-forms underscores just how dramatic the Cambrian explosion was. For over 3 billion years, the living world included little more than one-celled organisms such as bacteria and algae.[12] Then, beginning in the late Ediacaran period (about 555–570 million years ago), the first complex multicellular organisms appeared in the rock strata, including sponges and the peculiar Ediacaran biota discussed in Chapter 4.[13] This represented a large increase in complexity. Studies of modern animals suggest that the sponges that appeared in the late Precambrian, for example, probably required about ten cell types.[14]

Then 40 million years later, the Cambrian explosion occurred.[15] Suddenly the oceans swarmed with animals such as trilobites and anomalocaridids that probably required fifty or more cell types—an even greater jump in complexity. Moreover, as Valentine notes, measuring complexity differences by measuring differences in the number of cell types probably "greatly underestimate[s] the complexity differentials between bodyplans."[16]

One way to estimate the amount of new genetic information that appeared with the Cambrian animals is to measure the size of the genomes of modern representatives of the Cambrian groups and compare that to the amount of information in simpler forms of life. Molecular biologists have estimated that a minimally complex single-celled organism would require between 318,000 and 562,000 base pairs of DNA to produce the proteins necessary to maintain life.[17] More complex single cells might require upwards of a million base pairs of DNA. Yet to assemble the proteins necessary to sustain a complex arthropod such as a trilobite would need orders of magnitude more protein-coding instructions. By way of comparison, the genome size of a modern arthropod, the fruit fly *Drosophila melanogaster,* is approximately 140 million base pairs.[18] Thus, transitions from a single cell to colonies of cells to complex animals represent significant—and in principle measurable—increases in genetic information.

During the Cambrian period a veritable carnival of novel biological forms arose. But because new biological form requires new cell types, proteins, and genetic information, the Cambrian explosion of animal life also generated an explosion of genetic information unparalleled in the previous history of life.[19] (In Chapter 14, we'll see that building a new animal body plan also requires another type of information, not stored in genes, called epigenetic information.)

So can the neo-Darwinian mechanism explain the dramatic increase in genetic information that appears in the Cambrian explosion? Before addressing that question, it will help to define the concept of information and identify the kind of information that DNA contains.

BIOLOGICAL INFORMATION: SHANNON OR OTHERWISE?

Scientists typically recognize at least two basic types of information, functional (or meaningful) information and so-called Shannon information, which is not necessarily meaningful or functional. The distinction has arisen in part because of developments in a branch of applied mathematics known as information theory. During the late 1940s, mathematician Claude Shannon, working at the Bell Laboratories, developed a mathematical theory of information. Shannon equated the amount of information transmitted by a sequence of symbols or characters with the amount of uncertainty reduced or eliminated by the transmission of that sequence.[20]

Shannon thought that an event or communication that didn't eliminate much uncertainty was also not very informative. Consider an illustration. In the 1970s, when I was a teenager, if someone made a completely obvious statement, we would say, "Tell me something else I didn't know." Imagine that one of my classmates on the baseball team has just rushed up to breathlessly "inform" me that our team's star pitcher is planning to throw the ball to the catcher in the next game. Such a statement would earn the scornful reply, "Tell me something else I didn't know."

The obvious statement about the pitcher's intentions also illustrates why Shannon equated the elimination of uncertainty with the transmission of information. Since star pitchers who want to go on being star pitchers have no choice but to throw the ball across the plate to the catcher, the statement of my overwrought friend eliminated no previous uncertainty. It was not informative in the least. If, on the other hand, after days of speculation on campus about which one of our team's four pitchers the baseball coach would choose to pitch in the championship game, my friend had rushed up to me and revealed the identity of the starting pitcher, that would be different. In that case, he would have eliminated some significant uncertainty on my part with a decidedly informative statement.

Shannon's theory quantified the intuitive connection between reduced uncertainty and information by asserting that the *more* uncertainty an event or communication eliminated, the *more* information it conveyed. Imagine that after revealing to me the identity of the starting pitcher before the championship baseball game in the spring, he then also revealed to me the identity of the starting quarterback before the upcoming football season. Imagine as well that our baseball team had *four* equally competent pitchers and the football team had just *two* equally competent quarterbacks. Given these facts, my friend's decision to inform me of the identity of the starting pitcher eliminated *more* uncertainty than his decision to reveal to me the identity of the starting quarterback.

To assign precise quantitative measures of information, Shannon further linked the reduction of uncertainty and information to quantitative measures of probability (or *im*probability). Notice that in my illustration, the more informative communication reduced more uncertainty and also described a more *im*probable event. The probability of any one of the four pitchers being selected was 1 in 4. The probability of one of the quarterbacks being selected, given the same assumption, was only 1 in 2. The more improbable event eliminated more possibilities and more uncertainty. Thus, it conveyed more information.

Shannon applied these intuitions to quantify the amount of information present in sequences of symbols or characters stored in texts or codes or transmitted across communications channels. Thus, in his theory the presence of an English letter in a sequence of other such letters transmits more information than a single binary digit (a zero or one) in a section of computer code. Why? Again, the letter in the English alphabet reduces uncertainty among twenty-six possibilities, whereas a single binary digit reduces uncertainty among only two. The probability of any one character from the English alphabet occurring in a sequence of other such letters (disregarding the need for spaces and punctuation) is 1 in 26. The probability of either a zero or one arising in a sequence of binary characters is 1 in 2. In Shannon's theory the presence of the more improbable character conveys more information.

Yet even a binary alphabet can convey an unlimited amount of information because, in Shannon's theory, additional information is conveyed as improbabilities multiply. Imagine a grab bag of tiles with either zero or one etched onto each. Imagine someone producing a series of zeros and ones by reaching into the bag and placing them one by one on a game board. The probability of choosing a zero on the first pick is just 1 in 2. But the probability of choosing two consecutive zeros after placing the first back in the grab bag (and shaking the tiles) is 1 chance in 2×2, or 1 chance in 4. This is because there are four possible combinations of digits that could have been chosen—00, 01, 10, or 11. Similarly, the probability of producing any three-letter sequence as a result of choosing consecutively in this manner is 1 in $2 \times 2 \times 2$, or 1 in 2^3 (or 1 in 8). The improbability of any specific sequence of characters increases exponentially with the number of characters in the sequence. Thus, longer and longer sequences can generate larger and larger amounts of information even using a simple binary alphabet.

Information scientists measure such informational increases through a unit they call a *bit*. A bit represents the minimum amount of information that can be conveyed (or uncertainty reduced) by a single digit in a two-character alphabet.[21]

Biologists can readily apply Shannon's information theory to measure the amount of Shannon information in a sequence of DNA bases (or the sequence of amino acids in a protein) by assessing the probability of the sequence occurring and then converting that probability to an information measure in bits.[22] DNA conveys information, in Shannon's sense, in virtue of its containing long improbable arrangements of four chemicals—the four bases that fascinated Watson and Crick—adenine, thymine, guanine, and cytosine (A, T, G, and C). As Crick realized in formulating his sequence hypothesis, these nucleotide bases function as alphabetic or digital characters in a linear array. Since each of the four bases has an equal 1 in 4 chance of occurring at each site along the spine of the DNA molecule, biologists can calculate the probability, and thus the Shannon information, or what is technically known as the "information-carrying capacity," of any particular sequence *n* bases long. For instance, any particular sequence three bases long has a probability of 1 chance in $4 \times 4 \times 4$, or 1 chance in 64, of occurring—which corresponds to 6 bits of Shannon information. (Indeed, each base in a DNA sequence conveys 2 bits of information, since 1 in 4 is equal to 1 chance in 2×2.)

Yet the applicability of Shannon information theory to molecular biology has, to some degree, obscured a key distinction concerning the type of information that DNA possesses. Although Shannon's theory measures the *amount* of information in a sequence of symbols or characters (or chemicals functioning as such), it doesn't distinguish a meaningful or functional sequence from useless gibberish. For example:

"we hold these truths to be self-evident"
"ntnyhiznslhtgeqkahgdsjnfplknejmsed"

These two sequences are equally long and equally improbable if we imagine them being drawn at random. Thus, they contain the same amount of Shannon information. Yet clearly there is an important qualitative distinction between them that the Shannon measurement does not capture. The first meaningful sequence performs a communication function, while the second does not.

Shannon emphasized that the kind of information his theory described needs to be carefully distinguished from our common notions of information. As Warren Weaver, one of Shannon's close collaborators, made clear in 1949, "The word 'information' in this theory is used in a special mathematical sense that must not be confused with its ordinary usage."[23] By ordinary usage, Weaver, of course, was referring to the idea of meaningful or functional communication.

Webster's dictionary defines information as "the communication or reception of knowledge or intelligence." It also defines information as "the attribute inherent in, and communicated by, alternative sequences or arrangements of something that produce a specific effect." A sequence of characters possessing a large amount of Shannon information *may* convey meaning (as in an English text) or perform a function that "produces a specific effect" (as do both English sentences and computer codes, for example) or it may not (as would be the case with a meaningless pile of letters or a screen of scrambled computer code). In any case, Shannon's purely mathematical theory of information does not distinguish the presence of meaningful or functional sequences from *merely* improbable, though meaningless ones. It only provides a mathematical measure of the improbability—or information-carrying capacity—of a sequence of characters. In a sense, it provides a measure of a sequence's *capacity* to carry functional or meaningful information. It does not, and cannot, determine whether the sequence in question *does* convey meaning or generate a functionally significant effect.

Strands of DNA contain information-carrying capacity—something

Shannon's theory can measure.[24] But DNA, like natural languages and computer codes, also contains *functional* information.[25]

In languages such as English, specifically arranged characters convey functional information to conscious agents. In computer or machine code, specifically arranged characters (zeros and ones) produce functionally significant outcomes within a computational environment without a conscious agent receiving the meaning of the code inside the machine. In the same way, DNA stores and conveys functional information for building proteins or RNA molecules, even if it is not received by a conscious agent. As in computer code, the precise arrangement of characters (or chemicals functioning as characters) allows the sequence to "produce a specific effect." For this reason, I also like to use the term *specified information* as a synonym for functional information, because the function of a sequence of characters depends upon the *specific* arrangement of those characters.

And DNA contains specified information, not just Shannon information or information-carrying capacity. As Crick himself put it in 1958, "By information I mean the specification of the amino acid sequence in protein. . . . Information means here the *precise* determination of sequence, either of bases in the nucleic acid or on amino acid residues of the protein."[26]

THE MESSAGE AS THE MYSTERY

So if the origin of the Cambrian animals required vast amounts of new functional or specified information, what produced this information explosion? Since the molecular biological revolution first highlighted the primacy of information to the maintenance and function of living systems, questions about the origin of information have moved decidedly to the forefront of discussions about evolutionary theory. What's more, the realization that specificity of arrangement, rather than mere improbability, characterizes the genetic text has raised some challenging questions about the adequacy of the neo-Darwinian mechanism. Is it plausible to think that natural selection working on random mutations in DNA could produce the highly *specific* arrangements of bases necessary to generate the protein building blocks of new cell types and novel forms of life? Perhaps nowhere do such questions pose more of a challenge to neo-Darwinian theory than in discussions of the Cambrian explosion.

9

COMBINATORIAL INFLATION

Murray Eden (see Fig. 9.1), a professor of engineering and computer science at MIT, was accustomed to thinking about how to build things. But when he began to consider the importance of information to building living organisms, he realized something didn't add up. His critics said that he knew just enough biology to be dangerous. In retrospect, they were probably right.

In the early 1960s, just as molecular biologists had confirmed Francis Crick's famed sequence hypothesis, Eden began to think about the challenge of building a living organism. Of course, Eden wasn't contemplating building a living organism himself. Rather, he was thinking about what it would take for the neo-Darwinian mechanism of natural selection acting on random mutations to do the job. He wondered whether mutation and selection could generate the needed functional information.

To his way of thinking, *specificity* was a big part of the problem. Obviously, if DNA contained an improbable sequence of nucleotide bases in which the arrangement of bases does not matter to the function of the molecule, then random mutational changes in the sequence of bases would not have a detrimental effect on the function of the molecule. But, of course, sequence *does* affect function. Eden knew that in all computer codes or written text in which the specificity of sequence determines function, random changes in sequence consistently degrade function or meaning. As he explained, "No currently existing formal language can tolerate random changes in the symbol sequences which express its sentences. Meaning is almost invariably destroyed."[1] Thus, he suspected that the need for specificity in the arrangement of DNA bases made it extremely improbable that random mutations would generate new functional genes or proteins as opposed to degrading existing ones.

But how improbable? How difficult would it be for random mutations to generate, or stumble upon, the genetically meaningful or functional sequences needed to supply natural selection with the raw material—the genetic information and variation—it needed to produce new proteins, organs, and forms of life? Eden wasn't the only mathematician or scientist asking these questions. But the mathematically based challenge to evolutionary theory that he helped to initiate would indeed prove dangerous to neo-Darwinian orthodoxy.

THE WISTAR INSTITUTE CONFERENCE

During the early 1960s, Eden began discussing the plausibility of the neo-Darwinian theory of evolution with several MIT colleagues in math, physics, and computer science. As the discussion grew to include mathematicians and scientists from other institutions, the idea of a conference was born. In 1966, a distinguished group of mathematicians, engineers, and scientists convened a conference at the Wistar Institute in Philadelphia called "Mathematical Challenges to the Neo-Darwinian Interpretation of

FIGURE 9.1
Murray Eden. *Courtesy MIT Museum.*

Evolution." Prominent among the attendees were Marcel-Paul Schützenberger, a mathematician and physician at the University of Paris; Stanislaw Ulam, the codesigner of the hydrogen bomb; and Eden himself. The conference also included a number of prominent biologists, including Ernst Mayr, an architect of modern neo-Darwinism, and Richard Lewontin, at the time a professor of genetics and evolutionary biology at the University of Chicago.

Sir Peter Medawar, a Nobel laureate and the director of the North London Medical Research Council's laboratories, chaired the meeting. In his opening remarks, he said, "The immediate cause of this conference is a pretty widespread sense of dissatisfaction about what has come to be thought of as the accepted evolutionary theory in the English-speaking world, the so-called neo-Darwinian theory."[2]

For many, doubts about the creative power of the mutation and selection mechanism stemmed from the elucidation of the nature of genetic information by molecular biologists in the late 1950s and early 1960s.

The discovery that the genetic information in DNA is stored as a linear array of precisely sequenced nucleotide bases at first helped to clarify the nature of many mutational processes. Just as a sequence of letters in an English text might be altered either by changing individual letters one by one or by combining and recombining whole sections of text, so too might the genetic text be altered either one base at a time or by combining and recombining different sections of genes in various ways at random. Indeed, modern genetics has established various mechanisms of mutational change—not only "point mutations," or changes in individual bases, but also duplications, insertions, inversions, recombinations, and deletions of whole sections of the genetic text.

Although fully aware of this range of mutational options at nature's disposal, Eden argued at Wistar that such random changes to written texts or sections of digital code would inevitably degrade the function of information-bearing sequences, particularly when allowed to accumulate.[3] For example, the simple phrase "One if by land and two if by sea" will be significantly degraded by just a handful of random changes such as those in bold: "**I**ne if b**g** l**e**nd and two i**k** b**T** **N**ea." At the conference the French mathematician Marcel Schützenberger agreed with Eden's concerns about the effect of random alterations. He noted that if someone makes even a few random changes in the arrangement of the digital characters in a computer program, "we find that we have no chance (i.e., less than $1/10^{1000}$) even to see what the modified program would compute: it just jams."[4] Eden argued that much the same problem applied to DNA—that insofar as specific arrangements of bases in DNA function like digital code, random changes to these arrangements would likely efface their function, while attempts to generate completely new sections of genetic text by random means were likely doomed to failure.[5]

The explanation for this inevitable diminution in function is found in a branch of mathematics called combinatorics. Combinatorics studies the number of ways a group of things can be *combined* or arranged. At one level, the subject is fairly intuitive. If a thief slips round the corner of a dormitory after hours looking for a bike to steal, he will scan the bike rack for an easy target. If he spots a basic bicycle-style lock with only three dials of ten numbers each, and on the rack beside it one with five dials of ten numbers each, the thief wouldn't need a degree in mathematics to realize which one he should attempt to open. He knows that he would need to search fewer total possibilities with the three-dial lock.

A straightforward calculation supports his intuition. The simpler lock has only $10 \times 10 \times 10$, or 1000, possible combinations of digits—or what mathematicians refer to as "combinatorial" possibilities. The five-dial lock has $10 \times 10 \times 10 \times 10 \times 10$, or 100,000, combinatorial possibilities. With a lot of patience, the thief might elect to systematically work his way through the different combinations of digits on the simpler lock, knowing that at some point he will stumble across the correct combination. He shouldn't even bother with the five-dial lock, since making his way through all of the possible combinations on it would take 100 times as long. The five-dial lock simply has too many possibilities for the thief to have a reasonable chance of opening it by trial and error in the time available to him.

Several of the Wistar scientists noted that the mutation and selection mechanism faces a similar problem. Neo-Darwinism envisions new genetic information arising from random mutations in the DNA. If at any time from birth to reproduction the right mutation or combination of mutations accumulate in the DNA of cells involved in reproduction (whether sexual or asexual), then information for building a new protein or proteins will pass on to the next generation. When that new protein happens to confer a survival advantage on an organism, the genetic change responsible for the new protein will tend to be passed on to subsequent generations. As favorable mutations accumulate, the features of a population will gradually change over time.

Clearly, natural selection plays a crucial role in this process. Favorable mutations are passed on; unfavorable mutations are weeded out. Nevertheless, the process can only select variations in the genetic text that mutations have first produced. For this reason, evolutionary biologists typically recognize that mutation, not natural selection, provides the source of variation and innovation in the evolutionary process. As evolutionary biologists Jack King and Thomas Jukes put it in 1969, "Natural selection is the editor, rather than the composer, of the genetic message."[6]

And that was the problem, as the Wistar skeptics saw it: random mutations must do the work of composing new genetic information, yet the sheer number of possible nucleotide base or amino-acid combinations (i.e., the size of the combinatorial "space") associated with a single gene or protein of even modest length rendered the probability of random assembly prohibitively small. For every sequence of amino acids that generates a functional protein, there are a myriad of other combinations that don't. As the length of the required protein grows, the number of possible amino-acid combinations mushrooms exponentially. As this happens, the probability of ever stumbling by random mutation onto a functional sequence rapidly diminishes.

Consider another illustration. The two letters X and Y can be combined in four different two-letter combinations (XX, XY, YX and YY). They can be combined in eight different ways for three-letter combinations (XXX, XXY, XYY, XYX, YXX, YYX, YXY, YYY), sixteen ways for four-letter combinations, and so on. The number of possible combinations grows exponentially—2^2, 2^3, 2^4, and so on—as the number of letters in the sequence grows. Mathematician David Berlinski calls this the problem of "combinatorial inflation," because the number of possible combinations "inflates" dramatically as the number of characters in a sequence grows (see Fig. 9.2).

The combinations of bases in DNA are subject to combinatorial inflation of just this sort. The information-bearing sequences in DNA consist of specific arrangements of the four nucleotide bases. Consequently, there are four possible bases that could occur at each site along the DNA backbone and 4 × 4, or 4^2, or 16 possible two-base sequences (AA AT AG AC

TA TG TC TT CG CT CC CA GA GG GC GT). Similarly, there are $4 \times 4 \times 4$, or 4^3, or 64 possible three-base sequences. (I'll refrain from listing them all.) That is, increasing the number of bases in a sequence from 1 to 2 to 3 increases the number of possibilities from 4 to 16 to 64. As the sequence length continues to grow, the number of combinatorial possibilities cor-

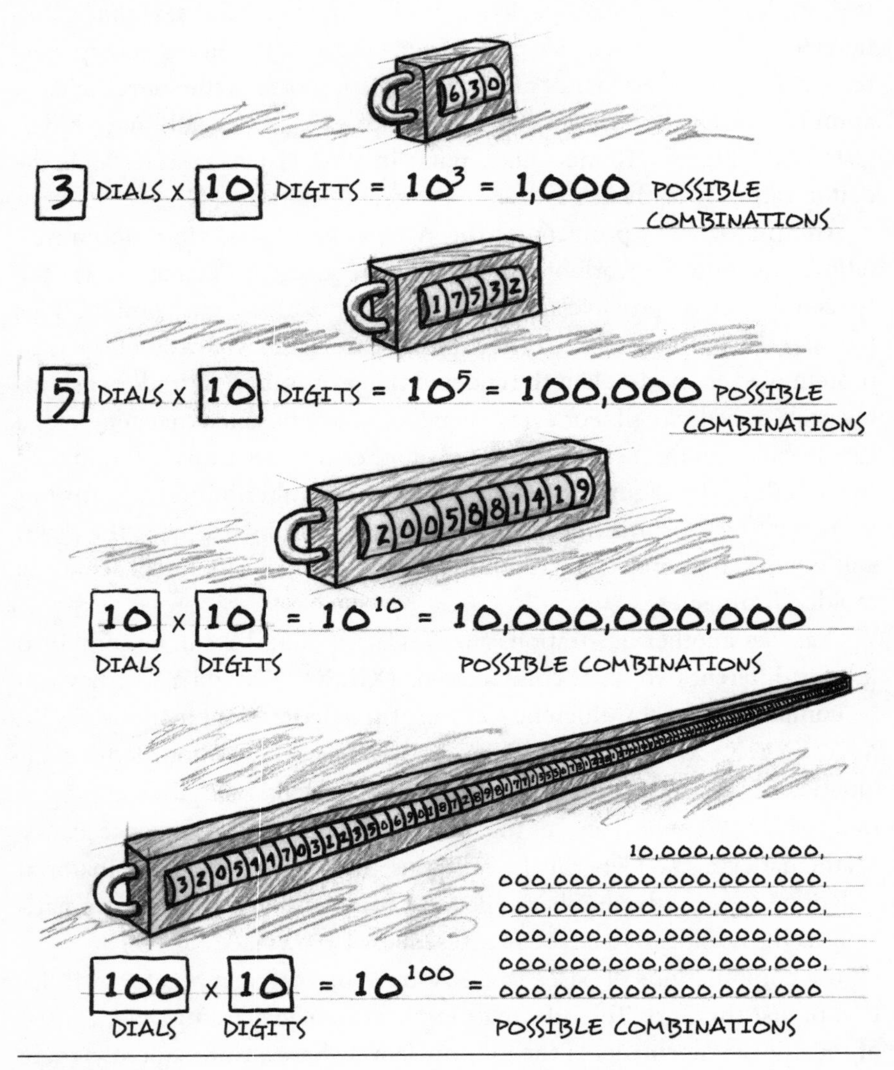

FIGURE 9.2
The problem of combinatorial inflation as illustrated by bike locks of varying sizes. As the number of dials on the bike locks increases, the number of possible combinations rises exponentially.

responding to sequences of increasing length inflates exponentially. For example, there are 4^{100}, or 10^{60}, possible ways of arranging one hundred bases in a row.

The amino-acid chains are also subject to such inflation. A chain of two amino acids could display 20^2, or 20×20, or 400 possible combinations, since each of the twenty protein-forming amino acids could combine with any one of that same group of twenty in the second position of a short peptide chain. With a three-amino-acid sequence, we're looking at 20^3, or 8,000, possible sequences. With four amino acids, the number of combinations rises exponentially to 20^4, or 160,000, total combinations, and so on.

Now, the number of combinatorial possibilities corresponding to a chain with four amino acids only marginally outstrips the combinatorial possibilities associated with the five-dial lock in my first illustration (160,000 vs. 100,000). It turns out, however, that many necessary, functional proteins in cells require far, far more than just four amino acids linked in sequence, and necessary genes require far, far more than just a few bases. Most genes—sections of DNA that code for a specific protein—consist of at least one thousand nucleotide bases. That corresponds to 4^{1000}—an unimaginably large number—possible base sequences of that length.

Moreover, it takes three bases in a group called a codon to designate one of the twenty protein-forming amino acids in a growing chain during protein synthesis. If an average gene has about 1000 bases, then an average protein would have over 300 amino acids, each of which are called "residues" by protein chemists. And indeed proteins typically require hundreds of amino acids in order to perform their functions. This means that an average-length protein represents just one possible sequence among an astronomically large number—20^{300}, or over 10^{390}—of possible amino-acid sequences of that length. Putting these numbers in perspective, there are only 10^{65} atoms in our Milky Way galaxy and 10^{80} elementary particles in the known universe.

That is what bothered Eden and other mathematically inclined scientists at Wistar. They understood the immensity of the combinatorial spaces associated with even single genes or proteins of average length. They realized that if the mutations themselves were truly random—that is, if they were neither directed by an intelligence nor influenced by the functional needs of the organism (as neo-Darwinism stipulates)—then the probability of the mutation and selection mechanism ever producing a new gene or protein could well be vanishingly small. Why? The mutations would have to generate, or "search" by trial and error, an enormous number of possibilities—far more than were realistic in the time available to the evolutionary process.

Eden pointed out in his Wistar presentation that the combinatorial space corresponding to an average-length protein (which he assumed to be about 250 amino acids long) is 20^{250}—or about 10^{325}—possible amino-acid arrangements. Did the mutation and selection mechanism have enough time—since the beginning of the universe itself—to generate even a small fraction of the total number of possible amino-acid sequences corresponding to a single functional protein of that length? For Eden, the answer was clearly no.

For this reason, Eden thought mutations had virtually no chance of producing new genetic information. He likened the probability of producing the human genome by relying on random mutations to that of generating a library of a thousand volumes by making random changes or additions to a single phrase in accord with the following instructions: "Begin with a meaningful phrase, retype it with a few mistakes, make it longer by adding letters [at random], and rearrange subsequences in the string of letters; then examine the result to see if the new phrase is meaningful. Repeat this process until the library is complete."[7] Would such an exercise have a realistic chance of succeeding, even granting it billions of years? Eden thought not.

In addition, Schützenberger emphasized that randomly cutting and pasting larger blocks of text, as evolutionary biologists often envision, would not make any appreciable difference to the efficacy of a random search of sequence space. Imagine a computer "mutating" at random the text of the play *Hamlet* either by individual-letter substitutions or by duplicating, swapping, inverting, or recombining whole sections of Shakespeare's text. Would such a computer simulation have a realistic chance of generating a completely different and equally informative text such as, say, *The Blind Watchmaker* by Richard Dawkins, even granting multiple millions of undirected mutational iterations?

Schützenberger didn't think so. He noted that making random changes "at the typographic level" in a computer program inevitably degrades its function whether those changes are made "by letters or by blocks, the size of the unit really does not matter."[8] Thus, he thought that a process of randomly shuffling blocks of text in any "typographic typology" would inevitably degrade meaning in much the same way that a series of individual-letter substitutions will.

Schützenberger insisted that the evolutionary process faced similar limitations. To him, it seemed extremely unlikely that random mutations of whatever sort would produce significant amounts of *novel and functionally specified* information within the time available to the evolutionary process.

Subsequent to the confirmation of Crick's sequence hypothesis, all present at Wistar understood that the entities that confer functional advantages on organisms—new genes and their corresponding protein products—constitute long linear arrays of precisely sequenced subunits, nucleotide bases in the case of genes and amino acids in the case of proteins. Yet, according to neo-Darwinian theory, these complex and highly specified entities must first arise and provide some advantage *before* natural selection can act to preserve them. Given the number of bases present in genes, and amino acids present in functional proteins, a large number of changes in the arrangement of these molecular subunits would typically have to occur before a new functional and selectable protein could arise. For even the smallest unit of functional innovation—a novel protein—to arise, many improbable rearrangements of nucleotide bases would need to occur before natural selection had anything new and advantageous to select.

Eden and others questioned whether mutations provided an adequate explanation for the origin of the genetic information necessary to build new proteins, let alone whole new forms of life. As physicist Stanislaw Ulam explained at the conference, the evolutionary process "seems to require many thousands, perhaps millions, of successive mutations to produce even the easiest complexities we see in life now. It appears, naïvely at least, that no matter how large the probability of a single mutation is, should it be even as great as one-half, you would get this probability raised to a millionth power, which is so very close to zero that the chances of such a chain seem to be practically non-existent."[9]

LOOKING FOR A LOOPHOLE

In his presentation at the conference, Eden himself acknowledged a possible way of resolving this dilemma. He suggested that it was at least possible that "functionally useful proteins are very common in this [combinatorial] space so that almost any polypeptide one is likely to find [as the result of mutation and selection] has a useful function."[10] Many neo-Darwinian biologists subsequently came to favor this possible solution. The solution was this: even though the size of the combinatorial space that mutations needed to search was enormous, *the ratio* of functional to nonfunctional base or amino-acid sequences in their relevant combinatorial spaces might turn out to be much higher than Eden and others had assumed. If that ratio turned out to be high enough, then the mutation and selection mechanism would frequently stumble onto novel genes and proteins and could easily leapfrog from one functional protein island to the next, with natural selection discarding the nonfunctional outcomes and seizing upon the rare (but not too rare) functional sequences.

As an electrical engineer who was used to working with computer code, Eden was intuitively disinclined to embrace this possibility. He noted that all codes and language systems can convey information precisely because they have rules of grammar and syntax. These rules ensure that not just any arrangement of characters will convey functional information. For this reason, functional sequences in working communications systems are typically surrounded in the larger combinatorial space by a multitude of nonfunctional sequences—sequences that don't obey the rules.

In known codes and language systems, functional sequences do indeed typically represent tiny islands of meaning amid a great sea of gibberish. Geneticist Michael Denton has shown that in English meaningful words and sentences are extremely rare among the set of possible combinations of letters of a given length, and they become proportionally rarer as sequence length grows.[11] The ratio of meaningful 12-letter words to 12-letter sequences is $1/10^{14}$; the ratio of meaningful 100-letter sentences to possible 100-letter strings has been estimated as $1/10^{100}$. Denton used these figures in 1985 to explain why random letter substitutions inevitably degrade meaning in English text after only a few changes and why the same thing might be true of the genetic text.

Given the alphabetic or "typographic" character of genetic information stored in DNA, Murray Eden and others at Wistar suspected that the same kind of problems would affect random mutational changes in DNA. It seemed logical that functional genes and proteins were also surrounded in their relevant combinatorial spaces by vast numbers of nonfunctional sequences—and, further, that the ratio of functional to nonfunctional sequences would also be exceedingly small.

Yet in 1966 none of the scientists on either side of the debates at Wistar knew how rare or common functional gene and amino-acid sequences are among the corresponding space of total possibilities. Do they occur with a frequency of 1 in 10, 1 in a million, or 1 in a million billion trillion? At the time, these questions could not be answered.

Most evolutionary biologists remained optimistic that the answer to this question would vindicate the neo-Darwinian model.[12] And some developments supported their confidence. During the late 1960s, molecular biologists learned that most of the functional roles performed by proteins are performed not just by one precise kind of protein, but by a wide variety, each with its own amino-acid sequence. This is unlike a bike lock, which has only one functional combination. Indeed, molecular biologists learned that though some amino acids at certain sites are absolutely essential for any particular protein to work, most sites tolerate amino-acid substitutions without loss of protein function. For many biologists, this suggested that mutation and selection had a reasonable chance of generating functional sequences of nucleotide bases or amino acids after all—that the ratio of functional to nonfunctional sequences was much higher than skeptics had anticipated.

How much higher? How much variability is allowed in the amino-acid sequences in proteins? Are there enough functional proteins within a relevant combinatorial space of possibilities to render a random mutational search for new proteins plausible?

When Denton compared linguistic and genetic text to explain the potential severity of the combinatorial inflation facing the neo-Darwinian mechanism, he noted that biologists still didn't know enough "to calculate with any degree of certainty the actual rarity of functional proteins." He concluded, however, that since future experiments surely would continue to deepen molecular biology's fund of knowledge, "it may be that before long quite rigorous estimates may be possible."[13]

IN SEARCH OF THE RATIO

Denton's prediction of imminent progress proved correct. During the late 1980s and early 1990s, Robert Sauer, a molecular biologist at MIT, performed a series of experiments that first attempted to measure the rarity of proteins within amino-acid sequence space.

Sauer's work exploited, for the first time, new technology that allowed for the systematic manipulation of gene sequences. Before the late 1970s, scientists typically used radiation and chemicals to produce mutant forms of DNA. Though these techniques sometimes paid off with dramatic results, such as mutant fruit flies with legs growing out of their heads (the famed *Antennapedia* mutation), they did not allow scientists to dictate or target any specific change to a sequence of bases in DNA. The treatments used simply replicated the conditions under which mutations occur naturally.

During the late 1970s and early 1980s, however, molecular biologists developed technologies for making customized synthetic DNA molecules. Robert Sauer used these techniques to make site-directed changes to DNA sequences of specific genes of known function and then to insert those variants into bacterial cells. He could then evaluate the effect of various targeted alterations to a DNA sequence on the function of their protein products within a bacterial cell culture.

Sauer's technique allowed him to begin to evaluate how many of the variant sequences, as a percentage of the total, still produced a functional form of the relevant protein (see Fig. 9.3). His initial results confirmed that proteins could indeed tolerate a variety of amino-acid substitutions at many of the sites in the protein chain. Yet his experiments also suggested that functional proteins might be incredibly rare among the space of all possible amino-acid sequences. Based on one set of mutagenesis experiments, Sauer and his colleagues estimated the ratio of functional to nonfunctional amino-acid sequences at about 1 to 10^{63} for a short protein of 92 amino acids in length.[14]

This result was in rough agreement with an earlier estimate by information theorist Hubert Yockey.[15] Yockey did not perform experiments to derive his estimate of the rarity of proteins in combinatorial sequence space. Instead, he used already published data to compare variants of the similar cytochrome c proteins (proteins involved in the biochemical pathways that generate energy in cells) in different species. He did this to see how much variability existed at each amino-acid site in molecules performing the same function with the same basic structure. Using this data about the allowable variability at each site, he estimated the probability of finding one of the allowable sequences among the total number of sequences corresponding to a cytochrome c protein 100 residues in length. He determined the ratio of functional to nonfunctional sequences to be about 1 to 10^{90} for amino-acid chains of this length.[16] So, although Sauer's experimentally derived results were numerically

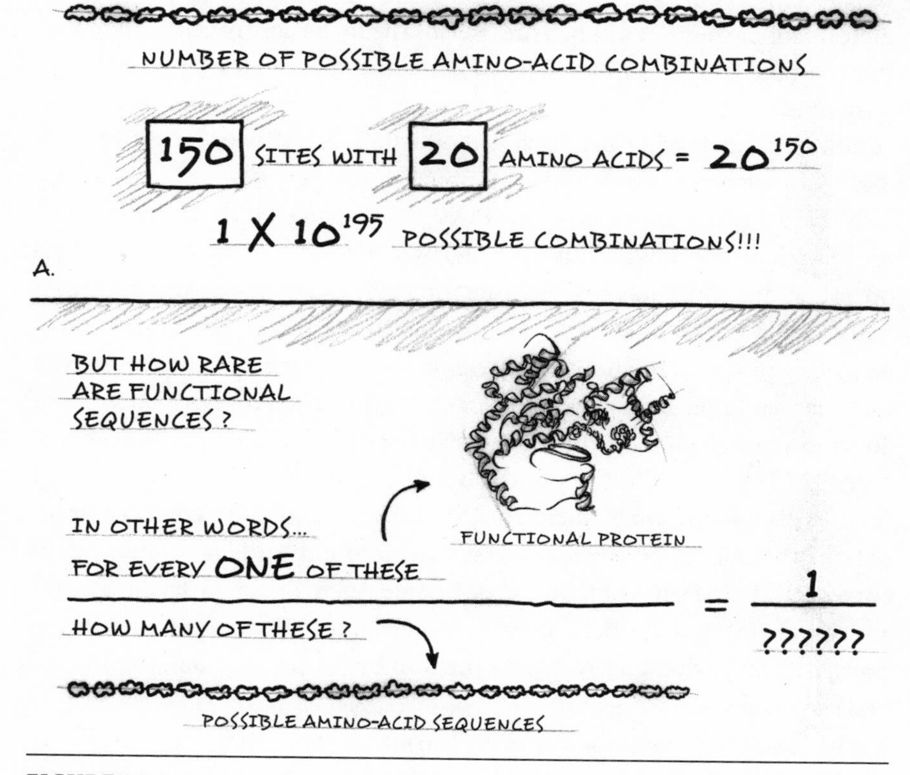

FIGURE 9.3
Figure 9.3a (top) depicts the problem of combinatorial inflation as it applies to proteins. As the number of amino acids necessary to produce a protein or protein fold grows, the corresponding number of possible amino-acid combinations grows exponentially. *Figure 9.3b (bottom)* poses graphically the question of the rarity of proteins in that vast amino-acid sequence space.

different from Yockey's, both approaches gave extremely low ratios suggesting that functional proteins are indeed rare in sequence space, even if proteins do admit significant variability in the specific amino acids present at various positions.

Taken at face value, Sauer's experiments appeared to yield contradictory conclusions. On the one hand, his results showed that many arrangements of amino acids could produce the same protein structure and function—that *numerous* amino-acid sequences populated amino-acid sequence space. On the other hand, the *ratio* of functional sequences to the total number of possible sequences corresponding to a sequence of roughly 100 amino acids appeared to be incredibly low, just 1 to 10^{63}.

Nevertheless, it's not hard to see how both of Sauer's seemingly contradictory conclusions could be true. Recall the locks confronting my hypothetical bike thief. Commercially manufactured bike locks typically have only one combination of digits that will allow them to be opened. The combination that will open a typical bike lock specifies one digit on each dial. No variability at any dial is allowed.

Now imagine a new kind of lock with three crucial differences from an ordinary lock. First, with this new alternative lock, there are four positions on every dial that may—in combination with other positions on other dials—open the lock. My bike thief would like this feature of this kind of lock, since it seems to allow more wiggle room at each dial. But he doesn't like the two other features of this lock. For one, each dial displays one of 20 letters rather than one of 10 numeric digits. Second, instead of 5 dials, there are 100 dials. On the upside, because 4 of the 20 letters on each of the 100 dials might work, there are 4^{100}, or a whopping 10^{60}, correct combinations that will open the lock. That's an astronomically large number of correct combinations. But, on the downside, there are 20 possible settings at each of 100 dials, which computes to 20^{100}, or 10^{130}, possible combinations, a number that totally dwarfs the number of correct combinations of dial settings.

The bike thief would be happy to learn that every dial has "only" four possible correct positions on each dial, but in a trial-and-error process, he still has only 1 chance in 5 (4 out of 20) of landing on a possibly functional dial position for any dial; and this 1-in-5 chance must be negotiated 100 times. In other words, the 1-in-5 chance must be multiplied by the

100 dials on the monster lock in order to arrive at the probability of the thief stumbling upon a functional combination on a given try. The odds are 1 chance in 5^{100} or—if we want to convert that to base 10—roughly 1 chance in 10^{70}. The odds are that slender because the functional combinations—numerous as they are—are dwarfed by the number of total combinations.

In the same way, Sauer established that though many different combinations of amino acids will produce roughly the same protein structure and function, the sequences capable of producing these functional outcomes are still *extremely* rare. He showed that for every functional 92-amino-acid sequence there are roughly another 10^{63} nonfunctional sequences of the same length. To put that ratio in perspective, the probability of attaining a correct sequence by random search would roughly equal the probability of a blind spaceman finding a single marked atom by chance among all the atoms in the Milky Way galaxy—on its face clearly not a likely outcome.[17]

UNCERTAIN SITUATION

Nevertheless, during the 1990s in the immediate wake of the publication of Sauer's results, the implications of his work for evolutionary theory were not entirely clear. Even in the scientific paper in which Sauer reported his work, the abstract summarizing his results emphasized the tolerance to amino-acid substitution that proteins allow. Consequently, scientists on both sides of the discussion about Darwinian evolution seized on different aspects of Sauer's findings either to support or challenge the plausibility of the neo-Darwinian account of the origin of genes and proteins.

Scientists sympathetic to neo-Darwinism emphasized the tolerance of proteins to amino-acid substitution; critics of the theory emphasized the rarity of proteins in sequence space. One scientist, Ken Dill, a biophysicist at University of California, San Francisco, cited Sauer's work to suggest that nearly any amino acid would work at any site in a protein chain, provided the amino acids in question exhibited the correct hydrophobicity (water-repelling) or hydrophilicity (water-attracting) properties.[18]

Yet, at least one scientist, Lehigh University biochemist Michael Behe, cited Sauer's quantitative estimate of the rarity of proteins as a decisive refutation of the creative power of the mutation and selection mechanism altogether.[19] So by the mid-1990s, though Sauer and his group had initiated a program of experimental research that addressed the key question that Murray Eden raised at Wistar, that question had still not been completely settled. Did the mutation and natural selection mechanism have a realistic chance of finding the new genes and proteins necessary to build, for example, a new Cambrian animal? Answering that would await an even more systematic and comprehensive experimental regime.

10

THE ORIGIN OF GENES AND PROTEINS

As a Ph.D. student in chemical engineering at the California Institute of Technology in the late 1980s, Douglas Axe (see Fig. 10.1) became interested in evolutionary theory after several fellow graduate students read the then-bestselling book by Richard Dawkins, *The Blind Watchmaker*. Axe's compatriots were quickly converted to zealous advocates of Dawkins's arguments and urged him to read the book for himself. Axe was impressed by the clarity of Dawkins's writing and illustrations, but he found his case for the creative power of natural selection and random mutations unpersuasive. Whether in the analogies he drew to animal breeding or the computer simulations he used to demonstrate the supposed ability of mutation and selection to generate new genetic information, Dawkins repeatedly smuggled in the very thing he insisted the concept of natural selection expressly precluded: the guiding hand of an intelligent agent.

He found Dawkins's computer simulation particularly interesting. In *The Blind Watchmaker*, Dawkins described how he had programmed a computer to generate the Shakespearean phrase: "Me thinks it is like a weasel."[1] Dawkins did this in order to simulate how random mutations and natural selection could generate new functional information.

Dawkins programmed the computer first to generate many separate strings (sequences) of English letters. He then programmed it to compare each string to the Shakespearean target phrase and select only the string that most closely resembled that target.[2] The program then generated variant versions of that newly selected string and compared those sequences to the target, selecting, again, only the one that most closely resembled the desired target. This eventually generated—after many iterations—a string that matched the target perfectly.

Axe recognized immediately the role that Dawkins's own intelligence had played. Not only did Dawkins provide the program with the information that he wanted it to generate ("Me thinks it is like a weasel"), he imbued the computer with a kind of foresight by directing it to compare the variant sequences of letters with the desired target. Axe realized that Dawkins's program did not simulate natural selection, which by definition is neither guided toward nor given information about a desired outcome generations in the future.

Axe began to wonder if there was some other way to assess the creative power of the mutation and natural selection mechanism, not with clever analogies or simplistic computer simulations, but with experimental and mathematical rigor.

Axe realized that Dawkins was right about one thing: the importance of genetic information. Like his fellow engineer Murray Eden, Axe's tendency to view biology as an engineer led him to ask whether selection and mutation could actually *build* new organisms. Axe's own research explored the connection between *process control* (a field of study in engineering) and *genetic regulation,* a sophisticated version of automated process control at work on a molecular scale inside living cells. Since cells use proteins to perform various feats of regulation, Axe was acutely aware that building

FIGURE 10.1
Douglas Axe. *Courtesy Brittnay Landoe.*

new organisms necessarily involved building new proteins, which would in turn require new genetic information.

But could mutation and selection generate the precise arrays of nucleotide bases necessary to build fundamentally new protein structures? Axe's interest in this question eventually led him to the scientific papers of Robert Sauer and the proceedings of a seemingly obscure 1966 Philadelphia conference called "Mathematical Challenges to the Neo-Darwinian Interpretation of Evolution."

UNRESOLVED ISSUES

As Axe read the papers that Sauer's research group had produced, he realized their importance as a first step toward answering the questions Murray Eden had raised at Wistar. If Sauer's quantitative measures of rarity held up, then Axe thought it obvious that mutation and selection could not adequately search a space that large. If, on the other hand, subsequent mutagenesis experiments overturned Sauer's work and showed that protein function was largely indifferent to changes in amino-acid sequence, then the number of functional sequences might be large enough that mutation and selection would have a good chance of finding new functional genes and proteins in a reasonable amount of time.

After completing his Ph.D., Axe made inquiries about doing postdoctoral research in a top research lab where he could address these unanswered questions. He was soon invited by Alan Fersht, a professor at the University of Cambridge and director of the Centre for Protein Engineering, part of the world-famous Medical Research Council (MRC) Centre at Cambridge, to join his research group.

The decision to accept Fersht's offer was an easy one. The star-studded history of the adjoining Laboratory for Molecular Biology (LMB), arguably the birthplace of molecular biology, included such luminaries as James Watson, Francis Crick, Max Perutz, John Kendrew, Sidney Brenner, and Fred Sanger. Starting in the chemistry department and then moving to the MRC Centre, Axe hoped to apply his research training to resolve the uncertainty that surrounded the interpretation of Sauer's results. Specifically, he wanted to eliminate what he saw as two sources of error in Sauer's method in order to get a more definitive estimation of the frequency of functional sequences in sequence space.

Axe thought, first, that Sauer's team might have *underestimated* the rarity of functional proteins. In their experiments, Sauer's group tested the tolerance of proteins to amino-acid substitution by changing amino acids at one or a few consecutive sites without making any other changes to other sites at the same time, much like a typist introducing an isolated typographic error in an otherwise accurately transcribed text. Not surprisingly, Sauer and his colleagues found that many sites along a protein chain could tolerate these isolated amino-acid substitutions, just as the reader of a text with only a few typos can often make out its meaning. Sauer's team *seemed* to be assuming that a similar tolerance would have emerged if they had changed many sites simultaneously.

Axe thought that this assumption ignored the importance of the larger context provided by the mostly unaltered protein. A single typographic error typically will not totally destroy the meaning of a section of English text, because of the surrounding context provided by the other words, as well as the correct letters in the altered word. That does not mean, however, that specificity of sequence doesn't matter. Instead, the meaning of a sentence with a typographic error can be discerned *only because the rest of the letters are specifically arranged* into meaningful words and phrases that provide a context for determining the meaning of the incorrectly spelled word. For just this reason, however, the meaning of a sentence is rapidly degraded if errors are allowed to accumulate at multiple sites.

Axe wondered if much the same could be true of genes and proteins. He wondered whether multiple, as opposed to single, position changes would quickly degrade function and whether a tolerance for substitutions at individual sites was itself context dependent—whether the tolerance for substitution at one site might depend upon having highly specific sequences at other sites. Thus, without questioning Sauer's experimental findings, Axe thought Sauer's result lent itself to misinterpretation. To many molecular biologists, it suggested that proteins could readily accommodate *many* simultaneous changes to their amino-acid sequences at many positions and remain functional.

As it turns out, Sauer recognized the potential for misinterpretation of his results. As he explained in the very paper in which he developed his quantitative estimate of rarity, "this calculation overestimates the number of functional sequences, since changes at individual positions are less likely to be independent of one another as more positions are allowed to vary."[3]

Another assumption in Sauer's approach had potentially the opposite effect—*exaggerating* the rarity of functional proteins. Axe thought the test that Sauer and his colleagues used to decide whether their mutant proteins were functional required a higher level of function than natural selection might require. Sauer and his team judged proteins with less than about 5–10 percent of the function seen in the natural protein to be nonfunctional. Yet Axe knew that even damaged enzymes with less than 5 percent of normal activity could add significantly more benefit than no enzymatic activity at all. Thus, from a neo-Darwinian point of view, the emergence of even such handicapped proteins might confer a selectable advantage on an organism. Axe thought that by rejecting as nonfunctional such mutated sequences, Sauer's team probably had introduced another estimation error. These competing errors made it hard to know if the estimate made by Sauer's team was too high or too low or whether perhaps they might neatly cancel each other out. To eliminate both sources of possible error, Axe carefully designed a new series of experiments.

THE IMPORTANCE OF FOLDS

Axe had a key insight that animated the development of his experimental program. He wanted to focus on the problem of the origin of new protein *folds* and the genetic information necessary to produce them as a critical test of the neo-Darwinian mechanism. Proteins comprise at least three distinct levels of structure:[4] primary, secondary, and tertiary, the latter corresponding to a protein fold. The specific sequence of amino acids in a protein or polypeptide chain make up its *primary structure*. The recurring structural motifs such as alpha helices and beta strands that arise from specific sequences of amino acids constitute its *secondary structure*. The larger folds or "domains" that form from these secondary structures are called *tertiary structures* (see Fig. 10.2).

Axe knew that as new life-forms arose during the history of life—in events such as the Cambrian explosion—many new proteins must also have arisen. New animals typically have new organs and cell types, and new cell types often call for new proteins to service them. In some cases new proteins, while *functionally* new, would perform their different functions with essentially the same fold or tertiary structure as earlier proteins. But more often, proteins capable of performing new functions require new

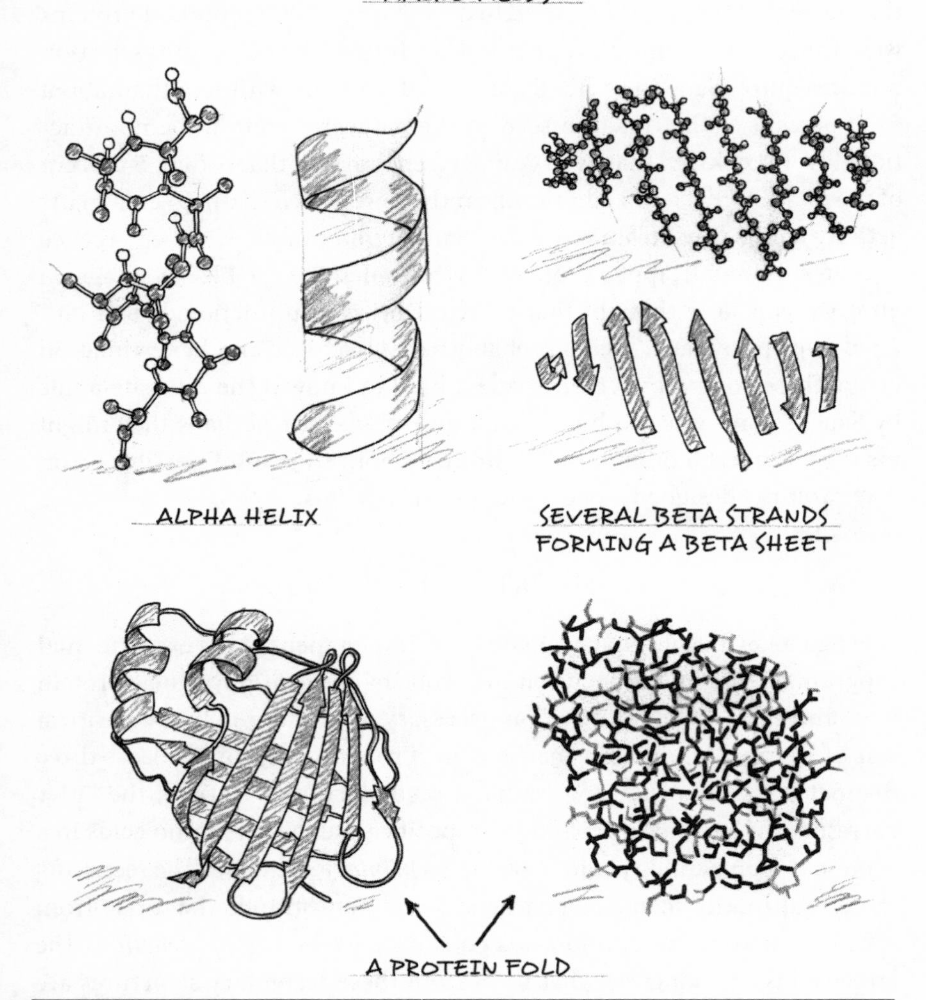

FIGURE 10.2
Different levels of protein structure. The first panel at the top shows the primary structure of a protein: a sequence of amino acids forming a polypeptide chain. The second panel depicts, in two different ways, two secondary structures: an alpha helix (*left*), and beta strands forming a beta sheet (*right*). The third panel at the bottom shows, in two different ways, a tertiary structure—that is, a protein fold.

folds to perform these functions. That means that explosions of new life-forms must have involved bursts of new protein folds as well.

The late geneticist and evolutionary biologist Susumu Ohno noted that Cambrian animals required complex new proteins such as, for example, lysyl oxidase in order to support their stout body structures. When these molecules originated in Cambrian animals, they also likely represented a completely novel folded structure unlike anything present in Precambrian forms of life such as sponges or one-celled organisms. Thus, Axe was convinced that explaining the kind of innovation that occurred during the Cambrian explosion and many other events in the history of life required a mechanism that could produce, at least, distinctly new protein folds.

He had another reason for thinking that the ability to produce novel protein folds provided a critical test for the creative power of the mutation and selection mechanism. As an engineer, Axe understood that building a new animal required innovation in form and structure. As a protein scientist, he understood that new protein folds could be viewed as the *smallest unit of structural innovation* in the history of life.

It follows that new protein folds represent the smallest unit of structural innovation that natural selection can select. Of course, natural selection can operate on smaller units of change—individual amino-acid changes that result in slight functional advantages or fitness gains, but not new folds, for example. But what if the functional or fitness gains that natural selection preserves and passes on never generate structural innovations? What if, instead, it only preserves slight differences in the sequence or function of proteins that confer an advantage without altering structure? Then, clearly, fundamental changes in the form of an organism will not occur. Building fundamentally new forms of life requires structural innovation. And new protein folds represent the smallest selectable unit of such innovation. Therefore, mutations must generate new protein folds for natural selection to have an opportunity to preserve and accumulate structural innovations. Thus, Axe realized that the ability to produce new protein folds represents a *sine qua non* of macroevolutionary innovation.

Could random mutations generate such novel protein folds? Axe realized that answering this question depended upon measuring the rarity of functional genes and proteins in sequence space and determining whether random genetic mutations would have enough opportunities to search the relevant sequence spaces within evolutionary time.

AXE'S INITIAL RESULTS

Axe read the paper by Sauer and his colleague John Reidhaar-Olson that estimated the proportion of functional protein sequences to be extremely low (1 in 10^{63}). He noticed that the authors chose not to emphasize this measure of rarity, however, but instead the variety of amino-acid substitutions that the protein under study could tolerate.

In their paper, Reidhaar-Olson and Sauer also repeated a then popular idea that the amino acids buried in the interior of a folded protein (forming what is known as the *hydrophobic core*) are most important for specifying the structure, while the arrangement of the exterior amino acids did not matter nearly as much.[5] They thought that the amino acids buried within folded proteins typically need only to be hydrophobic (water repelling), whereas the amino acids on the exterior, for the most part, need to be hydrophylic (water attracting). In fact, some protein scientists thought that these simple restrictions might be the whole story—that a functional protein fold might require nothing more than an appropriate arrangement of hydrophobic and hydrophylic amino acids in a given sequence.

At Fersht's lab in Cambridge, Axe conducted an experimental test of this idea, and surprised himself with the first result. In a paper he coauthored in the *Proceedings of the National Academy of Sciences* in 1996, he reported his findings. When he replaced the entire thirteen-residue hydrophobic core of a small enzyme with random combinations of other hydrophobic amino acids, a high fraction of the randomized proteins (about one-fifth) still performed their original function. This suggested that proteins were, perhaps, less susceptible to functional loss as a result of sequence changes than Axe had thought.

Next he focused on the exterior of proteins, randomizing portions of two different proteins' exteriors in much the same way that he had randomly changed the interior of one of them. This time his approach failed to produce any functional variants at all. Realizing that this seemed to contradict what Sauer and others had supposed, Axe decided to make only much more restrictive changes in the next trial. He replaced each exterior amino-acid residue only with its *most similar* amino-acid alternative. Nevertheless, both of the proteins that he studied still lost *all* function by the time he had replaced one-fifth of their exterior residues. Thus he concluded that the exterior parts of the proteins were much more susceptible to functional loss as a result of amino-acid changes than had been widely assumed.

In all this work, Axe designed his experiments to remedy the two sources of estimation error inherent in Sauer's method. First, by studying amino-acid changes in combination rather than in isolation, he determined that the surrounding context typically did influence whether a particular amino-acid change at a particular site caused functional loss. In other words, he discovered that the neglect of the influence of the surrounding context had the effect of exaggerating the tolerance to amino-acid changes at particular sites, as he (and Sauer) had suspected.

Second, the proteins that Axe chose for his study made it possible for function to be detected at much lower levels than was possible in Sauer's studies. For one protein Axe studied, the more sensitive test of function did indeed allow a greater proportion of single mutants to retain some function, with about 95 percent of the mutant proteins achieving the designation "active." This suggested that Sauer's less sensitive screen did contribute to another source of estimation error, this time in the opposite direction. Nevertheless, Axe's more sensitive screen also enabled him to establish that even though single mutations allow many proteins to retain some function, they still diminish or damage the function of the protein—often enough to ensure that they will be eliminated by the purifying effect of natural selection. Further, because of the extreme sensitivity of his test for function, Axe learned that any single mutation that failed his test was *single-handedly* destroying function. He determined that fully 5 percent of such changes did destroy protein function.

Overall, therefore, he showed that despite some allowable variability, proteins (and the genes that produce them) are indeed highly specified relative to their biological functions, especially in their crucial exterior portions. Axe showed that whereas proteins will admit some variation at most sites if the rest of the protein is left unchanged, multiple as opposed to single amino-acid substitutions consistently result in rapid loss of protein function. This was the case even when these changes occur at sites that allow variation when altered in isolation.[6] His new experiment also roughly confirmed Sauer's earlier quantitative assessment of the rarity of functional proteins, despite the estimation errors inherent in Sauer's method. Why? Because it appeared that Sauer's two estimation errors—ignoring context and using an insufficiently sensitive screen for function—did, in fact, roughly cancel each other out.

Despite these advances in understanding, Axe had not yet determined whether the greatly restricted picture of tolerance that his work had exposed would cause problems for the evolution of new protein folds. In order to answer that question he would need to obtain a more precise quantitative estimate of the rarity of proteins in sequence space.

Having developed a method that eliminated the main sources of estimation error in earlier mutagenesis experiments, Axe was now in a position to answer that question with unprecedented rigor. Once he did, he could determine whether *random* genetic changes would have enough opportunities—even on the scale of evolutionary time—to search the relevant sequence spaces for functional genes and proteins.

DON'T LEAVE THE FOLD

Of course, Axe understood that neo-Darwinists do not envision a completely random journey through nucleotide or amino-acid sequence space. They see *natural selection* acting to preserve useful mutational variations and to eliminate deleterious ones. Richard Dawkins, for example, likens an organism to a high mountain peak.[7] He compares climbing the sheer precipice up the front side of the mountain to building a new organism purely by chance—random mutations alone. He acknowledges that this approach up "Mount Improbable" will not succeed. Nevertheless, he asserts that there is a gradual slope up the back side of the mountain that can be climbed in small incremental steps. In his analogy, the back side up "Mount Improbable" corresponds to the process of natural selection acting on many small random changes in the genetic text. What chance alone cannot do, natural selection acting on random mutations can accomplish through the cumulative effect of many slight successive steps.

Yet Axe's experimental results presented a problem not only for scenarios involving random mutations acting alone, but also for scenarios envisioning selection and random mutation acting in concert. Further, his mutagenesis experiments cast doubt on each of the two scenarios by which evolutionary biologists might envision new protein folds (and the information necessary to produce them) arising as the result of the mutation and selection mechanism.

In theory, new genes capable of producing a new protein fold might arise either from (a) preexisting genes or from (b) nonfunctional sections of the genome. That is—to adapt Dawkins's visual analogy—mutation and natural selection could conceivably generate a new functional gene starting from either (a) another mountain peak (a different preexisting functional gene) or (b) from the valley floor (a nonfunctional section of the genome). Yet Axe's experimental results would show that the action of natural selection would not help solve the search problem confronting the mutation mechanism in either of these two cases. To see why, we need to understand a bit more about each of these two possible neo-Darwinian scenarios as well as Axe's subsequent experimental findings.

FROM PEAK TO PEAK

In the first case, evolutionary biologists might envision mutation and selection gradually altering a preexisting gene (and its protein product) to produce another functional gene (and a different protein product). This scenario involves moving metaphorically from one functional peak to another without dipping into a valley (a zone of diminished fitness or nonfunction).

Most evolutionary biologists reject this first scenario.[8] They do so because they recognize that mutations in preexisting genes will typically degrade functional genetic information. They know, too, that when genes lose function, natural selection will eventually eliminate the organisms that possess these genes. Genes that contribute to the healthy function of an organism, that have been mutated in such a way as to diminish that function, will be subject to what evolutionary biologists call "purifying selection." That is, natural selection will typically *eliminate* organisms possessing mutation-induced gene variants that diminish function or fitness. (When natural selection *preserves* genetic changes that enhance function or fitness, evolutionary biologists call that "positive selection.")

Axe's mutagenesis experiments confirmed these reasons for doubting the first of the possible neo-Darwinian scenarios, at least as an explanation for the origin of new protein folds. In work that he published in 2000, he showed that it is, indeed, exceedingly difficult to make extensive changes to functional amino-acid sequences without destabilizing a protein fold. Even best-case changes involving the most chemically similar amino acids in the exterior of proteins tended to destabilize protein folds.

In these experiments, Axe mutated a gene that produced a protein exhibiting a single fold and function. He found that as he altered this protein, multiple position changes in the exterior of the protein molecule quickly effaced or destroyed its function.[9] Yet to turn one protein with a distinctive folded structure into another with a completely novel structure and function requires specified changes at many, many sites—far more than Axe altered in his experiments.[10] The number of changes necessary to produce a new protein fold typically exceeds the number of changes that will result in functional loss. Given this, the probability of the evolutionary process successfully traversing a functional landscape from one functional peak to another—all the while escaping functional loss each step along the way—is extremely small, with the probability diminishing exponentially with each additional requisite change.[11] Indeed, by showing that functional proteins with distinct folds are far more sensitive to functional loss than protein scientists had previously assumed, Axe's experiments confirmed what most evolutionary biologists suspected—namely, that protein-to-protein (or functional gene–to–functional gene) evolution is a no-go where the mutation and selection mechanism must produce a new protein fold (see Fig. 10.3).

Axe had a more fundamental reason for considering the first evolutionary scenario implausible. Based on the physical principles of protein function, the vast majority of protein functions simply cannot be performed by unfolded proteins. In other words, stability of protein structure is a precondition of protein function. Destabilized protein folds not only lose the three-dimensional structures they need to perform functional tasks, they are also vulnerable to attack by other proteins called proteases that devour unfolded proteins or polypeptides in the cell.[12]

As one structure is degraded as the result of multiple sequence changes, it will *necessarily* lose structural stability, resulting in a catastrophic loss of function. Yet any diminution in the function of a protein will also diminish *fitness* in a way that will subject the protein (and its corresponding gene) to the purifying action of natural selection.[13] Indeed, according to the equations of population genetics, the standard mathematical expression of

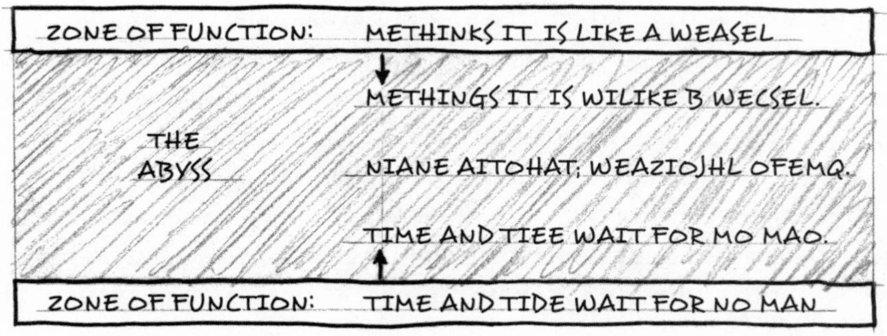

FIGURE 10.3
This figure illustrates why many evolutionary biologists reject the idea that genes and proteins under selection pressure will evolve into new functional genes and proteins. Since genes, like English sentences, contain sequence-specific functional information, multiple changes in the genetic text will inevitably degrade function (or fitness) long before a new functional sequence will arise—just as random changes in a meaningful English sentence will typically destroy meaning long before such changes produce a significantly different meaningful sentence.

neo-Darwinian theory, even slight losses in fitness will subject the disadvantageous traits that produce such losses to purifying selection, thus eliminating them. That means that even many protein sequences that retain a significant, though diminished, portion of their original function nevertheless will not survive the winnowing effects of the neo-Darwinian mechanism. Thus, the gradual transformation of one functional fold into another was a complete nonstarter.

Research performed at the European Molecular Biology Laboratory by molecular biologist Francisco Blanco has since confirmed this conclusion. Using site-directed mutagenesis, Blanco's team found that the sequence space between two naturally occurring protein domains is not continuously populated by folded or functional proteins. By sampling intermediate sequences between two sequences that do adopt different folds, Blanco found that the intermediate sequences "lack a well-defined three-dimensional structure." Thus, he concluded that "the appearance of a completely new fold from an existing one is unlikely to occur by evolution through a route of folded intermediate sequences."[14]

Thus, both experimental results and the physics of protein folding implied that random searches for novel proteins starting from preexisting protein-coding genes would result in functional loss long before a protein with a novel fold would emerge, as most evolutionary biologists already suspected. Although the first of the two possible evolutionary scenarios has the advantage of starting on a mountain peak—with a functional gene and protein—it also has a lethal disadvantage: randomly mutating the gene will soon destabilize a protein fold and/or generate nonfunctional intermediate sequences and structures long before a new gene (capable of generating a new fold) would arise. For this reason, this scenario involves not so much a climb up Mount Improbable, but a step out over Valley Impassable.

SCALING MOUNT IMPROBABLE

For all these reasons, like most evolutionary biologists, Axe thought the second neo-Darwinian scenario—in which new genes and proteins emerge from nonfunctional or neutral regions of the genome—provides a much more plausible means of producing the information necessary to construct novel protein folds. It was to this scenario that Axe turned his experimental energies.

In this scenario, neo-Darwinists envision new genetic information arising from sections of the genetic text that can vary freely without consequence to the organism. According to this scenario, noncoding sections of the genome or duplicated sections of coding regions undergo a protracted period of "neutral evolution"[15] in which alterations in nucleotide sequences have no discernible effect on the fitness of the organism. Functional genes and proteins gradually rise from a nonfunctional valley floor to a functional mountain peak—generating a new gene. Natural selection plays a role, but not until a new functional gene has arisen.

Evolutionary biologists typically picture this process beginning with a gene duplication event. Although several different mechanisms can generate gene duplicates in DNA,[16] the most common mechanism occurs during the crossing-over step of meiosis (a kind of cell division that produces sex cells, or gametes, in sexually reproducing organisms). During meiosis, homologous chromosomes swap segments of DNA. In a normal crossing-over event, corresponding chromosomal segments of equal size are swapped between the two homologous chromosomes, ensuring that both chromosomes experience no net gain or loss of genes. Sometimes, however, chromosomes swap genetic material of *unequal* length. When this happens, one chromosome (the one that gets the smaller piece) ends up losing some DNA, while the chromosome that receives the larger segment of genetic material ends up with a new stretch of chromosomal DNA—one that may include a gene or genes it already had. This results in *duplicate* copies of a gene on one chromosome.

When this occurs, one of the two genes may begin to vary—to experience mutations—without adversely affecting the function of the organism, while the other performs the original function. In the jargon of evolutionary biology, mutational changes in gene duplicates are "selectively neutral"—they initially provide no advantage or disadvantage to an organism or population. These gene-duplication events allow nature room to experiment safely. Unhelpful but harmless genetic novelties can be passed on to future generations, where additional mutations one day may render the evolving genetic material useful. Eventually, as mutational changes accumulate, a new gene sequence may arise in a new organism that can code for a novel protein fold and function. At that point, natural selection can favor the new gene and its protein product, preserving and passing it along to future generations—or so the story goes.

This scenario—which many evolutionary biologists now refer to as the "classical model" of gene evolution—has the advantage of allowing portions of the genome to vary freely through many generations, giving mutations many opportunities to "search" the space of possible base sequences without being punished for drifting into valleys of lost or diminished function.

But this scenario faces an overriding problem: the extreme rarity of sequences capable of forming stable folds and performing biological functions. Since natural selection does nothing to help *generate* new folded, functional sequences, but rather can only *preserve* such sequences once they have arisen, random mutations alone must search for the *exceedingly* rare folded and functional sequences within the vast sea of combinatorial possibilities.

And *that* is the big story associated with Axe's experiments. His research showed that folded, functional sequences of amino acids are indeed exceedingly rare within sequence space. After his initial round of experiments, Axe performed another series of site-directed mutagenesis experiments on a 150-amino-acid protein-folding domain within a β-lactamase enzyme and published the results in the *Journal of Molecular Biology*.[17] Recall that a folding domain is a portion of a larger protein that exhibits a distinctive fold. Since amino-acid chains must first fold into stable three-dimensional structures, Axe performed experiments that enabled him to estimate the frequency of sequences that will produce stable folds—any stable fold—before he estimated the frequency of sequences performing a specific (β-lactamase) function. His improved experimental method produced a precise quantitative result. He estimated (a) the number of 150-amino-acid-long sequences capable of folding into stable "function-ready" folded structures compared to (b) the whole set of possible amino-acid sequences of that length (recall Fig. 9.3). Based on his site-directed mutagenesis experiments, he determined that ratio to be a vanishingly small 1 in 10^{74}. In other words, for sequences 150 amino acids long, only 1 in 10^{74} sequences will be capable of folding into a stable protein.

For a sequence to achieve a protein fold is only a first step, however. A protein must be folded to be functional, but a folded protein is not necessarily a functional protein. And although sequences capable of forming stable protein folds are necessary to any significant evolutionary innovation, natural selection cannot select for the presence of a fold unless it also performs a function that confers a specific functional advantage on an organism. Thus, Axe also estimated (a) the number of proteins of modest length (150 residues) that perform *a specified function* via any folded structure compared to (b) the whole set of possible amino-acid sequences of that size. Based on his experiments and data about the number of stable folded proteins that exist, Axe estimated that ratio to be about 1 to 10^{77}. A telling conclusion follows from this experimental data: The probability of any given mutational trial generating (or "finding") a specific functional protein among all the possible 150 residue amino-acid sequences is 1 chance in 10^{77}—that is, one chance in one hundred thousand, trillion, trillion, trillion, trillion, trillion, trillion.

That is obviously an incredibly small probability, but is it small enough to justify rejecting the classical model of gene evolution? Or is it plausible to think that random mutations in the nonfunctional part of the genome could overcome these long odds to generate the genetic information necessary to produce a novel protein fold with a specific selectable function?

HOW MANY TRIALS?

When statisticians or scientists assess whether a chance hypothesis provides a plausible explanation for the occurrence of an event, they do not just evaluate the probability of that particular event occurring once; they evaluate the probability of the event occurring *given the number of opportunities it has to occur.*

For example, if our hypothetical bike thief from the previous chapter had enough time to try more than half (more than 500 of the 1000) total combinations of a three-dial bike lock, then the probability that he will stumble upon the right combination will exceed the probability that he will fail. In that case, it will be *more likely than not* that he will succeed in opening the lock by chance. In that case, the chance hypothesis—the hypothesis that he will succeed in opening the lock by chance—is more likely to be true than false. On the other hand, if just after he started trying to crack the lock he heard a security guard coming around the corner and only had time to explore a small fraction of the total number of possible combinations—far fewer than half—then it will be *much more likely than not* that he will *fail* to open the lock by chance. Consequently, anyone who knew his situation could conclude that the chance hypothesis is, in that case, much more likely to turn out false than true.

When statisticians or scientists assess the probability of an event occurring by chance, they often assess what is called a *conditional* probability. In deciding the plausibility of a chance hypothesis, they assess the probability of the event given or "*conditioned on*" what else we know, especially what else is known about the number of opportunities the event has to occur. And they refer to the number of opportunities an event has to occur as "the probabilistic resources."[18]

If the conditional probability of the chance hypothesis, given the number of opportunities it has to occur, is *less* than ½, then it is more likely than not that the event will *not* happen by chance. It will be viewed as *implausible*—more likely to be false than true. Conversely, if the conditional probability of the chance hypothesis, given the number of opportunities it has to occur, is *more* than ½, then it is more likely than not that the event in question *will* occur by chance. It will be deemed *plausible*—more likely to be true than false. And, of course, the smaller the conditional probability associated with a hypothesis, the *more* implausible the hypothesis—the more likely the chance hypothesis is to be false than true.

How then should we assess the chance hypothesis for the origin of biological information—in particular, the hypothesis that random mutations generated the information necessary to produce a novel protein fold with a selectable function? What is the conditional probability that such a folded protein could arise as the result of random mutations in duplicated non-functional sections of a genome? Axe realized that in order to answer that question he needed a way to estimate the number of opportunities that random mutations had for producing a new protein fold with a selectable function during the whole history of life on earth.

THE BIOLOGICAL UNIVERSAL PROBABILITY BOUND

Here, evolutionary theory itself provides the answer. Axe was interested in the number of times that mutations might have produced new sequences of bases in DNA that were capable of producing a new sequence of amino acids—one of the 10^{77} possible sequences in the relevant sequence space. Yet not every such sequence of bases that mutations might generate constituted a relevant trial. In theory, mutations may alter a gene many times during the life cycle of an organism. Nevertheless, natural selection can only act on the new sequence of bases that is actually passed on to offspring. It might seem that it would be difficult to quantify the number of mutational trials in each generation. But even if we think of mutations repeatedly shuffling and reshuffling the arrangement of bases during the life cycle of an organism, only those mutations in the genes (or DNA) in the *reproductive cells* of parent organisms can have any effect on the next generation. Since Axe wanted to know how many novel sequences capable of generating *a selectable* function might have arisen in the history of life, he only needed to concern himself with those sequences that could be transmitted during reproduction.

This meant that if he could estimate the total number of organisms that had lived during the history of life on earth and the number of new genes that mutations might produce and pass on to the next generation, he could establish an upper bound on the number of trials relevant to the evolutionary process.

Axe knew that the huge population sizes of prokaryotes like bacteria dwarf the population sizes of all other organisms combined. Thus, estimates for the size of the bacterial population—plus a smidge for everything else—would approximate the size of the number of organisms living at any given time. Based on the average length of time of a bacterial generation and the time since the first appearance of bacterial life on earth (3.8 billion years ago), scientists have estimated that a total of about 10^{40} organisms have lived on earth since life first appeared.[19] Axe made the assumption that each new organism received one new sequence of bases (one potential gene) capable of generating one of the possible amino-acid sequences in sequence space per generation.

This was an extremely generous assumption. Since mutations have to be quite rare for life to survive, most bacterial cells inherit an exact copy of their parent's DNA. Furthermore, the ones that differ from their parents are likely to carry a mutation that has already occurred many times in other cells. For these reasons, the actual number of new sequences sampled in the history of life is much lower than the total number of bacterial cells that have existed. Nevertheless, Axe assumed that one new gene per organism has been transmitted to the next generation. Thus, he used 10^{40} gene sequences as a liberal estimate of the total number of gene sequences (evolutionary trials) that have been generated to search sequence space in the history of life.

Even so, 10^{40} represents only a tiny fraction—1 ten trillion, trillion, trillion*th*—of 10^{77}. Thus, the conditional probability of generating a gene sequence capable of producing a novel protein fold and function is still only 1 in 10^{37}. This means that if every organism from the dawn of time had generated, by random mutation, one new base sequence in the sequence space of interest, that would amount to only one 10 trillion, trillion, trillionth of the sequences in that space—the space that needs to be searched. And, since conditional probability of a new gene arising in the manner envisioned by the classical model turns out to be almost unimaginably less than ½, the classical model turns out to be vastly more likely to be false than true. Thus, Axe concluded that a reasonable person should reject it. The probabilistic resources available to the classical model of gene evolution are simply far too small to tame 1 chance in 10^{77} (see Fig. 10.4).

To appreciate why this model fails, consider the following illustration. After the 1975 Steven Spielberg film *Jaws* became a big hit, one small-town motel advertised "Shark-Free Pool." Proponents of the second

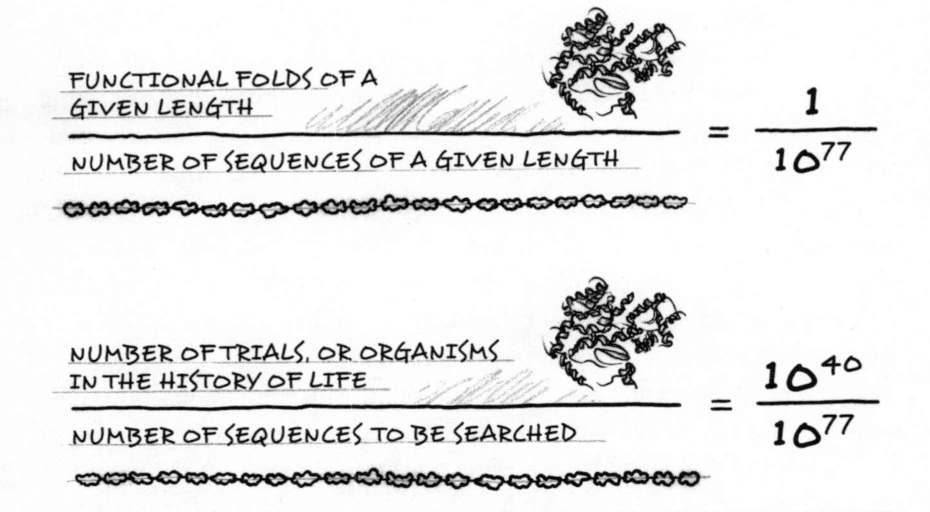

FIGURE 10.4
The top panel in this diagram represents the results of Axe's mutagenesis experiments showing the extreme rarity of functional proteins in sequence space. Based on his experiments Axe estimated that there are 10^{77} possible sequences corresponding to a specific functional sequence 150 amino acids long. The second panel shows that functional amino-acid sequences are extremely rare even in relation to the total number of opportunities the evolutionary process would have had to generate novel sequences (on the assumption that each organism that has ever lived during the history of life produced one such sequence per generation).

evolutionary scenario (gene duplication, followed by neutral evolution), envision an evolutionary pool where there are no consequences for mutational missteps—by analogy, a pool with no predators. But to extend the illustration, picture a predator-free pool the size of our galaxy. Now picture a blindfolded man dropped into the middle of it. He must swim to the far side, to the one spot on the edge of the pool where a ladder would give him a way out. He's safe from predators, but it will do him no good. He needs direction, some way of gauging his progress, and an immense amount of time. But he has none of these and so he will arrive at the ladder in neither a hundred years, nor a hundred billion. Similarly, in the classical model of gene evolution, random mutations must thrash about aimlessly in immense combinatorial space, a space that could not be explored by this means in the entire history of life on earth, let alone in the few million years of the Cambrian explosion.

PLATE 1 *(above)*. Mt. Maotian, near Chengjiang, China. The famous Chengjiang fossil site was discovered near the top of this mountain. *Courtesy Illustra Media (all).*

PLATE 2 *(left)*. The original discovery site of the first Chengjiang fossils. Today this area is a preserve.

PLATE 3 *(below)*. Looking down from Mt. Maotian, with the large Fuxian Lake in the distant foreground.

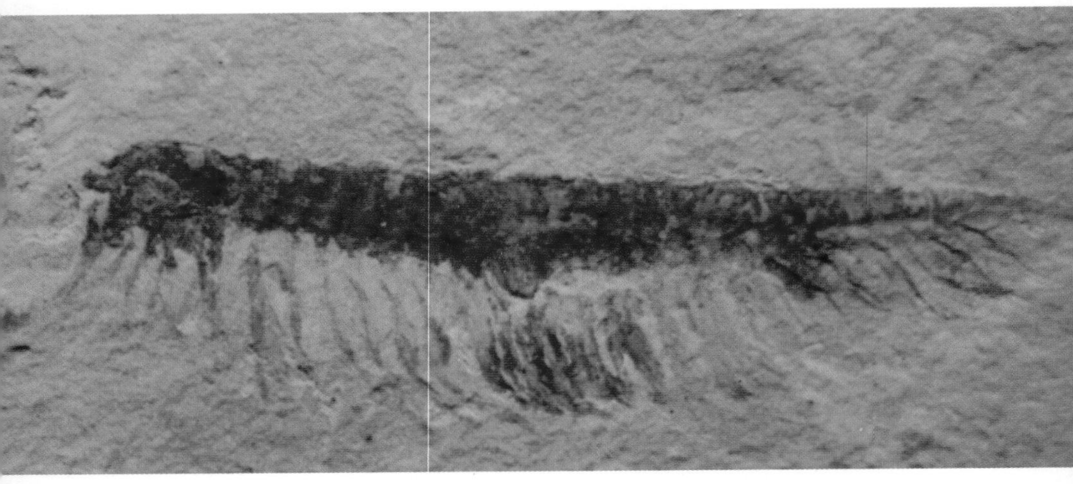

PLATE 4. The opabinid *Jianfengia multisegmentalis. Courtesy Paul Chien.* Unless otherwise noted, all fossils shown on this and subsequent color pages are from the Maotianshan Shale near Chengjiang, China.

PLATE 5. A fossilized comb jelly (a ctenophore). Species: *Maotianoascus octonarius. Courtesy J. Y. Chen.*

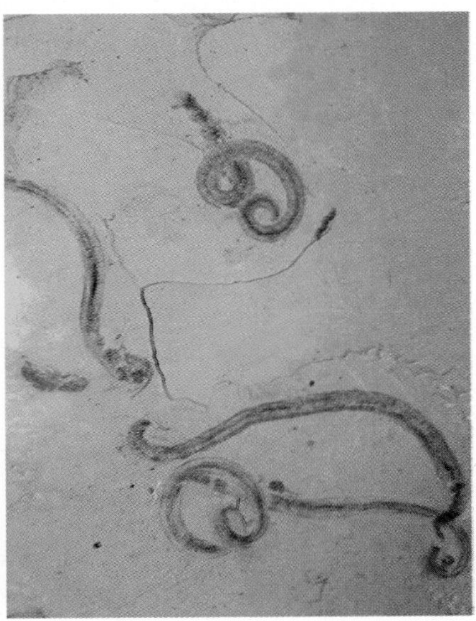

PLATES 6 & 7. Priapulid worms of the species *Maotianshania cylindrica. Courtesy Paul Chien and Illustra Media.*

PLATE 8 *(above)*. A lobopodian "worm with legs" of the genus *Microdictyon*.

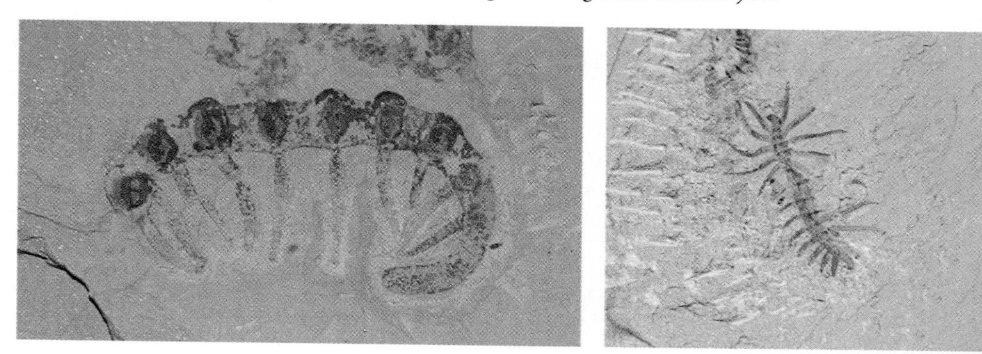

PLATE 9 *(above, left)*. Another lobopod worm. Species: *Microdictyon sinicum*.
PLATE 10 *(above, right)*. *Luolishania longicruris,* a rare lobopod of the Maotianshan Shale.
PLATE 11 *(below)*. The extinct and enigmatic *Vetulicolia*, thought to represent either an arthropod, a tunicate-like urochordate, or a distinct phylum.
Courtesy Illustra Media (all).

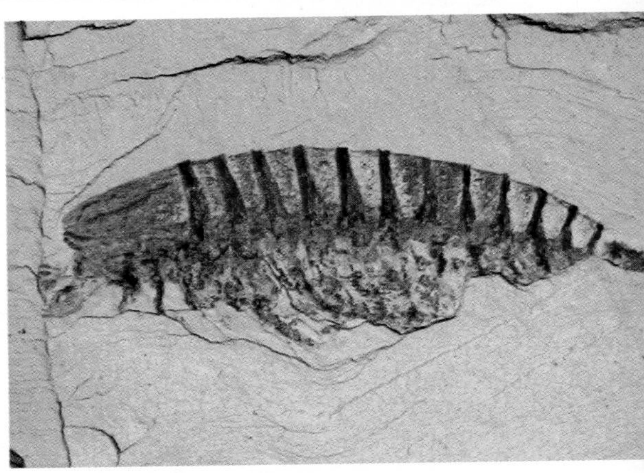

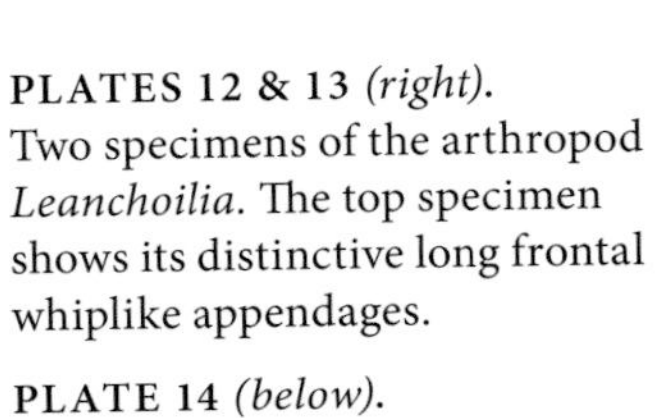

PLATES 12 & 13 *(right).*
Two specimens of the arthropod *Leanchoilia.* The top specimen shows its distinctive long frontal whiplike appendages.

PLATE 14 *(below).*
Chengjiangocaris longiformis (*below*), a rare arthropod known only from the Chengjiang biota. *Courtesy Illustra Media (all).*

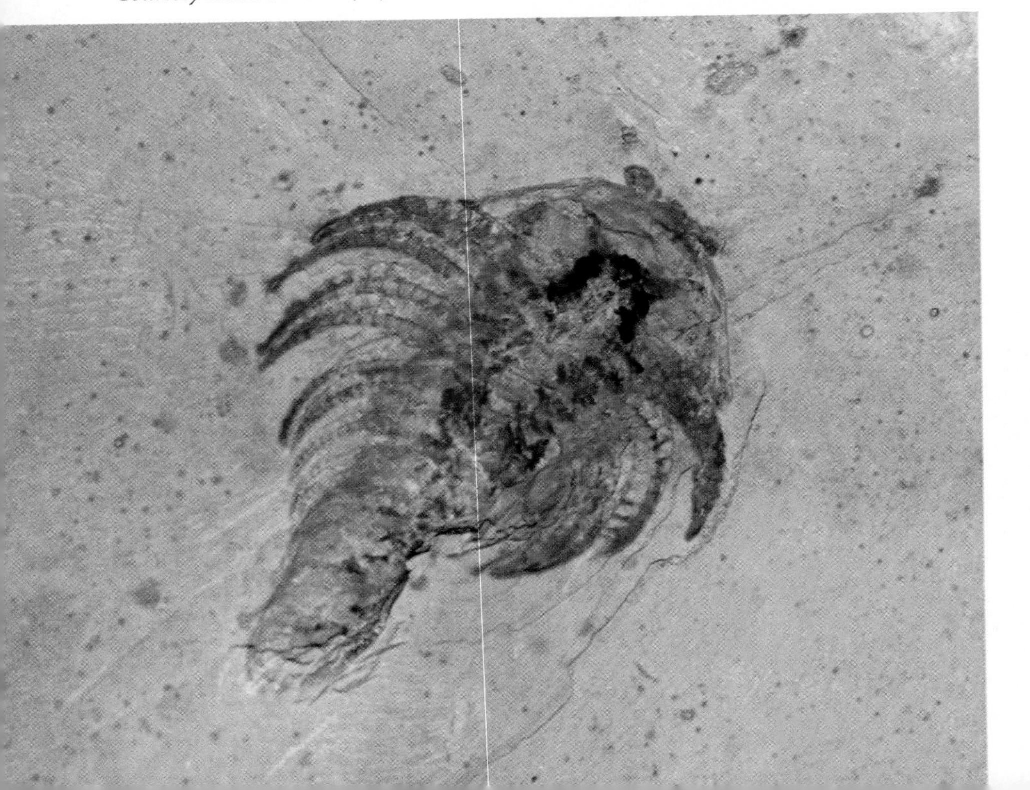

Two fossils from the dark-colored Burgess Shale, near the town of Field in British Columbia, Canada. Note the contrast in color to the fossils from the honey-colored Maotianshan Shale depicted on the previous and subsequent pages.

PLATE 15 *(above).* A fossil of the arthropod *Marrella* from the Burgess Shale. *Courtesy Illustra Media.*

PLATE 16 *(below).* The shrimp-like arthropod *Waptia. Courtesy Illustra Media.*

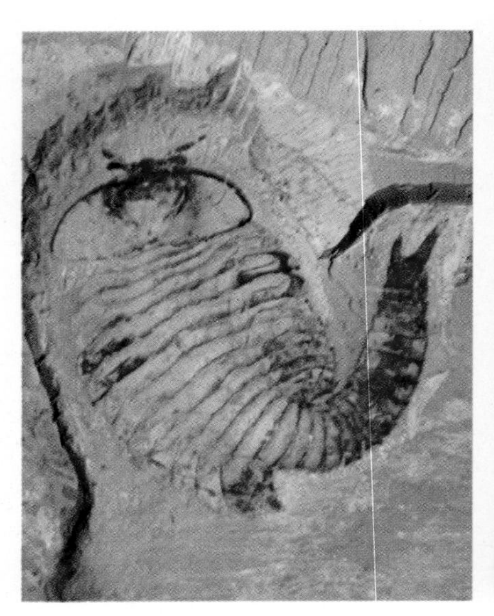

PLATE 17. The arthropod *Fuxianhuia protensa.* Note its exquisitely preserved eyes, antennae, segmented body, and forked tail. *Courtesy Illustra Media.*

PLATE 18. A fossil of the arthropod *Waptia,* from the honey-colored Maotianshan Shale. *Courtesy Paul Chien.*

PLATE 19. A fossil of an *Anomalocaris,* whose large eyes and jaws made it the fiercest predator in the Cambrian seas. Some reached a length of one meter. *Courtesy J. Y. Chen.*

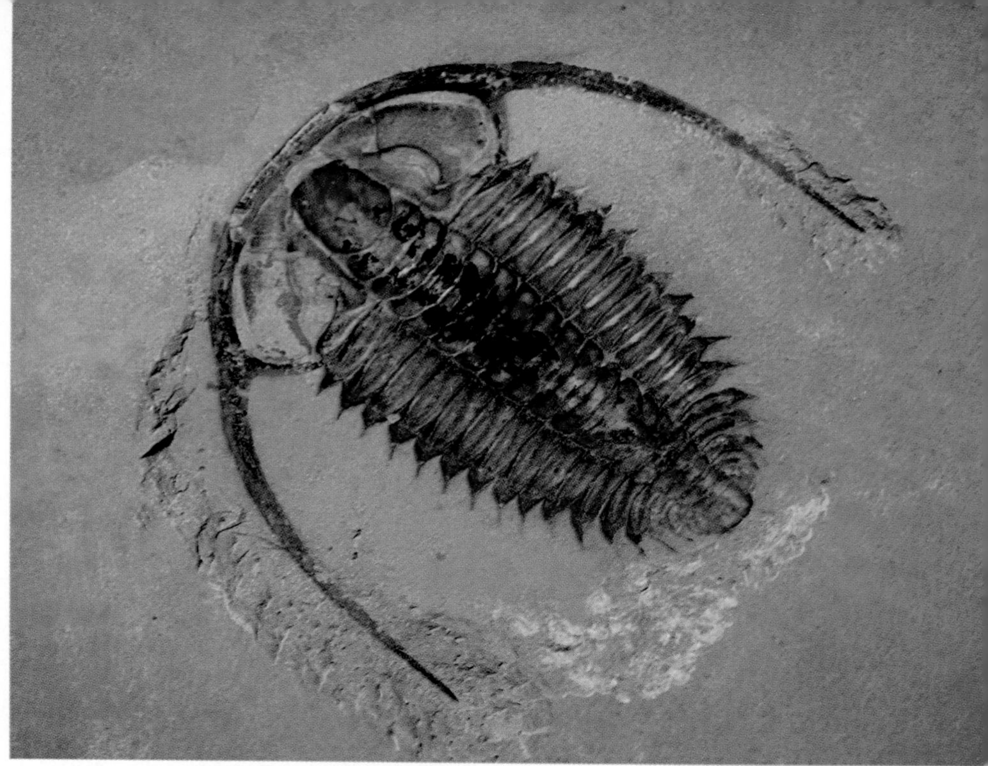

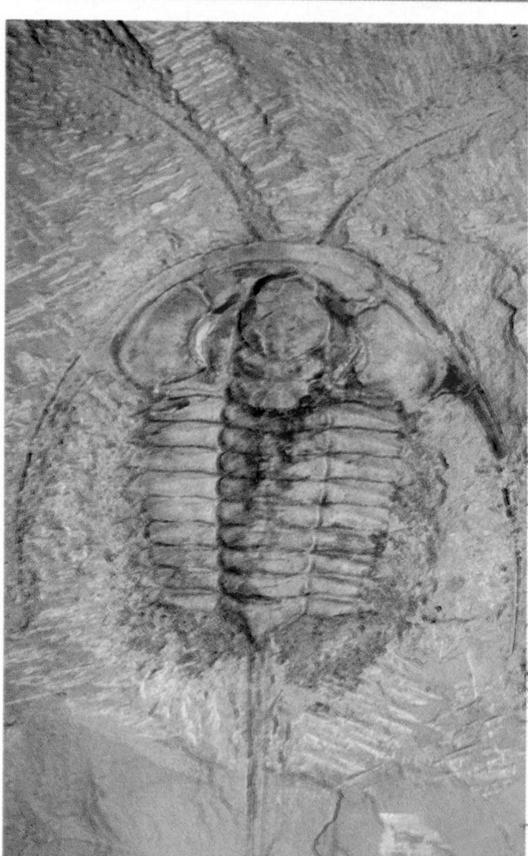

PLATES 20–22. A montage of beautifully preserved trilobites from the Maotianshan Shale, including a specimen of *Eoredlichia intermedia* (*left*). *Courtesy Illustra Media (all).*

PLATE 23. A specimen of the trilobite *Kuanyangia pustulosa* from the Maotianshan Shale near Chengjiang, China. *Courtesy Illustra Media.*

TO BUILD AN ANIMAL

Yet Axe's calculations only hint at the full problem for neo-Darwinian theory. By bending over backwards not to overstate the improbability of generating a new protein fold and by focusing narrowly on that one aspect of the challenge confronting evolutionary theory, his figures vastly *understate* the improbability of building a Cambrian animal. There are several reasons for this.

First, the Cambrian explosion as dated by fossil evidence took far less time than has elapsed since the origin of life on earth until the present (about 3.8 billion years).[20] Less time available for a given evolutionary transition means fewer generations of new organisms and fewer opportunities to generate new genes by which to search relevant sequence space. This makes it even harder to generate a new protein fold by chance in the relevant time period.

Second, bacteria are by far the most common type of organisms included in Axe's estimate of the total number of organisms that have lived on earth. Yet no one thinks that Cambrian animals evolved directly from bacteria. Nor does anyone think that the putative multicellular ancestors of the Cambrian forms would have been anywhere near as abundant as the bacterial populations that Axe used as the main basis of his estimate. A more realistic estimate for the number of possible animal ancestors would necessarily result in a much lower estimate for the number of gene sequences available for searching sequence space (corresponding to a single protein of modest length in a single Cambrian animal). Recall that, based on Axe's estimates, the probability of generating just one gene (for a new functional protein fold) from all the bacteria (and other organisms) that have ever lived on earth is just 1 in 10 trillion, trillion, trillion. Consequently, the mechanism for searching sequence space that Axe determined to be extremely implausible must be judged far more implausible as a mechanism for producing the Cambrian information explosion, since there were far fewer multicellular organisms present in the Precambrian period than there have been total organisms present in the entire history of life.

Third, building new animal forms requires generating far more than just one protein of modest length. New Cambrian animals would have required proteins much longer than 150 amino acids to perform necessary, specialized functions.[21] For example, as previously noted, many of these Cambrian animals needed the complex protein lysyl oxidase to support their stout body structures. In addition to a novel protein fold, these molecules (in living organisms) comprise over 400 precisely sequenced (nonrepeating) amino acids. Reasonable extrapolation from mutagenesis experiments done on shorter protein molecules suggests that the *im*probability of randomly producing functionally sequenced proteins of this length would be extremely unlikely to occur given the probabilistic resources (and duration) of the entire universe.[22]

The mutation and selection mechanism faces a related obstacle. The Cambrian animals exhibit structures that would have required many new *types* of cells, each requiring many novel proteins to perform their specialized functions. But new cell types require not just one or two new proteins, but coordinated systems of proteins to perform their distinctive cellular functions. The unit of selection in such cases ascends to the system as a whole. Natural selection selects for functional advantage, but no advantage accrues from a new cell type until a system of servicing proteins is in place. But that means random mutations must, again, do the work of information generation without the help of natural selection—and, now, not simply for one protein, but for many proteins arising together. Yet the odds of this occurring by chance alone are, of course, far smaller than the odds of the chance origin of a single new gene or protein—so small as to render the chance origin of the information needed to build a new cell type fantastically improbable (and implausible) given even the most optimistic estimates for the length of the Cambrian explosion.

Richard Dawkins has noted that scientific theories can rely on only so much "luck" before they cease to be credible.[23] But the second scenario, involving gene duplication and neutral evolution, by its own logic, precludes natural selection from playing a role in generating genetic information until after the fact. Thus, it relies entirely on "too much luck." The sensitivity of proteins to functional loss, the rarity of proteins within combinatorial sequence space, the need for long proteins to build new cell types and animals, the need for whole new *systems* of proteins to service new cell types, and the brevity of the Cambrian explosion relative to rates of mutation—all conspire to underscore the immense implausibility of any scenario for the origin of Cambrian genetic information that relies upon random variation alone, unassisted by natural selection.

Yet the classical model of gene evolution—which relies on neutral evolution—requires novel genes and proteins to arise, precisely, by random mutation alone. Adaptive advantage accrues *after* the generation of new functional genes and proteins. Natural selection cannot play a role *until* new functional information-bearing molecules have independently arisen. Thus, to return to Dawkins's imagery, evolutionary theorists envisioned the need to scale the steep face of a precipice of which there is effectively *no* gradually sloping back side, since the smallest increment of structural innovation in the history of life—a new protein fold—itself presents a formidable Mount Improbable.

By the way, Axe's later experiments establishing the extreme rarity of protein folds in sequence space also show *why* random changes to existing genes inevitably efface or destroy function before they generate fundamentally new folds or functions (scenario one). If only one out of every 10^{77} of the alternate sequences are functional, an evolving gene will *inevitably* wander down an evolutionary dead-end long before it can ever become a gene capable of producing a new protein fold. The extreme rarity of protein folds also entails their *isolation* from each other in sequence space.

A CATCH-22

Douglas Axe's results highlight an acute dilemma for neo-Darwinism, a "catch-22." On the one hand, if natural selection plays no role in generating new genes, as the idea of neutral evolution implies, then mutations alone must climb a Mount Improbable in a single leap—a situation that, given Axe's results and Dawkins's own logic, is probabilistically untenable. On the other hand, any model for the origin of genetic information that envisions a significant role for natural selection, by assuming a preexisting gene or protein under selective pressure, encounters other equally intractable difficulties. The evolving genes and proteins will range over a series of disadvantageous or nonfunctional intermediates that natural selection will not favor or preserve, but will, instead, eliminate. At that point, selection-driven evolution will cease, locking existing genes and proteins into place.

Thus, whether one envisions the evolutionary process beginning with a preexisting functional gene or a duplicated non-coding region of the genome, the results of mutagenesis experiments present a precise quantitative challenge to the efficacy of the neo-Darwinian mechanism. Indeed, our growing knowledge about the rarity and isolation of proteins and functional genes in sequence space implies that neither neo-Darwinian scenario for producing new genes is at all plausible. Thus, neo-Darwinism does not explain the Cambrian information explosion.

11

ASSUME A GENE

When I first heard that Douglas Axe had succeeded in making a rigorous estimate of the rarity of proteins in sequence space, I wondered what neo-Darwinists would say in response. Given the experimental rigor and mathematical precision of the work he reported in the *Journal of Molecular Biology* in 2004, and the long odds against mutation and selection ever finding a novel gene or functional protein, what *could* they say? That the probability of a successful search for new genes and proteins was higher than Axe's experiments suggested? That his methods or calculations were flawed? That no one else had gotten similar results? Since Axe's work confirmed other analyses and experiments, and since his paper had passed through the careful scrutiny of peer review, none of those responses seemed plausible. Yet defenders of the adequacy of the neo-Darwinian mechanism were far from admitting defeat, as I would soon find out.

The same year, I published a peer-reviewed scientific article about the Cambrian explosion and the problem of the origin of the biological information needed to explain it.[1] In the paper, I cited Axe's results and explained why the rarity of functional proteins in sequence space posed such a severe challenge to the adequacy of the neo-Darwinian mechanism. The article appeared in a biology journal, *Proceedings of the Biological Society of Washington,* published out of the Smithsonian Institution by scientists working for the Smithsonian's National Museum of Natural History (NMNH). Because the article also argued that the theory of intelligent design could help explain the origin of biological information (see Chapter 18), its publication created a firestorm of controversy.

Museum scientists and evolutionary biologists from around the country were furious with the journal and its editor, Richard Sternberg, for allowing the article to be peer-reviewed and then published. Recriminations followed. Museum officials took away Sternberg's keys, his office, and his access to scientific samples. He was transferred from a friendly to a hostile supervisor. A congressional subcommittee staff later investigated and found that museum officials initiated an intentional disinformation campaign against Sternberg in an attempt to get him to resign. His detractors circulated false rumors: "Sternberg has no degrees in biology" (actually he has two Ph.D.'s, one in evolutionary biology and one in systems biology); "He is a priest, not a scientist" (Sternberg is not a priest, but a research scientist); "He is a Republican operative working for the Bush campaign" (he was far too busy doing scientific research to be involved in political campaigns, Republican or otherwise); "He's taken money to publish the article" (not true); and so on. Eventually, despite the demonstrable falsehood of the charges, he was demoted.[2]

Major news stories about the controversy appeared in *Science, Nature, The Scientist,* and the *Chronicle of Higher Education*.[3] Then articles appeared in the mainstream press, including the *Washington Post* and the *Wall Street Journal*.[4] A major story aired on National Public Radio.[5] Sternberg himself even appeared on *The O'Reilly Factor.*

Despite the intense furor, there was no formal scientific response to my article: neither the *Proceedings* nor any other scientific journal published a scientific refutation. The members of the Council of the Biological Society of Washington who oversaw the publication of the journal insisted that they didn't want to dignify it by responding.

Eventually two scientists and a science education policy advocate—each associated with the National Center for Science Education, a group that lobbies for teaching evolution in the public schools—stepped forward. The three authors—geologist Alan Gishlick, education policy advocate Nicholas Matzke, and wildlife biologist Wesley R. Elsberry—published a response to my article on TalkReason.org, a prominent atheistic website.[6] Although the website's guidelines prohibit "ad hominem arguments," the rule was somewhat loosely enforced in the case of Gishlick, Matzke, and Elsberry's response, which they titled "Meyer's Hopeless Monster."

Gishlick, Matzke, and Elsberry attempted to refute my central argument by citing a scientific paper that they said solved the problem of the origin of genetic information. The paper, a scientific review essay titled "The Origin of New Genes: Glimpses from the Young and Old," had appeared in *Nature Reviews Genetics* in 2003. Gishlick, Matzke, and Elsberry asserted that this paper—coauthored by Manyuan Long, an evolutionary biologist at the University of Chicago, and several colleagues—was representative of an extensive "scientific literature documenting the origin of new genes."[7]

Other biologists echoed Gishlick, Matzke, and Elsberry's claim in the context of another public controversy. During the 2005 *Kitzmiller v. Dover* trial about an ill-advised attempt to require teachers in a Pennsylvania school district to read a statement about intelligent design, Brown University biologist Kenneth Miller cited Long's paper in his testimony. He said that it shows how new genetic information evolves. The judge in the case, John E. Jones, then cited Miller's testimony about Long's article in his own decision. Judge Jones asserted there are "more than three dozen peer-reviewed scientific publications showing the origin of new genetic information by evolutionary processes."[8] Elsewhere Matzke, along with biologist Paul Gross, stated that the paper by Long "reviews all the mutational processes involved in the origin of new genes and then lists dozens of examples in which research groups have reconstructed the genes' origins."[9] In their view, "Competent scientists know how new genetic information arises."[10]

But do evolutionary biologists really know this?

Let's take a closer look at the article that allegedly shows "how new genetic information arises."[11]

ONCE UPON A GENE

The oft-cited Long paper points to a variety of studies that purport to explain the evolution of various genes. These studies typically begin by taking a gene and then seeking to find other genes that are similar (or homologous) to it. They then seek to trace the history of slightly different homologous genes back to a hypothetical common ancestor gene (or genes). To do this, the studies survey databases of gene sequences looking for similar sequences in representatives of different taxonomic groups—often in closely related species. Some studies also attempt to establish the existence of a common ancestor gene on the basis of similar genes within the very same organism. They then typically propose evolutionary scenarios in which an ancestral gene duplicates itself,[12] and then the duplicate and the original evolve differently as the result of subsequent mutations in each gene.

Next, these scenarios invoke various kinds of mutations—duplication events, exon shuffling, retropositioning, lateral gene transfer, and subsequent point mutations—as well as the activity of natural selection (see Fig. 11.1). The evolutionary biologists conducting these studies *postulate* that modern genes arose as the result of these various mutational processes—processes that they envision as having shaped genes during a long evolutionary history. Since the information in modern genes is presumably different from the information in the hypothetical ancestor genes, they regard the mutational mechanisms that are allegedly responsible for these differences as the explanation for the *origin* of genetic information.

Upon closer examination, however, none of these papers demonstrate *how* mutations and natural selection could find truly novel genes or proteins in sequence space in the first place; nor do they show that it is reasonably probable (or plausible) that these mechanisms would do so in the time available. These papers *assume* the existence of significant amounts of *preexisting* genetic information (indeed, many whole and unique genes) and then *suggest* various mechanisms that might have slightly altered or fused these genes together into larger composites. At best, these scenarios "trace" the history of preexisting genes, rather than *explain* the origin of the original genes themselves (see Fig. 11.2).

This kind of scenario building can suggest potentially fruitful avenues of research. But an obvious error comes in mistaking a hypothetical scenario for either a demonstration of fact or an adequate explanation. None of the scenarios that the Long paper cites demonstrate the mathematical or experimental plausibility of the mutational mechanisms they assert as explanations for the origin of genes. Nor do they directly observe the presumed mutational processes in action. At best, they provide hypothetical, after-the-fact reconstructions of a few events out of a sequence of many supposed events, starting with the existence of a presumed common ancestor gene. But that gene itself does not represent a hard data point. It is inferred to have existed on the basis of the similarity of two or more other

FIGURE 11.1
Various types of mutations that are alleged to result in the modification of genes: exon shuffling, retropositioning, lateral gene transfer, and gene fusion.

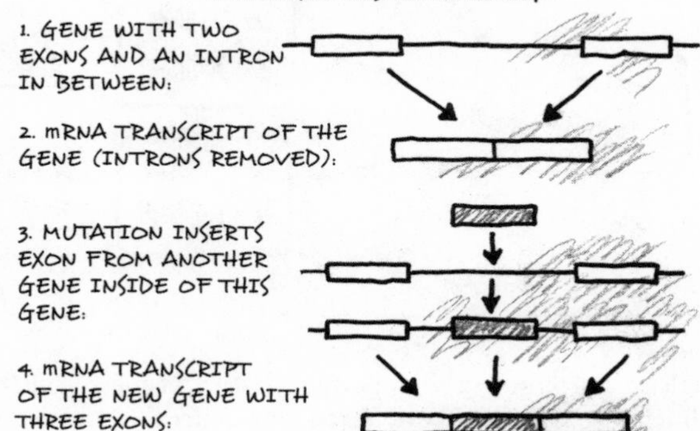

B. RETROPOSITIONING OF MESSENGER RNA TRANSCRIPT

1. TWO GENES, GENE A, AND GENE B, ARE NEXT TO ONE ANOTHER ON A CHROMOSOME:

2. THE mRNA TRANSCRIPT OF ANOTHER GENE, GENE C, IS RETROTRANSPOSED BETWEEN GENES A AND B:

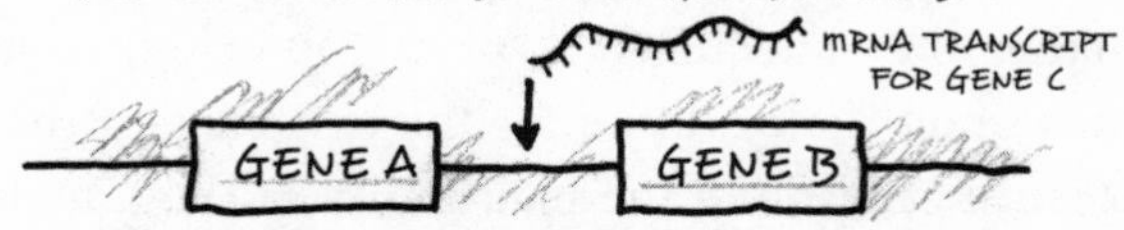

3. GENE C IS NOW INSERTED BETWEEN GENES A AND B ON THE CHROMOSOME.

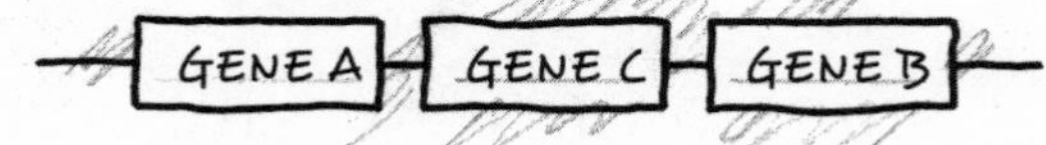

C. LATERAL GENE TRANSFER

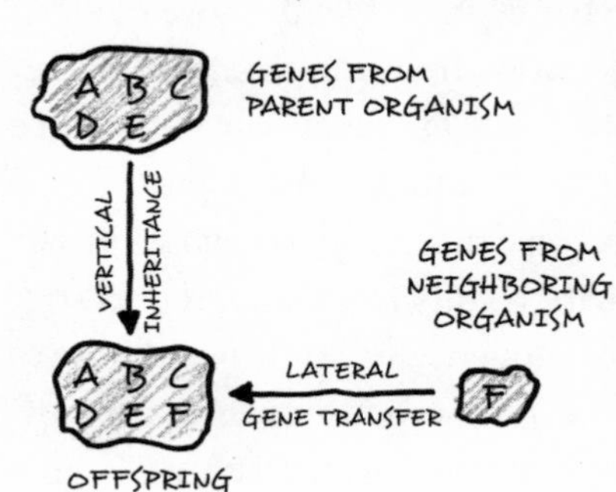

D. GENE FUSION (ONE MECHANISM)

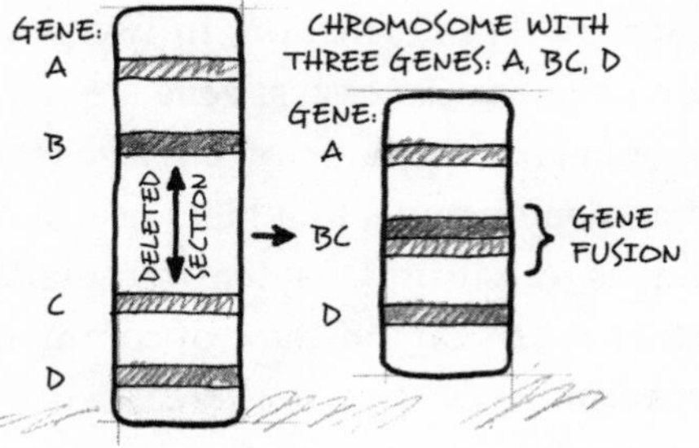

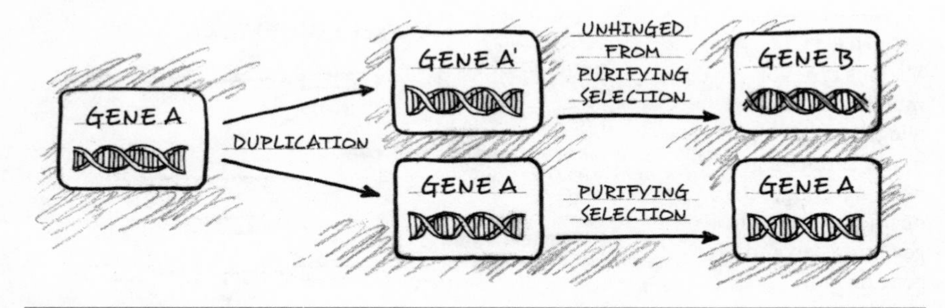

FIGURE 11.2
Depiction of how gene duplication and subsequent gene evolution might take place. While the gene on the bottom remains under selective pressure and cannot experience many mutations without loss of fitness or function (see Figure 10.3), the duplicated gene at the top can in theory vary without deleterious consequences to the organism.

existing genes, which are the only actual pieces of observational evidence upon which these often elaborate scenarios are based.

That these scenarios depend on various inferences and postulations doesn't, by itself, disqualify them from consideration. Nevertheless, whether they adequately explain the origin of genetic information depends upon the evidence for the existence of the entities they infer (the ancestral genes) and the plausibility of the mutational mechanisms they postulate. Let's look at both parts of these scenarios.

COMMON ANCESTOR GENES?

Nearly all of the scenarios developed in the papers that Long cites start with an inferred common ancestral gene from which two or more modern genes diverged and developed. These scenarios treat the similarity of sequence (the information) in two or more genes as unequivocal evidence for a common ancestral gene (see Fig. 11.2). As I noted in Chapters 5 and 6, standard methods of phylogenetic reconstruction *presuppose,* rather than demonstrate, that biological similarity results from shared ancestry. Yet, as we saw in Chapter 6, similarity of sequence by itself is not always an unequivocal indicator of common ancestry. Sometimes similarity appears between species where it cannot be explained by inheritance from a common ancestor (e.g., the similar forelimbs on moles and mole crickets) and, at the very least, there are other possible explanations for sequence similarity.

In the first place, similar gene sequences might have evolved independently on two parallel lines of descent starting with two different genes, as the hypothesis of convergent evolution asserts. Recent examples of convergent *genetic* evolution now abound in the literature of molecular and evolutionary biology.[13] For example, molecular biologists have discovered that both whales and bats use similar systems—involving similar genes and proteins—for echolocation. The striking similarity of these systems used in two otherwise disparate mammalian species has led biologists to posit the parallel evolution of echolocation, including the gene sequences and proteins that make it possible, from a common ancestor that did not possess this system.[14]

In addition, it is possible that similar genes might have been separately *designed* to meet similar functional needs in different organismal contexts. Viewed this way, similarity of sequence does not necessarily reflect descent with modification from a common ancestor, but could reflect *design* in accord with *common functional* considerations, constraints, or goals. I recognize, of course, that to this point I have not given any independent reasons for considering the design hypothesis, and that, as a hypothesis for sequence similarity by itself, intelligent design may not seem compelling. (For more compelling reasons to consider intelligent design, see Chapters 17 through 19.) Nevertheless, I mention both these other possible explanations for the similarity of gene sequences in order to demonstrate that sequence similarity does not *necessarily* indicate, or derive from, a common ancestral gene.

ORFAN GENES

Some genes, and the information-rich sequences they contain, most certainly cannot be explained by reference to the kind of scenarios that Long cites. All of these scenarios attempt to explain the origin of two similar genes by reference to descent with modification (via mutation) from common ancestral genes. Yet genomic studies are now turning up hundreds of thousands of genes in many diverse organisms that exhibit no significant similarity in sequence to any other known gene.[15] These "taxonomically restricted genes" or "ORFans" (for "open reading frames of unknown origin") now dot the phylogenetic landscape. ORFans have turned up in every major group of organisms, including plants and animals as well as both eukaryotic and prokaryotic one-celled organisms. In some organisms, as much as one-half of the entire genome comprises ORFan genes.[16]

Thus, even if it could be assumed that similar gene sequences always point to a common ancestor gene, these ORFan genes cannot be explained using the kind of scenarios that Long's article cites. Since ORFans lack sequence similarity to any known gene—that is, they have no known homologs in even distantly related species—it is impossible to posit a common ancestral gene from which a particular ORFan and its homolog might have evolved. Remember: ORFans, *by definition,* have no homologs. These genes are unique—one of a kind—a fact tacitly acknowledged by the increasing number of evolutionary biologists who attempt to "explain" the origin of such genes through *de novo* ("out of nowhere") origination.

Some might argue that as biologists map the sequence of more genomes and add more gene sequences to protein databases, homologs of these ORFans will eventually turn up, thus gradually eliminating the mystery surrounding the ORFan phenomenon. Yet to date the trend has gone in the opposite direction. As scientists have explored and sequenced more genomes, they have discovered more and more ORFans without finding anything like a corresponding number of homologs. Instead, the number of "unpaired" ORFan genes continues to grow with no sign of the trend reversing itself.[17]

THE PLAUSIBILITY OF THE MUTATIONAL PROCESSES

Even if evolutionary biologists could establish the existence of the common ancestral genes from which their scenarios begin, that would not establish the plausibility of a neo-Darwinian mechanism for generating genetic information from that ancestor. Moreover, the term "plausibility" in this context has a specific scientific and methodologically significant meaning. Studies in the philosophy of science show that successful explanations in historical sciences such as evolutionary biology need to provide "causally adequate" explanations—that is, explanations that cite a cause or mechanism capable of producing the effect in question. In *On the Origin of*

Species, Darwin repeatedly attempted to show that his theory satisfied this criterion, which was then called the *vera causa* (or "true cause") criterion. In the third chapter of the *Origin,* for example, he sought to demonstrate the causal adequacy of natural selection by drawing analogies between it and the power of animal breeding and by extrapolating from observed instances of small-scale evolutionary change over short periods of time.

In this, Darwin hewed to a principle of scientific reasoning that one of his scientific role models, the great geologist Charles Lyell, used as a guide for reasoning about events in the remote past. Lyell insisted that good explanations for the origin of geological features should cite "causes now in operation"—causes known from present experience to have the capacity to produce the effects under study.[18]

Do the scenarios developed by various evolutionary biologists cited in the Long review essay meet this criterion? Duplication mutations and various other modes of random mutational change along with natural selection clearly constitute "causes now in operation." No one disputes that. But have these processes demonstrated the capacity to produce the effect in question, namely, the genetic information necessary to structural innovation in the history of life? There are several good reasons to think that they have not.

BEGGING QUESTIONS

First, most of the mutational processes that evolutionary biologists invoke in the scenarios cited in the Long essay presuppose significant amounts of *preexisting* genetic information on *preexisting* genes or modular sections of DNA or RNA. The Long essay highlights seven main mutational mechanisms at work in the sculpting of new genes: (1) exon shuffling, (2) gene duplication, (3) retropositioning of messenger RNA transcripts, (4) lateral gene transfer, (5) transfer of mobile genetic units or elements, (6) gene fission or fusion, and (7) *de novo* origination (see Fig. 11.1). Yet each of these mechanisms, with the exception of *de novo* generation, begins with preexisting genes or extensive sections of genetic text. This preexisting functionally specified information is in some cases enough to code for the construction of an entire protein or a distinct protein fold. Moreover, these scenarios not only assume unexplained preexisting sources of biological information, they do so *without explaining or even attempting to explain* how any of the mechanisms they envision could have solved the combinatorial search problem described in Chapters 9 and 10.

A closer look at each of these mechanisms will show why scenarios that rely on them beg important questions about the origin of genetic information.

Advocates of exon shuffling envision modular sections of a genome randomly arranging and rearranging themselves to generate entirely new genes, not unlike rearranging whole paragraphs in an essay to generate a new article. In genomes that have regions that code for the production of proteins interspersed with regions that do not code for proteins, the term "exon" refers to a protein-coding region of the genome. These protein-coding regions of the genome are often interrupted by nonprotein-coding sections of the genome (called introns) that serve other functions, such as coding for the production of regulatory RNAs. In any case, exons store significant quantities of preexisting functionally specified information.

Though most proteins are encoded by multiple exons, a single exon may encode a substantial unit of protein structure, such as a functional protein fold—a fact that advocates of exon shuffling count in their own attempts to explain novel proteins. They assume that exons can be blindly shuffled and mixed around to form genes. Nevertheless, this mechanism cannot produce new protein folds. Either an exon is large enough that it already encodes a protein fold—in which case it's not creating a *new* fold—or it's too small, small enough that multiple exons must be combined in order to form a stable protein fold. In this latter case, other problems—in particular, something called adverse side-chain interaction—will preclude success, as we will see.

Evolutionary scenarios envisioning other mutational mechanisms also presuppose important sources of preexistent genetic information. Gene duplication, as the name implies, involves the production of a duplicate copy of a *preexisting* gene, already rich in functionally specified information. Retropositioning of messenger RNA transcripts occurs when an enzyme called reverse transcriptase takes a preexisting strand of messenger RNA and inserts its corresponding DNA sequence into a genome, also producing a duplicate of the coding portion of a preexisting gene. Lateral gene transfer involves transferring a preexisting gene from one organism (usually a bacterium) into the genome of another. The transfer of mobile genetic elements likewise occurs when preexisting genes enclosed in circular strands of DNA called plasmids enter one organism from another and eventually find themselves incorporated into a new genome. This process also mainly occurs in single-celled organisms. A similar process can occur in eukaryotes, where mobile genetic elements called transposons—often called "jumping genes"[19]—can hop from place to place in the genome. Gene fusion occurs when two adjacent preexisting genes, each rich with specified genetic information, link together after the deletion of intervening genetic material."[20]

Each of these six mutational mechanisms presupposes preexisting modules of specified genetic information. Some of these mutational mechanisms also depend upon sophisticated preexistent molecular machines such as the reverse transcriptase enzyme used in retropositioning or other complex cellular machinery involved in DNA replication. Since building these machines requires other sources of genetic information, scenarios that presuppose the availability of such molecular machines to assist in the cutting, splicing, or positioning of modular sections of genetic information clearly beg the question.

Overall, what evolutionary biologists have in mind is something like trying to produce a new book by copying the pages of an existing book (gene duplication, lateral gene transfer, and transfer of mobile genetic elements), rearranging blocks of text on each page (exon shuffling, retropositioning, and gene fusion), making random spelling changes to words in each block of text (point mutations), and then randomly rearranging the new pages. Clearly, such random rearrangements and changes will have no realistic chance of generating a literary masterpiece, let alone a coherent read. That is to say, these processes will not likely generate specificity of arrangement and sequence and, therefore, do not solve the combinatorial search problem. In any case, all such scenarios also beg the question. There is a big difference between shuffling and slightly altering preexisting sequence-specific modules of functional information and explaining how those modules came to possess information-rich sequences in the first place.

EVOLUTION EX NIHILO?

Long does cite at least one type of mutation that does not presuppose existing genetic information, the *de novo* origination of new genes. For example, one paper he discusses sought to explain the origin of a promoter region for a gene (the part of the gene that helps initiate the transcription of the gene's instructions) and found that "this unusual regulatory region did not really 'evolve.'" Instead, it somehow snapped into being: "It was aboriginal, created de novo by the fortuitous juxtaposition of suitable sequences."[21]

Many other papers invoke *de novo* origination of genes. Long mentions, for example, a study seeking to explain the origin of an antifreeze protein in an Antarctic fish that cites "*de novo* amplification of a short DNA sequence to spawn a novel protein with a new function."[22] Likewise, Long cites an article in *Science* to explain the origin of two human genes involved in neurodevelopment that appealed to "*de novo* generation of building blocks—single genes or gene segments coding for protein domains," where an exon spontaneously "originated from a unique noncoding sequence."[23] Other papers make similar appeals. A paper in 2009 reported "the *de novo* origin of at least three human protein-coding genes since the divergence with chimp[s]," where each of them "has no protein-coding homologs in any other genome."[24] An even more recent paper in *PLoS Genetics* reported "60 new protein-coding genes that originated *de novo* on the human lineage since divergence from the chimpanzee,"[25] a finding that was called "a lot higher than a previous, admittedly conservative, estimate."[26]

Another 2009 paper in the journal *Genome Research* was appropriately titled "Darwinian Alchemy: Human Genes from Noncoding RNA." It investigated the *de novo* origin of genes and acknowledged, "The emergence of complete, functional genes—with promoters, open reading frames (ORFs), and functional proteins—from 'junk' DNA would seem highly improbable, almost like the elusive transmutation of lead into gold that was sought by medieval alchemists."[27] Nonetheless, the article asserted without saying *how* that: "evolution by natural selection can forge completely new functional elements from apparently nonfunctional DNA—the process by which molecular evolution turns lead into gold."[28]

The presence of unique gene sequences forces researchers to invoke *de novo* origin of genes more often than they would like. After one study of fruit flies reported that "as many as ~12% of newly emerged genes in the *Drosophila melanogaster* subgroup may have arisen *de novo* from noncoding DNA,"[29] the author went on to acknowledge that invoking this "mechanism" poses a severe problem for evolutionary theory, since it doesn't really explain the origin of any of its "nontrivial requirements for functionality."[30] The author proposes that "preadaptation" might have played some role. But that adds nothing by way of explanation, since it only specifies when (before selection played a role) and where (in noncoding DNA), not how the genes in question first arose. Details about how the gene became "preadapted" for some future function is never explained. Indeed, evolutionary biologists typically use the term "*de novo* origination" to describe *unexplained* increases in genetic information; it does not refer to any known mutational process.

Taking stock, then, many of the mutational processes that Long cites either: (1) beg the question as to the origin of the specified information contained in genes or parts of genes, or (2) invoke completely unexplained *de novo* jumps—essentially evolutionary creation *ex nihilo* ("from nothing").

Thus, ultimately, the scenarios featured in Long's review essay do not *explain* the origin of the specified information in either genes or sections of genes. That would require a cause capable of solving the combinatorial inflation problem discussed in the previous chapters. But none of the scenarios discussed in Long's article even addresses this problem, let alone demonstrates the mathematical plausibility of the mechanisms they cite. Yet, Gishlick, Matzke, and Elsberry originally cited Long as a definitive refutation of my article—the one in which I argued that the rarity of genes and proteins in sequence space cast doubt on the power of selection and mutations to generate novel genetic information. Professor Miller, in his testimony at the celebrated Dover trial, even convinced a federal judge to affirm that Long had succeeded in demonstrating how genetic information originates in a celebrated legal ruling. Clearly, one cannot solve a problem or refute an argument by failing to address it.[31]

PROTEIN FOLDS: PLAUSIBLE BUT IRRELEVANT SCENARIOS

There is a second and closely related difficulty associated with the scenarios cited by Long. Typically, they do not even try to explain the origin of new protein *folds,* and few of them analyze genes different enough from each other that their protein products could, even conceivably, exemplify different folds. Instead, they usually attempt to explain the origin of homologous genes—genes that produce proteins with the same folded structure performing the same function or a closely related one.

For example, Long cites one study comparing the two genes *RNASE1* and *RNASE1B,* which code for homologous digestive enzymes.[32] The two proteins perform nearly the same function: breaking down RNA molecules in the digestive tracts of colobine leaf-eating monkeys, though each does its job at a slightly different optimal chemical pH. More important, given that the amino-acid sequences of the two enzymes are 93 percent identical, structural biologists would expect both enzymes to utilize the same protein fold to accomplish their closely related tasks.

Long also references a study of a gene that codes for a histone protein, Cid, in two closely related species of fruit flies—*Drosophila melanogaster* and *Drosophila simulans*. The study didn't try to explain the "origin" of the gene—it merely compared the gene in the two species, catalogued some minor differences between them, and asked how those differences arose. The study identified some two dozen nucleotide differences between the genes for Cid in the two species—only 17 of which might have changed an amino acid in the sequence out of 226 total amino acids in the Cid protein.[33] Such a slight (7.5 percent) difference would be extremely unlikely to translate into different protein *folds*. Indeed, natural sequences known to have different folds do not have anything like the correspondingly high degree (92.5 percent) of sequence identity. Instead, known natural sequences with this high level of sequence identity have the same fold.

Long also cites two studies of *FOXP2*, a gene involved in regulating gene expression in humans, chimps, other primates, and mammals. In humans and other mammals, this gene is involved in brain development.[34] Nevertheless, according to one study, the protein coded for by this gene in humans acquired just "two amino-acid changes on the human lineage"[35] during the entire course of its evolution from a common chimp-human ancestor—again, not likely a sufficient enough change to generate a new protein fold.

The Long review essay cites numerous scenarios of this type—scenarios attempting to explain the evolution of slight gene variants (and their similar proteins), not the origin of new protein folds. This is an important distinction because, as we saw in Chapter 10, new protein folds represent the smallest unit of selectable structural innovation, and much larger structural innovations in the history of life depend upon them. Explaining the origin of structural innovation requires more than just explaining the origin of variant versions of the same gene and protein or even the origin of new genes capable of coding for new protein functions. It requires producing enough genetic information—truly novel genes—to produce new protein folds.

Thus, even where these scenarios are plausible, they are not *relevant* to explaining the origin of the genetic information necessary to produce the kind of structural innovation that occurs in the Cambrian explosion (or many other events in the history of life).

PROTEIN FOLDS: RELEVANT BUT IMPLAUSIBLE SCENARIOS

In a few cases, the evolutionary scenarios cited in the Long paper appear to be attempts to explain genes that are different enough from each other that they could conceivably code for proteins with different folds. For example, Long discusses several papers that equate exon shuffling with the shuffling of protein domains. Recall that a protein domain is a stable "tertiary" protein structure or fold made of many smaller "secondary" structures such as alpha helices or beta strands (see Fig. 10.2). Many complex proteins have numerous domains, each exhibiting a unique fold or tertiary structure. One version of the exon-shuffling hypothesis assumes that each exon codes for a specific protein domain. It envisions random cutting and splicing—excising, shuffling, and recombination—of the exon portions of the genome, resulting in the modular rearrangement of genetic information. The resulting composite gene will then code for a new composite protein structure. As Long proposes, "Exon shuffling, which is also known as domain shuffling, often recombines sequences that encode various protein domains to create mosaic proteins."[36]

Of the mechanisms that Long discusses, exon shuffling (and the closely related idea of gene fusion) provides perhaps the most plausible means of generating new (composite) proteins.[37] Nevertheless, the idea that exon shuffling can explain the origin of the genetic information necessary to produce new protein folds or whole composite proteins is problematic for several reasons.

First, the exon-shuffling hypothesis seems to assume that each exon involved in the process codes for a protein domain that folds into a distinctive tertiary structure. To a protein scientist, a protein domain is equivalent to a protein fold, though distinct protein structures (folds) may be composed of several smaller domains (or folds). Thus, at the very least, the exon-shuffling hypothesis presupposes the prior existence of a significant amount of genetic information—enough information to build an independent protein domain or fold. As such, it fails as an explanation for the *origin* of protein folds and the information necessary to produce them.

Some advocates of exon shuffling, however, may be using the term "protein domain" in a slightly fuzzier way. They might be equating domains with smaller structural units such as fragments of a fold made of several units of secondary structure such as alpha helices or beta strands. Conceived this way, the exon-shuffling hypothesis would then entail the construction of a new protein structure by combining these smaller "fragments."

But in most cases, if the amino-acid chain that forms a domain is chopped into fragments, then the resulting isolated pieces would cease to retain their original shapes. Why? Because the three-dimensional shape of one small section of a protein is heavily dependent upon the overall structure and shape of the rest of the protein. Snip out a section or fragment, or synthesize a fragment in isolation from the rest of the protein, and a floppy amino-acid chain will result—one that has entirely lost its original shape, or ability to form a stable structure. Thus, this version of the exon-shuffling hypothesis lacks credibility because it incorrectly assumes that shapeless protein fragments can be mixed and matched in a modular fashion to form new stable, functional protein folds. Moreover, even if such shuffling were physically plausible, this version of the hypothesis would have another problem. It still presupposes unexplained functional information—in particular, the information necessary to specify, not just the smaller fragments, but also the information required to *arrange* these smaller units into stable folds, and ultimately functional proteins.

Second, since the exon-shuffling hypothesis assumes that each exon involved in the shuffling codes for a specific protein domain, it also assumes that exon boundaries correspond to the boundaries of protein domains or folds. In existing genes, however, exon boundaries do not typically correspond to the boundaries of folded domains within the larger proteins.[38] If the shuffling of exons explained how actual proteins had come into existence, then there should be a clear correspondence or correlation between exon boundaries within genes and the corresponding protein domains within larger composite structures (i.e., whole proteins). The absence of such a correspondence suggests that exon shuffling does not account for the origin of known compound protein structures.

Third, relying on exon shuffling to cobble together a new protein fold from smaller units of protein structure is physically implausible for another reason. To see why, we need to examine what a "side chain" is. All twenty protein-forming amino acids have a common backbone (made of nitrogen, carbon, and oxygen), but each one has a different chemical group called a side chain sticking out at roughly right angles from that backbone. The interactions between side chains determine whether secondary units of protein structure made from chains of amino acids will fold into larger stable three-dimensional folds.[39] Though many different sequences will generate secondary structures (alpha helices and beta strands), generating stable folds is much more difficult and requires much more specificity in the arrangement of amino acids and their side chains. Specifically, since the elements in smaller secondary structural units in proteins are surrounded by side chains, they cannot be combined into new folds unless the elements have the sequence specificity required for the side chains to complement one another.[40] That means smaller secondary structural units will rarely[41] fuse together to form stable tertiary structures or folds. Instead, attempts to form new folds from smaller units of structure repeatedly encounter adverse interactions *between the side chains* of the amino acids within units of secondary structure.

The need for extreme specificity in the sequential arrangements of amino acids, discussed in the previous chapter, means that the overwhelming majority of amino-acid sequences in units of secondary structure will not result in stable folds as these units of structure come into contact with each other. As discussed in Chapter 10, the extreme rarity of functional proteins (with stable folds) in sequence space ensures that the probability of finding a correct fold-stabilizing sequence will be astonishingly small. For this reason, even skilled protein scientists have struggled to design sequences that will produce stable protein folds.[42] Almost invariably the units of secondary structure that they attempt to combine or otherwise place into stable composite structures will not fold because of the interactions of their amino-acid side chains.[43] As molecular biologist Ann Gauger explains, "Thus, [alpha] helices and [beta] sheets are *sequence-dependent* structural elements within protein folds. You can't swap them around like Lego bricks."[44]

Nor is it an easy matter to simply find different sequences of amino acids that will stabilize folds from smaller secondary units of structure, again, because of the extreme rarity of functional (and folding) sequences within amino-acid sequence space. Generating specific sequences that will fold into stable structures, whether in the lab or during the history of life, requires solving the combinatorial inflation problem. Even small folds will require five or six units of secondary structure with 10 or so amino acids in each unit, that is, 60 or more precisely sequenced amino acids. Modest-size folds will require a dozen or more units of secondary structure and 150 to 200 specifically arranged amino acids in order to stabilize a fold. Larger protein folds will require many more secondary units and specifically arranged amino acids. Since, however, many mission-critical functions within even the simplest cell require many folds (of at least 150 amino acids) working in close coordination, the need to produce proteins of at least this length numerous times through the history of life cannot be avoided.

All this requires searching for a functional needle in a vast haystack of combinatorial possibilities. Recall that Douglas Axe estimated the ratio of needles (functional sequences) to strands of straw in the haystack (nonfunctional sequences) to be 1 to 10^{77} for sequences of modest-length (150 amino acids).

Of course, in naturally occurring proteins, the interactions between side chains in units of secondary structure do maintain stable folds. But these proteins, with their stable three-dimensional folded structures, depend upon exceedingly rare and precisely arranged sequences of amino acids. The question is not whether the combinatorial search problem necessary to produce stable protein folds has ever been solved, but whether a neo-Darwinian mechanism relying on random mutations (in this case random shuffling of exons) provides a plausible explanation for *how* it might have been solved.

The papers that Long cites give no reason to think that exon shuffling (or any other mutational mechanism) has solved this problem. The exon-shuffling hypothesis ignores the need for side-chain specificity, though the need for such specificity has repeatedly defeated attempts in the laboratory to build new proteins from units of secondary structure in the manner required.

But advocates of exon shuffling make no attempt to show how random rearrangements of protein domains—whether the domains are conceived of as fragments of a fold or whole folds—would solve the combinatorial problem. Nor do they challenge Sauer's or Axe's experimentally derived quantitative estimates of the rarity of functional genes or proteins. They do not challenge the probability calculations based on these estimates. And they do not show that a mechanism exists that can search amino-acid sequence spaces more effectively or efficiently than random mutation and selection. Nor do they demonstrate the efficacy of exon shuffling in a model system in the laboratory. Instead, basic considerations of protein structure imply the *implausibility* of exon shuffling as a means of generating the genetic information necessary to produce a new protein fold.[45] So, in the end, with few words and with apparent confidence, advocates of exon shuffling simply assert, as the Long paper does, that "exon shuffling often recombines sequences that encode various protein domains to create mosaic proteins."

WORD SALAD

The assertion of Long and his colleagues about exon shuffling, like many other statements about postulated mutational mechanisms, blurs the distinction between theory and evidence. Despite the authoritative tone of such statements, evolutionary biologists rarely directly observe the mutational processes they envision. Instead, they see patterns of similarities and differences in genes and then attribute them to the processes they postulate. Yet the papers that Long cites offer neither mathematical demonstration, nor experimental evidence, of the power of these mechanisms to produce significant gains in biological information.

In the absence of such demonstrations, evolutionary biologists have taken to offering what one biologist I know calls "word salad"—jargon-laced descriptions of unobserved past events—some possible, perhaps, but none with the demonstrated capacity to generate the information necessary to produce novel forms of life. This genre of evolutionary literature envisions exons being "recruited"[46] and/or "donated"[47] from other genes or from an "unknown source"[48]; it appeals to "extensive refashioning"[49] of genes; it attributes "fortuitous juxtaposition of suitable sequences"[50] to mutations or "fortuitous acquisition"[51] of promoter elements; it assumes that "radical change in the structure" of a gene is due to "rapid, adaptive evolution";[52] it asserts that "positive selection has played an important role in the evolution"[53] of genes, even in cases when the function of the gene under study (and thus the trait being selected) is completely unknown;[54] it imagines genes being "cobbled together from DNA of no related function (or no function at all)";[55] it assumes the "creation" of new exons "from a unique noncoding genomic sequence that fortuitously evolved";[56] it invokes "the chimeric fusion of two genes";[57] it explains "near-identical"[58] proteins in disparate lineages as "a striking case of convergent evolution";[59] and when no source material for the evolution of a new gene can be identified, it asserts that "genes emerge and evolve very rapidly, generating copies that bear little similarity to their ancestral precursors" because they are apparently "hypermutable."[60] Finally, when all else fails, scenarios invoke the "*de novo* origination" of new genes, as if that phrase—any more than the others just mentioned—constitutes a scientific demonstration of the power of mutational mechanisms to produce significant amounts of new genetic information.[61]

These vague narratives resemble nothing so much as the naming games of scholastic philosophers in the Middle Ages. Why does opium put people to sleep? Because it has a "dormative" virtue. What causes new genes to evolve so rapidly? Their "hypermutability" or perhaps their ability to undergo "rapid, adaptive evolution." How do we explain the origin of two similar genes in two separate, but otherwise widely disparate lineages? Convergent evolution, of course. What is convergent evolution? The presence of two similar genes in two separate, but otherwise widely disparate lineages. How does convergent evolution occur, given the improbability of finding even one functional gene in sequence space, let alone the same gene arising twice independently? No one knows exactly, but perhaps it was a "fortuitous juxtaposition of suitable sequences," or "positive selection," or "*de novo* origination." Need to explain two similar genes in more closely related lineages? Try "gene duplication," or "chimerical gene fusion," or "retropositioning," or "extensive refashioning of the genome," or some other scientific-sounding combination of words.

The vagueness of these scenarios raises serious questions about how scientists could regard them as decisive demonstrations or refutations of anything—let alone refutations of the kind of experimentally based, mathematically precise challenges to mutation and selection described in the previous chapter.

So despite the official pronouncement of a federal judge and claims of extensive "scientific literature documenting the origin of new genes," evolutionary biologists have not demonstrated *how* new genetic information arises, at least not in amounts sufficient to build protein folds, the crucial units of biological innovation. Biologists have not solved the problem of combinatorial inflation or refuted the precise quantitative argument against the creative power of the selection and mutation mechanism presented in the previous chapter (or in my 2004 article). Nor has anyone provided a compelling refutation of Douglas Axe's assessment of the rarity of genes and proteins on which that argument is based.

In fairness, neo-Darwinian biologists have mathematical models of their own—models indicating to them that nearly unlimited evolutionary change can occur under the right conditions. The assumption that these models, which are based on the equations of population genetics, accurately represent how much evolution can occur has left many evolutionary biologists confident in the creative power of various mutational mechanisms. But should they be?

In the next chapter, I will take up this question. As I do, I'll explain why evolutionary biologists have been, heretofore, untroubled by mathematical challenges to neo-Darwinism. I'll also show why that has begun to change as new developments in molecular genetics have introduced another formidable mathematical challenge to the creative power of the neo-Darwinian mechanism—a challenge that arises *from within the neo-Darwinian framework* and raises yet new questions about the causal adequacy of the neo-Darwinian mechanism.

12

COMPLEX ADAPTATIONS AND THE NEO-DARWINIAN MATH

University of Illinois biologist Tom Frazzetta knew the textbook story as well as anyone. According to neo-Darwinian theory, organisms with all their complex systems came into existence via natural selection acting on randomly arising, small-scale variations and mutations. As Frazzetta understood, this evolutionary mechanism necessarily transforms organisms gradually, with modifications parceled into increments "as a sort of continuous change, where one structural condition melts gradually into another."[1]

Frazzetta had his doubts, however. As an expert in functional biomechanics—studying how animals actually work—he had dissected the skulls of rare snakes found only on the island of Mauritius, in the Indian Ocean. These snakes, called bolyerines, are boa-like but have an anatomical specialization found in no other vertebrate. Their maxilla, the tooth-bearing bone of the upper jaw, is divided into *two* segments, linked by a flexible joint and serviced by many specialized nerves, extra bones, tissues, and differently arranged ligaments. This unique trait allows the snakes to bend the front half of their upper jaw backwards when they attack prey (see Fig. 12.1).

Could this complex system of bones, joints, tissues, and ligaments have evolved gradually? "A movable joint dividing the maxilla into two segments," observed Frazzetta, "seems to have either a presence or absence, with no intermediate to connect the two conditions."[2] That is, either the maxilla occurs as one bone (as it does in every other vertebrate) or as two segments with all the accompanying joints, bones, ligaments, and tissues necessary to make it work, as it does in the bolyerine snakes. No intermediate condition—a broken maxilla with two pieces of bone lacking the necessary joints, tissues, and ligaments, for example—appears viable. As Stephen Jay Gould asked of the same system, "How can a jawbone be half broken?"[3] Or as Frazzetta himself observed, "I thus find it difficult to envision a smooth transition from a single maxilla to the divided condition seen in bolyerines."[4] Yet because the intermediate forms would not be viable, building a bolyerine jaw would require all the necessary parts—the jointed maxilla, the adjoining ligaments, and the necessary muscles and tissues—arising together.

Yet the problem for neo-Darwinian theory, Frazzetta realized, extended well beyond the anatomical peculiarities of rare snakes. As a young evolutionary biology professor, he had studied complex features in a wide variety of species. He knew that almost any biological structure of interest—the inner ear, the amniotic egg, eyes, olfactory organs, gills, lungs, feathers, the reproductive, circulatory, and respiratory systems—possesses multiple necessary components. To change such systems requires altering each of the many independent parts upon which their functions are based. This cannot be done willy-nilly. For example, changing any of the three bones of the mammalian inner ear—the incus, stapes, or malleus—will perforce

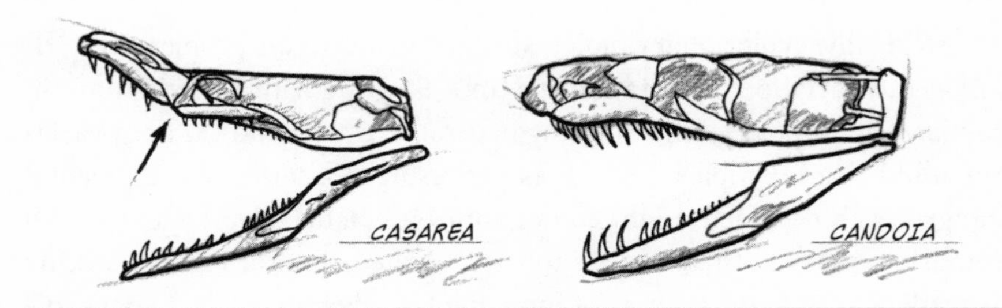

FIGURE 12.1
A complex adaptation: the jointed upper jaw of the bolyerine snake, made possible by its accompanying tendons, ligaments, and musculature. The other skull shows the single-boned jaw, found in other related snakes.

require corresponding changes in the other bones and in other parts of the ear as well, such as the tympanic membrane or the cochlea. Complex biological systems depend for their functions on tens or hundreds of such independent, yet jointly necessary parts. As the number of necessary components increases, the requisite number of coordinated changes increases too, rapidly driving up the difficulty of maintaining the functional integrity of the system while modifying any of its parts.

And that was the problem, as Frazzetta understood it. Any system that depends for its function on the coordinated action of many parts could not be changed gradually without losing function. But in the neo-Darwinian scheme of things, natural selection acts to preserve only functional advantages. Changes that result in death or reduced function will not be preserved. The integrated complexity of many biological systems thus imposes limitations on the evolutionary process—limitations that human engineers do not face when they design complex integrated systems. In 1975, Frazzetta wrote a minor classic entitled *Complex Adaptations in Evolving Populations* explaining this concern. He wrote:

> *When modifying the design of a machine, an engineer is not bound by the need to maintain a real continuity between the first machine and the modification. . . . But in evolution, transitions from one type to the next presumably involve a greater continuity by means of a vast number of intermediate types. Not only must the end product—the final machine—be feasible, but so must be all the intermediates. The evolutionary problem is, in a real sense, the gradual improvement of a machine while it is running!*[5]

Historically, evolutionary biologists tried to solve this problem one advantageous variation or mutation at a time. Starting with Darwin himself, they have attempted to explain how natural selection and random variation could build complex systems as the result of a series of incremental changes, each of which might confer some selectable advantage. Darwin famously employed this strategy to explain the origin of the eye, asking his readers to imagine a series of incremental, advantageous changes to a simple "nerve sensitive to light."[6]

As Frazzetta thought about the problem of explaining the origin of complex systems, he came to doubt both the classical and the modern

Darwinian accounts of such systems. Frazzetta acknowledged that he was influenced in part by the skepticism expressed by the Wistar "outsiders" (see Chapter 9). He admitted "revealing some hideous *personalia*" in confessing that he was attracted to the worries about neo-Darwinism expressed by Murray Eden and other Wistar skeptics.[7]

Frazzetta's concerns about the adequacy of the neo-Darwinian mechanism, like Eden's, turned on the growing appreciation of the nature and importance of genetic information. Though biologists then (as now) didn't fully understand how genetic information in DNA correlates or "maps" to these higher-level complex morphological structures, by 1975 they did know that many hundreds of genes can be involved in coding for a single complex integrated structure. Thus, altering the anatomical structure of the mammalian ear or the vertebrate eye, for example, would involve altering the genes that code for its constituents, which implies, most implausibly, that *multiple* coordinated mutations would occur virtually simultaneously.

As Frazzetta explained, "Phenotypic alteration of integrated systems requires an improbable coincidence of genetic (and hence, heritable phenotypic) modifications of a tightly specified kind."[8] Yet the extreme specificity of the fit of the components and the functional dependence of the whole system on this fit imply limits to allowable genetic change. Genetic change affecting any one of the necessary components, unless matched by many corresponding—and vastly improbable—genetic changes, will result in functional loss and often death. For this reason, as Frazzetta concluded, "We are still left with the unabating need to explain evolutionary changes in systems that have the operational integration characteristic of things we recognize as 'machines.'"[9] At the time, the doubts he expressed gained little traction in the evolutionary biology community, because neo-Darwinian evolutionary biologists assumed that mutation and selection had nearly unlimited creative power, enough to generate even complex systems of the kind described in Frazzetta's book.

The mathematical expression of neo-Darwinian theory, as represented in the equations of a subdiscipline of biology known as population genetics, seemed to confirm this conviction. Population genetics models how gene frequencies change as the result of processes such as mutation, genetic drift (neutral changes in the genome that natural selection neither favors nor eliminates), and natural selection. On the assumption that advantageous variations or traits will arise as the result of even single mutations, the mathematical models of population genetics describe how much evolutionary change can occur in a given period of time. These estimates are based upon, among others, three primary factors: mutation rates, effective population sizes, and generation times. When evolutionary biologists plug estimates for these factors into the equations of population genetics, their calculations seem to imply that standard evolutionary mechanisms could generate significant amounts of evolutionary change in many groups of organisms—even enough to build complex systems. As long as mutations generate a continuous supply of new traits, any system, however complex, can be built one trait at a time—trait upon trait—via the creative power of natural selection. Or so the story goes.

Confidence in these mathematical models (and their underlying assumptions) led many neo-Darwinists to disregard the need to give detailed accounts of the specific evolutionary pathways by which complex systems might have arisen. For example, in an evolutionary biology text widely used about the time Frazzetta first posed this challenge, evolutionary biologists Paul Ehrlich and Richard Holm advised:

> *One need not go into the details of the evolution of the bird's wing, the giraffe's neck, the vertebrate eye, the nest building of some fish, etc., as the selective origins of these and other structures and of behavioral patterns may be assumed to be basically the same in outline as those, such as industrial melanism, which have already been discussed. Even a slight advantage or disadvantage in a particular genetic change provides a sufficient differential for the operation of natural selection.*[10]

The phrase "sufficient differential for the operation of natural selection" refers to the equations of population genetics and one of the factors (the so-called selection coefficient) that determines how rapidly particular traits would be likely to disseminate through a population. The message was clear: the math tells the story; the biological details of the origin of complex systems don't matter.

The neo-Darwinian focus on mathematical modeling helps to explain why mainstream evolutionary biologists haven't worried about the problem of the origin of new genes and proteins or the problem of combinatorial inflation, discussed in Chapters 9 and 10. Many contemporary evolutionary biologists, like the founders of population genetics, assumed that some mechanism for building new genes already existed. Indeed, they assumed that new traits (and the genes for building them) can arise as the result of even single mutations (or a series of such mutations that each confer a small, incremental, selectable advantage). Thus, the mathematical expression of neo-Darwinian theory seemed to certify the plausibility of even large-scale evolutionary changes—again, provided these changes could occur one mutation at a time.

But what if there are systems in living organisms that *cannot* be built one mutation at a time, and instead must be built by simultaneous coordinated changes? What if building just a single new gene or protein requires such coordinated mutational changes? What if individual genes turned out to be *complex adaptations*?

Mathematical challenges of the kind first advanced at Wistar, and that Douglas Axe's experimental findings have exacerbated, initially did not dent confidence in the adequacy of neo-Darwinian explanations. Many evolutionary biologists have simply regarded mathematical challenges to the creative power of the mechanism, coming as they mostly do from scientists and engineers in other fields, as exotic or irrelevant.

That has begun to change. And it has begun to change in a way that has not only introduced a new mathematical challenge to the creative power of the neo-Darwinian mechanism, but also in a way that indirectly confirms Axe's key insight about the rarity of genes and proteins. In the last decade, developments in molecular genetics and population genetics have exposed a connection between the problem of the origin of new genes and proteins and the origin of complex adaptations, a connection first perceived by Tom Frazzetta back in 1975. As more biologists have recognized that connection, they too have begun to share Frazzetta's doubt.

POPULATION GENETICS AND THE ORIGIN OF GENETIC INFORMATION

The neo-Darwinian synthesis was formulated during the 1930s before the elucidation of the structure of DNA. Biologists at that time did not yet understand the nature, structure, or precise location of genetic information.[11] They did not associate genes with long strings of nucleotide bases along the spine of the DNA molecule. They did not think of genes as long sections of digital code stored in complex biomacromolecules. Instead, after Mendel, but before Watson and Crick, genes were defined operationally as those entities, associated with chromosomes, that produced specific visible or selectable anatomical traits, such as eye color or beak shape.

The architects of neo-Darwinism working in the 1930s reformulated evolutionary theory to emphasize the importance of mutations as the source of genetic variation. It followed, therefore, that the mutations—which they regarded as the source of heritable variation—must operate on genes. Not knowing the nature of genes, they also assumed that *a single mutation could alter a gene in such a way as to produce a new trait.*

The equations of population genetics are predicated upon this assumption. The rate of mutation thus emerges as an important factor in computing the amount of evolutionary change that can occur in any given population. If every individual mutation can produce a new, potentially selectable trait, then the rate at which such variation accumulates partially determines how much change can occur in a given time.

After 1953, biologists no longer conceived of the gene as an abstract entity. Watson and Crick showed that the gene had a definite locus and structure and that individual genes contain hundreds or thousands of precisely sequenced nucleotide bases, each functioning as digital characters in a larger instruction set. Consequently, biologists changed their understanding of mutations as well. Biologists came to understand mutations as something like typographic errors in long strings of digital code. As a result, many scientists began to realize that individual mutations were unlikely by themselves to produce new beneficial traits. Some scientists realized that mutations were instead overwhelmingly more likely to *degrade* the information contained in a gene than to produce a new function or trait, and that the accumulation of mutations would eventually and typically result in the *loss* of function.

This change of perspective called for an explanation of how mutations could generate new genes—an explanation that was provided beginning in the 1970s with the ideas of gene duplication, subsequent neutral evolution, and positive selection.

Though the theory of gene duplication played no formal role in the mathematical structure of population genetics, it did serve to buttress a critical assumption of the whole enterprise. After the 1950s, evolutionary biologists no longer assumed that single mutations would necessarily generate whole new traits. That left a critical assumption of population genetics essentially undefended. For many evolutionary biologists, the theory of gene duplication closed that conceptual gap. After the theory was formulated, many evolutionary biologists thought that a mechanism had been discovered by which sections of genetic text could accumulate multiple changes without compromising the fitness of an organism, thus ensuring the eventual production of new genes and a steady supply of new traits.

So when Frazzetta confronted evolutionary biology with the problem of complex adaptations in the mid-1970s, most neo-Darwinian biologists responded with a collective yawn, if they noticed it at all. His challenge was no challenge after all. So implied the math of population genetics—*provided its assumptions about the ease with which new mutations could generate new traits were valid.*

But were they valid? Could a series of separate mutations generate the new genes necessary to build new proteins and new traits, or did building genes require multiple coordinated mutations?

ARE GENES COMPLEX ADAPTATIONS?

Classically, Darwinian biologists have assumed that small, separate step-by-step changes could produce all biological structures and features, provided each change confers some survival or reproductive advantage. In his chapter in the 1909 anthology *Darwin and Modern Science*, the British geneticist William Bateson wryly described how this widespread assumption prevented evolutionary biologists from confronting the real difficulty of explaining the origin of complex adaptations:

> *By suggesting that the steps through which an adaptive mechanism arises are indefinite and insensible, all further trouble is spared. While it could be said that species arise by an insensible and imperceptible process of variation, there was clearly no use in tiring ourselves by trying to perceive that process. This labour-saving counsel found great favor.*[12]

One of the first prominent evolutionary biologists to consider the possibility that building new genes and proteins might require multiple coordinated mutations was John Maynard Smith. Maynard Smith worked as an aeronautical engineer during World War II, but then took up the formal study of evolutionary biology after the war. He eventually helped to found the University of Sussex, where he also served as a distinguished professor of biology until the mid-1980s.[13]

In 1970, Maynard Smith wrote an article in *Nature* responding to an earlier article by Frank Salisbury, a biologist from Utah State University. Salisbury had raised questions about whether random mutations could explain the specificity of the arrangement of nucleotide bases necessary to produce functional proteins. Salisbury worried, following discussions at Wistar, that the probability of random mutations generating functional arrangements of bases or amino acids was prohibitively low. According to Salisbury's calculations, "The mutational mechanism as presently imagined could fall short by hundreds of orders of magnitude of producing, in a mere four billion years, even a single required gene."[14]

To overcome this improbability, Maynard Smith proposed a model of protein evolution. While admitting that the origin of the first proteins remained a mystery, he suggested that one protein could evolve into another as the result of small incremental changes in amino-acid sequences, provided each sequence maintained some function at each step along the way. Maynard Smith compared protein-to-protein evolution to changing one letter in an English word in order to generate a different word (while at each step generating a different meaningful word). He used this example to convey how he thought protein evolution might work:

WORD ⟶ WORE ⟶ GORE ⟶ GONE ⟶ GENE

He explained:

> *The words [in this analogy] represent proteins; the letters represent amino acids; the alteration of a single letter corresponds to the simplest evolutionary step, the substitution of one amino acid for another; and the requirement of meaning corresponds to the requirement that each unit step in evolution should be from one functional protein to another.*[15]

As a self-professed "convinced Darwinist," Maynard Smith realized that natural selection and random mutation could only build new biological structures from preexisting structures *if* each intermediate structure along the way conferred some adaptive advantage. He thought that this requirement applied as much to the evolution of new genes and proteins as it did to the evolution of new phenotypic traits or larger-scale anatomical structures.[16]

Nevertheless, the essentially digital or alphabetic character of the genetic information that directs protein synthesis suggested a problem to Maynard Smith. How, he asked, could one gene or protein evolve into another if such a transformation required multiple simultaneous changes in the bases of the genetic text (or arrangement of amino acids)? If building new genes required multiple coordinated mutations, then the probability of generating a new gene or protein would drop precipitously, since such a transformation would require not just one improbable mutational event, but two or three or more, occurring more or less at once. Here's how he described the potential problem:

> *Suppose that a protein ABCD . . . exists, and that a protein abCD . . . would be favoured by selection if it arose. Suppose further that the intermediates aBCD . . . and AbCD . . . are nonfunctional. These forms would arise by mutation, but would usually be eliminated by selection before a second mutation could occur. The double step from abCD . . . to ABCD would thus be very unlikely to occur.*[17]

In Maynard Smith's view, the improbability associated with "double-step" or multiple-step coordinated mutations presented a significant *potential* problem for molecular evolution. In the end, however, he concluded that such mutations were so improbable that they *must* not have played a significant role in the evolution of novel structures. As he explained, "Such double steps . . . may occasionally occur, but are probably too rare to be important in evolution."[18]

For several decades, the problem he flagged receded into obscurity. As biochemist H. Allen Orr pointed out in 2005 in the journal *Nature Reviews Genetics,* "Although Maynard Smith's work appeared early in the molecular revolution," his ideas about problems facing protein evolution "were almost entirely ignored for two decades."[19] Thus, Orr noted that evolutionary biologists stopped thinking about molecular evolution as a consequence of adaptive changes at the amino-acid level. Not until the first decade of the twenty-first century would biologists confront the challenge of making a rigorous *quantitative* analysis of the plausibility of protein-to-protein evolution.

WAITING FOR COMPLEX ADAPTATIONS

In 2004, Lehigh University biochemist Michael Behe (see Fig. 12.2), introduced briefly at the end of Chapter 9, and University of Pittsburgh physicist David Snoke published a paper in the journal *Protein Science* that returned to the problem first described by Maynard Smith.[20] By this time, Behe had established himself as a prominent critic of neo-Darwinism by arguing that the neo-Darwinian mechanism did not provide an adequate explanation for the origin of functionally integrated "irreducibly complex" molecular machines. In his 2004 paper, Behe sought to extend his critique of neo-Darwinism by assessing its adequacy as an explanation for new genes and proteins. He and Snoke attempted to assess the plausibility of protein evolution in the case that it does indeed require multiple coordinated mutations. They applied standard neo-Darwinian modes of analysis derived from population genetics to make their evaluation. They considered the plausibility of the main neo-Darwinian model of gene evolution in which evolutionary biologists envision new genes arising by gene duplication and subsequent mutations in the duplicated gene.

Behe and Snoke assessed the plausibility of this model for multicellular organisms in the case that multiple (two or more) point mutations must occur *simultaneously* in order to generate a new selectable gene or protein. Whereas Maynard Smith saw the need for multiple coordinated mutations as a *potential* problem, one that ultimately needn't trouble evolutionary biologists, Behe and Snoke argued that evolutionary biologists *do* need to worry about it, and they quantified its severity.

Behe and Snoke first noted that many proteins, as a condition of their function, require unique *combinations* of amino acids interacting in a coordinated way. For example, ligand binding sites on proteins—places where small molecules bind to large proteins to form larger functional complexes—typically require a combination of several amino acids. Behe and Snoke argued that in such cases the combinations of amino acids would have to arise in a coordinated fashion since the capacity for ligand binding depends on all the necessary amino acids being present together. In support of this inference, they cited an authoritative textbook, *Molecular Evolution*, by University of Chicago evolutionary biologist Wen-Hsiung Li. In it, Li notes that evolving ligand binding capacity in proteins such as hemoglobin may require "many mutational steps,"[21] even though the first steps on the way to

FIGURE 12.2
Michael Behe. *Courtesy Laszlo Bencze.*

building such capacity would confer no selective advantage. As Li explains, "Acquiring a new function may require many mutational steps, and a point that needs emphasis is that the early steps might have been selectively neutral [non-advantageous] because the new function might not be manifested until a certain number of steps had already occurred."[22]

Behe and Snoke point out that this observation implies that a series of separate mutations could not generate a ligand binding function in a protein that previously did not have this capacity, since individual amino-acid changes would initially confer no selectable advantage on the protein lacking this function. Instead, evolving ligand binding capability would require multiple *coordinated* mutations. Behe and Snoke make a similar argument about the requirements for the evolution of protein-to-protein interactions. They note that for proteins to interact with each other in specific ways, typically at least several individually necessary amino acids must be present in combination in each protein, again, suggesting the need for multiple coordinated mutations.

SO MANY CHANGES, SO LITTLE TIME

Behe and Snoke used the principles of population genetics to assess the likelihood of various numbers of coordinated mutational changes occurring in a given period of time. They asked: Is it probable that there was enough time in evolutionary history to generate coordinated mutations? If so, how many coordinated mutations is it reasonable to expect in a period of time given various population sizes, mutation rates, and generation times? Then, for different combinations of these various factors, they assessed how long it would typically take to generate two or three or more coordinated mutations. They determined that generally the probability of multiple mutations arising in close (functionally relevant) coordination to each other was "prohibitively" low—it would likely take an immensely long time, typically far longer than the age of the earth.

THE POWERBALL LOTTERY—POPULATION GENETICS MADE EASY

Before going on, it might be helpful to understand a bit more about how the equations and principles of population genetics can be used to calculate what evolutionary biologists call "waiting times," the expected time that it will take for a given trait to arise by various evolutionary processes. In his book *The Edge of Evolution,* Michael Behe illustrates these principles using a charming analogy to the Powerball lottery game that many American state governments use to raise money.

To win at Powerball, contestants must purchase tickets with six numbers that match the numbers printed on six balls drawn from two drums. Five of the balls are selected from a drum containing 59 white balls, numbered 1 through 59. A sixth red ball, the so-called power ball, is chosen out of a drum of 35 red balls numbered 1 to 35. To win the jackpot—which can exceed $100 million—a player must purchase a ticket listing all six of the chosen numbers in any order. The Powerball website lists the probability of matching all six balls at roughly 1 in 175 million. Depending on how many tickets have been purchased and how frequently drawings occur, it may take a very long time for someone to win.

Behe asked his readers first to consider how long it will take, on average, to generate a lottery ticket with the winning numbers. He notes that knowing the probability of drawing such a winning ticket isn't sufficient. The calculation *also* requires knowing how often drawings occur and how many tickets are sold. As Behe explains: "If the odds of winning are one in a hundred million, and if a million people play every time, then it will take on average about a hundred drawings for someone to win." If there are about a hundred drawings per year, with a million people playing per drawing, "then it would take about a year before someone won. But if there were only one drawing per year, on average it would take a century to hit the jackpot."[23] More frequent drawings produce shorter waiting times. Less frequent drawings tend to require longer waiting times. Similarly, more players will decrease the average time necessary to produce a winner, while fewer players result in longer waits.

Similar mathematical principles apply when calculating the expected waiting times for the evolution of biological features by mutation and selection. Biologists first need to assess the complexity of the system—or its inverse, the improbability of the feature occurring. As in Powerball, however, knowing the probability of an event by itself does not allow someone to calculate how long it will likely take for that event to occur. Such a calculation requires also knowing the size of the population (equivalent to how many people are playing Powerball) *and* how frequently new genetic sequences arise (equivalent to how frequently drawings are held).

In Powerball, a new sequence of numbers arises in every drawing. But when organisms reproduce, they do not always generate a new sequence of nucleotide bases in their individual genes. For this reason, to calculate the rate at which new sequences arise in living organisms requires knowing two factors: the generation time and the mutation rate. More rapid rates of mutation and/or shorter generation times will increase the rate at which new genetic sequences arise, resulting in shorter waiting times. Slower rates of mutation and/or longer times between generations produce longer waiting times. Also, as in Powerball, the number of "players" is important. Larger populations generate new genetic sequences more frequently than smaller ones and, thus, *decrease* expected waiting times. Smaller populations reduce the rate at which new sequences are generated, *increasing* waiting times.

Now, under the rules of Powerball, you can "win" without picking all six numbers correctly—you just won't win the entire jackpot.[24] If you pick just the red "power ball" correctly, you win $4. Pick three white balls correctly, and you win $7. If you correctly pick the numbers of four white balls, you win $100. Guess all five white balls correctly (but not the red ball), and you can win a cool $1 million.

With each additional ball necessary to secure a new level of winnings, the probability of winning decreases exponentially, while the values of the prizes increase dramatically. The Powerball website lists the probability of

NUMBER OF BALLS WHICH MUST BE GUESSED CORRECTLY	ODDS OF WINNING	PRIZE
1 (RED)	1 IN 55.41	$4
3 (ALL WHITE)	1 IN 360.14	$7
4 (ALL WHITE)	1 IN 19,087.53	$100
5 (4 WHITE, 1 RED)	1 IN 648,975.96	$10,000
5 (ALL WHITE)	1 IN 5,153,632.65	$1,000,000
6 (5 WHITE, 1 RED)	1 IN 175,223,510.00	JACKPOT

FIGURE 12.3
A chart showing the probability of winning and the corresponding payouts for different combinations of balls in the Powerball lottery game.

winning a $4 prize at just 1 in 55, the probability of winning $1 million dollars at roughly 1 in 5 million, and the probability of winning the jackpot as 1 in 175 million (see Fig. 12.3).[25]

Neo-Darwinists have long assumed that biological evolution works something like matching one number in Powerball. In their view, natural selection acts to reward or preserve small but *relatively* probable changes in gene sequences—like winning the small but more likely $4 prize in Powerball over and over again. They assume the mutation and selection mechanism doesn't depend on winning extremely unlikely "prizes" (like the whole Powerball jackpot) all at once.

But what if, to produce a functional advantage at the genetic level, the mutation and selection mechanism had to generate the biological equivalent of all six (or more) correct balls in the Powerball lottery with no reward for guessing a smaller number of balls correctly first? Clearly, the probability of this would be extremely small. And the *waiting time* for winning such a lottery could become prohibitively long.

BACK TO THE BIOLOGY

That brings us back to Behe and Snoke's conclusion. In their 2004 paper, they argued that generating a single new protein will often require many improbable mutations occurring at once. They took into account the improbability of multiple functionally necessary mutations appearing together—the equivalent of needing to get a Powerball ticket matching several numbers to win any money at all. Then they sought to determine how long it would take and/or how large the population sizes would need to be to generate a new gene via multiple coordinated mutational changes—the genetic equivalent of the "jackpot scenario."

Behe and Snoke found that *if* generating a new gene required multiple coordinated mutations, then the waiting time would grow exponentially with each additional necessary mutational change. They also assessed how population sizes affected how long it would take to generate new genes, *if* multiple coordinated mutations were necessary to produce those genes. They found, not surprisingly, that just as larger populations diminished expected waiting times, smaller populations dramatically increased them.

More important, they found that even if building a new gene required just two coordinated mutations, the neo-Darwinian mechanism would likely either require huge population sizes or extremely long waiting times or both. If coordinated mutations were necessary, then evolution at the genetic level faced a catch-22: for the standard neo-Darwinian mechanism to generate just two coordinated mutations, it typically either needed unreasonably long waiting times, times that exceeded the duration of life on earth, or it needed unreasonably large population sizes, populations exceeding the number of multicellular organisms that have ever lived. To get population sizes that were reasonable, they had to have waiting times that were unreasonable. To get waiting times that were reasonable, they had to have population sizes that were unreasonable. As they put it, either way the "numbers seem prohibitive."[26]

Behe and Snoke found that mutation and selection could generate two coordinated mutations in a mere 1 million generations, a reasonable length of time given the age of the earth. But that was only in a population of 1 trillion or more multicellular organisms, *a number that exceeds the size of the effective breeding populations of practically all individual animal species that have lived at any given time.*[27] Conversely, they found that mutation and selection could generate two coordinated mutations in a population of only 1 million organisms, but only if the mechanism had 10 billion generations at its disposal. Yet on the assumption that each multicellular organism lived only one year, 10 billion generations computes to 10 billion years—more than twice the age of the earth. This is clearly an unreasonable length of time to wait for the emergence of a single gene, let alone more significant evolutionary innovations.

Behe and Snoke did, however, find one tiny "sweet spot" in which a gene requiring only two coordinated mutations could arise (see Fig. 12.4). Such a gene could conceivably arise from 1 billion organisms in a "mere" 100 million generations. Since many more than 1 billion multicellular organisms have lived on earth during its history and since multicellular life on earth has existed for more than 500 million years, these numbers offer (assuming, again, one year per generation) the prospect of enough time and organisms to generate one new gene—if only two coordinated mutations are necessary. (Of course, if the population evolving a two-mutation trait had fewer than 1 billion organisms, then the waiting times again increased to unreasonable lengths.)

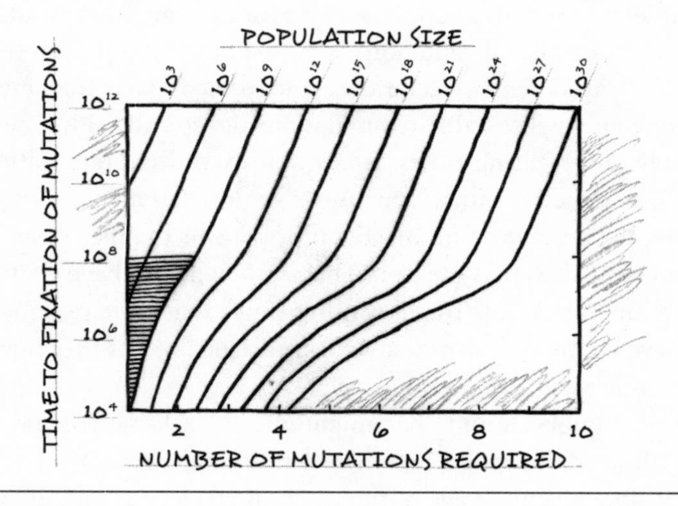

FIGURE 12.4

This diagram shows the population sizes and times (measured in number of generations) required to produce a gene or trait if building that gene or trait requires multiple coordinated mutations. The shaded gray area shows the "sweet spot"—population sizes and available time sufficient to generate the coordinated mutations necessary to produce a new gene. Note that any multi-mutation feature requiring more than two mutations could not, in all probability, evolve by gene duplication and subsequent coordinated mutation in a population of multicellular organisms, however large. Note also that for most normal population sizes and reasonable generation times, even evolving two mutations lies beyond the reach of gene duplication, mutation, and selection. *Courtesy John Wiley and Sons and The Protein Society.*

Nevertheless, these numbers only apply to the case in which only two coordinated mutations are necessary to build a new gene. Behe and Snoke found that *if* generating a new functional gene or trait required more than two coordinated mutations, then excessively long waiting times were necessary regardless of the size of the population. If three or more coordinated mutations were necessary, their calculations generated no "sweet spots" at all. Thus, they concluded that "the mechanism of gene duplication and point mutation alone would be ineffective, at least for multicellular . . . species."[28]

In summary, Behe and Snoke applied the principles derived from population genetics to evaluate the creative power of the standard neo-Darwinian model of gene evolution. They showed that the standard model encounters clear probabilistic limits *if* the structures it needs to build require more than two coordinated mutations in multicellular eukaryotic organisms.

THE EDGE OF EVOLUTION AND ITS CRITICS

Behe and Snoke are well-known critics of the creative power of the neo-Darwinian mechanism, so their conclusion might seem suspect to some observers. Nevertheless, evolutionary biologists attempting to defend the creative power of the neo-Darwinian mechanism have inadvertently confirmed Behe and Snoke's conclusions.

Two recent scientific publications tell this story. First, in 2007, Michael Behe published a book, *The Edge of Evolution,* amplifying the results of his 2004 paper with David Snoke. Using public-health data about a genetic trait—resistance to the antimalarial drug chloroquine in the one-celled organism that causes malaria—Behe provided another line of evidence and argument to support the conclusion that multiple coordinated mutations *are* often necessary to produce even minor genetic adaptations.

Based on public-health data, Behe determined that resistance to chloroquine only arises once in every 10^{20} malaria-causing cells. Behe inferred, by working the problem backwards, that the trait probably required multiple—though not necessarily coordinated—mutations to develop. He called this trait a "chloroquine complexity cluster," or a "CCC."[29] Behe wanted to explore what he called the "edge of evolution," the limits to the creative power of mutation and selection at the genetic level. Having established that this trait could arise by random mutation in a reasonably short period of time, he wondered how much time would be required to produce traits of greater complexity in populations of various sizes.

He asked his readers to consider a hypothetical genetic trait twice as complex as a CCC cluster—a feature requiring the origin of two *coordinated* traits, each as complex as a CCC. In other words, Behe wondered how long it would take to develop a hypothetical trait that required two genetic changes as complex as a chloroquine complexity cluster, *if* both changes had to occur together in the same organism—in a coordinated fashion—in order to produce the trait. He then showed, using the principles of population genetics, that multi-mutation traits of that complexity—the molecular equivalent of two coordinated CCCs—would require many more organisms or vastly more time than was reasonable given the history of life. Remember the Powerball lottery: waiting times increase *exponentially* with each additional coordinated change or winning element needed. Behe showed, for example, that if 10^{20} organisms were required to obtain one CCC, then the square of that amount—10^{40} organisms—would be required to evolve a trait that required two *coordinated* CCCs before providing any advantage.[30] But, as we saw in Chapter 10, only 10^{40} total organisms have ever existed on earth, implying that the entire history of the earth would barely provide enough opportunities to generate a trait of this complexity.[31]

Similarly, Behe reasoned that for organisms in smaller population sizes, developing a trait of twice the complexity of a CCC would require immensely long waiting times. He also determined that exceedingly long waiting times are typically required to generate even less complex genetic adaptations in smaller populations.

Behe showed that the problem of coordinated mutations was particularly acute for longer-lived organisms with small population sizes—organisms such as mammals or, more specifically, human beings and their presumed prehuman ancestors. Behe estimated, based upon relevant mutation rates, known human population sizes, and generation times, the time required for two coordinated mutations to occur in the hominid line. He calculated that producing even such a modest evolutionary change would require many hundreds of millions of years. Yet, humans and chimps are thought to have diverged from a common ancestor only 6 million years ago. Behe's calculation implied that the neo-Darwinian mechanism does not have the capacity to generate even two coordinated mutations in the time available for human evolution—and thus does not explain how humans arose.

Here the story gets really interesting. Soon after the publication of *The Edge of Evolution,* two Cornell University mathematical biologists, Rick Durrett and Deena Schmidt, both defenders of neo-Darwinism, attempted to refute Behe's conclusion by making their own calculations. Their paper, "Waiting for Two Mutations: With Applications to Regulatory Sequence Evolution and the Limits of Darwinian Evolution," also applied a model based upon population genetics to calculate the amount of time necessary to generate two coordinated mutations in the hominid line. Although they calculated a shorter waiting time then Behe did, their result nevertheless underscored the implausibility of relying on the neo-Darwinian mechanism to generate coordinated mutations during the relevant evolutionary timescale. Their calculation suggested that it would take not several hundred million years, but "only" 216 million years to generate and fix two coordinated mutations in the hominid line—more than thirty times the amount of time available to produce humans and chimps and all their distinctive complex adaptations and differences from their inferred common ancestor.

In seeking to refute Behe, Durrett and Schmidt inadvertently confirmed his main contention. As they acknowledged, their calculation implies that generating two or more coordinated mutations is "very unlikely to occur on a reasonable timescale."[32] In sum, calculations performed by both critics *and defenders* of neo-Darwinian evolution now reinforce the same conclusion: *if* coordinated mutations are necessary to generate new genes and proteins, then the neo-Darwinian math itself, as expressed in the principles of population genetics, establishes the implausibility of the neo-Darwinian mechanism.

TESTING THE CO-OPTION OPTION

But *does* generating novel genes and proteins require coordinated mutations? Behe and Snoke *inferred* as much based upon an undisputed fact of molecular biology: many proteins rely on sets of amino acids acting in close coordination in order to perform their functions. In addition, in *The Edge of Evolution,* Behe argued on functional grounds that many complex biological systems *would* require coordinated adaptive mutations since in these systems, the absence of even one or a few gene products (proteins or traits) will cause them to lose function. Behe specifically showed that several molecular machines within cells (such as the cilium and intraflagellar transport system, and the bacterial flagellar motor[33]) require the coordinated interaction of multiple protein parts in order to maintain their function. Nevertheless, in making this argument, Behe did not address an alternative idea about the pathway by which new genes and proteins might have evolved, and thus did not establish conclusively that new genes and proteins themselves represent complex adaptations.

Some neo-Darwinists have proposed a model of protein evolution known as "co-option." In this model, a protein that performs one function is transformed, or "co-opted" to perform some other function. This model envisions new features requiring multiple "mutations" arising in a step-by-step way to produce some protein, call it "Protein B," from some other protein that lacked those features, call it "Protein A." In proposing a series of single separate mutations, advocates of co-option acknowledge that the initial individual amino-acid changes, the first few steps in evolution, from Protein A, the protein lacking the multisite feature, would not allow Protein A to perform the function of Protein B. Nevertheless, they propose that these initial changes might have allowed Protein A to perform some *other* advantageous function, thus making it selectable and preventing protein evolution from terminating due to diminution or loss of its initial function. Eventually, as mutations continued to generate new proteins with slightly different functions, they would have generated a protein close enough in sequence and structure that just one or a very few additional changes would suffice to convert it into Protein B.

Aware of these imaginative scenarios, Douglas Axe and his colleague, molecular biologist Ann Gauger (see Fig. 12.5), now working together at

FIGURE 12.5
Ann Gauger. *Courtesy Laszlo Bencze.*

the Biologic Institute in Seattle, decided to put them to an ingenious experimental test.[34] In so doing, they sought to determine whether the evolution of new multisite features does indeed typically require multiple *coordinated* mutations, or instead whether such a feature could arise by co-option.

Axe and Gauger scoured protein databases looking for proteins that are as similar as possible in sequence and structure, but that nevertheless perform different functions. They identified two proteins that meet those criteria. One of these proteins (Kbl_2) is needed for breaking down an amino acid called threonine, and the other ($BioF_2$) is needed for building a vitamin called biotin. (See Fig. 12.6.)

Gauger and Axe realized that if they could transform Kbl_2 into a protein performing the function of $BioF_2$ with just one or very few coordinated amino-acid changes, then that might demonstrate (depending upon how few) that the two proteins were close enough in sequence that a conversion in function of the kind envisioned by co-option advocates is plausible in evolutionary time. What's more, because they knew the difficulty scientists have had in showing any real change of protein function to be feasible, a positive result would suggest that they had at last discovered a functional gap that one or very few mutations could plausibly jump—as co-option envisioned.

If, however, they found that *many* coordinated mutational changes were needed, then that could establish—depending upon *how many* were needed—that the Darwinian mechanism could not accomplish the functional jump from A to B in a reasonable time. That would imply that an even greater degree of structural similarity between proteins would be needed

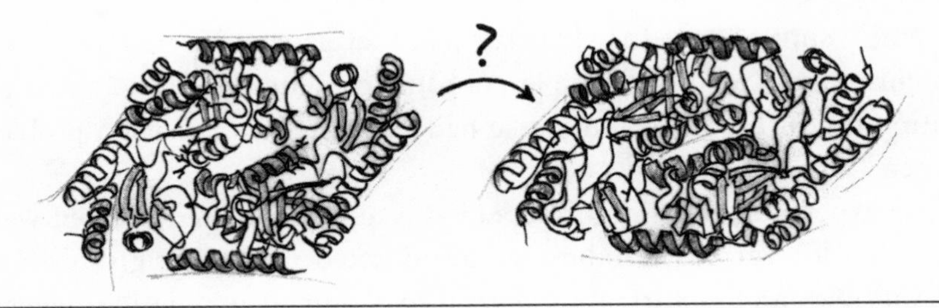

FIGURE 12.6
The proteins, Kbl_2 (*left*) and $BioF_2$ (*right*) are enzymes that use similar catalytic mechanisms to accelerate different chemical reactions in the bacterium *E. coli. Courtesy Ann Gauger and Douglas Axe.*

for the co-option hypothesis to be plausible. Having carefully examined the structural similarities between members of a large class of structurally similar enzymes, they knew that Kbl_2 and $BioF_2$ were about as close in sequence and structure as any two known proteins that performed different functions. Thus, if it turned out that converting one protein function into the other required many coordinated mutations—more than could be expected to occur in a reasonable time—then the outcome of their experiment would have devastating implications for standard accounts of protein evolution. If proteins that perform two different functions have to be even more similar than Kbl_2 and $BioF_2$ in order for mutational changes to convert the function of one to the other, then for all practical purposes co-option would not work. There simply aren't many known jumps that small.

Axe and Gauger first identified those amino-acid sites that were most likely, if mutated, to cause a change from Kbl_2 function to $BioF_2$ function. They then systematically mutated those sites individually and in groups involving various amino-acid combinations. Their results were unambiguous. They found that they could not induce, with either one or a small number of amino acids, the change in function they sought. In fact, they found that they could not get Kbl_2 to perform the function of $BioF_2$, even if they mutated larger numbers of amino acids in concert—that is, even if they made many more coordinated mutations than could plausibly occur by chance in all of evolutionary history.

Although their attempts to convert Kbl_2 to perform the function of $BioF_2$ failed, their experiment did not. It allowed them to establish experimentally for the first time that the co-option hypothesis of protein evolution lacks credibility—simply too many coordinated mutations would be required to convert one protein function to another, even in the case of extremely similar proteins. That implied that generating new genes and proteins would require multiple coordinated mutations, and thus, the waiting times that Behe and Snoke had calculated *do* present a problem for neo-Darwinian theory.

The experimental work also enabled Axe to calculate expected waiting times for various numbers of coordinated mutations given different variables and factors. Axe developed a refined population-genetics mathematical model to calculate various waiting times. His results roughly confirmed the previous calculations of Behe and Snoke. He found, for example, that if he took into account the probable fitness cost to an organism of carrying unnecessary gene duplicates (as was necessary to give the evolution of a new gene a reasonable chance), that the probable waiting time for even three coordinated mutations exceeded the duration of life on earth.

He therefore effectively determined an *upper bound* of two for the number of coordinated mutations that could be expected to occur in a duplicate gene during the history of life on earth (taking into account the negative effects of carrying gene duplicates in the evolutionary process). He also calculated six coordinated mutations as an upper bound, neglecting the fitness cost of carrying gene duplicates. Nevertheless, in their experiments, he and Gauger could not induce a functional change in a single gene with *more* than six coordinated mutations. So, even that more generous—and, again, unrealistically generous upper bound—does little to render the co-option hypothesis credible. Indeed, Axe and Gauger's experiments showed that the smallest realistically conceivable step exceeded what was plausible given the time available to the evolutionary process. In their words, "evolutionary innovations requiring that many changes . . . would be extraordinarily rare, becoming probable only on timescales much longer than the age of life on earth."

WHAT IT ALL MEANS

By showing the implausibility of the co-option model of protein evolution and the need for multiple coordinated mutations in order to generate multisite features in proteins, Axe and Gauger confirmed that genes and proteins *themselves* represent complex adaptations—entities that depend upon the coordinated interaction of multiple subunits that must arise as a group to confer any functional advantage.

The need for coordinated mutations means that evolutionary biologists cannot just assume that mutations will readily generate new genes and traits, as neo-Darwinists have long presupposed. Indeed, by applying mathematical models based on the standard principles of population genetics to the questions of the origin of genes themselves, Behe and Snoke, Durrett and Schmidt (inadvertently), Axe and Gauger, and other biologists[35] have recently shown that generating the number of multiple coordinated mutations needed to produce even one new gene or protein is unlikely to occur within a realistic waiting time. Thus, these biologists establish the implausibility of the neo-Darwinian mechanism as a means of generating new genetic information.

There is one other aspect of this. The body of work published between 2004 and 2011 also provides additional confirmation of Axe's research showing the rarity of genes and proteins in sequence space. In fact, that research helps to explain *why* such long waiting times are necessary. If functional sequences are rare in sequence space, it stands to reason that finding them by purely random and undirected means will take a long time. Moreover, waiting times increase exponentially with each additional necessary mutation. Thus, long waiting times for the production of new functional genes and proteins is exactly what we should expect if indeed functional genes and proteins are rare, and if coordinated mutations are necessary to produce them. Thus, the various experiments and calculations performed between 2004 and 2011 indirectly confirm Axe's earlier conclusion about the rarity of functional genes and proteins and supply further evidence that the neo-Darwinian mechanism cannot generate the information necessary to build new genes, let alone a new form of animal life, in the time available to the evolutionary process.

THE MATH AND THE MECHANISM

There is a concluding irony in all this. The researchers calculating waiting times for the appearance of complex adaptations have in each case done so using models based on the core principles of population genetics, the mathematical expression of neo-Darwinian theory. In a real sense, therefore, the neo-Darwinian math is itself showing that the neo-Darwinian mechanism cannot build complex adaptations—including the new information-rich genes and proteins that would have been necessary to build the Cambrian animals. To adapt a metaphor that Tom Frazzetta might appreciate, the snake has eaten its own tail.

13

THE ORIGIN OF BODY PLANS

Rarely have the implications of a Nobel Prize–winning scientific discovery received so little notice. Of course, the discovery itself received great acclaim. But the deeper meaning was another matter.

Starting in the autumn of 1979, at the European Molecular Biology Laboratory in Heidelberg, two venturesome young geneticists, Christiane Nüsslein-Volhard and Eric Wieschaus (see Fig. 13.1), generated thousands of mutations to investigate the genomes of tens of thousands of fruit flies (species: *Drosophila melanogaster*). They hoped to get them to divulge the secrets of embryological development. In technical jargon, Nüsslein-Volhard and Wieschaus performed "saturation mutagenesis" experiments. After feeding male flies the potent mutation-causing chemical (i.e., mutagen) ethyl methane sulphonate (EMS), Nüsslein-Volhard and Wieschaus bred the males with virgin females. They then examined the offspring larvae for visible defects.

In generating many thousands of mutants, thereby "saturating" the *Drosophila* genome, Nüsslein-Volhard and Wieschaus induced variations in the small subset of genes that specifically regulate embryonic development. These regulatory genes normally control the expression of many other genes that build the fly embryo, progressively subdividing it into regions that will become the head, thorax, and abdomen of the adult fly. The EMS mutagen disrupts DNA replication, thereby mutating genes. These mutations affect the process of development, leaving visible defects in the fly larvae. By observing the damaged larvae, Wieschaus and Nüsslein-Volhard inferred how specific genes regulate the development of different parts of the fly body plan. In essence, Wieschaus and Nüsslein-Volhard reverse-engineered the fly's genome to determine the function of the different genes, including the regulatory genes crucial to fly development.[1]

The thoroughness and novelty of the "Heidelberg screens" (as the experiments came to be known) and their importance for revealing the mechanisms of regulatory control during animal embryogenesis won the attention of the Nobel committee. In 1995, the committee awarded the Nobel Prize in Medicine or Physiology to Nüsslein-Volhard and Wieschaus. "This work was revolutionary," University of Cambridge geneticist Daniel St. Johnston explained, "because it was the first mutagenesis in any multicellular organism that attempted to find most or all of the mutations that affect . . . the essential patterning genes that are used throughout development."[2]

That's the story as it is usually told. And it's correct, as far as it goes. But the mutant fruit flies obtained by Nüsslein-Volhard and Wieschaus tell another story—one less widely known, but one containing important clues for the unsolved mystery of the origin of animal body plans.

Wieschaus himself alluded to these clues in a memorable interaction at the 1982 meeting of the American Association for the Advancement of Science (AAAS). After a session on the processes of macroevolution in which Wieschaus had presented a paper, one audience member asked him what he meant by the term "strong" as he used it to describe the mutations he and Nüsslein-Volhard had induced in flies. Wieschaus explained with a laugh that "strong" certainly did not mean *alive*. Without exception, the mutants he studied perished as deformed larvae long before achieving reproductive age. "No, dead *is* dead," he joked, "and you can't be more dead."[3]

Another questioner then asked Wieschaus about the implications of his findings for evolutionary theory. Here Wieschaus responded more soberly, wondering aloud about whether his collection of mutants offered any insights into how the evolutionary process could have constructed novel body plans. "The problem is, we think we've hit all the genes required to specify the body plan of *Drosophila*," he said, "and yet these results are obviously not promising as raw materials for macroevolution. The next question then, I guess, is what *are*—or what *would be*—the right mutations for major evolutionary change? And we don't know the answer to that."[4]

Thirty years later, developmental and evolutionary biologists still don't know the answer to that question. At the same time, mutagenesis experiments—on fruit flies as well as on other organisms such as nematodes (roundworms), mice, frogs, and sea urchins—have raised troubling questions about the role of mutations in the origin of animal body plans. If mutating the genes that regulate body-plan construction destroy animal forms as they develop from an embryonic state, then how do mutations and selection build animal body plans in the first place?

The neo-Darwinian mechanism has failed to explain the generation of new genes and proteins needed for building the new animal forms that arose in the Cambrian explosion. But even if mutation and selection could generate fundamentally new genes and proteins, a more formidable problem remains. To build a new animal and establish its body plan, proteins need to be organized into higher-level structures. In other words, once new proteins arise, something must arrange them to play their parts in distinctive cell types. These distinctive cell types must, in turn, be organized to form distinctive tissues, organs, and body plans. This process of organization occurs during embryological development. Thus, to explain how animals are actually built from smaller protein components, scientists must understand the process of embryological development.

FIGURE 13.1
Figure 13.1a (left): Christiane Nüsslein-Volhard. *Courtesy Wikimedia Commons, user Rama. Figure 13.1b (right):* Eric F. Wieschaus.

THE ROLE OF GENES AND PROTEINS IN ANIMAL DEVELOPMENT

As much as any other subdiscipline of biology, developmental biology has raised disquieting questions for neo-Darwinism. Developmental biology describes the processes, called *ontogeny,* by which embryos develop into mature organisms. Within the past three decades the field has dramatically advanced our understanding of how body plans arise during ontogeny. Much of this new knowledge has come from studying so-called model systems—organisms that biologists can easily mutate in the lab, such as the fruit fly *Drosophila* and the nematode *Caenorhabditis elegans.*

Although the exact details of animal development can vary in bewildering ways depending on the species, all animal development exemplifies a common imperative: start with one cell, end with many different cells. In most animal species, development begins with the fertilized egg. Once the egg divides into its daughter cells, becoming an embryo, the organism begins heading toward a well-defined target, namely, an adult form that can reproduce. Arriving at that distant target requires the embryo to produce many specialized cell types, in the correct positions and at the right time.

Cell differentiation involves coordinating the expression of specific genes in space and time, as the number of cells, taking on their different roles, rises from one to two to four to eight, doubling and doubling until it reaches thousands, millions, and even trillions, depending on the species. The number of cell divisions and the total number of cells reflects the number of different cell types the adult needs. This in turn requires producing different proteins for different cell types.

For example, the specialized digestive proteins that service the cells lining the adult intestine differ from proteins expressed in a neuron found in the nerve tract of a limb. They must differ because each performs dramatically different functions. So, during development, the appropriate genes must be turned on, or "up-regulated," and turned off, or "down-regulated," to ensure the production of the correct protein products at the right time and in the right cell types.

Specific proteins play active roles in regulating the expression of genes for building other proteins. The protein actors playing these coordinating roles are known as transcriptional regulators (TRs) or transcription factors (TFs). TRs (or TFs) usually bind directly to specific sites in DNA, either preventing (*repressing*) or enabling (*activating*) the transcription of specific genes into RNA. TRs or TFs convey instructions about which genes to turn on or turn off. Their three-dimensional geometries exhibit characteristic DNA-binding features, including a specific domain of 61 amino acids that wraps around the DNA double helix. Other transcription factors include the zinc finger and leucine zipper motifs that also bind to DNA. Transcriptional regulators and factors are themselves controlled by complex circuits and signals transmitted by other genes and proteins, the overall complexity and precision of which is breathtaking.

Painstaking genetic research—performed by Nüsslein-Volhard and Wieschaus and many other developmental biologists[5]—has uncovered many of the key embryonic regulatory genes that help switch cells into their differentiated adult types. This research also uncovered a profound difficulty cutting to the very core of the neo-Darwinian view of life.

EARLY-ACTING BODY-PLAN MUTATIONS AND EMBRYONIC LETHALS

To create significant changes in the forms of animals requires attention to timing. Mutations in genes expressed late in the development of an animal will affect relatively few cells and architectural features. That's because by late in development the basic outlines of the body plan have already been established.[6] Late-acting mutations therefore cannot cause any significant or heritable changes in the form or body plan of the whole animal. Mutations that are expressed early in development, however, may affect many cells and could conceivably produce significant changes in the form or body plan, especially if these changes occur in key regulatory genes.[7] Thus, mutations that are expressed early in the development of animals have probably the only realistic chance of producing large-scale macroevolutionary change.[8] As evolutionary geneticists Bernard John and George Miklos explain, "macroevolutionary change" requires changes in "very early embryogenesis."[9] Former Yale University evolutionary biologist Keith Thomson concurs: only mutations expressed early in the development of organisms can produce large-scale macroevolutionary change.[10]

Yet from the first experiments by geneticist T. H. Morgan systematically mutating fruit flies early in the twentieth century until today, as many model species have been subjected to mutagenesis, developmental biology has shown that mutations affecting body-plan formation expressed early in development inevitably damage the organism.[11] (See Fig. 13.2, for examples.) As one of the founders of neo-Darwinism geneticist R. A. Fisher noted, such mutations are "either definitely pathological (most often lethal) in their effects," or they result in an organism that cannot survive "in the wild state."[12]

Normal development in any animal can be represented as an expanding network of decisions, where the earliest (upstream) decisions have greater impact than those occurring later. Regulatory genes and their DNA-binding protein products help to control this unfolding network of decisions—such that if regulatory proteins are altered or destroyed by mutation, the effects cascade downstream into the whole developmental process. The earlier the failure, the more widespread the destruction. Geneticist Bruce Wallace explains why early-acting mutations are thus overwhelmingly likely to disrupt animal development. "The extreme difficulty encountered," he observes, "when attempting to transform one organism into another . . . still functional one lies in the difficulty in resetting a number of the many controlling switches in a manner that still allows for the individual's orderly (somatic) development."[13]

Nüsslein-Volhard and Wieschaus discovered this problem in experiments performed on fruit flies after their first Nobel Prize–winning efforts. In these later experiments they studied protein molecules that influence the organization of different types of cells early in the process of embryological development. These molecules, called "morphogens," including one called Bicoid, are critical to establishing the fruit fly's head-to-tail axis. They found that when these early-acting, body-plan-affecting molecules are perturbed, development shuts down. When mutations occur in the gene that codes for Bicoid, the resulting embryos die[14]—as they do in all other known cases in which mutations occur early in the regulatory genes that affect body-plan formation.

There are good functional reasons for this, familiar to us from the logic of other complex systems. If an automaker modifies a car's paint color or seat covers, nothing else needs to be altered for the car to operate, because the normal function of the car does not depend upon these features. But if an engineer changes the length of the piston rods in the car's engine, and does not modify the crankshaft accordingly, the engine won't run. Similarly, animal development is a tightly integrated process in which various proteins and cell structures depend upon each other for their function, and later events depend crucially on earlier events. As a result, one change early in the development of an animal will require a host of other coordinated changes in separate but functionally interrelated developmental processes and entities downstream.[15] This tight functional integration helps explain why mutations early in development inevitably result in embryonic death and why even mutations expressed somewhat later in development commonly leave organisms crippled.

Looking more closely at a specific experimental result of this kind further illuminates the problem. A mutation in the regulatory *Ultrabithorax* gene (expressed midway in the development of a fly) produces an extra pair of wings on a normally two-winged creature. Although an extra set of wings may sound like a useful piece of equipment, it's not at all. This "innovation" results in a crippled insect that cannot fly because it lacks,

FIGURE 13.2
Examples of deleterious macromutations produced by experiments on fruit flies, including the "short wings," "curly wings," "eyeless," and *Antennapedia* mutants.

among other things, a musculature to support the use of its new wings. Because the developmental mutation was not accompanied by the many other coordinated developmental changes that were needed to make the wings useful, the mutation is decidedly harmful.

This problem has led to what Georgia Tech geneticist John F. McDonald has called a "great Darwinian paradox."[16] He notes that the genes that are obviously variable within natural populations seem to affect only minor aspects of form and function—while those genes that govern major changes, the very stuff of macroevolution, apparently do not vary or vary only to the detriment of the organism. As he puts it, "Those [genetic] *loci* that are obviously variable within natural populations do not seem to lie at the basis of many major adaptive changes, while those *loci* that seemingly do constitute the foundation of many if not most major adaptive changes are not variable within natural populations."[17] In other words, the kind of mutations the evolutionary process would need to produce new animal body plans—namely, *beneficial* regulatory changes expressed early in development—don't occur. Whereas, the kind that it doesn't need—viable genetic mutations in DNA generally expressed late in development—do occur. Or put more succinctly, the kind of mutations we need for major evolutionary change we don't get; the kind we get we don't need.

My Discovery Institute colleague Paul Nelson (see Fig. 13.3), a philosopher of biology who specializes in evolutionary theory and developmental biology, summarizes the challenge to neo-Darwinism posed by animal development as three premises:

1. Animal body plans are built in each generation by a stepwise process, from the fertilized egg to the many cells of the adult. The earliest stages in this process determine what follows.

2. Thus, to evolve any body plan, mutations expressed early in development must occur, must be viable, and must be stably transmitted to offspring.

3. Such early-acting mutations of global effect on animal development, however, are those *least likely* to be tolerated by the embryo and, in fact, never have been tolerated in any animals that developmental biologists have studied.

FIGURE 13.3
Paul Nelson. *Courtesy Paul Nelson.*

Nelson came to appreciate the depth of the problem posed by these facts after many years of discussion with two members of his University of Chicago Ph.D. committee, evolutionary biologist Leigh Van Valen (1935–2010) and evolutionary theorist and philosopher of biology William Wimsatt. Van Valen, famous for his "Red Queen hypothesis" about the need for organisms to continue to evolve in order to maintain fitness, was passionately interested in the mechanisms of macroevolution. Wimsatt originated the theory of "generative entrenchment," an account of the "causal asymmetries" at work in complex systems, including those responsible for animal development.[18] Both acknowledged to Nelson that the scientific literature offers no examples of viable mutations affecting early animal development and body-plan formation (Premise 3, on previous page) and also that the macroevolution of novel animal form requires just such early-acting mutations (Premise 2, on previous page). Nevertheless, both Van Valen and Wimsatt remained committed to the descent of animal forms from a common ancestor via some kind of undirected mutations. Nelson argues, however, that those premises strongly imply that the neo-Darwinian mechanism does not—and indeed cannot—provide an adequate mechanism for producing new animal body plans. As he has told me: "If the only kind of mutations that can conceivably produce enough morphological change to alter whole body plans never causes *beneficial* and *heritable* changes, then it is difficult to see how mutation and selection could ever produce new body plans in the first place."[19]

Thus, he concludes:

Research on animal development and macroevolution over the last thirty years—research done from within the neo-Darwinian framework—has shown that the neo-Darwinian explanation for the origin of new body plans is overwhelmingly likely to be false—and for reasons that Darwin himself would have understood.

Indeed, Darwin himself insisted that "nothing can be effected" by natural selection, "unless favorable variations occur."[20] Or as Danish evolutionary biologist Søren Løvtrup succinctly explains: "Without variation, no selection; without selection, no evolution. This assertion is based on logic of the simplest kind. . . . Selection pressure as an evolutionary agent becomes void of sense unless the availability of the proper mutations is assumed."[21] Yet the "proper" kind of mutations—the mutations that produce *favorable* changes to *early-acting,* body-plan–shaping, regulatory genes—do not occur.

Microevolutionary change is insufficient; macromutations—large-scale changes—are harmful. This paradox has beset Darwinism from its inception, but discoveries about the genetic regulation of development in animals have made this paradox more acute and cast serious doubt on the efficacy of the modern neo-Darwinian mechanism as an explanation for the new body plans that arise in the Cambrian period.

DEVELOPMENTAL GENE REGULATORY NETWORKS

Another line of research in developmental biology has revealed a related challenge to the creative power of the neo-Darwinian mechanism. Developmental biologists have discovered that many gene products (proteins and RNAs) needed for the development of specific animal body plans transmit *signals* that influence the way individual cells develop and differentiate themselves. Additionally, these signals affect how cells are organized and interact with each other during embryological development. These signaling molecules influence each other to form circuits or networks of *coordinated* interaction, much like integrated circuits on a circuitboard. For example, exactly *when* a signaling molecule gets transmitted often depends upon when a signal from another molecule is received, which in turn affects the transmission of still others—all of which are coordinated and integrated to perform specific time-critical functions. The coordination and integration of these signaling molecules in cells ensures the proper differentiation and organization of distinct cell types during the development of an animal body plan. Consequently, just as mutating an individual regulatory gene early in the development of an animal will inevitably shut down development, so too will mutations or alterations in the whole network of interacting signaling molecules destroy a developing embryo.

No biologist has explored the regulatory logic of animal development more deeply than Eric Davidson, at the California Institute of Technology. Early in his career, collaborating with molecular biologist Roy Britten, Davidson formulated a theory of "gene regulation for higher cells."[22] By "higher cells" Davidson and Britten meant the differentiated, or specialized, cells found in any animal after the earliest stages of embryological development. Davidson observed that the cells of an individual animal, no matter how varied in form or function, "generally contain identical genomes."[23] During the life cycle of an organism, the genomes of these specialized cells express only a small fraction of their DNA at any given time and produce different RNAs as a result. These facts strongly suggest that some animal-wide system of genetic control functions to turn specific genes on and off as needed throughout the life of the organism—and that such a system functions during the development of an animal from egg to adult as different cell types are being constructed.

When they proposed their theory in 1969, Britten and Davidson acknowledged that "little is known . . . of the molecular mechanisms by which gene expression is controlled in differentiated cells."[24] Nevertheless, they deduced that such a system must be at work. Given: (1) that tens or hundreds of specialized cell types arise during the development of animals, and (2) that each cell contains the same genome, they reasoned (3) that some control system must determine which genes are expressed in different cells at different times to ensure the differentiation of different cell types from each other—some system-wide regulatory logic must oversee and coordinate the expression of the genome.[25]

Davidson has dedicated his career to discovering and describing the mechanisms by which these systems of gene regulation and control work during embryological development. During the last two decades, research in genomics has revealed that *non*protein-coding regions of the genome control and regulate the timing of the expression of the protein-coding regions of the genome. Davidson has shown that the *non*protein-coding

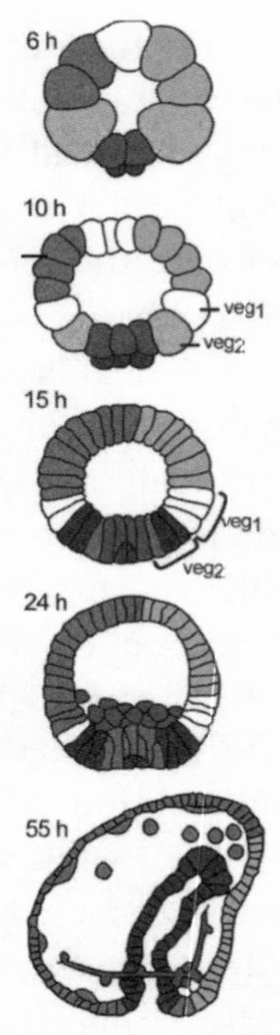

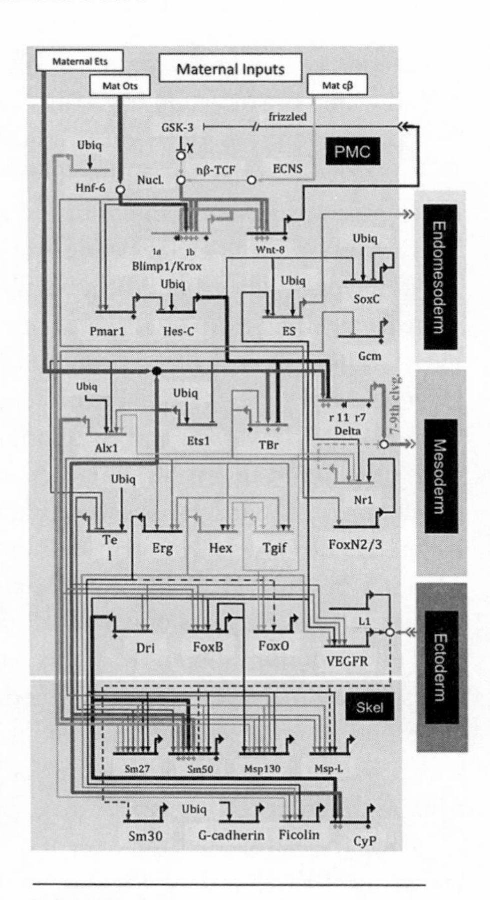

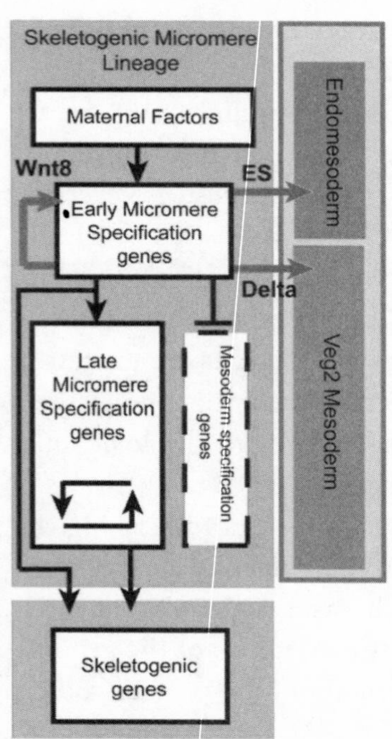

FIGURE 13.4
Developmental gene regulatory networks (dGRNs) and development in the purple sea urchin, *Strongylocentrotus purpuratus. Figure 13.4a (top, left):* shows the actual embryo, starting at 6 hours and progressing through cell division to 55 hours, when the larval skeleton appears. *Figure 13.4b (bottom, left):* depicts the major classes of genes involved in specifying the larval skeleton. *Figure 13.4c (top, right):* shows the detailed genetic circuitry implicated in the overall "gene regulatory network" ("GRN") that controls the construction of the larval skeleton. *Courtesy National Academy of Sciences, U.S.A.*

regions of DNA that regulate and control gene expression and the protein-coding regions of the genome together function as circuits. These circuits, which Davidson calls "developmental gene regulatory networks" (or dGRNs) control the embryological development of animals.

On arriving at Caltech in 1971, Davidson chose the purple sea urchin, *Strongylocentrotus purpuratus,* as his experimental model system. The biology of *S. purpuratus* makes it an attractive laboratory subject: the species occurs abundantly along the Pacific coast, produces enormous quantities of easily fertilized eggs in the lab, and lives for many years.[26] Davidson and his coworkers pioneered the technology and experimental protocols required to dissect the sea urchin's genetic regulatory system.

The remarkable complexity of what they found needs to be depicted visually. Figure 13.4a shows the urchin embryo as it appears six hours after development has begun (top left of diagram). This is the 16-cell stage, meaning that four rounds of cell division have already occurred (1 → 2 → 4 → 8 → 16). As development proceeds in the next four stages, both the number of cells and the degree of cellular specialization increases, until, at 55 hours, elements of the urchin skeleton come into focus. Figure 13.4b shows, corresponding to these drawings of embryo development, a schematic diagram with the major classes of genes (for cell and tissue types) represented as boxes, linked by control arrows. Last, Figure 13.4c shows what Davidson calls "the genetic circuitry" that turns on the specific biomineralization genes that produce the structural proteins needed to build the urchin skeleton.[27]

This last diagram represents a developmental gene regulatory network (or dGRN), an integrated network of protein and RNA-signaling molecules responsible for the differentiation and arrangement of the specialized cells that establish the rigid skeleton of the sea urchin. Notice that, to express the biomineralization genes that produce structural proteins that make the skeleton, genes far upstream, activated many hours earlier in development, must first play their role.

This process does not happen fortuitously in the sea urchin but via highly regulated and precise control systems, as it does in all animals. Indeed, even one of the simplest animals, the worm *C. elegans,* possessing just over 1,000 cells as an adult, is constructed during development by dGRNs of remarkable precision and complexity. In all animals, the various dGRNs direct what Davidson describes as the embryo's "progressive increase in complexity"—an increase, he writes, that can be measured in "informational terms."[28]

Davidson notes that, once established, the complexity of the dGRNs as integrated circuits makes them stubbornly resistant to mutational change—a point he has stressed in nearly every publication on the topic over the past fifteen years. "In the sea urchin embryo," he points out, "disarming any one of these subcircuits produces some abnormality in expression."[29]

Developmental gene regulatory networks resist mutational change because they are organized hierarchically. This means that some developmental gene regulatory networks control other gene regulatory networks, while some influence only the individual genes and proteins under their control. At the center of this regulatory hierarchy are the regulatory networks that specify the axis and global form of the animal body plan during development. These dGRNs cannot vary without causing catastrophic effects to the organism.

Indeed, there are no examples of these deeply entrenched, functionally critical circuits varying at all. At the periphery of the hierarchy are gene regulatory networks that specify the arrangements for smaller-scale features that can sometimes vary. Yet, to produce a new body plan requires altering the axis and global form of the animal. This requires mutating the very circuits that do not vary without catastrophic effects. As Davidson emphasizes, mutations affecting the dGRNs that regulate body-plan development lead to "catastrophic loss of the body part or loss of viability altogether."[30] He explains in more detail:

> *There is always an observable consequence if a dGRN subcircuit is interrupted. Since these consequences are always catastrophically bad, flexibility is minimal, and since the subcircuits are all interconnected, the whole network partakes of the quality that there is only one way for things to work. And indeed the embryos of each species develop in only one way.*[31]

ENGINEERING CONSTRAINTS

Davidson's findings present a profound challenge to the adequacy of the neo-Darwinian mechanism. Building a new animal body plan requires not just new genes and proteins, but new dGRNs. But to build a new dGRN from a preexisting dGRN by mutation and selection necessarily requires altering the preexisting developmental gene regulatory network (the very kind of change that, as we saw in Chapter 12, cannot arise without multiple coordinated mutations). In any case, Davidson's work has also shown that such alterations inevitably have catastrophic consequences.

Davidson's work highlights a profound contradiction between the neo-Darwinian account of how new animal body plans are built and one of the most basic principles of engineering—the principle of constraints. Engineers have long understood that the more functionally integrated a system is, the more difficult it is to change any part of it without damaging or destroying the system as a whole. Davidson's work confirms that this principle applies to developing organisms in spades. The system of gene regulation that controls animal-body-plan development is exquisitely integrated, so that significant alterations in these gene regulatory networks inevitably damage or destroy the developing animal.[32] But given this, how could a new animal body plan, and the new dGRNs necessary to produce it, ever evolve gradually via mutation and selection from a preexisting body plan and set of dGRNs?

Davidson makes clear that no one really knows: "contrary to classical evolution theory, the processes that drive the small changes observed as species diverge cannot be taken as models for the evolution of the body plans of animals."[33] He elaborates:

> *Neo-Darwinian evolution . . . assumes that all process works the same way, so that evolution of enzymes or flower colors can be used as current proxies for study of evolution of the body plan. It erroneously assumes that change in protein-coding sequence is the basic cause of change in developmental program; and it erroneously assumes that evolutionary change in body-plan morphology occurs by a continuous process. All of these assumptions are basically counterfactual. This cannot be surprising, since the neo-Darwinian synthesis from which these ideas stem was a premolecular biology concoction focused on population genetics and . . . natural history, neither of which have any direct mechanistic import for the genomic regulatory systems that drive embryonic development of the body plan.*[34]

NOW AND THEN

Eric Davidson's work, like that of Nüsslein-Volhard and Wieschaus, highlights a difficulty of obvious relevance to the Cambrian explosion. Typically, paleontologists understand the Cambrian explosion as the geologically sudden appearance of new *forms* of animal life. Building these forms requires new developmental programs—including both new early-acting regulatory genes *and* new developmental gene regulatory networks. Yet if neither early-acting regulatory genes nor dGRNs can be altered by mutation without destroying existing developmental programs (and thus animal form), then mutating these entities will leave natural selection with nothing favorable to select and the evolution of animal form will, at that point, terminate.

Darwin's doubt about the Cambrian explosion centered on the problem of missing fossil intermediates. Not only have those forms not been found, but the Cambrian explosion itself illustrates a profound engineering problem that fossil evidence does not address—the problem of building a new form of animal life by gradually transforming one tightly integrated system of genetic components and their products into another. Yet, in the next chapter, we will see that an even more formidable problem remains.

14

THE EPIGENETIC REVOLUTION

In 1924, two German scientists, Hans Spemann and Hilda Mangold, reported an intriguing experiment, the significance of which could not have been fully appreciated at the time, three decades before the discovery of the information-bearing properties of DNA. Using microsurgery, Spemann and Mangold excised a portion of a newt embryo and transplanted that portion into another developing newt embryo.[1]

They achieved a startling result. The second embryo produced two bodies, each with a head and tail, joined together at the belly, not unlike Siamese twins. Yet despite dramatically altering the anatomy of the embryo, Spemann and Mangold did not alter its DNA.

Their experiment later suggested a radical possibility: that something in addition to DNA profoundly influences the development of animal body plans. Other experiments suggested as much. In the 1930s and 1940s, American biologist Ethel Harvey showed experimentally that sea urchin embryos could undergo development up to about 500 cells after removal of their nuclei—in other words, *without* their nuclear DNA.[2] In the 1960s, Belgian scientists chemically blocked the transcription of DNA into RNA in amphibian embryos and found that the embryos could still develop to the point of containing several thousand cells.[3] In the 1970s, Canadian biologists showed that a frog embryo could undergo early development without its nucleus if the cell division apparatus from a sea urchin was injected into the egg.[4]

None of these results indicate that embryos can develop fully without DNA. In every case, DNA was eventually necessary to complete embryological development. Yet these results suggest that DNA is not the whole story, that other sources of information are playing important roles in directing at least the early stages of animal development.

ABOVE AND BEYOND: EPIGENETIC INFORMATION

In 2003, MIT Press published a groundbreaking collection of scientific essays titled *Origination of Organismal Form: Beyond the Gene in Developmental and Evolutionary Biology,* edited by two distinguished developmental and evolutionary biologists, Gerd Müller, of the University of Vienna, and Stuart Newman, of New York Medical College. In their volume, Müller and Newman included a number of scientific articles describing recent discoveries in genetics and developmental biology—discoveries suggesting that genes alone do not determine the three-dimensional form and structure of an animal. Instead, many of the scientists in their volume reported that so-called epigenetic information—information stored in cell structures, but not in DNA sequences—plays a crucial role. The Greek prefix *epi* means "above" or "beyond," so epigenetics refers to a source of information that lies beyond the genes. As Müller and Newman explain in their introduction, "Detailed information at the level of the gene does not serve to explain form."[5] Instead, as Newman explains, "epigenetic" or "contextual information" plays a crucial role in the formation of animal "body assemblies" during embryological development.[6]

Müller and Newman not only highlighted the importance of epigenetic information to the formation of body plans during development; they also argued that it must have played a similarly important role in the origin and evolution of animal body plans in the first place. They concluded that recent discoveries about the role of epigenetic information in animal development pose a formidable challenge to the standard neo-Darwinian account of the origin of these body plans—perhaps the most formidable of all.

In the introductory essay to their volume, Müller and Newman list a number of "open questions" in evolutionary biology, including the question of the origin of Cambrian-era animal body plans and the origin of organismal form generally, the latter being the central topic of their book. They note that though "the neo-Darwinian paradigm still represents the central explanatory framework of evolution," it has "no theory of the generative."[7] In their view, neo-Darwinism "completely avoids [the question of] the origination of phenotypic traits and of organismal form."[8] As they and others in their volume maintain, neo-Darwinism lacks an explanation for the origin of organismal form precisely because it cannot explain the origin of epigenetic information.

I first learned about the problem of epigenetic information and the Spemann and Mangold experiment while driving to a private meeting of Darwin-doubting scientists on the central coast of California in 1993. In the car with me was Jonathan Wells (see Fig. 14.1), who was then finishing a Ph.D. in developmental biology at the University of California at Berkeley. Like some others in his field, Wells had come to reject the (exclusively) "gene-centric" view of animal development and to recognize the importance of nongenetic sources of information.

By that time, I had studied many questions and challenges to standard evolutionary theories arising out of molecular biology. But epigenetics was new to me. On our drive, I asked Wells why developmental biology was so important to evolutionary theory and to assessing neo-Darwinism. I'll never forget his reply. "Because" he said, "that's where the whole theory is going to unravel."

In the years since, Wells has developed a powerful argument against the adequacy of the neo-Darwinian mechanism as an explanation for

FIGURE 14.1
Jonathan Wells. *Courtesy Laszlo Bencze.*

the origin of animal body plans. His argument turns on the importance of epigenetic information to animal development. To see why epigenetic information poses an additional challenge to neo-Darwinism and what exactly biologists mean by "epigenetic" information, let's examine the relationship between biological form and biological information.

FORM AND INFORMATION

Biologists typically define "form" as a distinctive shape and arrangement of body parts. Organismal forms exist in three spatial dimensions and arise in time—in the case of animals during development from embryo to adult. Animal form arises as material constituents are constrained to establish specific arrangements with an identifiable three-dimensional shape or "topography"—one that we would recognize as the body plan of a particular type of animal. A particular "form," therefore, represents a highly specific arrangement of material components among a much larger set of possible arrangements.

Understanding form in this way suggests a connection to the notion of information in its most theoretically general sense. As I noted in Chapter 8, Shannon's mathematical theory of information equated the amount of information transmitted with the amount of uncertainty reduced or eliminated in a series of symbols or characters. Information, in Shannon's theory, is thus imparted as some options, or possible arrangements, are excluded and others are actualized. The greater the number of arrangements excluded, the greater the amount of information conveyed. Constraining a set of possible material arrangements, by whatever means, involves excluding some options and actualizing others. Such a process generates information in the most general sense of Shannon's theory. It follows that the constraints that produce biological form also impart in*form*ation, even if this information is not encoded in digital form.

DNA contains not only Shannon information but also *functional* or *specified* information. The arrangements of nucleotides in DNA or of amino acids in a protein are highly improbable and thus contain large amounts of Shannon information. But the function of DNA and proteins depends upon extremely *specific* arrangements of bases and amino acids.

Similarly, animal body plans represent, not only highly improbable, but also highly specific arrangements of matter. Organismal form and function depend upon the precise arrangement of various constituents as they arise during, or contribute to, embryological development. Thus, the specific *arrangement* of the other building blocks of biological form—cells, clusters of similar cell types, dGRNs, tissues, and organs—also represent a kind of specified or functional information.

In Chapter 8, I noted that the ease with which Shannon's information theory applies to molecular biology has sometimes led to confusion about the kind of information contained in DNA and proteins. It may have also created confusion about the places that specified information might reside in organisms. Perhaps because the information-carrying capacity of the gene can be so easily measured, biologists have often treated DNA, RNA, and proteins as the sole repositories of biological information. Neo-Darwinists have assumed that genes possess all the information necessary to specify the form of an animal. They have also assumed that mutations in genes will suffice to generate the new information necessary to build a new form of animal life.[9] Yet if biologists understand organismal form as resulting from constraints on the possible arrangements of matter at many levels in the biological hierarchy—from genes and proteins, to cell types and tissues, to organs and body plans—then biological organisms may well exhibit many levels of information-rich structure. Discoveries in developmental biology have confirmed this possibility.

BEYOND GENES

Many biologists no longer believe that DNA directs virtually everything happening within the cell. Developmental biologists, in particular, are now discovering more and more ways that crucial information for building body plans is imparted by the form and structure of embryonic cells, including information from both the unfertilized and fertilized egg.

Biologists now refer to these sources of information as "epigenetic."[10] Spemann and Mangold's experiment is only one of many to suggest that something beyond DNA may be influencing the development of animal body plans. Since the 1980s, developmental and cell biologists such as Brian Goodwin, Wallace Arthur, Stuart Newman, Fred Nijhout, and Harold Franklin have discovered or analyzed many sources of epigenetic information. Even molecular biologists such as Sidney Brenner, who pioneered the idea that genetic programs direct animal development, now insist that the information needed to code for complex biological systems vastly outstrips the information in DNA.[11]

DNA helps direct protein synthesis. Parts of the DNA molecule also help to regulate the timing and expression of genetic information and the synthesis of various proteins within cells. Yet once proteins are synthesized, they must be arranged into higher-level systems of proteins and structures. Genes and proteins are made from simple building blocks—nucleotide bases and amino acids, respectively—arranged in specific ways. Similarly, distinctive cell types are made of, among other things, systems of specialized proteins. Organs are made of specialized arrangements of cell types and tissues. And body plans comprise specific arrangements of specialized organs. Yet the properties of individual proteins do not fully determine the organization of these higher-level structures and patterns.[12] Other sources of information must help arrange individual proteins into systems of proteins, systems of proteins into distinctive cell types, cell types into

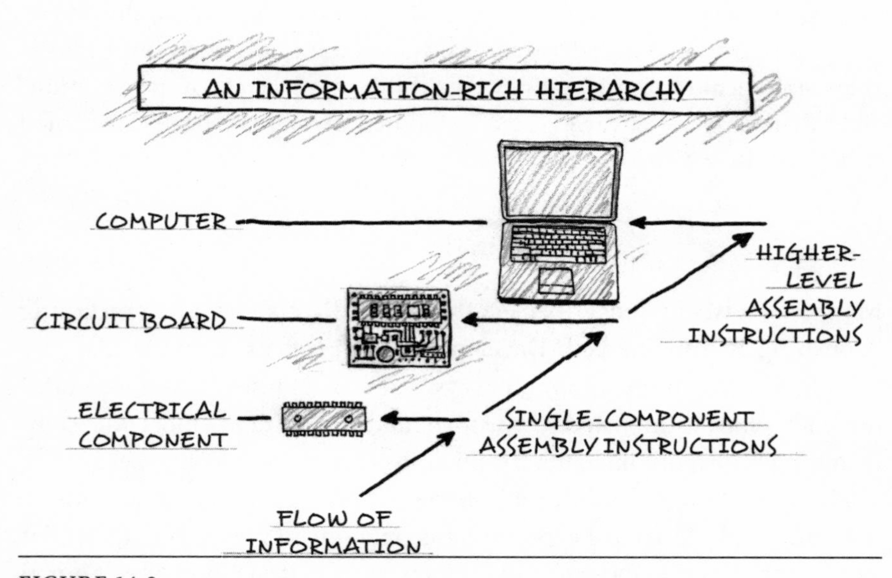

FIGURE 14.2
The hierarchical layering or arrangement of different sources of information. Note that the information necessary to build the lower-level electronic components does not determine the arrangement of those components on the circuit board or the arrangement of the circuit board and the other parts necessary to make a computer. That requires additional informational inputs.

tissues, and different tissues into organs. And different organs and tissues must be arranged to form body plans.

Two analogies may help clarify the point. At a construction site, builders will make use of many materials: lumber, wires, nails, drywall, piping, and windows. Yet building materials do not determine the floor plan of the house or the arrangement of houses in a neighborhood. Similarly, electronic circuits are composed of many components, such as resistors, capacitors, and transistors. But such lower-level components do not determine their own arrangement in an integrated circuit (see Fig. 14.2).

In a similar way, DNA does not by itself direct how individual proteins are assembled into these larger systems or structures—cell types, tissues, organs, and body plans—during animal development.[13] Instead, the three-dimensional structure or spatial architecture of embryonic cells plays important roles in determining body-plan formation during embryogenesis. Developmental biologists have identified several sources of epigenetic information in these cells.

CYTOSKELETAL ARRAYS

Eukaryotic cells have internal skeletons to give them shape and stability. These "cytoskeletons" are made of several different kinds of filaments including those called the "microtubules." The structure and location of the microtubules in the cytoskeleton influence the patterning and development of embryos. Microtubule "arrays" within embryonic cells help to distribute essential proteins used during development to specific locations in these cells. Once delivered, these proteins perform functions critical to development, but they can only do so if they are delivered to their correct locations with the help of preexisting, precisely structured microtubule or cytoskeletal arrays (see Fig. 14.3). Thus, the precise arrangement of microtubules in the cytoskeleton constitutes a form of critical structural information.

These microtubule arrays are made of proteins called tubulin, which are gene products. Nevertheless, like bricks that can be used to assemble many different structures, the tubulin proteins in the cell's microtubules are identical to one another. Thus, neither the tubulin subunits, nor the genes that produce them, account for the differences in the shape of the microtubule arrays that distinguish different kinds of embryos and developmental

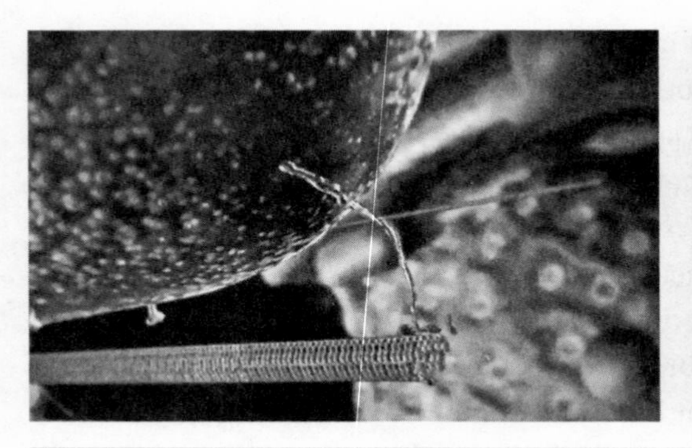

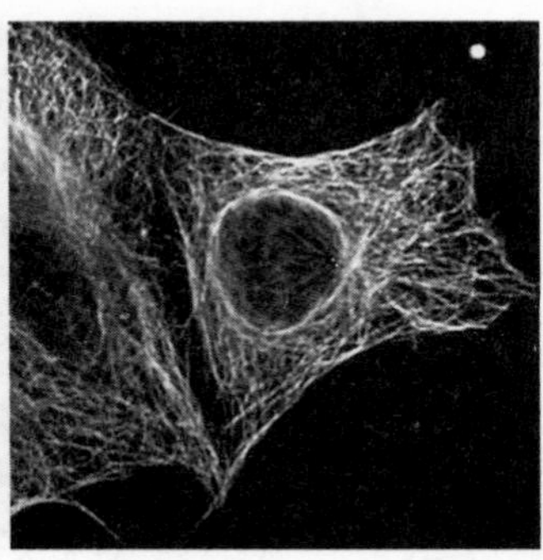

FIGURE 14.3
Figure 14.3a (left) shows a still shot from an animation of microtubule (at the bottom of the image) made of tubulin proteins. *Courtesy Joseph Condeelis. Figure 14.3b (right)* shows a microscopic image of a large section of cytoskeleton made of many microtubules (and other elements) inside the cell in cross section. *Courtesy The Company of Biologists and Journal of Cell Science.*

pathways. Instead, the structure of the microtubule array itself is, once again, determined by the location and arrangement of its subunits, not the properties of the subunits themselves. Jonathan Wells explains it this way: "What matters in [embryological] development is the shape and location of microtubule arrays, and the shape and location of a microtubule array is not determined by its units."[14] For this reason, as University of Colorado cell biologist Franklin Harold notes, it is impossible to predict the structure of the cytoskeleton of the cell from the characteristics of the protein constituents that form that structure.[15]

Another cell structure influences the arrangement of the microtubule arrays and thus the precise structures they form and the functions they perform. In an animal cell, that structure is called the centrosome (literally, "central body"), a microscopic organelle that sits next to the nucleus between cell divisions in an undividing cell. Emanating from the centrosome is the microtubule array that gives a cell its three-dimensional shape and provides internal tracks for the directed transport of organelles and essential molecules to and from the nucleus.[16] During cell division the centrosome duplicates itself. The two centrosomes form the poles of the cell-division apparatus, and each daughter cell inherits one of the centrosomes; yet the centrosome contains no DNA.[17] Though centrosomes are made of proteins—gene products—the centrosome structure is not determined by genes alone.

MEMBRANE PATTERNS

Another important source of epigenetic information resides in the two-dimensional patterns of proteins in cell membranes.[18] When messenger RNAs are transcribed, their protein products must be transported to the proper locations in embryonic cells in order to function properly. Directed transport involves the cytoskeleton, but it also depends on spatially localized targets in the membrane that are in place before transport occurs. Developmental biologists have shown that these membrane patterns play a crucial role in the embryological development of fruit flies.

Membrane Targets

For example, early embryo development in the fruit fly *Drosophila melanogaster* requires the regulatory molecules Bicoid and Nanos (among others). The former is required for anterior (head) development, and the latter is required for posterior (tail) development.[19] In the early stages of embryological development, nurse cells pump Bicoid and Nanos RNAs into the egg. (Nurse cells provide the cell that will become the egg—known as the oocyte—and the embryo with maternally encoded messenger RNAs and proteins.) Cytoskeletal arrays then transport these RNAs through the oocyte, where they become attached to specified targets on the inner surface of the egg.[20] Once in their proper place—but only then—Bicoid and Nanos play critical roles in organizing the head-to-tail axis of the developing fruit fly. They do this by forming two gradients (or differential concentrations), one with Bicoid protein most concentrated at the anterior end and another with Nanos protein most concentrated at the posterior end.

Insofar as both of these molecules are RNAs—that is, gene products—genetic information plays an important role in this process. Even so, the information contained in the *bicoid* and *nanos* genes does not by itself ensure the proper function of the RNAs and proteins for which the genes code. Instead, preexisting membrane targets, already positioned on the inside surface of the egg cell, determine where these molecules will attach and how they will function. These membrane targets provide crucial information—spatial coordinates—for embryological development.

Ion Channels and Electromagnetic Fields

Membrane patterns can also provide epigenetic information by the precise arrangement of ion channels—openings in the cell wall through which charged electrical particles pass in both directions. For example, one type of channel uses a pump powered by the energy-rich molecule ATP to transport three sodium ions out of the cell for every two potassium ions that enter the cell. Since both ions have a charge of plus one (Na+, K+), the net difference sets up an electromagnetic field across the cell membrane.[21]

Experiments have shown that electromagnetic fields have "morphogenetic" effects—in other words, effects that influence the form of a developing organism. In particular, some experiments have shown that the targeted disturbance of these electric fields disrupts normal development in ways that suggest the fields are controlling morphogenesis.[22] Artificially applied electric fields can induce and guide cell migration. There is also evidence that direct current can affect gene expression, meaning internally generated electric fields can provide spatial coordinates that guide embryogenesis.[23] Although the ion channels that generate the fields consist of proteins that may be encoded by DNA (just as microtubules consist of subunits encoded by DNA), their pattern in the membrane is not. Thus, in addition to the information in DNA that encodes morphogenetic proteins, the spatial arrangement and distribution of these ion channels influences the development of the animal.

The Sugar Code

Biologists know of an additional source of epigenetic information stored in the arrangement of sugar molecules on the exterior surface of the cell membrane. Sugars can be attached to the lipid molecules that make up the membrane itself (in which case they are called "glycolipids"), or they can be attached to the proteins embedded in the membrane (in which case they are called "glycoproteins"). Since simple sugars can be combined in many more ways than amino acids, which make up proteins, the resulting cell surface patterns can be enormously complex. As biologist Ronald Schnaar explains, "Each [sugar] building block can assume several different positions. It is as if an A could serve as four different letters, depending on whether it was standing upright, turned upside down, or laid on either of its sides. In fact, seven simple sugars can be rearranged to form hundreds of thousands of unique words, most of which have no more than five letters."[24]

These sequence-specific information-rich structures influence the arrangement of different cell types during embryological development. Thus, some cell biologists now refer to the arrangements of sugar molecules as the "sugar code" and compare these sequences to the digitally encoded information stored in DNA.[25] As biochemist Hans-Joachim Gabius notes, sugars provide a system with "high-density coding" that is "essential to allow cells to communicate efficiently and swiftly through complex surface interactions."[26] According to Gabius, "These [sugar] molecules surpass amino acids and nucleotides by far in information-storing capacity."[27] So the precisely arranged sugar molecules on the surface of cells clearly represent another source of information independent of that stored in DNA base sequences.

NEO-DARWINISM AND THE CHALLENGE OF EPIGENETIC INFORMATION

These different sources of epigenetic information in embryonic cells pose an enormous challenge to the sufficiency of the neo-Darwinian mechanism. According to neo-Darwinism, new information, form, and structure arise from natural selection acting on random mutations arising at a very low level within the biological hierarchy—within the genetic text. Yet both body-plan formation during embryological development and major morphological innovation during the history of life depend upon a specificity of arrangement at a much higher level of the organizational hierarchy, a level that DNA alone does not determine. If DNA isn't wholly responsible for the way an embryo develops—for body-plan morphogenesis—then DNA sequences can mutate indefinitely and still not produce a new body plan, regardless of the amount of time and the number of mutational trials available to the evolutionary process. Genetic mutations are simply the wrong tool for the job at hand.

Even in a best-case scenario—one that ignores the immense improbability of generating new genes by mutation and selection—mutations in DNA sequence would merely produce new *genetic* information. But building a new body plan requires *more* than just genetic information. It requires both genetic and *epigenetic* information—information by definition that is not stored in DNA and thus cannot be generated by mutations to the DNA. It follows that the mechanism of natural selection acting on random mutations in DNA cannot by itself generate novel body plans, such as those that first arose in the Cambrian explosion.

GENE-CENTRIC RESPONSES

Many of the biological structures that impart important three-dimensional spatial information—such as cytoskeletal arrays and membrane ion channels—are made of proteins. For this reason, some biologists have insisted that the genetic information in DNA that codes for these proteins does account for the spatial information in these various structures after all. In each case, however, this exclusively "gene-centric" view of the location of biological information—and the origin of biological form—has proven inadequate.

First, in at least the case of the sugar molecules on the cell surface, gene products play no direct role. Genetic information produces proteins and RNA molecules, not sugars and carbohydrates. Of course, important glycoproteins and glycolipids (sugar-protein and sugar-fat composite molecules) are modified as the result of biosynthetic pathways involving networks of proteins. Nevertheless, the genetic information that generates the proteins in these pathways only determines the function and structure of the individual proteins; it does not specify the coordinated interaction between the proteins in the pathways that result in the modification of sugars.[28]

More important, the *location* of specific sugar molecules on the exterior surface of embryonic cells plays a critical role in the function that these sugar molecules play in intercellular communication and arrangement. Yet their location is not determined by the genes that code for the proteins to which these sugar molecules might be attached. Instead, research suggests that protein *patterns* in the cell membrane are transmitted directly from parent membrane to daughter membrane during cell division rather than as a result of gene expression in each new generation of cells.[29] Since the sugar molecules on the exterior of the cell membrane are attached to proteins and lipids, it follows that their position and arrangement probably result from membrane-to-membrane transmission as well.

Consider next the membrane targets that play a crucial role in embryological development by attracting morphogenetic molecules to specific places on the inner surface of the cell. These membrane targets consist largely of proteins, most of which are mainly specified by DNA. Even so, many "intrinsically disordered"[30] proteins fold differently depending on the surrounding cellular context. This context thus provides epigenetic information. Further, many membrane targets include more than one protein, and these multiprotein structures do not automatically self-organize to form properly structured targets.[31] Finally, it is not only the molecular structure of these membrane targets, but also their specific *location* and distribution that determines their function. Yet the location of these targets on the inner surface of the cell is not determined by the gene products out of which they are made any more than, for example, the locations of the bridges across the River Seine in Paris are determined by the properties of the stones out of which they are made.

Similarly, the sodium-potassium ion pumps in cell membranes are indeed made of proteins. Nevertheless, it is, again, the location and distribution of those channels and pumps in the cell membrane that establish the contours of the electromagnetic field that, in turn, influence embryological development. The protein constituents of these channels do not determine where the ion channels are located.

Like membrane targets and ion channels, microtubules are also made of many protein subunits, themselves undeniably the products of genetic information. In the case of microtubule arrays, defenders of the gene-centric view do not claim that individual tubulin proteins determine the structure of these arrays. Nevertheless, some have suggested that other proteins, or suites of proteins, acting in concert could determine such higher-level form. For example, some biologists have noted that so-called helper proteins—which are gene products—called "microtubule associated proteins" (MAPs) help to assemble the tubulin subunits in the microtubule arrays.

Yet MAPs, and indeed many other necessary proteins, are only part of the story. The locations of specified target sites on the interior of the cell membrane also help to determine the shape of the cytoskeleton. And, as noted, the gene products out of which these targets are made do not determine the location of these targets. Similarly, the position and structure of the centrosome—the microtubule-organizing center—also influences the structure of the cytoskeleton. Although centrosomes are made of proteins, the proteins that form these structures do not entirely determine their location and form. As Mark McNiven, a molecular biologist at the Mayo Clinic, and cell biologist Keith Porter, formerly of the University of Colorado, have shown, centrosome structure and membrane patterns *as a whole* convey three-dimensional structural information that helps determine the structure of the cytoskeleton and the location of its subunits.[32] Moreover, as several other biologists have shown, the centrioles that compose the centrosomes replicate independently of DNA replication: daughter centrioles receive their form from the overall structure of the mother centriole, not from the individual gene products that constitute them.[33]

Additional evidence of this kind comes from ciliates, large single-celled eukaryotic organisms. Biologists have shown that microsurgery on the cell membranes of ciliates can produce heritable changes in membrane patterns without altering the DNA.[34] This suggests that membrane patterns (as opposed to membrane constituents) are impressed directly on daughter cells. In both cases—in membrane patterns and centrosomes—form is transmitted from parent three-dimensional structures to daughter three-dimensional structures directly. It is not entirely contained in DNA sequences or the proteins for which these sequences code.[35]

Instead, in each new generation, the form and structure of the cell arises as the result of *both* gene products and the preexisting three-dimensional structure and organization inherent in cells, cell membranes, and cytoskeletons. Many cellular structures are built from proteins, but proteins find their way to correct locations in part because of preexisting three-dimensional patterns and organization inherent in cellular structures. Neither structural proteins nor the genes that code for them can alone determine the three-dimensional shape and structure of the entities they build. Gene products provide necessary, but not sufficient, conditions for the development of three-dimensional structure within cells, organs, and body plans.[36] If this is so, then natural selection acting on genetic variation and mutations alone cannot produce the new forms that arise in the history of life.

EPIGENETIC MUTATIONS

When I explain this in public talks, I can count on getting the same question. Someone in the audience will ask whether mutations could alter the structures in which epigenetic information resides. The questioner wonders if changes in epigenetic information could supply the variation and innovation that natural selection needs to generate new form, in much the same way that neo-Darwinists envision *genetic* mutations doing so. It's a reasonable thing to ask, but it turns out that mutating epigenetic information doesn't offer a realistic way of generating new forms of life.

First, the structures in which epigenetic information inheres—cytoskeletal arrays and membrane patterns, for example—are much larger than individual nucleotide bases or even stretches of DNA. For this reason, these structures are not vulnerable to alteration by many of the typical sources of mutation that affect genes such as radiation and chemical agent.

Second, to the extent that cell structures can be altered, these alterations are overwhelmingly likely to have harmful or catastrophic consequences. The original Spemann and Mangold experiment did, of course, involve forcibly altering an important repository of epigenetic information in a developing embryo. Yet the resulting embryo, though interesting and illustrative of the importance of epigenetic information, did not stand a chance of surviving in the wild, let alone reproducing.

Altering the cell structures in which epigenetic information inheres will likely result in embryo death or sterile offspring—for much the same reason that mutating regulatory genes or developmental gene regulatory networks also produces evolutionary dead ends. The epigenetic information provided by various cell structures is critical to body-plan development, and many aspects of embryological development depend upon the precise three-dimensional placement and location of these information-rich cell structures. For example, the specific function of morphogenetic proteins, the regulatory proteins produced by master regulatory (*Hox*) genes, and developmental gene regulatory networks (dGRNs) all depend upon the location of specific, information-rich, and preexisting cell structures. For this reason, altering these cell structures will in all likelihood damage *something* else crucial during the developmental trajectory of the organism. Too many different entities involved in development depend for their proper function upon epigenetic information for such changes to have a beneficial or even neutral effect.

In Chapter 16 I will examine several new theories of evolution, including one known as "epigenetic inheritance." We'll see that there are some additional difficulties associated with the idea that mutations in epigenetic structures can produce significant evolutionary innovation.

DARWIN'S GROWING ANOMALY

With the publication of *On the Origin of Species* in 1859, Darwin advanced, first and foremost, an explanation for the origin of biological form. At the time, he acknowledged that the pattern of appearance of the Cambrian animals did not conform to his gradualist picture of the history of life. Thus, he regarded the Cambrian explosion as primarily a problem of incompleteness in the fossil record.

In Chapters 2, 3, and 4, I explained why the problem of fossil discontinuity exemplified by the Cambrian forms has, since Darwin's time, only intensified. Yet clearly a more fundamental problem now afflicts the whole edifice of modern neo-Darwinian theory. The neo-Darwinian mechanism does not account for either the origin of the genetic or the *epigenetic* information necessary to produce new forms of life. Consequently, the problems posed to the theory by the Cambrian explosion remain unsolved. But further, the central problem that Darwin set out to answer in 1859, namely the origin of animal form in general, remains unanswered—as Müller and Newman in particular have noted.[37]

Contemporary critics of neo-Darwinism acknowledge, of course, that preexisting forms of life can *diversify* under the twin influences of natural selection and genetic mutation. Known microevolutionary processes can account for small changes in the coloring of peppered moths, the acquisition of antibiotic resistance in different strains of bacteria, and cyclical variations in the size of Galápagos finch beaks. Nevertheless, many biologists now argue that neo-Darwinian theory does not provide an adequate explanation for the origin of new body plans or events such as the Cambrian explosion.

For example, evolutionary biologist Keith Stewart Thomson, formerly of Yale University, has expressed doubt that large-scale morphological changes could accumulate by minor changes at the genetic level.[38]

Geneticist George Miklos, of the Australian National University, has argued that neo-Darwinism fails to provide a mechanism that can produce large-scale innovations in form and structure.[39] Biologists Scott Gilbert, John Opitz, and Rudolf Raff have attempted to develop a new theory of evolution to supplement classical neo-Darwinism, which, they argue, cannot adequately explain large-scale macroevolutionary change. As they note:

> *Starting in the 1970s, many biologists began questioning its [neo-Darwinism's] adequacy in explaining evolution. Genetics might be adequate for explaining microevolution, but microevolutionary changes in gene frequency were not seen as able to turn a reptile into a mammal or to convert a fish into an amphibian. Microevolution looks at adaptations that concern the survival of the fittest, not the arrival of the fittest. As Goodwin (1995) points out, "the origin of species—Darwin's problem—remains unsolved."*[40]

Gilbert and his colleagues have tried to solve the problem of the origin of form by invoking mutations in genes called *Hox* genes, which regulate the expression of other genes involved in animal development—an approach that I will examine in Chapter 16.[41] Notwithstanding, many leading biologists and paleontologists—Gerry Webster and Brian Goodwin, Günter Theissen, Marc Kirschner, and John Gerhart, Jeffrey Schwartz, Douglas Erwin, Eric Davidson, Eugene Koonin, Simon Conway Morris, Robert Carroll, Gunter Wagner, Heinz-Albert Becker and Wolf-Eckhart Lönnig, Stuart Newman and Gerd Müller, Stuart Kauffman, Peter Stadler, Heinz Saedler, James Valentine, Giuseppe Sermonti, James Shapiro and Michael Lynch, to name several—have raised questions about the adequacy of the standard neo-Darwinian mechanism, and/or the problem of evolutionary novelty in particular.[42] For this reason, the Cambrian explosion now looks less like the minor anomaly that Darwin perceived it to be, and more like a profound enigma, one that exemplifies a fundamental and as yet unsolved problem—the origination of animal form.

PART THREE

AFTER DARWIN, WHAT?

15

THE POST-DARWINIAN WORLD AND SELF-ORGANIZATION

The year 2009 marked the 150th anniversary of the publication of *The Origin of Species*. In that year, the renowned Cambrian paleontologist Simon Conway Morris published an essay in the journal *Current Biology* titled "Walcott, the Burgess Shale and rumours of a post-Darwinian world," assessing the current state of evolutionary biology. "Everywhere elsewhere in the *Origin* the arguments slide one by one skillfully into place, the towering edifice rises, and the creationists are left permanently in its shadow," he wrote. "But not when it comes to the seemingly abrupt appearance of animal fossils."[1] Instead, unresolved problems exposed by the Cambrian explosion have, in Conway Morris's view, "opened the way to a post-Darwinian world."[2] The evidence we reviewed in the previous sections of the book—evidence *for* a real, rather than merely apparent, explosion of animal form in the fossil record, and *against* the neo-Darwinian mechanism as an explanation for the origin of form and information—may help to explain why biology has begun to enter such a world.

Moreover, any doubts that at least some biologists have begun to embrace a post-Darwinian perspective should have been laid to rest in the summer of 2008, when sixteen influential *evolutionary* biologists met for a private conference at the Konrad Lorenz Institute in Altenberg, Austria. The scientists, whom the science media later dubbed the "Altenberg 16,"[3] met to explore the future of evolutionary theory. These biologists had many different, and sometimes conflicting, ideas about how new forms of life might have evolved. But all were united by the conviction that the neo-Darwinian synthesis had run its course and that new evolutionary mechanisms were needed to explain the origin of biological form. As paleontologist Graham Budd, who was in attendance, explained, "When the public thinks about evolution, they think about [things like] the origin of wings. . . . But these are things that evolutionary theory has told us little about."[4]

Of course, explaining the origin of form is precisely what has made the Cambrian explosion so mysterious. In Chapter 7, in discussing the idea of punctuated equilibrium, I quoted Cambrian paleontologists James Valentine and Douglas Erwin, who concluded exactly that. They argued that neither punctuated equilibrium nor neo-Darwinism has accounted for the origin of new body plans and that, consequently, biology needs a new theory to explain "the evolution of novelty."[5]

The Altenberg 16 sought to address this challenge. Since the conference, and for nearly two decades preceding it, many evolutionary biologists have been working to formulate new theories of evolution, or at least new ideas about evolutionary mechanisms with more creative power than mutation and natural selection alone. Each of these new theories attempts to answer the increasingly urgent question: After Darwin—or neo-Darwinism—*what?*

THE NEO-DARWINIAN TRIAD

The neo-Darwinian mechanism rests on three core claims: first, that evolutionary change occurs as the result of random, minute variations (or mutations); second, that the process of natural selection sifts among those variations and mutations, such that some organisms leave more offspring than others (differential reproduction) based on the presence or absence of certain variations; and third, favored variations must be *inherited* faithfully in subsequent generations of organisms, thus causing the population in which they reside to change or evolve over time.[6] Biologists Marc Kirschner and John Gerhart call these three elements—variation, natural selection, and heritability—the "three pillars" of neo-Darwinian evolution.[7]

Those evolutionary biologists who now doubt orthodox neo-Darwinian theory typically question or reject one or more of the elements of this neo-Darwinian triad (see Fig. 16.1). Eldredge and Gould questioned Darwinian gradualism, which led them to reject the idea that mutational change occurs in minute increments (i.e., the first element of the neo-Darwinian triad just mentioned). Other evolutionary biologists have since rejected other core elements of the neo-Darwinian mechanism and sought to replace them with other mechanisms or processes. This chapter will examine a new class of post-neo-Darwinian evolutionary models that attempt to explain the origin of biological form by deemphasizing the role of random mutations. These models instead emphasize the importance of "self-organizational" laws or processes to the evolution of biological form.

SELF-ORGANIZATIONAL MODELS

Well before the Altenberg 16 convened, a significant number of evolutionary theorists had already begun to explore alternatives to the neo-Darwinian synthesis. Punctuated equilibrium was one such alternative. But as scientific criticisms of that theory began to mount during the 1980s and 1990s, a group of scientists associated with a think tank in New Mexico, the Santa Fe Institute, developed a new theoretical approach. They called it "self-organization."

Whereas neo-Darwinism explains the origin of biological form and structure as the consequence of natural selection acting on random mutations, self-organizational theorists suggest that biological form often arises (or "self-organizes") spontaneously as a consequence of the laws of nature (or "laws of form"). Natural selection, they theorize, acts to preserve this spontaneously arising order. They think spontaneous self-organizing order, not random genetic mutations, typically provides the ultimate source of new biological form. Thus, they deemphasize two of the three parts of the classical neo-Darwinian triad: random mutations and, to a lesser extent, natural selection.

In 1993, the most prominent scientist associated with the Santa Fe Institute, former University of Pennsylvania biochemist Stuart Kauffman (see Fig. 15.1), released an eagerly awaited treatise, *The Origins of Order: Self-Organization and Selection in Evolution*.[8] Kauffman articulated a trenchant

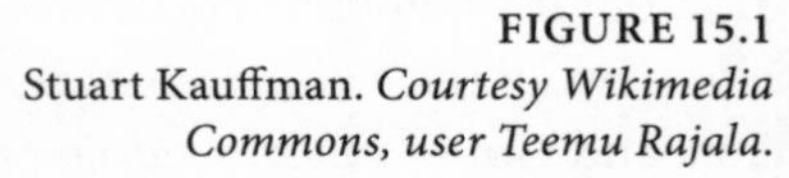

FIGURE 15.1
Stuart Kauffman. *Courtesy Wikimedia Commons, user Teemu Rajala.*

critique of the creative power of the mutation and selection mechanism, emphasizing some of the criticisms described here in previous chapters. Kauffman advanced a comprehensive alternative theory to account for the emergence of new form. In addition, he advanced a specific proposal for explaining the Cambrian explosion.[9]

Kauffman notes that the development of animal body plans involves two phases: cell differentiation and body-plan morphogenesis (cell organization). He explores the possibility that self-organizational processes at work today—specifically in cell differentiation and body-plan formation—might help explain how new animal forms originated in the past.

Kauffman proposes, first, that gene regulatory networks in animal cells—genes that regulate other genes—influence cell differentiation. They do this by generating predictable "pathways of differentiation,"[10] patterns by which one type of cell will emerge from another over the course of embryological development as cells divide. For example, early in embryological development, one type of cell (call it cell type "A") will divide and give rise to two other types of cells (call them types "B" and "C"), which will eventually generate cell types "D" and "E," and "F and G," respectively, and many other cell types as the process continues. Kauffman suggests that these pathways of differentiation "may reflect self-organizing features of complex genomic regulatory networks."[11] In other words, networks of regulatory genes in embryonic cells determine the pathways by which cells divide and differentiate. Since these patterns of cell differentiation may be *determined* by regulatory genes, Kauffman regards them as the inevitable by-products of self-organizational processes. Moreover, since "pathways of cell differentiation [have been] present in all multicellular organisms presumably since the Precambrian,"[12] he suggests that self-ordering properties "inherent in a wide class of genomic regulatory networks"[13] played a significant role in the origin of the animal forms.

Kauffman makes a similar case for the importance of self-organizational processes during body-plan morphogenesis, the second phase of animal development. This phase involves not so much the differentiation of one cell type from another, but the arrangement and organization of different cell types into the distinct tissues and organs that jointly constitute various animal body plans.

Kauffman again points to known processes of body-plan development and suggests that they could have played an important role in the formation of the first animal body plans. He cites the importance of the structural or "positional"[14] information in cells and cell membranes as the crucial determinants of how different cell types are organized into different animal forms. I discussed the importance of such "epigenetic" information to animal development in Chapter 14 and explained why it poses a problem for neo-Darwinian theory. By recognizing the importance of such information, Kauffman also rejects the neo-Darwinian assumption that a "genetic program" entirely determines animal development. He further regards the patterns of development that result from this positional information as evidence of self-ordering tendencies in matter and the existence of laws of biological form.

Do these self-ordering tendencies or laws of form, if they exist, explain the origin of animal body plans and the information necessary to build them? They don't.

Self-Organization and Epigenetic Information

To see why let's look first at how Kauffman attempts to explain the epigenetic "positional" information that directs the organization of cells in the second phase of animal development. Kauffman attempts to explain this "positional" information by offering an entirely hypothetical and, ultimately, question-begging proposal. He invokes an idea sketched out in the 1940s by the famous English mathematician Alan Turing.[15] Turing proposed that specific arrangements of cells in animal development might ultimately derive from the diffusion and specific arrangement of crucial molecules—presumably something like the morphogen proteins present in embryonic cells. (Recall that morphogens, or morphogen proteins, influence cell differentiation and organization during animal development.) Rather than attributing the distribution of these morphogenic proteins to preexisting genetic and epigenetic information in cells, as occurs during development in modern animals, Turing postulated that the distribution of these molecules might have originated in the first place independently of such information as the result of simple chemical reactions. He imagined one molecule producing both a copy of itself ("autocatalyzing") *and,* in addition, producing a different molecule as well. Then he envisioned one of these molecules inhibiting the production of the other, thus allowing, through repeated cycles, the production of more and more of one molecule and less and less of the other. Turing thought the resulting nonuniformities in the patterns of distribution of those molecules would eventually result in nonuniform patterns in the distributions of different cells, possibly resulting in different animal forms.

Kauffman expanded upon this proposal as a way of understanding how crucial positional information might organize as the result of chemical interactions of different molecules. Nevertheless, his proposal suffers from an obvious difficulty: it lacks any chemical or biological specificity. In explaining the proposal, Kauffman does not mention any specific chemicals or proteins that would behave in the way he envisions. Instead, he describes the behavior of hypothetical molecules that he labels with the indistinct monikers "X" and "Y." More important, Kauffman offers no evidence that chemicals interacting in the way he envisions could create specific *biologically relevant* configurations or distributions of morphogen proteins—apart, that is, from the processes that generate specifically arranged distributions of these proteins in *preexisting information-rich* embryonic cells today.

Instead, it is inherently implausible to think that the specificity necessary to coordinate the movements and arrangements of the billions or trillions of cells present in adult animal forms could be established by the interactions of one or two simple chemicals, even if they formed autocatalytic cycles. Kauffman himself seems tacitly to acknowledge the difficulty of generating biological specificity from the reactions of chemicals alone. He notes, in critique of his own model, that patterns of molecular diffusion produced by chemical autocatalysis would depend crucially upon "the initial conditions."[16] In other words, getting a biologically relevant information-rich arrangement of morphogenic proteins would require starting with a very specific (presumably information-rich) arrangement of autocatalyzing molecules.

Kauffman encounters this same problem in attempting to explain the origin of the first life as the result of autocatalytic reactions starting from a prebiotic soup. In *The Origins of Order*, he acknowledges that generating an autocatalytic, or self-reproducing, set of molecules—a crucial step in his origin-of-life scenario—would require "high molecular specificity"[17] in the initial set of peptides or RNA molecules. In other words, it would require specificity of *arrangement* and structure, that is to say, functional information.

Self-Organization and Genetic Information

And what about the specifically *genetic* information necessary to the earlier phase of animal development? Does Kauffman's self-organizational theory explain the origin of the "genetic regulatory networks" necessary to cell differentiation? Again, it does not. Instead, in an even more obvious way, it begs the question of the origin of these regulatory networks. Indeed, though Kauffman discusses cell differentiation as a kind of "self-ordering" or self-organizational process, he acknowledges that the predictable pathways of differentiation that characterize this process *derive from preexisting* gene regulatory networks. As Kauffman notes, the spontaneous ordering tendencies in cell differentiation are "*inherent* in a wide class of genomic regulatory networks."[18] Indeed, the genetic information in the gene regulatory networks does not *come from* self-ordering processes of cell differentiation. Instead, cell differentiation, to the extent that it can be properly described as "self-ordering," *results from* preexisting genetic sources of information. Thus, the self-organizational process that Kauffman cites cannot *explain* the origin of genetic information, because it derives from it, as Kauffman's own description reveals.

In a later book, *At Home in the Universe: The Search for the Laws of Self-Organization and Complexity*, Kauffman does offer computer simulations of two "model systems"[19] that seek to explain, at least in principle, how genetic information might have self-organized. In one example, he describes a system of buttons connected by strings.[20] The buttons represent novel genes or proteins and the strings represent self-organizational forces of attraction between the proteins. Kauffman suggests that when the complexity of this system reaches a critical threshold (as represented by the number of buttons and strings), new modes of organization might arise in the system "for free"[21]—without intelligent guidance—similar to the way that water spontaneously changes to ice or vapor under specific conditions.

Kauffman asks his readers to imagine a system of many interconnected lights. Each light can flash in a variety of states—on, off, twinkling, and so on. Since each light can adopt more than one possible state, the system may adopt a vast number of possible states. Further, in his system, rules determine how past states influence future states. Kauffman asserts that, as a result of these rules, the system would, if *properly tuned*, eventually produce a kind of order in which a few basic patterns of light activity recur with greater than random frequency. Since these patterns represent a small portion of the total number of possible states in which the system can reside, Kauffman suggests that self-organizational laws might similarly find highly improbable biological outcomes—perhaps even functional sequences of bases or amino acids within a much larger sequence space of possibilities.[22]

It's not hard to see why these simulations also would fail to account for the origin of the new genes and proteins needed to produce the Cambrian animals. In both of his examples Kauffman *presupposes* significant sources of preexisting information. In his buttons and strings simulation, he intends the buttons to represent proteins, themselves the result of preexisting genetic information. Where did that information come from? Kauffman doesn't say, but it is an essential part of what needs explanation in the history of life. Similarly, in his light system, the order that allegedly arises "for free"—that is, apart from an intelligent input of information—only does so if, as Kauffman acknowledges, the programmer "tunes" the system to keep it from either (a) generating an excessively rigid order or (b) devolving into chaos.[23] This tuning presumably involves an intelligent programmer selecting certain parameters and excluding others—that is, inputting information. In fact, in summarizing the import of this illustration, Kauffman insists that it shows how the "orderliness of the cell, long attributed to the honing of Darwinian evolution, seems instead likely to arise from *the dynamics of the genome network*,"[24] that is, from preexisting—unexplained—sources of genetic information.

In addition, Kauffman's model systems are not analogous to biological systems because they are not constrained by functional considerations. A system of interconnected lights governed by preprogrammed rules may well settle into a small number of patterns within a much larger space of possibilities. But since these patterns have no function and need not meet any functional requirements, they have no specificity analogous to that in the genes of actual organisms. Kauffman's model systems do not produce sequences or systems characterized by *specified* complexity or functional information. They produce modules of repetitive order distributed in an aperiodic manner, yielding mere complexity (i.e., information only in the Shannon sense).[25] Getting a law-governed system to generate repetitive patterns of flashing lights, even with a certain amount of variation, is interesting, but not biologically relevant. A system of lights flashing "Vote for Jones," on the other hand, would model a biologically relevant outcome, at least, if such a functional sequence of letters arose without intelligent agents programming the system with equivalent amounts of functionally specified information.

Kauffman on the Cambrian

Kauffman also proposes a specific self-organizational mechanism to explain some aspects of the Cambrian explosion. According to Kauffman, new Cambrian animals emerged through "long-jump" mutations that established new body plans in a discrete rather than gradual fashion.[26] He recognizes that mutations affecting early development are almost inevitably harmful.[27] Thus he concludes that body plans, once established, will not change, whatever subsequent evolution may occur. This keeps his proposal consistent with a top-down pattern in the fossil record in which higher taxa (and the body plans they represent) appear first, only later to be followed by the multiplication of lower taxa representing variations within those original body designs.

Even so, Kauffman's proposal begs the most important question: *What produces the new Cambrian body plans in the first place?* By invoking "long-jump mutations," he identifies no specific self-organizational process that can produce such changes. Moreover, he concedes a principle that undermines his own proposal. As noted above, Kauffman acknowledges that mutations early in development are almost inevitably deleterious. Yet developmental biologists know that these are the only kind of mutations that have a realistic chance of producing large-scale evolutionary change—the big jumps that Kauffman invokes. Though Kauffman repudiates the neo-Darwinian reliance upon random mutations, he must invoke the most implausible kind of random mutation to provide a self-organizational account of the new Cambrian body plans.

DEVELOPMENTAL TOOLKITS AND SELF-ORGANIZATIONAL PROCESSES

More recently, another advocate of self-organization, Stuart Newman, a cell biologist at the New York Medical College, has published several papers suggesting that self-organizational processes can help explain the origin of body plans. In a paper in the volume produced from the Altenberg 16 conference, Newman develops a model that resembles Kauffman's, but one that offers more biological specificity.[28]

Newman, like Kauffman, invokes self-organizational processes. But Newman sees these processes acting dynamically and *in coordination* with a genetic "toolkit." His model emphasizes the importance of a highly conserved (i.e., similar) set of regulatory genes in all the major Cambrian taxa. In his view, this common "developmental genetic toolkit"[29] has been used "to generate animal body plans and organ forms for more than half a billion years"[30] since the inception of the animal kingdom.

But if all the animal taxa have the same toolkit, why are the various forms of animals and higher metazoan taxa so different from one another? For Newman, the answer to this question requires understanding how self-organizing processes influence the interaction of cells during development and how they cause genes to acquire different functions affecting the interactions of cells.

For example, he attributes the emergence of multicellularity to cells acquiring the capacity "to remain attached to one another after dividing."[31] This capacity in turn derives not from generating new genes and proteins (as neo-Darwinism would assume). Instead, it derives from the repurposing of old genes and proteins in response to specific self-organizational (and epigenetic) processes such as the "physical force of adhesion."[32] Newman proposes, further, that once the first multicellular organisms had arisen, they would have "set the stage for additional physical processes to come into play"[33]—processes that could alter the expression and function of still other genes in the developmental genetic toolkit, resulting in wholly new and different body plans. As Newman explains, "The phenomenon of multicellularity opened up possibilities for these molecules to become involved in the molding of bodies and organs."[34]

Newman envisions new animal body plans resulting from different cells sticking to each other in different configurations because of different forces of attraction between molecules on the surface of cells and because of different patterns of distribution of crucial molecules within cells. He calls these self-organizational forces and factors "dynamical patterning modules" (or DPMs).[35] Figure 15.2 shows some typical ways that cells cluster together or arrange themselves as the result of these

DYNAMICAL PATTERNING MODULES (DPMs)

DPM NAME	WHAT IT DOES	WHAT IT LOOKS LIKE
ADH (ADHESION)	CAUSES ADHESION AMONG A GROUP OF UNAGGREGATED CELLS, ALLOWING FOR MULTICELLULARITY.	
LAT (LATERAL INHIBITION)	TAKES A GROUP OF AGGREGATED CELLS AND ALLOWS CO-EXISTENCE OF ALTERNATIVE CELL STATES WITHIN THE GROUP.	
DAD (DIFFERENTIAL ADHESION)	TAKES AN AGGREGATED GROUP OF DIFFERENT TYPES OF CELLS AND ALLOWS SEPARATION OF CELLS INTO MULTILAYERED TISSUES.	
POL_A (POLARITY, APICAL-BASAL)	TAKES A GROUP OF AGGREGATED CELLS AND CAUSES FORMATION OF INTERIOR CAVITIES.	
POL_B (POLARITY, PLANAR)	TAKES AN AGGREGATED GROUP OF CELLS AND CAUSES ELONGATION OF TISSUES WITHIN A PLANE.	
OSC (CHEMICAL OSCILLATION)	TAKES AN ELONGATED TISSUE AND CHEMICALLY INDUCES OSCILLATION OF CELL PATTERNS, ALLOWING SEGMENTATION OF A BODY PLAN.	

FIGURE 15.2
Dynamical Patterning Modules (DPMs), showing the different ways that, according to biologist Stuart Newman, cells can stick to one another ("aggregate") and form structures during animal development.

self-organizational forces. Newman lists many "dynamical patterning modules," or self-organizational forces, responsible for the spontaneous emergence of these different cell clusters, including "adhesion, shape and surface polarization, switching between alternative biochemical states, biochemical oscillation, and the secretion of diffusible and nondiffusible factors."[36]

To get a handle on what Newman has in mind, think of cells as Lego blocks. There are many different ways of connecting Lego blocks, depending on the shape of the blocks and the pattern of raised bumps and indentations on the blocks. These patterns allow small groups of Legos to be arranged into different modular structures: cubes, walls, circular rings, and so forth. Each of these smaller modular structures can then be combined to make many different larger structures, from airplanes and skyscrapers to submarines and castles. In a similar way, Newman suggests that different forces of adhesion between cells and different patterns of molecular diffusion within and between cells will generate many different patterns or motifs of multicellular organization, which in turn function as modular elements that can be combined in various ways to make diverse animal forms.[37]

Do these self-organizational processes account for the origin of animal body plans in the Cambrian explosion or the information necessary to produce new animal forms? Again, they do not. Instead, Newman, like Kauffman, either fails to offer an adequate mechanism for generating crucial sources of biological information, or he begs the question by presupposing the existence of various sources of information.

ASSUME A TOOLKIT

In the first place, Newman obviously *presupposes* the existence of a "developmental genetic toolkit," that is, a whole set of genes, including regulatory genes, that help to direct the development of animal body plans. Where does this genetic information come from? He doesn't specify, though presumably he might be assuming the neo-Darwinian mechanism somehow produced the genetic information in the toolkit. If so, he leaves his model vulnerable to the criticisms outlined in Chapters 9 through 12. He certainly does not cite any specifically self-organizational process to explain the origin of the genetic toolkit. He also incorrectly seems to presuppose that the genes present in this common toolkit provide all the *genetic* information necessary to specify individual body plans. But this overlooks a host of recent findings showing that individual species within specific taxa often require genes for development that are specific to those species and taxa.[38] Thus, these genes would not have been present in a *common* metazoan toolkit of the kind Newman postulates.

Second, Newman does not account for the origin of the information necessary to organize modular arrangements or groups of cells into whole animal body plans. The forces at work in his dynamical patterning modules explain, at best, only the arrangements of small groups of cells, not the arrangements of those modular cell clusters into tissues, organs, and whole body plans.

Think, again, of arranging Lego blocks. There are many ways of arranging small numbers of Lego blocks. These various arrangements form common structural motifs such as: two blocks stuck together at right angles; several curved blocks forming circular rings, stacked blocks forming hollow squares or walls or cube-like shapes; blocks arranged as prisms or cylinders; flat layers of blocks stacked two bumps thick or three bumps thick or more. Though these structural elements stick together because of interactions between the bumps and indentations on each block, those bumps and indentations themselves do not specify any particular larger structure—a castle or an airplane, for example—because each motif may be combined or recombined with many other structural motifs in numerous different ways. The shape and properties of the modular elements do not dictate the type of larger structure that must be built from them. Instead, to build a particular structure, the modular elements must be arranged in particular ways. And since there are many possible ways to arrange these modular elements, only one or a few of which will result in a desired structure, every Lego set includes a blueprint with step-by-step instructions—in other words, additional information.

In a similar way, producing a body plan from the different types of cell clusters generated by Newman's dynamical patterning modules (DPMs), would also require additional information. Newman does not account for this information. He correctly highlights the way certain recurrent motifs for organizing groups of cells seem to form spontaneously as the result of physical interactions between individual cells (his DPMs). He does not, however, establish that these groups of cells must arrange themselves into specific tissues, organs, or body plans in response to any known physical process or law. Instead, it seems entirely possible that these modular elements (cell clusters) have many "degrees of freedom" and can be arranged in innumerable ways. If so, then some additional information—an overall organismal blueprint or set of assembly instructions—would need to direct the arrangement of these modular elements. Newman does not consider this possibility. Nor does he cite any law-like self-organizational process that would eliminate the need for such information to direct animal development.

There is yet a further problem with Newman's proposal. Even the capacity for cells to self-organize into dynamical patterning modules probably derives from prior unexplained sources of information. Newman's DPMs undoubtedly form as the result of interactions between molecules on the surface of cells, and as the result of chemical gradients between cells—with the specific configuration and properties of those molecules determining the exact structure of the individual DPMs. In that sense, the DPMs do, of course, self-organize, but clearly the specific ways that cells typically cluster together will depend upon highly *specific* and *complex* forces of interaction between the molecules and groups of molecules on the surface of these cells. Many of the molecules that contribute to these interactions are no doubt proteins, obvious products of genetic information. But, in addition, cell-to-cell interactions are affected by the *arrangement* of proteins and other molecules on the surface of cells (such as the sugars in the sugar code, see Chapter 14) as well as by the *arrangement of structures* made of proteins. But these molecular arrangements are, in turn, specified by either preexisting genetic or, more probably, *epigenetic* sources of information and structure. Thus, Newman's analysis shows that the self-ordering tendencies (or biological laws of form), to the extent they exist, depend upon preexisting sources of biological *information*. Newman, again, does not explain where that information comes from.

Newman does emphasize how epigenetic sources of information affect the expression and the function of gene products during the process of animal development. He notes that different gene products may perform different functions depending upon the organismal context in which they find themselves. But Newman does not explain with any specificity how the genes in the common toolkit acquire different functions in response to self-organizational processes or where the epigenetic information comes from that determines those functions. Yet, clearly context-dependent gene expression depends upon a host of other *preexisting* epigenetic sources of information and structure.

Recall, for example, the discussion of cell membrane targets from Chapter 14. These targets provide an important source of epigenetic information by influencing the positioning of crucial morphogenic proteins. Yet, the arrangements of the targets on the cell membrane do not self-organize as the result of simple chemical interactions between the proteins out of which they are made[39]—that is, the proteins do not determine the location of the membrane targets on the interior of the cell. Instead, the location and the structure of membrane targets are transmitted from parent membrane to daughter membrane, a process that transmits *preexisting* epigenetic structural information from the parent membrane. Newman does not attempt to explain the origin of this information or structure by any known self-organizational process. Instead, the evidence indicates that interior and exterior membrane targets, cytoskeletal arrays, the sugar code, and many other sources of epigenetic structural information do not self-organize as the result of physical interactions between their respective molecular subunits.[40]

ORDER VS. INFORMATION

Self-organizational theorists face, in addition, a conceptual distinction that has cast doubt on the relevance of their theories to biological systems. Self-organizational theorists seek to explain the origin of "order" in living systems by reference to purely physical or chemical processes (or laws describing those processes). But what needs to be explained in living systems is not mainly order in the sense of simple repetitive or geometric patterns. Instead, what requires explanation is the adaptive complexity and the information, genetic and epigenetic, necessary to build it.

Yet advocates of self-organization fail to offer examples of either biological information or complex anatomical structures arising from physics and chemistry alone. They either point, as Newman and Kauffman do, to embryological development unfolding predictably as the result of *preexisting* information-rich gene products, cell membranes, and other cell structures. Or they offer examples of purely physical and chemical processes generating a kind of order that has little relevance to the features of living systems that most need explanation.

In the latter case, self-organizational theorists often point to simple geometric shapes or repetitive forms of order arising from or being modified by purely physical or chemical processes. They suggest that such order provides a model for understanding the origin of biological information or body-plan morphogenesis.[41] Self-organizational theorists have cited crystals, vortices, and convection currents (or stable patterns of flashing lights) to illustrate the supposed power of physical processes to generate "order for free." Crystals of salt do form as the result of forces of attraction between sodium and chloride ions; vortices can result from gravitational and other forces acting on water in a draining bathtub; convection currents do emerge from warm air (or molten rock) rising in enclosed spaces. And some molecules found in living systems do adopt highly ordered structures and recognizable geometric shapes as the result of the physical interactions of their constituent parts alone. Nevertheless, the type of order evident in these molecules or physical systems has nothing to do with the specific "order" of arrangement—the information or specified complexity—that characterizes the digital code in DNA and other higher-level information-rich biological structures.

This is easiest to see in the case of the information encoded in DNA and RNA. Some of what follows may be familiar from my discussion in Chapter 8, but it bears repeating. The bases in the coding region of a section of DNA or in an RNA transcript are typically arranged in a nonrepetitive or aperiodic way. These sections of genetic text display what information scientists call "complexity," not simple "order" or "redundancy."

To see the difference between order and complexity consider the difference between the following sequences:

Na-Cl-Na-Cl-Na-Cl-Na-Cl
AZFRT< MPGRTSHKLKYR

The first sequence, describing the chemical structure of salt crystals, displays what information scientists call "redundancy" or simple "order." That's because the two constituents, Na and Cl (sodium and chloride), are highly ordered in the sense of being arranged in a simple, rigidly repetitive way. The sequence on the bottom, by contrast, exhibits complexity. In this randomly generated string of characters, there is no simple repetitive pattern. Whereas the sequence on the top could be generated by a simple rule or computer algorithm, such as "Every time Na arises, attach a Cl to it, and vice versa," no rule shorter than the sequence itself could generate the second sequence.

The information-rich sequences in DNA, RNA, and proteins, by contrast, are characterized not by either simple order or *mere* complexity, but instead by "specified complexity." In such sequences, the irregular and unpredictable arrangement of the characters (or constituents) is critical to the function that the sequence performs. The three sequences below illustrate these distinctions:

Na-Cl-Na-Cl-Na-Cl-Na-Cl (Order)
AZFRT < MPGRTSHKLKYR (Complexity)
Time and tide wait for no man (Specified complexity)

What does all this have to do with self-organization? Simply this: the law-like, self-organizing processes that generate the kind of order present in a crystal or a vortex do not also generate complex sequences or structures; still less do they generate *specified* complexity, the kind of "order" present in a gene or functionally complex organ.

Laws of nature by definition describe repetitive phenomena—order in that sense—that can be described with differential equations or universal "if-then" statements. Consider, for example, these informal expressions of the law of gravity: "All unsupported bodies fall" or "*If* an elevated body is left unsuspended, *then* it will fall." These statements represent reasonably accurate law-like descriptions of natural gravitational phenomena precisely because we have *repeated* experience of unsupported bodies falling to the earth. In nature, repetition provides grist for lawful description.

The information-bearing sequences in protein-coding DNA and RNA molecules do not exhibit such repetitive "order," however. As such, these sequences can be neither described nor explained by reference to a natural law or law-like "self-organizational" process. The kind of non-repetitive "order" on display in DNA and RNA—a precise sequential "order" necessary to ensure function—is not the kind that laws of nature or law-like self-organizational processes can—in principle—generate or explain.

Otherwise, the nucleotide bases would repeat rigidly—such as ACACACACACACACACAC—in a way that would not allow DNA to store or convey specified information. A curious feature of the chemistry of DNA allows any one of the four nucleotide bases to attach to any site on the interior backbone of the DNA molecule. This chemical indeterminacy makes it possible for DNA and RNA to store any one of a virtually unlimited number of different arrangements of nucleotide bases—in effect, to encode any genetic message. But this indeterminacy also categorically defies explanation by deterministic law-like forces of chemical attraction. And because forces of attraction do not determine the sequence of nucleotide bases in DNA or RNA, the *origin* of the specific arrangement of the bases—the information in DNA and RNA—cannot be attributed to self-organizing forces of attraction either.

Hubert Yockey, a leading innovator in the application of information theory to molecular biology, first recognized the problems associated with invoking self-organization to explain the origin of biological information. These theories fail, he argued, for two reasons. First, they do not distinguish order from information. And, second, the information in the DNA molecule does not derive from law-like forces of attraction.[42] As he explained in 1977: "Attempts to relate the idea of order . . . with biological organization or specificity must be regarded as a play on words that cannot stand careful scrutiny. Informational macromolecules can code genetic messages and therefore can carry information because the sequence of bases or residues is affected very little, if at all, by [self-organizing] physicochemical factors."[43]

Much the same thing is true of many vital sources of *epi*genetic information. The forces of attraction between constituent proteins in membrane targets or cytoskeletal arrays, for example, do not determine the structure or location of these epigenetic structures and the positional information they provide. The *origin* of these structures cannot be attributed to self-organizing forces of attraction either. Instead, in each case, information-rich epigenetic structures are generated from preexisting sources of epigenetic information.

Thus, self-organizational theories explain well what does not mainly need explaining in biology, namely, repetitive or simple geometric forms of order. Self-organizational theorists do cite structures that might have self-organized. But these examples are typically extremely modest in scope. They include repeating patterns of atoms in crystals; simple geometric figures; patterns of lines, triangles, and streaks; vortices; spiral wave currents; and simple shapes that glide across computer screens.[44] None exhibit the specified complexity that characterizes the digital information in DNA and RNA or the complex arrangements of proteins, cells, tissues, and organs necessary to build a functioning form of animal life.

NATURAL MAGIC OR TRUE CAUSE?

In 2007, I participated in a private meeting of evolutionary biologists and other scientists who shared the conviction that a new theory of biological origins is now needed. In attendance were several prominent advocates of the self-organization approach. During the meeting, these scientists presented intriguing analogies from physics and chemistry to show how order might have arisen "for free"—that is, without intelligent guidance—in the biological realm. Yet the order they described in these analogies seemed to have no direct relevance to the complexity—indeed the specified complexity—of genes or cell membranes or animal body plans. Other scientists at the conference challenged the advocates of self-organization to cite known processes that could produce biologically relevant form and information.

Near the end of the meeting one advocate of self-organization privately acknowledged to me the validity of these critiques, admitting that, for now, "self-organization is really more of a slogan than a theory." Stuart Kauffman, perhaps attempting to make a virtue of the necessity of accepting this explanatory deficit, has recently celebrated the self-organizational perspective for embracing what he calls "natural magic." In a lecture at MIT, he concluded: "Life bubbles forth in a natural magic beyond the confines of entailing law, beyond mathematization."[45] He went on to explain that one benefit of the self-organizational perspective is that it allows us to be "reenchanted" with nature and to "find a way beyond modernity."[46]

Since the beginning of modern science, scientists have championed a commonsense principle of scientific reasoning known as the *vera causa* principle. This principle holds that explaining a particular phenomenon or event requires identifying a "true cause" that is known from experience to have the power to produce the event or phenomenon in question. The early modern scientists affirmed this principle as one of the key aspects of a scientific approach to understanding nature. This stood in opposition to the magical thinking that had gone before in which people attributed powers to nature that they had never observed it manifesting. As the scientific revolution matured, endeavors such as alchemy, for example, were eventually rejected precisely because the alchemists could not identify a cause that could effect the transformation they were seeking to demonstrate.

Self-organizational theories have clearly failed to provide a *vera causa* for the origin of biologically relevant forms of "order"—the functional complexity and specified information present in living systems. Instead, they either beg the question as to the ultimate origin of biological information or point to physical and chemical processes that do not produce the specified complexity that characterizes actual animals.

Viewed in this light, Kauffman's recent discussion of natural magic and calls for a "reenchantment" with nature sound less like a bold new initiative to reconcile science and spirituality (which is what he intended) than a tacit admission that self-organizational theories have failed to identify *known* physical and chemical processes capable of generating the form and information present in actual living systems. Indeed, after years of attempting to solve the problem of the origin of form, Kauffman's recent ruminations about "natural magic" sound a lot like an admission that a profound mystery remains.

16

OTHER POST-NEO-DARWINIAN MODELS

When Stephen Jay Gould was first wrestling with the question of how new forms of animal life could have arisen so quickly in the fossil record, he considered many possible mechanisms of change. In the famed 1980 paper in which he declared neo-Darwinism "effectively dead,"[1] he didn't just propose allopatric speciation and species selections as new evolutionary mechanisms. He also granted a rehearing to a long discredited idea. Specifically, he argued that large-scale "macromutations" might generate significant innovations in form relatively quickly.[2]

In the 1930s and 1940s, this idea had been associated with University of California at Berkeley geneticist Richard Goldschmidt. Aware of the many discontinuities in the fossil record, Goldschmidt envisioned radical transformations in the form of animals arising in even one generation as the result of such large-scale mutations. He endorsed, for instance, the view of the German paleontologist Otto Schindewolf (1896–1971) that "the first bird hatched from a reptilian egg" and, thus, in Goldschmidt's words, "that the many missing links in the paleontological record are sought for in vain because they have never existed."[3] If a bird hatched directly from a reptilian egg as the result of heritable, large-scale mutations, then such a sudden leap or "saltation" would obviously leave no fossil intermediates behind.

Neo-Darwinists rejected this idea as biologically implausible in the extreme. They argued that changing so many functionally integrated anatomical and physiological systems so quickly would inevitably result in deformed mutants, not different integrated systems of organs constituting a whole new animal.[4] Goldschmidt's macromutations, they contended, would produce not what Goldschmidt called "hopeful monsters," but "hopeless monsters"—that is, nonviable organisms.[5]

Though Gould wanted to reconsider a role for large-scale mutations, he carefully disassociated his proposal from Goldschmidt's much ridiculed idea. Instead, he suggested that the mutations affecting genes in animal development might generate larger increments of morphological innovation than the mutations that affected other genes. These "developmental mutations," he thought, might generate modular parts of biological systems in a short time—without needing to generate whole new forms of life in a single generation. He offered, as an example, the possibility that the gill arch bones of ancient jawless fish, though not the whole fish, might have arisen in one step as the result of a developmental macromutation. Gould explained: "I do not refer to the saltational origin of entire new designs, complete in all their complex and integrated features. . . . Instead, I envisage a potential saltational origin for the essential features of key adaptations."[6]

In response to heavy criticism from neo-Darwinists, Gould later downplayed the role of such larger-scale developmental mutations in the theory of punctuated equilibrium. Other evolutionary biologists, however, took his idea as an inspiration and developed theories that emphasize such developmental mutations as a driving force in macroevolution. Evolutionary theorists and developmental biologists such as Rudolf Raff, Sean B. Carroll, and Wallace Arthur have developed a subdiscipline of biology known as evolutionary developmental biology, or "evo-devo" for short. The evolutionary developmental biologists have since formulated alternative models that challenge a key aspect of the neo-Darwinian triad. Whereas neo-Darwinism envisions new form arising as the result of slow, incremental accumulations of minor mutations, evolutionary developmental biologists argue that mutations affecting genes involved in animal development can cause large-scale morphological change and even whole new body plans.

This chapter will not only examine the ideas of evolutionary developmental biologists, but three of the other most prominent alternatives to neo-Darwinism, some proposed by members of the Altenberg 16 (see Fig. 16.1). Each of these alternatives emphasizes certain elements of the "triad" at the expense of others. Whereas the self-organizational alternatives that I discussed in the last chapter emphasize the role of law-like processes over random mutations, these other new theories reaffirm the importance of mutations, though each also reconceptualizes how mutations act. One approach falls under the rubric of "evo-devo" and conceives of mutations producing modifications in *larger increments.* Another, the neutral theory of evolution, sees mutations acting *absent selection.* Another, neo-Lamarckian "epigenetic inheritance," envisions heritable alterations in *epigenetic* information influencing the future course of evolution. Still another, called "natural genetic engineering," affirms that *nonrandom* genetic rearrangements drive evolutionary innovation.[7]

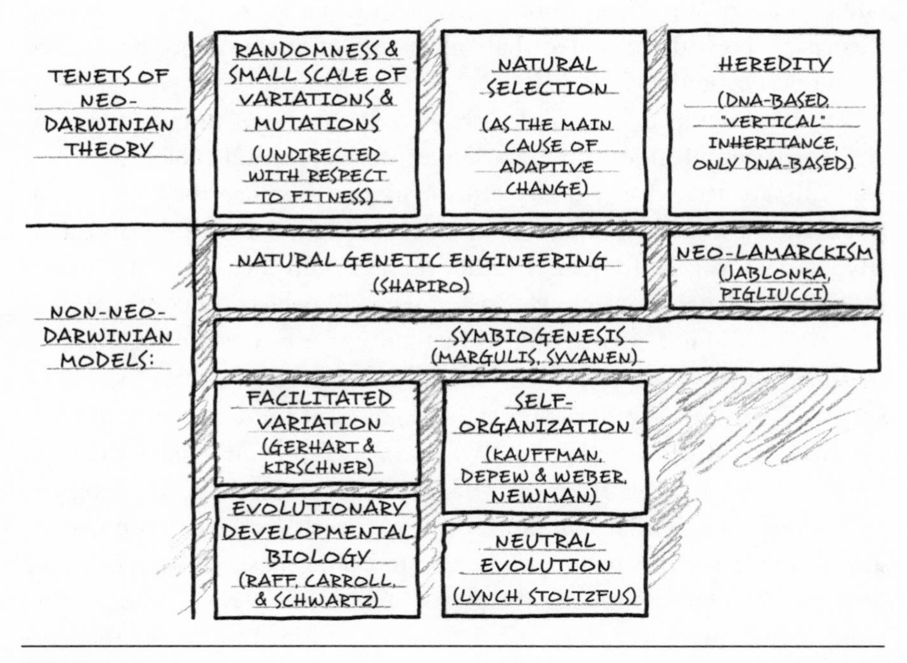

FIGURE 16.1
The tenets of neo-Darwinian orthodoxy and the different ways that various non-Darwinian models of evolution deviate from those tenets. The boxes representing new evolutionary models are positioned under the headings of the neo-Darwinian tenets that they challenge.

ne of these proposals solves the twin problems of the origin
formation and whether, therefore, it might also help to re-
ery of the Cambrian explosion.[8]

EVO-DEVO AND ITS PROPOSALS

The neo-Darwinian synthesis has long emphasized that large-scale macroevolutionary change occurs as the inevitable by-product of the accumulation of small-scale "microevolutionary" changes within populations. The consensus in support of this idea began to fray in evolutionary biology during the early 1970s, when young paleontologists such as Gould, Niles Eldredge, and Steven Stanley realized that the fossil record did not show a pattern of gradual "micro-to-macro" change. In 1980, at a now famous symposium on macroevolution at the Field Museum in Chicago, the rebellion burst into full view, exposing what developmental biologist Scott Gilbert called "an underground current in evolutionary theory" among theorists who had concluded that "macroevolution could not be derived from microevolution."[9]

At the conference, paleontologists who doubted the "micro-to-macro" consensus found allies among younger developmental biologists. They were dissatisfied with neo-Darwinism in part because they knew that population genetics, its mathematical expression, sought only to quantify changes in gene frequency rather than explain the origin of genes or novel body plans. Thus, many developmental biologists thought that neo-Darwinism did not offer a compelling theory of *macro*evolution.[10]

To formulate a more robust theory, many developmental biologists, such as Rudolf Raff, a developmental biologist at the University of Indiana and one of the founders of "evo-devo," urged evolutionary theorists to incorporate insights from their discipline.[11] For example, developmental biologists know that mutations expressed early in the development of animals are necessary to alter body-plan morphogenesis. Thus, they argue that these mutations must have played a significant role in generating whole new animal forms during the history of life. They assert that this understanding of developmental processes is crucial to understanding animal evolution. Some evo-devo advocates such as Sean B. Carroll and Jeffrey Schwartz have pointed specifically to homeotic (or *Hox*) genes—

master regulatory genes that affect the location, timing, and expression of other genes—as entities capable of producing such large-scale change in animal form.[12] These evo-devo advocates have broken with classical neo-Darwinism primarily in their understanding of the size or increment of mutational change.

MAJOR BUT NOT VIABLE, VIABLE BUT NOT MAJOR

Despite the enthusiasm surrounding the field, evo-devo fails, and for an obvious reason: its main proposal, that early-acting developmental mutations can cause stably heritable, large-scale changes in animal body plans, contradicts the results of one hundred years of mutagenesis experiments.[13] As we saw in Chapter 13, the experiments of scientists such as Nüsslein-Volhard and Wieschaus have shown definitively that early-acting body-plan mutations invariably generate embryonic lethals—dead animals incapable of further evolution. The results of these experiments have generated the dilemma for evolutionary biologists that geneticist John McDonald aptly described as the "great Darwinian paradox." Recall that McDonald noted that early-acting regulatory mutations do not produce viable alterations in form that will persist in populations, as evolution absolutely requires. Instead, these mutations are eliminated immediately by natural selection because of their invariably destructive consequences. On the other hand, later-acting mutations can generate viable changes in the features of animals, but these changes do not affect global animal architectures. This generates a dilemma: major changes are not viable; viable changes are not major. In neither case do the kinds of mutation that actually occur produce viable major changes of the kind necessary to build new body plans.

In 2007, I coauthored a textbook with several colleagues titled *Explore Evolution*. In it, we explained this "either/or" ("major-not-viable, viable-not-major") dilemma and suggested that it posed a challenge to theories that rely on the mutation and selection mechanism to explain the origin of major morphological changes.[14] The National Center for Science Education (NCSE)—an influential activist group that opposes allowing students to learn about scientific criticisms of evolutionary theory—challenged our critique. They charged that our textbook "fails

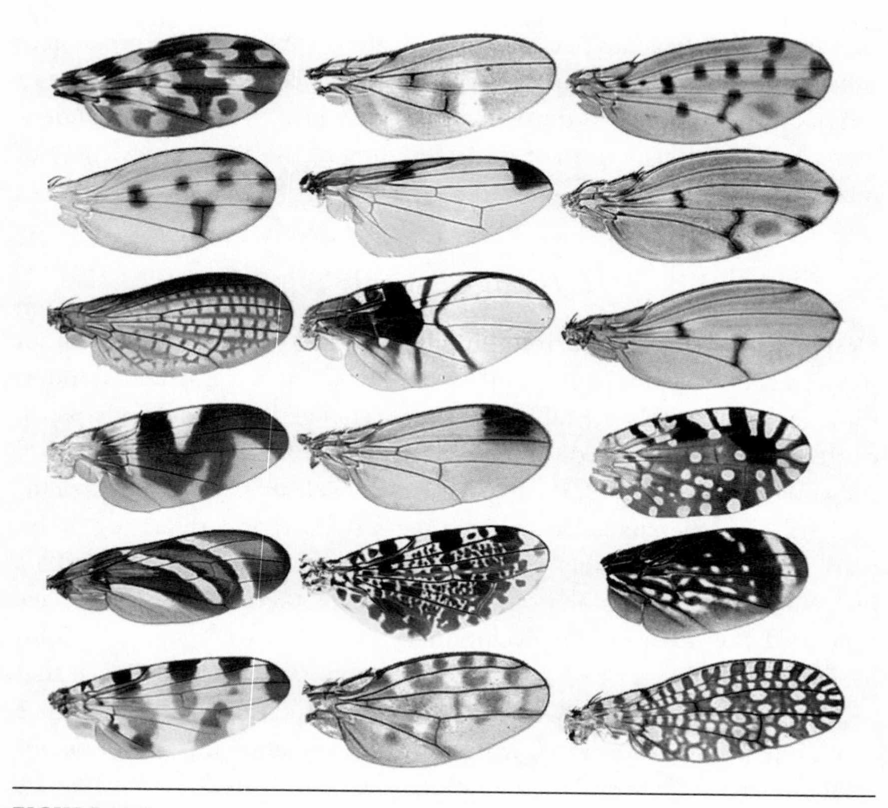

FIGURE 16.2
Superficial changes in insect wing coloration thought to be caused by mutations in *cis*-regulatory elements. Such examples show that mutations that affect development and that also result in viable offspring tend to be minor. *Courtesy National Academy of Sciences, U.S.A.*

to acknowledge the extensive research on mutations in DNA sequences that do not encode proteins, but which have important morphological effects."[15] In other words, they claimed that some viable mutations do produce major large-scale changes.

The NCSE cited papers from the "evo-devo" literature claiming that a type of mutation in the regulatory regions of the genome, "*cis*-regulatory" regions, have been shown to produce large-scale changes in winged insects. According to the NCSE, mutations in these *cis*-regulatory elements (or CREs) are "considered by many evolutionary biologists to have the greatest potential for generating evolutionary change."[16] What's more, they insisted that "mutations in CREs play an important role in morphological evolution."[17] The NCSE cited a paper in the *Proceedings of the Na-*

tional Academy of Sciences by three developmental biologists, Benjamin Prud'homme, Nicolas Gompel, and Sean B. Carroll.[18]

The paper did not show what the NCSE claimed, however. It did assert that changes in regulatory DNA produce "both relatively modest morphological differences among closely related species and more profound anatomical divergences among groups at higher taxonomical levels."[19] But the study only showed how changes in the *cis*-regulatory elements in fruit fly DNA might have affected the *coloration* of wing spots in several different types of flying insects. It did not report any significant change in the *form or body plan* of these insects. Instead, the study highlighted a clear case of a viable mutation generating merely a minor or superficial change (see Fig. 16.2).

Not surprisingly, many evolutionary biologists recognize that such regulatory mutations do not explain the evolution of new body plans. For example, Hopi Hoekstra, of Harvard University, and Jerry Coyne, two traditional neo-Darwinists, have published an article reviewing various evo-devo proposals in the journal *Evolution*. They note, "Genomic studies lend little support to the *cis*-regulatory theory" of evolutionary change.

They also argue, tellingly, that most *cis*-regulatory mutations result in the *loss* of genetic and anatomical traits, including a famous case in which evolutionary biologists attributed the loss of pelvic spines in stickleback fish to mutations in *cis*-regulatory elements.[20] Yet, as they argue, "supporting the evo-devo claim that *cis*-regulatory changes are responsible for morphological innovations requires showing that promoters are important in the evolution of *new* traits, not just the losses of old ones." Hoekstra and Coyne conclude, "There is no evidence at present that *cis*-regulatory changes play a major role—much less a pre-eminent one—in adaptive evolution."[21] Given their commitment to neo-Darwinism, it's fair to assume that Hoekstra and Coyne probably did not intend, in making this argument, to refute the NCSE's criticism of our textbook *Explore Evolution*. Nevertheless, science, like politics, sometimes makes for strange bedfellows.

WHAT ABOUT *HOX* GENES?

When biology students hear my colleague Paul Nelson describe the "great Darwinian paradox" (see Chapter 13) in public lectures on university campuses, they often ask, "What about *Hox* genes?" Recall that *Hox* (or homeotic) genes regulate the expression of other protein-coding genes during the process of animal development. Some biologists have likened them to the conductor of an orchestra who plays the role of coordinating the contributions of the players. And because *Hox* genes affect so many other genes, many evo-devo advocates think that mutations in these genes can generate large-scale changes in form.

For example, Jeffrey Schwartz, at the University of Pittsburgh, invokes mutations in *Hox* genes to explain the sudden appearance of animal forms in the fossil record. In his book *Sudden Origins,* Schwartz acknowledges the discontinuities in the fossil record. As he notes, "We are still in the dark about the origin of most major groups of organisms. They appear in the fossil record as Athena did from the head of Zeus—full-blown and raring to go, in contradiction to Darwin's depiction of evolution as resulting from the gradual accumulation of countless infinitesimally minute variations."[22]

What resolves this mystery? Schwartz, an evo-devo advocate, reveals his answer: "A mutation affecting the activity of a homeobox [*Hox*] gene can have a profound effect—such as turning . . . larval tunicates into the first chordates. Clearly, the potential homeobox genes have for enacting what we call evolutionary change would seem to be almost unfathomable."[23]

But can mutations in *Hox* genes transform one form of animal life—one body plan—into another? There are several reasons to doubt that they can.

First, precisely *because Hox* genes coordinate the expression of so many other different genes, experimentally generated mutations in *Hox* genes have proven harmful. William McGinnis and Michael Kuziora, two biologists who have studied the effects of mutations on *Hox* genes, have observed that in fruit flies "most mutations in homeotic [*Hox*] genes cause fatal birth defects."[24] In other cases, the resulting *Hox* mutant phenotype, while viable in the short term, is nonetheless markedly less fit than the wild type. For example, by mutating a *Hox* gene in a fruit fly, biologists have produced the dramatic *Antennapedia* mutant, a hapless fly with legs growing out of its head where the antennae should be (see Fig. 16.3).[25] Other *Hox* mutations have produced fruit flies in which the balancers (tiny structures behind wings that stabilize the insect in flight, called "halteres") are transformed into an extra pair of wings.[26] Such mutations alter the structure of the animal, but not in a beneficial or permanently

FIGURE 16.3
Photograph of an *Antennapedia* mutant with a pair of legs growing out of its head, where antennae would normally develop. Such examples show that mutations that occur early in animal development and that also produce major changes typically result in less fit offspring—in this case offspring that cannot reproduce. *Courtesy Elsevier, Inc.*

heritable way. The *Antennapedia* mutant cannot survive in the wild; it has difficulty reproducing, and its offspring die easily. Similarly, fruit-fly mutants sporting an extra set of wings lack the musculature to make use of them and, absent their balancers, cannot fly. As Hungarian evolutionary biologist Eörs Szathmáry notes with cautious understatement in the journal *Nature,* "macromutations of this sort [i.e., in *Hox* genes] are probably frequently maladaptive."[27]

Second, *Hox* genes in all animal forms are expressed *after* the beginning of animal development, and well after the body plan has begun to be established. In fruit flies, by the time that *Hox* genes are expressed, roughly 6,000 cells have already formed, and the basic geometry of the fly—its anterior, posterior, dorsal, and ventral axes—is already well established.[28] So *Hox* genes don't determine body-plan formation. Eric Davidson and Douglas Erwin have pointed out that *Hox* gene expression, although necessary for correct *regional* or local differentiation within a body plan, occurs much later during embryogenesis than global body-plan specification itself, which is regulated by entirely different genes. Thus, the primary origin of animal body plans in the Cambrian explosion is not merely a question of *Hox* gene action, but of the appearance of much deeper control elements—Davidson's "developmental gene regulatory networks" (dGRNs).[29] And yet, as we saw in Chapter 13, Davidson argues that it is extremely difficult to alter dGRNs without damaging their ability to regulate animal development.

Third, *Hox* genes only provide information for building proteins that function as switches that turn other genes on and off. The genes that they regulate contain information for building proteins that form the parts of other structures and organs. The *Hox* genes themselves, however, do not contain information for building these structural parts. In other words, mutations in *Hox* genes do not have all the *genetic* information necessary to generate new tissues, organs, or body plans.

Nevertheless, Schwartz argues that biologists can explain complex structures such as the eye just by invoking *Hox* mutations alone. He asserts that "[t]here are homeobox genes for eye formation and that when one of them, the *Rx* gene in particular, is activated in the right place and at the right time, an individual has an eye."[30] He also thinks that mutations in *Hox* genes help arrange organs to form body plans.

In a review of Schwartz's book, Eörs Szathmáry finds Schwartz's reasoning deficient. He too notes that *Hox* genes don't code for the proteins out of which body parts are made. It follows, he insists, that mutations in *Hox* genes cannot by themselves build new body parts or body plans. As he explains, "Schwartz ignores the fact that homeobox genes are selector genes. They can do nothing if the genes regulated by them are not there."[31] Though Schwartz says he has "marveled" at "the importance of homeobox genes in helping us to understand the basics of evolutionary change,"[32] Szathmáry doubts that mutations in these genes have much creative power. After asking whether Schwartz succeeds in explaining the origin of new forms of life by appealing to mutations in *Hox* genes, Szathmáry concludes, "I'm afraid that, in general, he does not."[33]

Nor, of course, do *Hox* genes possess the *epigenetic* information necessary for body-plan formation. Indeed, even in the best of cases mutations in *Hox* genes still only alter *genes*. Mutations in *Hox* genes can only generate new *genetic* information in DNA. They do not, and cannot, generate *epi*genetic information.

Instead, epigenetic information and structures actually determine the function of many *Hox* genes, and not the reverse. This can be seen when the same *Hox* gene (as determined by nucleotide sequence homology) regulates the development of different anatomical features found in different phyla. For instance, in arthropods the *Hox* gene *Distal-less* is required for the normal development of jointed arthropod legs. But in vertebrates a homologous gene (e.g., the *Dlx* gene in mice) builds a different kind of (nonhomologous) leg. Another homologue of the *Distal-less* gene in echinoderms regulates the development of tube feet and spines—anatomical features classically thought not to be homologous to arthropod limbs, nor to limbs of tetrapods.[34] In each case, the *Distal-less* homologues play different roles determined by the higher-level organismal context. And since mutations in *Hox* genes do not alter higher-level epigenetic contexts,[35] they cannot explain the origin of the novel epigenetic information and structure that establishes the context and that is necessary to building a new animal body plan.[36]

NEUTRAL OR NONADAPTIVE EVOLUTION

Michael Lynch, a geneticist at Indiana University, has offered a different mechanism of evolutionary change, and a different explanation for the origin (or growth) of the genome as well as the origin of anatomical novelty. Lynch proposes a neutral or "non-adaptive" theory of evolution in which natural selection plays a largely insignificant role. His theory is based on contrasting observations about the features and strength of evolutionary mechanisms at work in populations of different sizes.

He observes, first, that in general, the larger the population of organisms, the lower the mutation rate and (in sexually reproducing eukaryotes) the higher the rate of genetic recombination. He notes that the genomes of organisms in larger populations (such as those of bacteria and unicellular eukaryotic organisms) tend to be smaller and more streamlined—meaning they have fewer intervening nonprotein-coding sequences (i.e., introns). Most important, he notes that in large populations, natural selection tends to be relatively effective in eliminating deleterious mutations and fixing beneficial ones, whereas the process of genetic drift (the tendency for gene variants to be lost through random processes) plays a relatively less significant role.

By contrast, Lynch observes that small populations—which would include almost all animal groups—are characterized by higher mutation rates and lower rates of genetic recombination. They also tend to have large genomes with a lot of nonprotein-coding DNA—introns, pseudogenes, transposons and various repetitive DNA elements—as well as gene duplicates. Lynch argues that in these small populations, natural selection tends to be weak—unable to remove mildly deleterious mutations or to fix mildly beneficial ones efficiently. As Lynch summarizes, "Three factors (low population sizes, low recombination rates and high mutation rates) conspire to reduce the efficiency of natural selection with increasing organism size."[37] Consequently, nonprotein-coding elements are not removed from the genome, but instead tend to accumulate, causing the genomes of organisms living in small populations to grow—even though these sequences may be neutral or even deleterious. Moreover, in small populations, "neutral" processes such as random mutation, genetic recombination, and genetic drift predominate in their effects over natural selection.

What does all this have to do with the origin of animals and the Cambrian explosion? Evolutionary biologists think that the ancestral groups of the Cambrian animals would likely have existed in relatively small populations. Lynch argues that in small populations, animal genomes will *inevitably grow* over time as nonprotein-coding sections of DNA (as well as gene duplicates) accumulate due to the weakness of natural selection. He thinks these neutral mutations drive the evolution and growth of genomic and phenotypic complexity in animals. In short, Lynch attempts to explain the expansion of the genome and the origin of anatomical complexity as the result of neutral, non-adaptive processes of genetic accretion, rather than as an adaptive process involving natural selection acting on random mutations. As he states, "Many of the unique complexities of the eukaryotic gene arose by semi-neutral processes with little, if any, direct involvement of positive selection."[38]

In his work, Lynch has also advanced a powerful mathematical critique of the efficacy of the neo-Darwinian mechanism. He has argued that natural selection plays a lesser role in shaping the features of evolving populations than many evolutionary theorists have previously assumed—especially in the case of relatively small populations. Lynch has argued, instead, that random environmental factors—an organism being in the right place at the right time (near an abundant food source, for example) or in the wrong place at the wrong time (in a drought-stricken region or near an erupting volcano, for example)—will play a more important role in determining reproductive success than variations in the fitness of organisms within the same population.

COUNTERINTUITIVE APPROACH

Lynch develops his mathematical critique of the creative power of natural selection based on the principles of population genetics. Nevertheless, it does not follow from his analysis showing the weakness of natural selection that neutral processes alone are sufficient to build new functional genes and proteins. Nor does it follow that neutral processes alone can account for the many complex anatomical systems that require new sources of genetic (and epigenetic) information for their construction. Indeed, it may seem counterintuitive, at least from a neo-Darwinian point of view, to think that the accumulation of random mutations alone can accomplish what neo-Darwinists have long invoked *both* mutations and natural selection to do. In effect, Lynch's theory attempts to explain the origin of anatomical complexity by reference to what would seem on its face to be a less—not a more—potent mechanism than the one offered by neo-Darwinism. Could such a counterintuitive theory be correct?

Perhaps, but as a comprehensive theory of how biological information and anatomical complexity arises, Lynch's neutral theory leaves much to be desired.

In the first place, Lynch's theory offers no explanation for some of the crucial molecular machinery present in eukaryotes—machinery that is necessary to rendering his mechanism for the accumulation and subsequent expression of genetic information credible. Recall that Lynch thinks that small populations of multi-cellular organisms in particular would have inevitably accumulated many insertional genomic elements. But for the functional information in these growing genomes to be expressed, the cell must have some way of excising the non-functional randomly accreting genetic elements—at least, until some of them mutate to the point that they contribute to producing *functional* genes and proteins.

Extant eukaryotic organisms depend on a sophisticated molecular machine called a spliceosome—a machine that excises introns and fuses (the portions of the genome that code for proteins) before [illegible]ssion takes place. "This large complex," observes cell biologist [illegible]elissa Jurica, is "composed of over 150 individual proteins" and several structural RNAs, and thus "may indeed deserve the moniker 'the most complicated macromolecular machine in the cell.'"[39]

So where do spliceosomes, and the genes necessary to produce them, come from? Lynch doesn't say, though he recognizes, of course, the importance of this molecular machinery to gene expression and to his scenario. As he explains, "The problem is that introns are inside genes and get transcribed to mRNA but then have to be spliced out perfectly. If you're one nucleotide off, you get a dead transcript."[40] Nevertheless, Lynch's theory presupposes, but does not explain, the origin of the genetic information necessary to produce the spliceosomes that perform this function. He certainly does not explain the origin of these massive multi-protein, multi-RNA complexes by reference to any neutral evolutionary process. Nor can he, since his theory of genomic accretion and expression *presupposes* the existence of precisely such intricate machines. Instead, as my colleague Paul Nelson has put it rather colorfully, "to get Lynch's theory of genomic accretion up and running, a great deal of complicated molecular machinery must be rolled in from offstage."

Of course, it could be argued that these machines and systems arose much earlier with the origin of the eukaryotic cell as the result of selection-driven evolution in the large populations of simpler unicellular organisms in which, according to Lynch's theory, natural selection played a more significant role. Nevertheless, Lynch does not make that argument—and for good reason. Most evolutionary biologists today recognize the origin of the eukaryotic cell as a *completely* unsolved problem—unexplained by either neutral or adaptive theories of evolution.[41]

Of course, insofar as these molecular machines are present in even one-celled eukaryotic organisms, they would have arisen, presumably, well before the origin of animals. Thus, explaining their origin is not, strictly speaking, directly relevant to explaining the Cambrian explosion. Nevertheless, Lynch's inability to account for their origin reflects directly on the credibility of his theory—at least insofar as it seeks to offer a *comprehensive* account of the mechanisms by which biological information and complexity arise during the history of life.

Drifting In and Out

In any case, there are good reasons to doubt that Lynch's neutral mechanism could generate the novel biological information and form necessary to explain the origin of animals, even granting the prior existence of the molecular machinery (in small populations of eukaryotic organisms) that his scenario requires.

First, Lynch assumes a false gene-centric view of the origin of biological form. As he writes: "Most of the phenotypic diversity that we perceive in the natural world *is directly attributable* to the peculiar structure of *the eukaryotic gene*."[42] His view overlooks the crucial role of *epigenetic* information and structure in the origin of animal form discussed in Chapter 14 and, therefore, does nothing to explain its origin.

Second, neutral processes such as genetic drift do not favor beneficial mutations, and thus do not *fix*, with any *efficiency*, those mutation-induced *genetic* traits in small populations.[43] Natural selection, as we saw in Chapter 10, is something of a double-edged sword. On the one hand, natural selection helps to fix beneficial traits in a population. On the other hand, natural selection also makes it difficult for functional genes to vary widely without being eliminated. Neutral theories of evolution attempt to avoid the latter problem by invoking gene duplication and other processes that can add nonfunctional sequences to the genome—sequences that are unaffected, at least initially, by selective pressure. In so doing, however, these theoretical formulations significantly diminish the role of natural selection as a mechanism that can fix beneficial mutations in place once they have arisen. Thus, in all neutral theories, including Lynch's, any beneficial mutations that arise and begin to drift through a population, can just as readily—without a significant influence from natural selection to impede it—drift out of a population as well. This limitation vastly increases the time it will take for neutral processes to fix beneficial genetic changes in a population. Both skeptics and proponents of neo-Darwinism have recognized this deficiency in Lynch's model.[44]

Third, and most important, Lynch's theory not only fails to account for the *fixation* of new genes and traits in small populations, it also fails to account for their *origin*. Lynch's mechanism of neutral mutational change does envision the *addition* of brute genomic complexity as the result of the accretion of preexisting genetic elements (introns, transposons, pseudogenes and gene duplicates). Nevertheless, the addition of these elements does not generate any novel *functional* (or *specified*) genetic information. Instead, it merely transfers preexisting genetic sequences from one organismal context where those sequences may have performed a function, to another where they likely will not. Indeed, the point of neutral theory is to postulate the addition of genetic elements that, initially, do not perform crucial functions such that they can experience mutations without deleterious consequence to the organism. Lynch himself assumes that these added elements will not perform functions in their new context, which is why he envisions the need for spliceosomes to excise them, at least initially.

Instead, for Lynch's theory to explain the origin of new *and functional* genes and proteins (and the anatomical complexities that depend on them), his theory would have to solve the problem of combinatorial inflation discussed in Chapter 10. He would have to show that purely random mutations could efficiently search the relevant combinatorial space of possible sequences corresponding to a given novel functional gene or protein.

Nevertheless, Lynch does not even address the problem of combinatorial inflation or the closely related problem of the rarity of genes and proteins in sequence space. He provides no experimental evidence that recombination and/or mutation (given genetic drift) will actually produce *functional* or *specified* genetic complexity. Instead, the examples he provides are entirely hypothetical. In addition, he offers no reason to think that the probability of a successful search for functional genes or proteins would be any higher (i.e., more likely to occur) than the probabilities calculated in Chapter 10. He does not, therefore, answer the challenge of the problem of combinatorial inflation and the rarity of functional genes and proteins in sequence space.

Lynch does provide, perhaps, a more detailed characterization than other neutral theories of where neutral, nonadaptive processes must predominate. Nevertheless, he does not show that such processes—random genetic mutations unhinged from natural selection—are sufficient to generate novel functional genes and proteins, let alone complex anatomical novelties requiring the origin of many such genes and proteins. Instead, as Axe's experimental results have shown, random mutations of whatever kind will not generate enough trials to render probable (or plausible) a successful search of the sequence space corresponding to a given *functional* gene or protein.

LYNCH AND WAITING TIMES

Lynch does argue in one paper that neutral evolutionary processes can generate new complex adaptations—adaptations requiring multiple coordinated mutations—within realistic waiting times. In particular, writing in a recent paper with colleague Adam Abegg of St. Louis University, he argues that "conventional population genetic mechanisms" such as random mutation and genetic drift can cause the "relatively rapid emergence of specific complex adaptations."[45] Lynch makes two specific claims in this regard. First, he claims that in large populations, *arbitrarily* complex adaptations can occur if the mutational intermediates are neutral in their effects on the organism. That is, Lynch purports to show that in large microbial populations, complex adaptations requiring a virtually *unlimited* number of mutations can occur within realistic waiting times. According to Lynch, this can occur provided that each mutation in a series of mutations has neutral (but not deleterious) effects on the organism. Second, Lynch argues that even though it generally takes longer to build complex traits in small populations, such traits can still evolve within realistic waiting times provided, again, that the mutational intermediates are neutral in their effects. In fact, he concludes that "the elevated power of both random genetic drift and mutation may enable the acquisition of complex adaptations in multicellular species at rates that are not appreciably different from those achievable in enormous microbial populations."[46]

Though Lynch makes these assertions in the context of a densely mathematical scientific article, the significance of his claims, if true, can hardly be overstated. In essence, he claims that his population genetics–based mathematical model shows that purely random mutations and genetic drift can generate extremely complex adaptations in realistic waiting times—that his neutral evolutionary theory *solves* the problem of complex adaptations and long expected waiting times discussed in Chapter 12.

But some things are just too good to be true, and it turns out that Lynch and Abegg made a subtle but fundamental mathematical error in coming to their conclusion. Appropriately, perhaps, the first person to demonstrate that Lynch's incredible claim was problematic was Douglas Axe. Although Axe could see that much of the math in Lynch's paper with Abegg was correct, Axe suspected from his own calculations and experiments that they had made some crucial error. In the end, he traced Lynch and Abegg's claims to two erroneous equations, both of which were based on an erroneous assumption. In essence, Lynch and Abegg assumed that organisms will acquire a given complex adaptation by traversing a *direct* path to the new anatomical structure. Each mutation would build on the previous one in the most efficient manner possible—with no setbacks, false starts, aimless wandering, or genetic degradation—until the desired structure or system (or gene) is constructed. Thus, they formulated an undirected model of evolutionary change, and one that assumes, moreover, that there is no mechanism available (such as natural selection) that can lock in potentially favorable mutational changes on the way to some complex advantageous structure. Nevertheless, they calculated the waiting times required to produce such structures *as if* a process for locking in potentially advantageous changes did exist, and as if their undirected and purely random mechanism was in some way directed to these functionally propitious outcomes. As Axe notes in a trenchant mathematical critique of Lynch and Abegg's argument, "Of all the possible evolutionary paths a population can take, the analysis of Lynch and Abegg considers only those special paths that lead directly to the desired end—the complex adaptation."[47]

Yet nothing in Lynch's neutral model ensures that potentially advantageous mutations will remain in place while other mutations accrue. As Axe explains, "Productive changes cannot be 'banked', whereas Equation 2 [one of Lynch's equations] presupposes that they can."[48] Instead, Axe shows, mathematically, that degradation (the fixation of mutational changes that make the complex adaptation *less* likely to arise) will occur much more rapidly than constructive mutations, causing the expected waiting time to increase exponentially.

The illustration I developed in Chapter 10 to explain the problem facing neutral models of gene evolution may help illuminate the mistaken assumption underlying Lynch and Abegg's calculations. Recall from that chapter that my hypothetical blindfolded man dropped in the middle of an enormous shark-free pool did not face any predators (by analogy the purifying effects of natural selection). But he still faced the problem of finding the ladder at the edge of the enormous body of water (by analogy, the need to search an enormous number of possible mutational paths and possible sequences to find the rare functional ones). Now suppose someone were to calculate how long on average it would take for the blindfolded man to swim to the ladder at the edge of the enormous pool. If someone

simply divided the distance to the ladder at the ed
maximum speed that man could swim, he or she
cally optimistic estimate of the severity of the prol
tunate swimmer. Why? Because calculating the pr
this manner would overlook the main problem the man
man does not know *where* the ladder is or *how* to get there. Nor does he have any way to gauge his progress.

Thus, any realistic estimate of how long it will actually take him to swim to the ladder—as opposed to an estimate of the theoretically fastest route possible—must take into account his probable aimless wandering, fits and starts, swimming in circles and drifting in various directions. Similarly, Lynch and Abegg fail to reckon in their calculation on the random, undirected and, literally, aimless nature of the mechanism that they propose. Instead, they mistakenly assume that neutral processes of evolution will make a beeline for some specific complex adaption. In fact, these processes will—in all probability—also wander aimlessly in a vast sequence space of neutral, functionless possibilities with nothing to direct them, or preserve them in any forward progress they happen to make, toward the rare and isolated islands of function represented by complex adaptations. For this reason, Lynch vastly underestimates the waiting times required to generate complex adaptations and, therefore, does not solve the problem of the origin of genes and proteins or any other complex adaptation. Instead, Axe shows in his own mathematical model, which accompanies his critique of Lynch, that the waiting times problem is just as severe as he (Axe) had previously calculated.

NEO-LAMARCKIAN EPIGENETIC INHERITANCE

The third element of the neo-Darwinian triad concerns the transmission and inheritance of genetic information. Not surprisingly, a new version of evolutionary theory questions the neo-Darwinian understanding of heredity as well.

Darwin himself lacked an accurate theory of how features of organisms are transmitted from one generation to the next. He thought that changes in organisms that occurred during their lifetimes, as the result of the "use and disuse" of different organs and anatomical systems, would be transmitted to offspring through reproduction.[49] In this respect, his y of inheritance resembled that of an earlier evolutionary theorist, n-Baptiste de Lamarck (1744–1829), who also believed in the inheri-ance of acquired characteristics.

Lamarckian mechanisms, although unsupported by any evidence at the time, came to play an increasingly important role in Darwin's thinking, as criticisms of natural selection caused Darwin to place more weight on the direct influence of the environment in evolutionary change. Indeed, by the sixth edition of the *Origin* (1872), Darwin specifically emphasized the importance of these modes of inheritance.[50]

But with the rediscovery of Mendel's laws in 1900 and the identification of chromosomes as the material entity responsible for the transmission of inheritance, Lamarckian theories of inheritance fell out of favor. Following the rise of the neo-Darwinian synthesis, most evolutionary biologists came to regard the gene as the locus of all heritable change in the organism. And after 1953, biologists equated the gene with specifically arranged nucleotide bases within the DNA molecule.

Recently, however, as more biologists have recognized that some biological information—epigenetic information—resides in structures outside of DNA, interest has grown in the possibility that these nongenetic sources of information may influence the course of evolution. The discovery that epigenetic information can be altered and directly inherited independently of DNA has attracted further attention. This discovery has, in turn, led to the formulation of a contemporary "neo-Lamarckian"[51] theory that envisions changes in the nongenetic structures of an organism affecting subsequent generations.

Today, prominent defenders of neo-Lamarckism include Eva Jablonka, of Tel Aviv University, and Massimo Pigliucci, of the City University of New York. Lamarck, of course, knew nothing about the role of genes and believed that inheritance of acquired characteristics was an important driving force in evolution. Modern neo-Lamarckians, fully apprised of the reality of genetic inheritance, nevertheless think that nongenetic sources of information and structure may play some role in the evolution of biological form. According to Jablonka, neo-Lamarckism "allow[s] evolutionary possibilities denied by the 'Modern Synthesis' version of evolutionary theory, which states that variations are blind, are genetic (nucleic acid–based), and that saltational events do not significantly contribute to evolutionary change."[52]

Jablonka has collected several categories of evidence in support of what she calls "epigenetic inheritance systems." In the first place, in some single-celled organisms (such as *E. coli* and yeast) environmentally induced changes in metabolic pathways can be transmitted to the next generation independently of any changes in the cell's DNA. Second, she notes that structural information mediating organismal form (and function) does pass from parent to offspring independently of DNA, via membranes and other three-dimensional cellular patterns.

Third, she discusses the process of DNA methylation—a process in which special enzymes attach a methyl group (CH_3) to nucleotide bases within the double helix. Processes like this can alter gene regulation and chromatin structure. Jablonka notes that the changes produced by processes that alter gene regulation are often transmitted to subsequent generations of cells without any changes to DNA base sequences. Finally, she cites a process called "RNA-mediated" epigenetic inheritance, a recently discovered phenomenon. Here, small RNAs, again acting in concert with special enzymes, affect gene expression and chromatin structure, and these modifications appear to be heritable independently of genes.

Can any of these mechanisms help to explain the origin of animal form in the Cambrian explosion? Not really.

By its nature, macroevolution requires *stable*—meaning permanently heritable—changes. But Jablonka's evidence shows that where nongenetic inheritance occurs in animals, it involves structures that either (a) do not change (such as membrane patterns and other persistent templates of structural information), or (b) do not *persist* over more than several generations. And neither case generates significant evolutionary innovation in animal form. Instead, for directional evolutionary change to occur in a population of organisms, changes must not only be heritable, but permanent. Stability—the irreversible and enduring heritability of traits—is a logically inescapable requirement for any theory of evolution. This is precisely what "descent with modification" means.

And here Jablonka's evidence for *stable* nongenetic inheritance is equivocal at best, as she readily admits. Reviewing Jablonka's assembled data for animals reveals no case where an induced epigenetic change persisted permanently in any population. The heritability of such changes is transient, lasting (depending on the species in question) from a few generations up to forty.

Jablonka candidly addresses this lack of evidence for stability, noting, "We believe that epigenetic variants in every locus in the eukaryotic genome can be inherited, *but in what manner, for how long, and under what conditions, has yet to be qualified.*"[53] Consequently, despite its intriguing aspects, the evolutionary significance of neo-Lamarckian epigenetic inheritance remains uncertain or, in Jablonka's own words, "inevitably, somewhat speculative."[54]

NATURAL GENETIC ENGINEERING

University of Chicago geneticist James Shapiro has formulated another post-Darwinian perspective on how evolution works that he calls "natural genetic engineering." Shapiro has developed an understanding of evolution that takes account of the integrated complexity of organisms as well as the importance of nonrandom mutations and variations in the evolutionary process.

He observes that organisms within a population often modify themselves in response to different environmental challenges. He cites evidence showing that when populations are challenged by environmental stresses, signals, or triggers, organisms do not generate mutations or make genetic changes randomly, that is, without respect to, or unguided by, their survival needs. Instead, they often respond to environmental stresses or signals by inducing mutations in a directed or regulated way. As he explains, "The continued insistence on the random nature of genetic change by evolutionists should be surprising for one simple reason: empirical studies of the mutational process have inevitably discovered patterns, environmental influences, and specific biological activities at the roots of novel genetic structures and altered DNA sequences."[55]

The depth of Shapiro's challenge to orthodox neo-Darwinism is profound. He rejects the randomness of novel variation that Darwin himself emphasized and that neo-Darwinian theorists throughout the twentieth century have reaffirmed.[56] Instead, he favors a view of the evolutionary process that emphasizes preprogrammed adaptive capacity or "engineered" change, where organisms respond in a "cognitive" way to environmental influences, rearranging or mutating their genetic information in regulated ways to maintain viability.

As an example, Shapiro notes that—contrary to the neo-Darwinian assumption that "DNA alterations are accidental"[57]—all organisms possess sophisticated cellular systems for proofreading and repairing their DNA during its replication. He notes that these systems are "equivalent to a quality-control system in human manufacturing," where the "surveillance and correction" functions represent "cognitive processes, rather than mechanical precision."[58]

As an example of regulated mutation, Shapiro observes that in response to environmental assault—UV damage from sunlight or the presence of an antibiotic, for instance—bacteria activate what is known as the "SOS response" system. This system makes use of specialized error-prone DNA polymerases, normally left unexpressed, that are synthesized and set into action, allowing the population to generate a much wider range of genetic variation than usual. Bacterial cells regulate this process using a DNA-binding protein known as LexA, which normally represses the error-prone polymerases. When the SOS system is activated by environmental damage, the production of LexA first drops dramatically, allowing expression of the error-prone polymerases, but then rises, which "ensures that as soon as DNA repair occurs . . . LexA [will] reaccumulate and repress the SOS genes."[59] This system allows cells to "replicate DNA that carries unrepaired damage,"[60] keeping their essential replication machinery moving past a stall, in the absence of which the bacterium would die.

An analogy may help to illustrate what the cell is doing when confronted with an environmental challenge. Imagine a military unit, a combined armor and infantry battalion, crossing an open plain. Suddenly, the battalion falls under a fierce, unrelenting enemy artillery barrage, wounding many of its soldiers. To keep the wounded alive until the barrage ceases or reinforcements arrive, the commander instructs certain members of the unit with destructive skills to disassemble (in military jargon, "cannibalize") a few of the tanks to provide temporary armored cover from further incoming shells. His order tells them, however, to cease their tank-modifying actions as soon as the barrage ends. That is, the unit as a whole tolerates "damage" to some of its equipment to save as many of its members as possible.

In the same way, while at one level their "error-prone" role may appear counterintuitive, these mutation-generating DNA polymerases of the

SOS system actually constitute essential hardware in the cell's defensive armory.[61] From Shapiro's perspective, this survival strategy does not exemplify Darwinian randomness, but rather sophisticated preprogramming, an "apparatus that even the smallest cells possess" to maintain viability.[62] What's more, the carefully regulated expression of the SOS response provides evidence that cells employ the system only when needed.[63]

In addition to ramping up their mutation rates in specific sections of the genome, cells may also change the way they express the genetic information that they already carry, expressing some genes that were previously unexpressed and suppressing others. Organisms in populations under particular stresses may retrieve and access modular elements of genetic information stored in disparate locations on the genome or even on different chromosomes. Cells will then assemble, or concatenate, those modular elements to form a new gene or RNA-transcript capable of directing the synthesis of a novel protein or proteins that can help the organism survive.

Shapiro argues that these and other kinds of directed, rather than random, genetic changes and responses to stimuli occur under "algorithmic control." He describes the cell as "a powerful real-time distributed computing system"[64] implementing various "if-then" subroutines. This emphatically challenges one of the three key elements of the neo-Darwinian triad: the claim that mutations and variations occur in a strictly random way.

During the last fifteen years, Shapiro has published a series of fascinating papers about the newly discovered capacities of cells to direct or "engineer" the genetic changes they need to remain viable in a range of environmental conditions. His work represents a promising avenue of new biological research, bringing insight into how the cell's information-processing system modifies and directs the expression of its genetic information in real time in response to different signals. Shapiro's work also provides new insights into how observable evolutionary changes occur in living populations.

Could it, then, also provide a solution to the problem of the origin of the information necessary to build an animal body plan? It could, except for one question that Shapiro's otherwise brilliant characterization of how organisms modify themselves doesn't address.

Where does the programming come from that accounts for the "pre-programmed adaptive capacity" of living organisms? If, as James Shapiro argues, natural selection and exclusively random mutations don't produce this information-rich pre-programming, then what did? In the next chapters, I'll propose an answer to precisely this question.

17

THE POSSIBILITY OF INTELLIGENT DESIGN

The owner of a remote island estate has been murdered while out riding. When the local sheriff arrives, he learns there are several obvious suspects: the volatile gamekeeper, the owner of a neighboring estate with whom the murder victim has had a long-running feud, and the estate owner's estranged wife, who had been living on the island in a small mother-in-law cottage. The sheriff quickly learns the basic facts of the case. The victim was found dead, facedown on the beach, with his horse standing nearby. Any one of the three suspects could have taken a rifle, from an unlocked shed at the edge of the property. All were healthy enough to have hiked to the scene of the crime. Each of them has a motive. And none has an alibi.

But as the investigation unfolds additional facts come to light. Most importantly, when the coroner arrives, he determines that although the victim was shot in the stomach and then his head was harshly bludgeoned by the butt of the rifle, these injuries served merely to conceal the bullet wound that actually killed the estate owner. The man was dead when he hit the ground. What killed him was a perfect shot entering the head just behind the right ear, exactly where an expert marksman would place a bullet. Moreover, ballistics shows that this bullet came from a different gun altogether from the one stored in the shed, a weapon likely fired from quite a distance.

The sheriff then returns to the list of suspects and, one by one, eliminates them. Abundant evidence shows that none of the three prime suspects is a particularly good shot, much less a world-class marksman. The landowner's estranged wife has a shaky hand and no experience with firearms. The volatile gamekeeper has extremely poor eyesight. And the neighboring landowner turns out to have an alibi after all—as well as a broken arm, which would have prevented him from holding the kind of rifle from which the bullet was fired. There is, however, one other person living on the estate, though not even the other suspects suspect him. He is the victim's loyal and longtime personal assistant, a timorous older man much beloved by both the family and the other servants. No one wants to consider him as a possible suspect. But is it possible that he could have had something to do with the crime after all? Might an unexpected suspect—indeed "the butler"—have done it?

Clearly, standard evolutionary theory has reached an impasse. Neither neo-Darwinism nor a host of more recent proposals (punctuated equilibrium, self-organization, evolutionary developmental biology, neutral evolution, epigenetic inheritance, natural genetic engineering) have succeeded in explaining the origin of the novel animal forms that arose in the Cambrian period. Yet all these evolutionary theories have two things in common: they rely on strictly material processes, and they also have failed to identify a cause capable of generating the information necessary to produce new forms of life.

This raises a question. Is it possible that a different or unexpected kind of cause might provide a more adequate explanation for the origin of the new *form* and *information*—as well as the other distinctive features—present in the Cambrian explosion? In particular, is it possible that intelligent design—the purposeful action of a conscious and rational agent—might have played a role in the Cambrian explosion?

INTRODUCING INTELLIGENT DESIGN

When the case for intelligent design is made, it's often hard to get contemporary evolutionary biologists to see why such an idea should even be considered or why discussions of design should play any role in biology at all. Though many biologists now acknowledge serious deficiencies in current strictly materialistic theories of evolution, they resist considering alternatives that involve intelligent guidance, direction, or design.

Much of this resistance seems to come simply from not understanding what the theory of intelligent design is. Many evolutionary biologists see intelligent design as a religiously based idea—a form of biblical creationism. Others think the theory denies all forms of evolutionary change. But contrary to media reports, intelligent design is not a biblically based idea, but instead an evidence-based theory about life's origins—one that challenges some, but not all, meanings of the term "evolution."

Perhaps the best way to explain the theory of intelligent design is to contrast it with the specific aspect of the theory of Darwinian evolution that it directly challenges. Recall from our opening discussion in Chapter 1 that the term "evolution" has many different meanings and that Darwin's theory of evolution by natural selection affirmed several of them: first, change over time; second, universal common descent; and third, the creative power of natural selection acting on random variations. In affirming this third meaning of evolution, both classical Darwinism and modern neo-Darwinism also affirm what neo-Darwinist Richard Dawkins has called the "blind watchmaker" hypothesis. This hypothesis holds that the mechanism of natural selection acting on random genetic variations (and mutations) can produce not just new biological form and structure, but also the *appearance* of design in living organisms.[1]

Darwin argued for this idea in *The Origin of Species* as well as in his letters. Recall the sheep breeding illustration from Chapter 1 where I described how both intelligent human breeders and environmental change (a series of bitterly cold winters) might produce an adaptive advantage in a population of sheep. During the nineteenth century, biologists regarded the adaptation of organisms to their environment as one of the most powerful pieces of evidence of design in the living world. By observing that natural selection had the power to produce such adaptations, Darwin not only affirmed that his mechanism could generate significant biological change, but that it could explain *the appearance of design*—without invoking the activity of an actual designing intelligence. In doing so, he sought to refute the design hypothesis by providing a materialistic explanation for the origin of *apparent design* in living organisms. Modern neo-Darwinists also affirm that organisms look as if they were designed. They also affirm the sufficiency of an unintelligent natural mechanism—mutation and natural selection—as an explanation for this appearance. Thus, in both Darwinism, and neo-Darwinism, the selection/variation (or selection/mutation) mechanism functions as a kind of "designer substitute." As the late Harvard evolutionary biologist Ernst Mayr explains: "The real core of Darwinism . . . is the theory of natural selection. This theory is so important for the Darwinian because it permits the explanation of adaptation, the 'design' of the natural theologian, by natural means."[2] Or as another prominent evolutionary biologist, Francisco Ayala, has put it succinctly, natural selection explains "design without a designer."[3]

Other contemporary neo-Darwinian biologists including Richard Dawkins, Francis Crick, and Richard Lewontin have also emphasized that biological organisms only *appear* to have been designed.[4] They recognize that many biological structures—whether the chambered nautilus, the compound eye of a trilobite, the electrical system of the mammalian heart, or numerous molecular machines—attract our attention because the sophisticated organization of such systems is reminiscent of our own designs. Dawkins has noted, for example, that the digital information in DNA bears an uncanny resemblance to computer software or machine code.[5] He explains that many aspects of livings systems "give the appearance of having been designed for a purpose."[6]

Nevertheless, neo-Darwinists regard that appearance of design as entirely illusory, as did Darwin himself, because they think that purely mindless, materialistic processes such as natural selection and random mutations can produce the intricate designed-like structures in living organisms. In this view, natural selection and random mutation mimic the powers of a designing intelligence without themselves being intelligently directed or guided in any way.

That's where the theory of intelligent design comes into play. Intelligent design challenges the idea that natural selection and random mutation (and other similarly undirected materialistic processes) can explain the most striking appearances of design in living organisms. Instead, it affirms that there are certain features of living systems that are best explained by the design of an actual intelligence—a conscious and rational agent, a mind—as opposed to a mindless, materialistic process. The theory of intelligent design does *not* reject "evolution" defined as "change over time" or even universal common ancestry, but it does dispute Darwin's idea that the cause of major biological change and the appearance of design are wholly blind and undirected.

Nor does the theory seek to insert into biology an extraneous religious concept. Intelligent design addresses a key scientific question that has long been part of evolutionary biology: Is design real or illusory? Indeed, part of what Darwin set out to explain was precisely the appearance of design. With current materialistic evolutionary theories now failing to explain many of the most striking appearances of design in the Cambrian animals, including the presence of digital information as well as other complex adaptations, the possibility emerges that these appearances of design may not be *just* appearances after all. The Darwinian formulation of evolutionary theory in opposition to the design hypothesis,[7] coupled with the inability of neo-Darwinian and other materialistic theories to account for salient appearances of design, would seem logically to reopen the possibility of actual (as opposed to apparent) design in the history of animal life.

Either life arose as the result of purely undirected material processes or a guiding or designing intelligence played a role. Advocates of intelligent design favor the latter option and argue that living organisms look designed because they really were designed. Design proponents argue that living systems exhibit telltale indicators of prior intelligent activity that justify this claim, indicators that make intelligent design *scientifically detectable* from the evidence of the living world.

But that, for many evolutionary biologists, is precisely the rub. Because they think of intelligent design as a religiously based idea, they understand that people might want to affirm the intelligent design of life as part of their religious beliefs—but not as a consequence of scientific evidence. Indeed, most evolutionary biologists don't see how the idea of intelligent design could contribute to a scientific explanation of life's origins, nor do they see how intelligent design could ever be *detected* or *inferred* scientifically from evidence in nature. Exactly how *would* researchers justify such an inference?

MY STORY

When I left for my graduate studies in England in 1986, I was asking a similar set of questions. At that time, I wasn't thinking about the scientific legitimacy of the intelligent design hypothesis as an explanation for the origin of animals. Instead, I wanted to know if intelligent design could help explain the origin of life itself. My questions eventually led me to learn about a distinctive method of *historical* scientific inquiry. That discovery led me to a method of reasoning that allows for the detection or inference of past causes, including *intelligent* causes.

A year earlier, in 1985, I had met one of the first contemporary scientists to revive the idea that intelligent design might have played a causal role in the origins of life. Chemist Charles Thaxton (see Fig. 17.1) had recently published a book, *The Mystery of Life's Origin*. His coauthors were polymer scientist and engineer Walter Bradley and geochemist Roger Olsen. Their book received acclaim as a groundbreaking critique of current theories of chemical evolution. They showed that attempts to explain the origin of the first living cell from simpler nonliving chemicals had failed and that these theories had specifically failed to explain the origin of the information necessary to produce the first life.

But it was in the book's epilogue that the three scientists proposed a radical alternative. There they suggested that the information-bearing properties of DNA might point to the activity of a designing intelligence—to the work of a mind, or an "intelligent cause" as they put it.[8] Drawing on the analysis of the British-Hungarian physical chemist Michael Polanyi, they argued that chemistry and physics alone could not produce the information in DNA any more than ink and paper alone could produce the information in a book. Instead, they argued that our uniform experience suggests a cause-and-effect relationship between intelligent activity and the production of information.[9]

FIGURE 17.1
Charles Thaxton. *Courtesy Charles Thaxton.*

At the time the book appeared, I was working as a geophysicist for an oil company in Dallas where Thaxton happened to live. I met him at a scientific conference and became intrigued by his work. Over the next year, I began dropping by his office to discuss his book and the radical idea he was developing about DNA.

The first part of Thaxton's argument made sense to me. Experience does indeed seem to affirm that (specified or functional) information typically arises from the activity of intelligent agents, from minds as opposed to mindless, material processes. When a "tweet" appears on your smart phone's Twitter feed (if you're into that kind of thing), it clearly originated first in the mind of a person who created a Twitter account, scripted the "tweet," and then sent it out across the Internet. Information does arise from minds.

But Thaxton went further. He acknowledged that most branches of science didn't consider intelligent activity as an explanation because, he thought, intelligent agents don't usually generate repeatable or predictable phenomena and because they are difficult to study under controlled laboratory conditions. Nevertheless, Thaxton argued that scientists might propose an intelligent cause as a positive *scientific* explanation for some events in the past, as part of a special mode of scientific inquiry he called *origins science*. He noted that scientific disciplines such as archaeology, evolutionary biology, cosmology, and paleontology often infer the occurrence of singular, nonrepeatable events and that the methods used to make such inferences could help scientists identify positive indicators of intelligent causes in the past as well.

Here I wasn't initially so sure. Thaxton's ideas about a distinctive method of science concerned with origins, or at least with the past generally, seemed intuitively plausible. After all, evolutionary biologists and paleontologists do seem to use a method of investigation different from that employed by laboratory chemists. Nevertheless, I wasn't exactly sure what those methods were, how they were different from those used in other sciences, and whether using them in any way justified considering intelligent design as a scientific hypothesis.

So the next year when I left Dallas, Texas, for Cambridge, England, to pursue my studies in the history and philosophy of science, I had a lot on my mind. Is there a distinctive method of historical scientific inquiry? If so, does that method of reasoning and investigation justify a scientific reformulation of the design hypothesis? In particular, does the intuitive connection between information and the prior activity of a designing intelligence justify a *positive* (historical) scientific inference to intelligent design? Does it make intelligent design detectable?

HISTORICAL SCIENTIFIC METHOD AND THE DESIGN HYPOTHESIS

In my research, I discovered that historical scientists often do make inferences with a distinctive logical form. This type of inference is known technically as an *abductive* inference.[10] During the nineteenth century, American logician C. S. Peirce characterized this mode of reasoning and distinguished it from two better-known forms, inductive and deductive reasoning. He noted that in inductive reasoning, general rules are inferred from particular facts, whereas in deductive reasoning, general rules are applied to particular facts in order to deduce specific outcomes. In abductive reasoning, however, inferences are often made about *past* events or causes based on present clues or facts.[11]

To see the difference between these three types of inference, consider the following argument forms:

Inductive argument:

A_1 is B.

A_2 is B.

A_3 is B.

A_4 is B.

A_n is B.

All A's are B.

Deductive argument:

MAJOR PREMISE: If A has occurred, then B will follow as a matter of course.

MINOR PREMISE: *A has occurred.*

CONCLUSION: Hence, B will follow as well.

Abductive argument:

MAJOR PREMISE: If A occurs, then B would be expected as a matter of course.

MINOR PREMISE: *The surprising fact B is observed.*

CONCLUSION: Hence, there is reason to suspect that A has occurred.

Note the difference between deductive and abductive forms of inference. In deduction, the minor premise affirms the *antecedent* variable ("A"), while the conclusion deduces the *consequent* variable ("B"), an *anticipated* outcome. In this sense, deductive inferences look forward to something that will happen in the future. A classic illustration of deductive reasoning has this character:

MAJOR PREMISE: All men are mortal.

MINOR PREMISE: *Socrates is a man.*

CONCLUSION: Therefore, Socrates is a mortal (i.e., he *will* die).

In an abductive argument, the minor premise affirms the *consequent* variable ("B") and its conclusion infers the *antecedent* variable ("A")—the variable referring to something that went *before,* either logically or temporally. Abductive reasoning, thus, often affirms a *past* occurrence. For this reason, forensic or historical scientists such as geologists, paleontologists, archaeologists, and evolutionary biologists often use abductive reasoning to infer *past* conditions or causes from *present* clues. As Stephen Jay Gould notes, historical scientists typically "infer history from its results."[12]

For example, a geologist might reason as follows:

MAJOR PREMISE: If a mudslide occurred, we would expect to find felled trees.

MINOR PREMISE: *We find evidence of felled trees.*

CONCLUSION: Therefore, we have reason to think that a mudslide may have occurred.

In the deductive form, if the premises are true, the conclusion follows with certainty. The logic of the abductive arguments is different, however.

Abductive arguments do not produce certainty, but instead merely plausibility or possibility. To see why, consider the following variation of the preceding abductive argument:

MAJOR PREMISE: If a mudslide occurred, we would expect to find felled trees.

MINOR PREMISE: *We find felled trees.*

CONCLUSION: Therefore, a mudslide occurred.

or symbolically:

MAJOR PREMISE: If MS, then FT.

MINOR PREMISE: *FT.*

CONCLUSION: Therefore, MS.

Notice that unlike the first version of the abductive argument in which the conclusion was stated tentatively ("We have reason to think that a mudslide may have occurred"), in this version the conclusion is affirmed definitively ("A mudslide *occurred*"). Obviously, this latter form of argument has a problem. It does not follow that, because the trees have fallen, a mudslide *necessarily* occurred. The trees may have fallen for some other reason. A hurricane may have blown them down; perhaps an ice storm occurred and the trees fell under the weight of accumulating ice; or loggers may have cut them down. In logic, affirming the consequent variable of a minor premise (with certainty) constitutes a formal fallacy—a fallacy that derives from the failure to acknowledge that more than one cause (or antecedent) might produce the same evidence (or consequent).

Even so, the presence of downed timber *might* indicate that a mudslide has occurred. Thus, amending the above argument to conclude: "We have reason to think that a mudslide *may have* occurred" does not commit a fallacy. Even if we may not affirm the consequent with certainty, we may affirm it as a possibility. This is precisely what abductive reasoning does. It provides a reason for considering that a hypothesis—and often a hypothesis about the past—might be true, even if one cannot affirm the hypothesis (or conclusion) with certainty.[13]

THE METHOD OF MULTIPLE COMPETING HYPOTHESES

To address this limitation in abductive reasoning and to make it possible to strengthen inferences about the past, the nineteenth-century geologist Thomas Chamberlain developed a form of reasoning he called "the method of multiple working hypotheses."[14] Historical and forensic scientists employ this method when more than one cause or hypothesis can account for the same evidence. They use it to adjudicate between competing hypotheses by comparing them to see which one *best* explains not just one piece of evidence but, usually, a wider class of relevant facts.

For example, consider how this method of reasoning was used to establish the hypothesis of continental drift as the best explanation for a wide range of geological observations. During the early 1900s, a German geologist and meteorologist named Alfred Wegener became fascinated with the way the African and South American continents fit together on the map like pieces of a jigsaw puzzle.[15] He proposed that the continents had once been fused together as a single giant continent that he called "Pangea," which later separated and drifted apart.

Initially, many geologists ridiculed Wegener's idea. They thought that—given the vast distances separating the continents—the matching shapes were most likely just a coincidence. Wegener's critics dismissed his theory of continental drift as "delirious ravings," "Germanic pseudo-science," or a "fairy tale."[16] But Wegener cited other evidence that he thought continental drift could explain that the coincidence hypothesis could not. He noted that fossil forms discovered on the east coast of South America matched those on the west coast of Africa in corresponding places and sedimentary strata. This fact seemed *too* coincidental to him to be explained away by chance alone. Nevertheless, other geologists attempted to explain matching fossil forms an ocean apart not as the result of the movement of the continents, but instead as the result of the migration of flora and fauna—either across oceans or over ancient land bridges.[17] This introduced a third hypothesis into the mix, one that, in conjunction with the coincidence hypothesis, could explain each of the same facts that Wegener's hypothesis could.

Later, however, an additional set of facts came to light—one that helped scientists to decide between the competing hypotheses. During World War II, the United States Navy surveyed the seafloor topography and measured the earth's magnetic field across the oceans. These magnetic surveys showed parallel stripes of magnetized rock, each with the same polarity on either side of mountain ridges running down the middle of the ocean floor at equal distances from the mid-oceanic mountain ranges.[18] Geologists also learned that magma was continually seeping out at the middle of these mid-oceanic mountain ranges. They discovered that as the magma cools, it "acquires" a characteristic magnetic signature reflecting the polarity of the earth's magnetic field at that location at the time of its cooling. When ships towing sensitive magnetometers measured this "remanent magnetization," scientists learned that the magnetization of the seafloor alternated between sections of "normal" and "reverse" polarity as the magnetometer was towed away from a mid-ocean ridge in each direction. This led to the discovery of a famous symmetrical "piano key" pattern on each side of a mid-ocean ridge, seen in Figure 17.2.

To explain this symmetrical pattern of alternating magnetism, geologists proposed that the magnetic stripes were formed as the result of the seafloor spreading away from the mid-ocean ridge as magma was extruded and cooled in the presence of earth's changing magnetic field—in other words, that the continents were literally drifting apart. This hypothesis not only explained the symmetrical pattern of magnetic stripes but

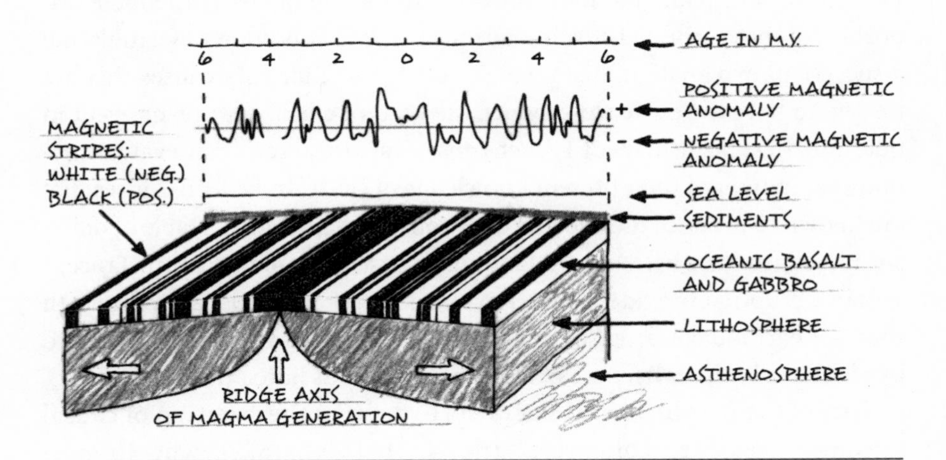

FIGURE 17.2
This diagram shows the symmetrical pattern of alternating magnetic stripes of either "normal" or "reversed" polarity on either side of a mid-ocean ridge. Because this "piano key" pattern could only be explained by plate tectonics and seafloor spreading, it contributed to the widespread acceptance of those theories within contemporary geology.

also other relevant evidence. Although the other hypotheses could explain (or explain away) the fit of the continents and/or the similar pattern of fossilization across the oceans, only continental drift (driven by seafloor spreading) could explain the magnetic seafloor stripes *and* these other pieces of evidence. Consequently, as the result of its superior explanatory power, a decisive case for continental drift was soon established, strengthening a merely plausible abductive inference about the past movement of the continents by showing that this inference provided the best (and only adequate) explanation of *all* the relevant facts.[19]

Contemporary philosophers of science such as Peter Lipton have called this method of reasoning "inference to the best explanation."[20] Scientists often use this method when trying to explain the origin of an event or structure from the past. They compare various hypotheses to see which would, if true, best explain it.[21] They then provisionally affirm the hypothesis that best explains the data as the one that is most likely to be true.

Obviously, saying, "The best explanation is the one that best explains the facts or that best explains the most facts," begs an important question. What does it mean to explain something well or best?

As it happens, historical scientists have developed criteria for deciding which explanation, among a group of competing possible hypotheses, provides the best explanation for some event in the remote past. The most important of these criteria is "causal adequacy." As a condition of formulating a successful explanation, historical scientists must identify causes that are known to have the power to produce the kind of effect, feature, or event in question. In seeking to identify such causes, historical scientists evaluate hypotheses in light of their present knowledge of cause and effect. Causes that are known to produce the effect in question (or are thought capable of doing so) are judged to be better candidates than those that are not. For instance, a volcanic eruption provides a better explanation for an ash layer in the earth than an earthquake or a flood, because eruptions have been observed to produce ash layers, whereas earthquakes and floods have not.

One of the first historical scientists to develop the criterion of causal adequacy was the geologist Charles Lyell (1797–1875), who in turn influenced Charles Darwin. Darwin read Lyell's magnum opus, *The Principles of Geology*, during his voyage on the *HMS Beagle* and employed Lyell's principles of reasoning in *On the Origin of Species*. The subtitle of Lyell's book summarized his central methodological principle: *Being an Attempt to Explain the Former Changes of the Earth's Surface, by Reference to Causes Now in Operation* (1830–1833). Lyell argued that when scientists seek to explain events in the past, they should not invoke some unknown type of cause, the effects of which we have not observed. Instead, they should cite causes that are known from our uniform experience to have the power to produce the effect in question.[22] Historical scientists should cite presently acting causes, that is, "*causes now in operation.*" This was the idea behind his uniformitarian method and its famous dictum: "The present is the key to the past." According to Lyell, our *present* experience of cause and effect should guide our reasoning about the causes of *past* events. Darwin adopted this methodological principle as he sought to demonstrate that natural selection qualified as a *vera causa,* that is, a true, known, or actual cause of significant biological change.[23] In other words, he sought to show that natural selection was "causally adequate" to produce the effects he was trying to explain.

THE ONLY KNOWN CAUSE

Both philosophers of science and leading historical scientists have emphasized causal adequacy as the key criterion by which competing hypotheses are judged. But philosophers of science have insisted that assessments of explanatory power lead to conclusive inferences only when there is *just one known cause* for the effect or evidence[24] (see Fig. 17.3) in question. If there are many causes that can produce the same effect, then the presence of the effect does not definitively establish the cause. When scientists know of only one cause for a given effect, however, they can infer that cause and yet avoid the fallacy of affirming the consequent—the error of ignoring other possible causes with the power to produce the same effect.[25] In that case, they can infer or detect a *uniquely* plausible past cause from the clues that are left behind.

This can happen in one of two ways. First, historical scientists might focus their investigation on a *single* fact (in isolation) for which only one cause happens to be known. In such a case, they can quickly and decisively infer the cause from the effect alone—without risk of affirming the consequent, because no other known cause produces the same effect. For example, because a volcanic eruption is the only known cause of a

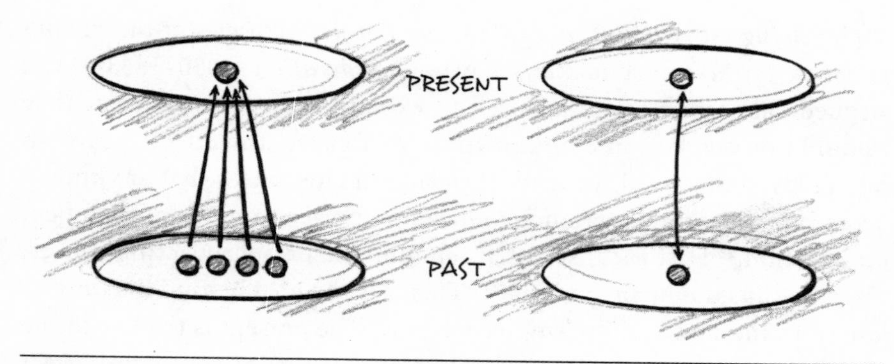

FIGURE 17.3
Schematic of the logical problem of retrodiction. Whether it is possible to reconstruct the past definitively or not depends upon whether there is a single cause or condition that gives rise to a present state or whether there are many possible past causes or conditions that give rise to a given present state. The diagram on the left portrays an information-destroying situation in which *many* past causes (or conditions) correspond to a given present state. The diagram on the right portrays an information-preserving situation in which only one past cause (or condition) corresponds to a present state. Adapted from Sober, *Reconstructing the Past*, 4.

volcanic ash layer, the presence of such a layer at an archeological site strongly indicates the prior eruption of a volcano.

In other cases where historical scientists encounter evidence for which there are *many* known causes, they will often broaden their investigation beyond an initial fact or set of facts. In such cases, they will use the strategy described above (as part of the method of multiple competing hypotheses), by looking for additional evidence until they find a piece for which there is only one known cause. They can then compare the explanatory power of the competing hypotheses. Using this strategy, historical scientists will choose the proposed cause with the demonstrated power to produce *all* the relevant evidence, including the new fact or piece of evidence for which there is only one known cause. For example, the discovery of the symmetrical pattern of ocean-floor magnetism on opposite sides of a mid-oceanic ridge allowed for a comparison of the explanatory power of the three hypotheses under consideration, leaving only seafloor spreading as a causally adequate explanation of *all* the relevant facts.

Such an approach often allows historical scientists to pick out a piece of evidence (from some combination of effects) for which there is only one known (or theoretically plausible) cause, thus making it possible to establish a past cause decisively. Though this strategy involves looking at a wider class of facts than the first strategy, the logical status of the inferences involved is the same. In each case, the presence of a fact (either standing on its own or in combination with other facts) for which only one cause is known allows historical scientists to make a definitive inference about the causal history in question without committing the fallacy of affirming the consequent. Logically, if a postulated cause is known to be a *necessary* condition or cause of a given event or effect, then historical scientists can validly infer that condition or cause from the presence of the effect. If it's true that where there is smoke there is always first fire, then the presence of smoke wafting up over a distant mountain range decisively indicates the prior presence of a fire on the other side of the ridge.

HISTORICAL INFERENCE AND INTELLIGENT DESIGN

What does all this have to do with the Cambrian explosion?

Quite a lot. In my investigation of the historical scientific method, I found that whether they always realize it or not, historical scientists typically use the method of inference to the best explanation. They make abductive inferences about past causes from present clues, evidence, or effects. This later suggested to me that if there were features of the Cambrian explosion or the Cambrian animals that would be "expected as a matter of course" if an intelligent designer had played a role in that event, then it was at least possible to formulate the hypothesis of intelligent design as a historical (abductive) scientific inference. An advocate of intelligent design could reason in a standard historical scientific way:

MAJOR PREMISE: If intelligent design played a role in the Cambrian explosion, then feature (X) known to be produced by intelligent activity would be expected as a matter of course.

MINOR PREMISE: *Feature (X) is observed in the Cambrian explosion of animal life.*

CONCLUSION: Hence, there is reason to suspect that an intelligent cause played a role in the Cambrian explosion.

Of course, a historical scientist would only be justified in making such an abductive inference to the past activity of an intelligent cause if "feature X" is evident in the Cambrian explosion and if intelligent design is known to produce "feature X." Moreover, just because the Cambrian explosion may exhibit some feature or features for which intelligent design is *a* known cause does not mean that intelligent design was necessarily *the actual* cause (or the best explanation) of those features. Only if the Cambrian event and animals exhibit features for which intelligent design is the *only* known cause may a historical scientist make a decisive inference to a past intelligent cause.

We are left with two crucial questions. Are there in fact such features present in the record of the Cambrian explosion or in the animals that arise in it—features that are known from our experience to be produced by intelligent causes such that they would justify making a tentative abductive inference to intelligent design? Are there also perhaps features of the Cambrian event that are known from our experience to be produced by intelligent causes, and *only* intelligent causes, justifying a more definitive inference to past intelligent activity as the *best* explanation for the relevant evidence? Might "the butler" have done it after all?

18

SIGNS OF DESIGN IN THE CAMBRIAN EXPLOSION

Well-crafted mystery novels, like real-world crime investigations, unfold with a distinctive logic. There is a death to be explained and, at the start, an indefinitely large universe of possible causes. That universe can be made smaller, narrowing to the one true cause, as more and more clues come to light. Those clues typically come in two forms: *positive* evidence, or indicators of what *likely* happened (e.g., .38 caliber shell casings on the ground and bullet wounds in a body) and *negative* evidence, or indicators of what could *not* have happened.

Let's say that the local sheriff who discovered the body of the estate owner (from my illustration in the previous chapter), did so as he was making his rounds on a dirt road that makes a close approach to the beach at the end of the estate where the owner died. And let's say that, as a result, the sheriff just happened to find the body soon after the murder had taken place. Let's suppose, further, that the sheriff had the presence of mind to immediately measure the temperature of the body only to find that the victim was still warm, indeed almost as warm as a living person. Clearly, in this situation, the sheriff would conclude that the victim had just died. At that point in the investigation, a physical regularity would govern the sheriff's reasoning, one that tells a lot about who *didn't* commit the murder. Following death, the human body cools to the surrounding temperature at a known rate. So, making allowances for vehicular transport, whoever committed the murder could *not* have gone beyond a certain distance from the remote estate at the time the body was found.

These facts would immediately provide a rock-solid alibi to the vast majority of humanity, anyone located safely outside that radius when the body was discovered. Of course, calling this information a *negative* clue is really only a convention of naming. "Negative" and "positive" refer to how we conceive of the *implications* of a fact, but not to the fact itself: the evidence, after all, is what it is. Even so, facts both exclude and allow competing possible hypotheses. As they accumulate, they typically paint a picture, a profile, of the actual cause of the event in need of explanation. Thus, when we say "the body temperature of the deceased rules out the 7 billion people who were well beyond the radius set by the cooling rate," we could equally well have said, "the body temperature implicates some person within 30 miles of the estate when the sheriff arrived," a population of possible suspects much smaller than when we started.

As I have described the many attempts to explain the scientific enigma motivating this book, that mystery has, in one sense, progressively deepened. As more and more attempts to explain the Cambrian explosion of animal life have failed, the evidence that these various competing theories fail to explain may be considered a set of negative clues—evidence that effectively precludes certain possible causes or explanations. I've already explained why the received version of evolutionary theory, neo-Darwinism, fails to account for the explosion of information and form in the Cambrian period. I've also examined more recent evolutionary theories and shown why they too fail to explain key aspects of the evidence. To this point, then, much of the evidence has returned a *negative* verdict. It has told us a lot about what, in all probability, did *not* cause the Cambrian explosion. But, as in our hypothetical murder case, an accumulating body of evidence that makes one set of explanations less and less plausible may also begin to paint a picture of an alternative cause and the true explanation.

PROFILE OF THE SUSPECT

Long before detectives know the actual identity of a suspect, they will often compose a profile of the person they are seeking. One leading paleontologist has used this strategy to begin to draw a bead on the cause responsible for the Cambrian explosion. Douglas Erwin has dedicated his career to solving the problem of the origin of animal body plans (see Fig. 18.1). Trained at the University of California by James Valentine, another Cambrian expert, Erwin has worked closely over the past decade with Eric Davidson, whom we first met in Chapter 13, trying to determine what happened to cause dozens of novel body plans to appear—and appear rapidly—in the Cambrian period.

Both Erwin and Davidson have now ruled out standard neo-Darwinian theory—vehemently in Davidson's case. He says that the standard theory "gives rise to lethal errors."[1] But Erwin and Davidson go further. They have assembled what is, in essence, a clue sheet—a list of key evidences that must be explained. Using that list, they have begun to sketch, at least in outline, a profile of the cause behind the Cambrian explosion.

On the positive side of the ledger, they conclude that this cause must have several attributes in order to explain key facts about the fossil record as well as what it takes to build animals. In particular, the cause must be capable of generating a top-down pattern of appearance; it must be capable of generating new biological form relatively rapidly; and it must be capable of constructing, not merely modifying, complex integrated genetic circuits (specifically, the developmental gene regulatory networks discussed in Chapter 13).

On the negative side, Davidson and Erwin rule out both observed microevolutionary processes and postulated macroevolutionary mechanisms (such as punctuated equilibrium and species selection) as explanations for the origin of the key features of the Cambrian explosion. They insist that

FIGURE 18.1
Douglas Erwin. *Courtesy UPHOTO/Cornell University.*

the requirements for constructing animal body plans *de novo* "cannot be accommodated by microevolutionary [or] macroevolutionary theory."[2]

In Chapter 13, I discussed their reason for coming to this conclusion: developmental gene regulatory networks, once in place, cannot be perturbed (or mutated) without "catastrophic"[3] consequences to the developing animal. Thus, fundamentally new gene regulatory networks (dGRNs) cannot evolve gradually from preexisting dGRNs, if those evolutionary changes require perturbing the deepest nodes of the earlier dGRNs. Yet building new dGRNs capable of producing new animals requires precisely such fundamental alterations in preexisting dGRNs. But then how would new regulatory networks ever arise? Davidson and Erwin insist that no current theory of evolution explains the origin of these systems. Thus, they conclude that the cause of the Cambrian explosion is not described by *any* currently proposed theory of micro- or macroevolution.

In saying this, Erwin emphasizes the uniqueness of the innovations that occurred in the Cambrian explosion. He explains: "Unlike later events, the most significant developmental events of the Cambrian radiation involved the proliferation of cell types, developmental hierarchies and epigenetic cascades."[4] Consequently, he concludes, "The crucial difference between the developmental events of the Cambrian and subsequent events is that the former involved the *establishment* of these developmental patterns, not their *modification*."[5] For this reason, Erwin denies that the central event of the Cambrian explosion—the origin of novel body plans—has any parallel to currently observed biological processes. Rather, he insists that *the events of the past were fundamentally different*—that profound asymmetries exist between evolution then, and evolution now.[6] Thus, he amplifies his denial of the sufficiency of current evolutionary theory by adding one additional attribute, albeit a negative one, to his portrait of "the suspect": the cause responsible for generating the new animal forms, whatever it was, *must have been unlike any observed biological process operating in actual living populations today.*

PROFILING A CAUSE

All of this raises an obvious question. Could the negative clues that increasingly disconfirm materialistic evolutionary theories also be positive indicators of a different kind of cause—perhaps even an intelligent cause?

By sketching the profile of the kind of cause needed to explain the origin of animal life, Davidson, Erwin, and many other evolutionary biologists may have, inadvertently, rendered the idea of intelligent design a bit less inconceivable. To see why, let's quickly review Erwin and Davidson's profile of the suspect. They have concluded that the cause of the origin of the new animal forms in the Cambrian explosion must be capable of:

- generating new form rapidly
- generating a top-down pattern of appearance
- constructing, not merely modifying, complex integrated circuits

They have also concluded that this cause is:

- not described by any currently proposed theory of micro- or macroevolution
- unlike any observed biological process operating in actual living populations today

Erwin and Davidson, no friends of intelligent design, have sketched a partial profile of an adequate cause as befits their particular interest in the importance of gene regulatory networks (Davidson) and fossil discontinuity (Erwin). But other evolutionary biologists have contributed to this picture as well. Simon Conway Morris marvels at "the uncanny ability of evolution to navigate to the appropriate solution through immense 'hyperspaces' of biological possibility."[7] As a result, he argues that evolution may in some way be "channeled" toward propitious functional and/or structural end points—without specifying any known evolutionary mechanism that can so direct evolution to such end points.[8] James Shapiro proposes a mechanism of evolutionary change that relies on preprogrammed adaptive capacity—without explaining where such preprogramming comes from.[9] Several of the new evolutionary theories discussed in the previous chapters presuppose, but do not explain, the existence of both genetic and epigenetic forms of information, highlighting the need for a cause capable of generating such information in the first place.

Erwin and Davidson have made a bold start with their clue list ruling out neo-Darwinism. But the evidence explored in these pages suggests additional attributes that need to be added to their profile of the actual cause of the Cambrian explosion. Our previous investigations have suggested that building an animal requires specified or functional information and that any explanation for the origin of the Cambrian animals must identify a cause capable of generating:

- digital information
- structural (epigenetic) information
- functionally integrated and hierarchically organized *layers* of information

Still, do any or even all of these clues add up to a reason for considering that an alternative kind of cause—a designing intelligence—might have played a role in the origin of animal life?

They do. As it turns out, each of the features of the Cambrian animals and the Cambrian fossil record that constitute negative clues—clues that render neo-Darwinism and other materialistic theories inadequate as causal explanations—also happen to be features of systems known from experience to have arisen as the result of intelligent activity. In other words, standard materialistic evolutionary theories have failed to identify an adequate mechanism or cause *for precisely those attributes* of living forms that we know from experience *only intelligence*—conscious rational activity—is capable of producing. That suggests, in accord with the method of historical scientific reasoning elucidated in the previous chapter, the possibility of making a strong historical inference to intelligent design as the best explanation for the origin of those attributes.

Let's have a look at each of these features of the Cambrian event, starting with key features of the Cambrian animals themselves, to see how they might point to the past activity of a designing intelligence, thereby making intelligent design scientifically detectable.

THE CAMBRIAN INFORMATION EXPLOSION

We have seen that building a Cambrian (or any other) animal would require vast new, functionally specified digital information. Moreover, the presence of such digitally encoded information in DNA presents, at least, a striking *appearance of design* in all living organisms. As Richard Dawkins observes, for example, "The machine code of the genes is uncannily computer-like."[10] Similarly, biotechnology pioneer Leroy Hood refers to the information stored in DNA as "digital code" and describes it in terms reminiscent of computer software.[11] And as we have seen, Microsoft's Bill Gates notes: "DNA is like a computer program but far, far more advanced than any software ever created."[12]

Yet we've also seen that neither neo-Darwinism nor any other materialistic evolutionary model or mechanism explains the origin of the genetic information (the digital code) necessary to produce the Cambrian animals or even the simplest structural innovations that they exhibit. Could this—from a materialistic point of view—*unexplained* appearance of design point instead to actual *intelligent* design?

I think it does. But to explain why, I need to tell a bit more about the "evolution" of my own thinking on the matter.

After learning about how historical scientists make inferences about the causes of events in the remote past, I first applied these methods of reasoning to the question of the origin of the information necessary to produce the first living cell. My book *Signature in the Cell* used the method of multiple competing hypotheses (or inference to the best explanation) to evaluate the "causal adequacy" of proposed explanations for the ultimate origin of biological information. I showed that chemical evolutionary models (whether based upon chance, physical-chemical necessity, or the combination of the two) failed to identify a cause capable of producing the digital information in DNA and RNA. Yet we do know of a cause that has demonstrated the causal power to produce digital code. That cause is intelligent agency. Since intelligent agency is the only cause known to be capable of generating information (at least starting from nonliving chemicals), intelligent design offers the best explanation for the origin of the information necessary to produce the first organism.

The case for intelligent design in *Signature* was carefully limited as a challenge to *chemical* evolution. Many evolutionary biologists acknowledge that chemical evolutionary theory has failed to account for the origin of the first life. Many cite its inability to account for the origin of biological information as one of the main reasons for that failure. Moreover, because they do not think that natural selection could have played a significant role in evolution until *after* the first self-replicating organisms had arisen, most evolutionary biologists also think that explaining the origin of information in a prebiotic context is much more difficult than explaining the origin of new information in already living organisms.

For this reason, in *Signature* I did not try to argue that intelligent design might help explain the origin of the information necessary to account for the origin of new animals from simpler preexisting forms of life. That would have required a separate demonstration showing the inadequacy of natural selection and mutation as a mechanism for generating new genetic information in already living organisms. This book—in Chapters 9–14—has provided that demonstration. These chapters show how neo-Darwinism fails to explain the origin of genetic information—at least, in amounts necessary to build a new protein fold. Chapters 15 and 16 showed, in addition, that the other main materialistic evolutionary theories also fail to account for the information necessary to build new forms of animal life. These theories presuppose, rather than explain, the origin of the information necessary for structural innovation in the history of life. And since the Cambrian explosion of animal life is an explosion of information and structural innovation, that raises a question. Is it possible that this increase of biological information not only represents evidence against materialistic theories of biological evolution, but also positive evidence *for* intelligent design?

A Cause Now in Operation

It does. Intelligent agents, due to their rationality and consciousness, have demonstrated the power to produce specified or functional information in the form of linear sequence-specific arrangements of characters. Digital and alphabetic forms of information routinely arise from intelligent agents. A computer user who traces the information on a screen back to its source invariably comes to a *mind*—a software engineer or programmer. The information in a book or inscription ultimately derives from a writer or scribe. Our experience-based knowledge of information flow confirms that systems with large amounts of specified or functional information invariably originate from an intelligent source. The generation of functional information *is* "habitually associated with conscious activity."[13] Our uniform experience confirms this obvious truth.

It also suggests, therefore, that intelligent design meets the key "causal adequacy" requirement of a good historical scientific explanation. Certainly, intelligence is a "cause now in operation" capable of generating functional or specified information in a digital form. As I write this, my mind is generating specified information. Intelligent agents generate information in the form of software code, ancient inscriptions, books, encrypted military codes, and much else. And since we know of no "presently acting" materialistic cause that also generates large amounts[14] of specified information (especially in a digital or alphabetic form), only intelligent design meets the causal adequacy requirement of a historical scientific explanation. In other words, our uniform experience of cause and effect shows that intelligent design is *the only known cause* of the origin of large amounts of functionally specified digital information. It follows that the great infusion of such information in the Cambrian explosion points decisively to an intelligent cause.

Intelligent design stands alone as an explanation for the origin of genetic information for another reason: purposive agents have just those necessary powers that natural selection lacks as a condition of its causal adequacy. We have seen that natural selection lacks the ability to generate novel information precisely because it can only act *after* new functional information has arisen. Natural selection can favor new proteins and genes, but only after they perform some function (influencing reproductive output). The job of generating new functional genes, proteins, and systems of proteins therefore falls entirely to random mutations. Yet without functional criteria to guide a search through the space of possible sequences, random variation is probabilistically doomed. What is needed is not just a source of variation (i.e., the freedom to search a space of possibilities) or a mode of selection that can operate after the fact of a successful search, but instead a means of selection that (a) operates during a search—*before* success—and that (b) is guided by information about or knowledge of a functional target.

Demonstration of this requirement has come from an unlikely quarter: genetic algorithms. Genetic algorithms are programs that allegedly simulate the creative power of mutation and selection. Richard Dawkins, Bernd-Olaf Küppers, and others have developed computer programs that putatively simulate the production of genetic information by mutation and natural selection.[15] Yet these programs succeed only by the illicit expedient of providing the computer with a "target sequence" and then treating proximity to *future* function (i.e., the target sequence), not actual present function, as a selection criterion. As mathematician David Berlinski shows, genetic algorithms need something akin to a "forward-looking memory" in order to succeed.[16] Yet such foresighted selection has no analogue in nature. In biology, where differential survival depends upon maintaining function, natural selection cannot occur before new functional sequences arise. Natural selection lacks foresight; the process, as evolutionary theorists Rodin and Szathmáry note, works strictly "'in the present moment,' right here and right now . . . lacking the foresight of potential future advantages."[17]

What natural selection lacks, intelligent design—purposive, goal-directed selection—provides. Rational agents can arrange both matter and symbols with distant goals in mind. They also routinely solve problems of combinatorial inflation. In using language, the human mind routinely "finds" or generates highly improbable linguistic sequences to convey an intended or *pre*conceived idea. In the process of thought, functional objectives precede and constrain the selection of words, sounds, and symbols to generate functional (and meaningful) sequences from a vast ensemble of meaningless alternative possible combinations of sound or symbol.[18] Similarly, the construction of complex technological objects and products, such as bridges, circuit boards, engines, and software, results from the application of goal-directed constraints.[19] Indeed, in all functionally integrated complex systems where the cause is known by experience or observation, designing engineers or other intelligent agents applied constraints on the possible arrangements of matter to limit possibilities in order to produce improbable forms, sequences, or structures. Rational agents have repeatedly demonstrated the capacity to constrain possible outcomes to actualize improbable but initially unrealized future functions. Repeated experience affirms that intelligent agents (minds) uniquely possess such causal powers.

Analysis of the problem of the origin of biological information, therefore, exposes a deficiency in the causal powers of natural selection and other undirected evolutionary mechanisms that corresponds precisely to powers that agents are uniquely known to possess. Intelligent agents have foresight. Such agents can determine or select functional goals *before* they are physically instantiated. They can devise or select material means to accomplish those ends from among an array of possibilities. They can then actualize those goals in accord with a *pre*conceived design plan or set of functional requirements. Rational agents can constrain combinatorial space with distant information-rich outcomes in mind. The causal powers that natural selection lacks—by definition—are associated with the attributes of consciousness and rationality—with purposive intelligence. Thus, by invoking intelligent design to overcome a vast combinatorial search problem and to explain the origin of new specified information, contemporary advocates of intelligent design are not positing an arbitrary explanatory element unmotivated by a consideration of the evidence. Instead, we posit an entity possessing precisely the causal powers that a key feature of the Cambrian explosion—the explosive increase in specified information—requires as a condition of its production and explanation.

INTEGRATED CIRCUITRY: DEVELOPMENTAL GENE REGULATORY NETWORKS

Keep in mind, too, that animal forms have more than just genetic information. They also need tightly integrated networks of genes, proteins, and other molecules to regulate their development—in other words, they require developmental gene regulatory networks, the dGRNs that Eric Davidson has so meticulously mapped over the course of his career. Developing animals face two main challenges. First, they must produce different types of proteins and cells and, second, they must get those proteins and cells to the right place at the right time.[20] Davidson has shown that embryos accomplish this task by relying on networks of regulatory DNA-binding proteins (called transcription factors) and their physical targets. These physical targets are typically sections of DNA (genes) that produce other proteins or RNA molecules, which in turn regulate the expression of still other genes.

These interdependent networks of genes and gene products present a striking appearance of design. Davidson's graphical depictions of these dGRNs look for all the world like wiring diagrams in an electrical engineering blueprint or a schematic of an integrated circuit, an uncanny resemblance Davidson himself has often noted. "What emerges, from the analysis of animal dGRNs," he muses, "is almost astounding: a network of logic interactions programmed into the DNA sequence that amounts essentially to a hardwired biological computational device."[21]

These molecules collectively form a tightly integrated network of signaling molecules that function as an integrated circuit. Integrated circuits in electronics are systems of individually functional components such as transistors, resistors, and capacitors that are connected together to perform an overarching function. Likewise, the functional components of dGRNs—the DNA-binding proteins, their DNA target sequences, and the other molecules that the binding proteins and target molecules produce and regulate—also form an integrated circuit, one that contributes to accomplishing the overall function of producing an adult animal form.

Yet, as explained in Chapter 13, Davidson himself has made clear that the tight functional constraints under which these systems of molecules (the dGRNs) operate preclude their gradual alteration by the mutation and selection mechanism. For this reason, neo-Darwinism has failed to explain the origin of these systems of molecules and their functional integration. Like advocates of evolutionary developmental biology, Davidson himself favors a model of evolutionary change that envisions mutations generating large-scale developmental effects, thus perhaps bypassing nonfunctional intermediate circuits or systems. Nevertheless, neither proponents of "evo-devo," nor proponents of other recently proposed materialistic theories of evolution, have identified a mutational mechanism capable of generating a dGRN or anything even remotely resembling a complex integrated circuit. Yet, in our experience, complex integrated circuits—and the functional integration of parts in complex systems generally—are known to be produced by intelligent agents—specifically, by engineers. Moreover, intelligence is the *only* known cause of such effects. Since developing animals employ a form of integrated circuitry, and certainly one manifesting a tightly and functionally integrated system of parts and subsystems, and since intelligence is the only known cause of these features, the necessary presence of these features in developing Cambrian animals would seem to indicate that intelligent agency played a role in their origin (see Fig. 13.4).

THE HIERARCHICAL ORGANIZATION OF GENETIC AND EPIGENETIC INFORMATION

In addition to the information stored in individual genes and the information present in the *integrated networks* of genes and proteins in dGRNs, animal forms exemplify *hierarchical* arrangements or layers of information-rich molecules, systems, and structures. For example, developing embryos require *epi*genetic information in the form of specifically arranged (a) membrane targets and patterns, (b) cytoskeletal arrays, (c) ion channels, and (d) sugar molecules on the exterior of cells (the sugar code). As noted in Chapter 13, much of this epigenetic information resides in the structure of the maternal egg and is inherited directly from membrane to membrane independently of DNA.

This three-dimensional structural information *interacts* with other information-rich molecules and systems of molecules to ensure the proper development of an animal. In particular, epigenetic information influences the proper positioning and thus the function of regulatory proteins (including DNA-binding proteins), messenger RNAs, and various membrane components. Epigenetic information also influences the function of developmental gene regulatory networks. Thus, information at a higher structural level in the maternal egg helps to determine the function of both whole networks of genes and proteins (dGRNs) and individual molecules (gene products) at a lower level within a developing animal. Genetic information is necessary to specify the arrangement of amino acids in a protein or bases in an RNA molecule. Similarly, dGRNs are necessary to specify the location and/or function of many gene products. And, in a similar way, epigenetic information is necessary to specify the location and determine the function of lower-level molecules and systems of molecules, including the dGRNs themselves.

Furthermore, the role of epigenetic information provides just one of many examples of the hierarchical arrangement (or layering) of information-rich structures, systems, and molecules within animals. Indeed, at every level of the biological hierarchy, organisms require specified and highly improbable (information-rich) arrangements of lower-level constituents in order to maintain their form and function. Genes require specified arrangements of nucleotide bases; proteins require specified arrangements of amino acids; cell structures and cell types require specified arrangements of proteins or systems of proteins; tissues and organs require specific arrangements of specific cell types; and body plans require specialized arrangements of tissues and organs. Animal forms contain information-rich lower-level components (such as proteins and genes). But they also contain information-rich *arrangements* of those components (such as the arrangement of genes and gene products in dGRNs or proteins in cytoskeletal arrays or membrane targets). Finally, animals also exhibit information-rich arrangements of *higher-level systems and structures* (such as the arrangements of specific cell types, tissues, and organs that form specific body plans).

The highly specified, tightly integrated, hierarchical arrangements of molecular components and systems within animal body plans also suggest intelligent design. This is, again, because of our experience with the features and systems that intelligent agents—and only intelligent agents—produce. Indeed, based on our experience, we know that intelligent human agents have the capacity to generate complex and functionally specified arrangements of matter—that is, to generate specified complexity or specified information. Further, human agents often design information-rich hierarchies, in which both individual modules and the arrangement of those modules exhibit complexity and specificity—specified information as defined in Chapter 8. Individual transistors, resistors, and capacitors in an integrated circuit exhibit considerable complexity and specificity of design. Yet at a higher level of organization, the specific arrangement and connection of these components within an integrated circuit requires additional information and reflects further design (see Fig. 14.2).

Conscious and rational agents have, as part of their powers of purposive intelligence, the capacity to design information-rich parts and to organize those parts into functional information-rich systems and hierarchies. We know of no other causal entity or process that has this capacity. Clearly, we have good reason to doubt that mutation and selection, self-organizational processes, or any of the other undirected processes cited by other materialistic evolutionary theories, can do so. Thus, based upon our present experience of the causal powers of various entities and a careful assessment of the efficacy of various evolutionary mechanisms, we can infer intelligent design as the best explanation for the origin of the hierarchically organized layers of information needed to build the animal forms that arose in the Cambrian period.

LOCATION, LOCATION, LOCATION

There is another remarkable aspect of the hierarchical organization of information in animal forms. Many of the same genes and proteins play very different roles, depending upon the larger organismal and informational context in which they find themselves in different animal groups.[22] For example, the same gene (*Pax-6* or its homolog, called *eyeless*), helps to regulate the development of the eyes of fruit flies (arthropods) and those of squid and mice (cephalopods and vertebrates, respectively). Yet arthropod eyes exemplify a completely different structure from vertebrate or cephalopod eyes. The fruit fly possesses a compound eye with hundreds of separate lenses (ommatidia), whereas both mice and squid employ a camera-type eye with a single lens and retinal surface. In addition, although the eyes of squid and mice resemble each other optically (single lens, large internal chamber, single retinal surface), they focus *differently.* They undergo completely different patterns of development and utilize different internal structures and nerve connections to the visual centers of the brain. Yet the *Pax-6* gene and its homologs play a key role in regulating the construction of all three of these different adult sensory structures. Moreover, evolutionary and developmental biologists have found that this pattern of "same genes, different anatomy" recurs throughout the bilaterian phyla, for features as fundamental as appendages, segmentation, the gut, heart, and sense organs (see Fig. 18.2).[23]

This pattern contradicts the expectations of textbook evolutionary theory. Neo-Darwinism predicts that disparate adult *structures* should be produced by different *genes.* This prediction follows directly from the neo-Darwinian assumption that all evolutionary (including anatomical) transformations begin with mutations in DNA sequences—mutations that are fixed in populations by natural selection, genetic drift, or other evolutionary processes. The arrow of causality flows one way from genes

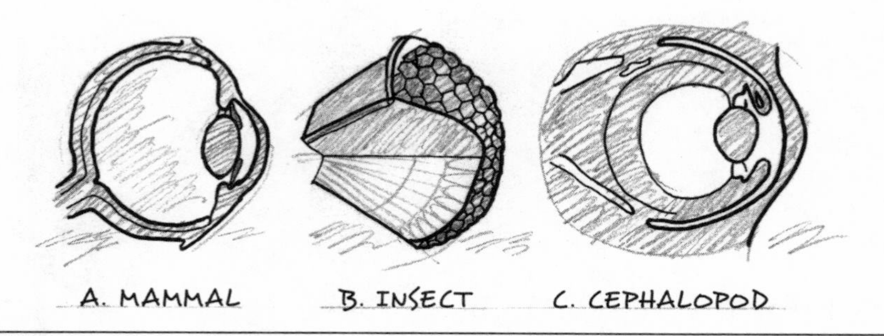

FIGURE 18.2
The same genes can be used in different animals to produce dramatically different structures, contradicting neo-Darwinian expectations.

(DNA) to development to adult anatomy. Thus, if biologists observe different animal forms, it follows that they should expect that *different genes* will specify those forms during animal development. Given the profound differences between the fruit-fly compound eye and the vertebrate camera eye, neo-Darwinian theory *would not* predict that the "same" genes would be involved in building different eyes in arthropods and chordates.[24]

Many leading evolutionary theorists have acknowledged this problem. University of Wisconsin evo-devo researcher Sean B. Carroll has noted that the neo-Darwinian prediction of similar genes producing similar structures is "entirely incorrect."[25] Stephen Jay Gould described the discovery of the polyfunctional role of similar genes as "explicitly unexpected" and "discombobulat[ing] the confident expectations of orthodox theory."[26]

The theory of intelligent design suggests a solution to the problem—a solution familiar to us from the construction and operation of our own artifacts. Figure 18.3 shows a general-purpose switching transistor. These electronic components can be used to help build many electronic systems, from a computer to a microwave oven to a radio. And the exact functional role that the transistor will play will be governed by the system in which it finds itself. (One must, of course, make allowances for the particular specifications of the transistor itself; a transistor cannot function as a battery).

Nowhere, however, is this feature of polyfunctional modularity more intuitively clear than in our use of natural languages, such as English. To illustrate this, my colleague Paul Nelson once "disassembled" the last forty-four words of Abraham Lincoln's Gettysburg Address (see Fig. 18.4)

FIGURE 18.3
A general purpose transistor—an example of a component that can be used to perform different functions in different designed systems. *Courtesy iStockphoto.com/S230.*

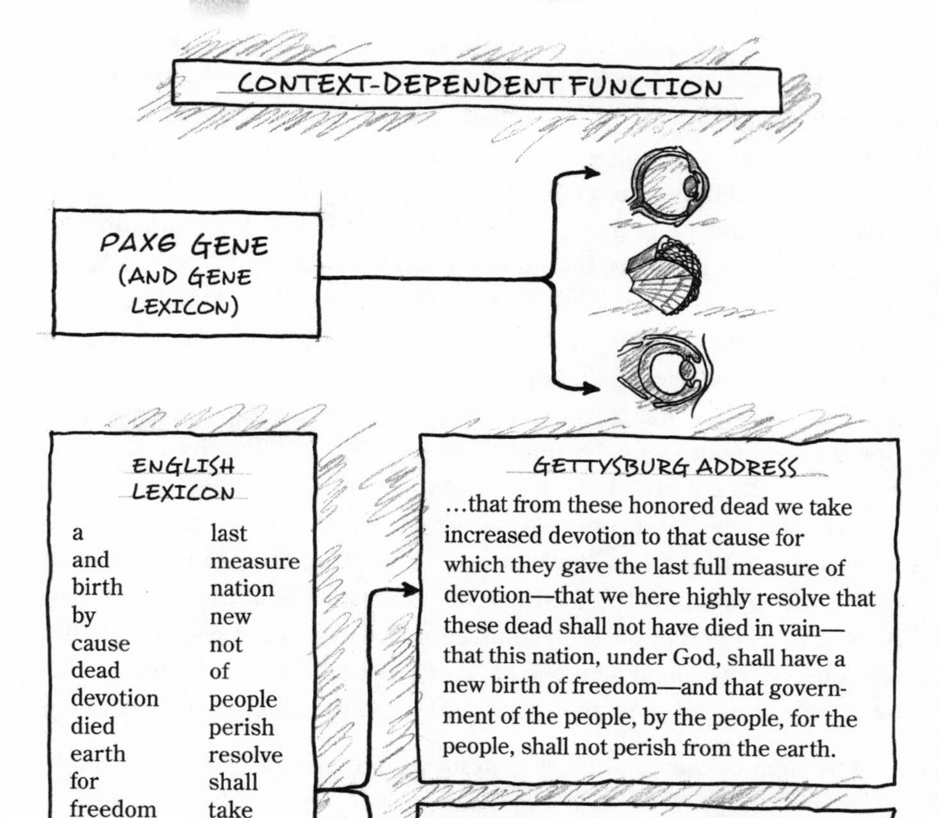

FIGURE 18.4

This comparison of Lincoln's Gettysburg Address and an "Anarchist's Manifesto" shows how the same modular elements (words) can perform different functions depending on their surrounding context, just as many genes do in biological systems. *Courtesy Paul Nelson.*

into a lexicon. Using the same words, at roughly the same frequency, he then wrote an "Anarchist's Manifesto," with a meaning diametrically opposite to that of Lincoln's. What had changed were *not* the words—that is, the lower-level modules. Rather, the higher-level context, or the system as a whole, differed. Lincoln wrote with one meaning in mind, the anarchist with another—and that made all the difference to the functions of the elements, or modules, within the respective systems. Nelson calls this context dependency "the organismal context principle" when it occurs in biology and he likens it to the context dependence of words in language or low-level parts in a technological system.

He also argues that intelligent design provides a compelling explanation for the presence of polyfunctional modularity in living systems. Why? Not only is the polyfunctionality of genetic modules unexpected in a neo-Darwinian view, it is a common feature of intelligently designed systems. As Nelson and Jonathan Wells note, "An intelligent cause may reuse or redeploy the same module in different systems, without there necessarily being any material or physical connection between those systems."[27] They also observe that intelligent agents "can generate identical patterns independently" and put them to different uses in different systems of parts:

> *If we suppose that an intelligent designer constructed organisms using a common set of polyfunctional genetic modules—just as human designers, for instance, may employ the same transistor or capacitor in a car radio or a computer, . . . then we can explain why we find the "same" genes expressed in the development of what are very different organisms. . . . A particular gene, employed for its DNA-binding properties, finds its functional role in a higher-level system whose ultimate origin was intelligently caused.*[28]

Wells and Nelson go on to explain that "the overall system, not the gene itself" determines the ultimate functional significance of the lower level modules, just as it does in all human technological or communication systems. Certainly, both the computer software and computer hardware (integrated circuits) exhibit this feature—what might be called "context-dependent, polyfunctional modularity." Similarly, in information-rich texts—such as the Gettysburg Address or the Anarchist's Manifesto—humans convey different meanings with the same low-level modules (words) depending upon their surrounding context. Experience shows that when we know how systems possessing this feature arose, invariably they arose by intelligent design.

FEATURES OF THE PRE-CAMBRIAN–CAMBRIAN FOSSIL RECORD

Intelligent design not only helps to explain many key features of the Cambrian animals themselves; it also helps to explain many otherwise anomalous features of the Cambrian fossil record.

An Inverted Cone: Disparity Preceding Diversity

As discussed in Chapter 2, the fossil record shows a "top-down" pattern in which phyla-level morphological disparity appears first followed only later by species-level diversity. Major innovations in body plans precede minor variations on basic designs.[29] This "inverted cone of diversity" also suggests intelligent design.

Neo-Darwinism seeks to explain the origin of novel body plans by starting with simpler body plans and gradually assembling animals with more complex body plans via the gradual accumulation of small successive material variations. Thus, neo-Darwinism employs a "bottom-up" mode of causation. With a bottom-up approach, small-scale diversification should eventually produce large-scale morphological disparity—differences in body plan. The "bottom-up" metaphor thus describes a kind of self-assembly in which the gradual production of the material parts eventually generates the organization of the whole. This suggests in turn that the parts stand causally prior to the organization of the whole. As I have argued, however, this approach encounters both paleontological and biological difficulties: the fossil record leaves no evidence of such a process and the morphological innovation and transformations that it requires are, in any case, biologically implausible.

But if a bottom-up approach fails, perhaps a "top-down" approach will succeed. "Top-down" causation begins with a basic architecture, blueprint, or plan and then proceeds to assemble parts in accord with it. The blueprint stands causally prior to the assembly and arrangement of the parts.

But where could such a blueprint come from? One possibility involves a mental mode of causation. Intelligent agents often conceive of plans prior to their material instantiation—that is, the *pre*conceived design of a blueprint often precedes the assembly of parts in accord with it. An observer touring the parts section of a General Motors plant will see no direct evidence of a prior blueprint for GM's new models, but will perceive the basic design plan immediately upon observing the finished product at the end of the assembly line. Designed systems, whether automobiles, airplanes, or computers, invariably manifest a design plan that preceded their first material instantiation. But the parts do not generate the whole. Rather, an idea of the whole directed the assembly of the parts.

This form of causation can certainly explain the pattern in the fossil record. As new species appear in the Cambrian, they manifest completely novel, morphologically disparate, and functionally integrated body plans. Thus, although the fossil record does not directly establish the existence of a prior mental plan or blueprint, such a plan could certainly explain, or be inferred from, the top-down pattern of fossil evidence. In other words, if the body plans of the Cambrian animals did arise as the result of a "top-down" mode of causation involving a *pre*conceived design plan, we would expect, based on our experience of complex designed systems, to find precisely the pattern of evidence that we do see in the fossil record. Further, materialistic "bottom-up" models of causation fail to explain this same pattern of fossil evidence. Thus, intelligent design provides a better explanation of this feature of the Cambrian fossil record than do competing materialistic evolutionary theories.

The design hypothesis can also explain why smaller-scale diversity arises *after*, not before, morphological disparity in the fossil record or, to put it more poetically, why the basic themes of life precede the variation on those themes. Complex designed systems have a fundamental functional integrity that makes their alteration difficult. For this reason, we should not expect gradual mechanisms of change to produce new body plans or alter them fundamentally after they have arisen. We might, however, expect to find *variations* on these basic themes within the functional limits established by a basic architecture or body plan. Fundamentally new forms of organization require design from scratch. For example, airplanes did not arise gradually or incrementally from automobiles. Nevertheless, new innovations often accrete to novel designs provided the fundamental organizational plan is not altered.

Since the invention of the automobile, all cars have included the same basic structural and functional elements, including a motor, at least three (and usually four) wheels, a carriage with seats for passengers, a structure connecting the wheels to the carriage, a steering wheel and column (or analogous mechanism), and a means of translating energy generated by the motor to the wheels. These are minimal requirements, of course; many cars have used axles to connect the wheels, though some have not and a "stretch" limousine may need additional axles or wheels. Indeed, though many new variations on the original model have arisen *after* the invention of the basic automobile design, all exemplify this same basic design. Interestingly, we also observe this pattern in the fossil record. The major animal body plans appear first instantiated by a single (or very few) species or genera. Then, later, many other varieties arise with many new features, yet all still exhibit the same basic body plan.

Experience shows a hierarchical relationship between functionally *necessary* and functionally *optional* features in designed systems. An automobile cannot function without a motor or steering mechanism; it can function with or without twin I-beam suspension, antilock brakes, or "stereo surround sound." This distinction between functionally necessary and optional features suggests the possibility of future innovation and variation on basic design plans, even as it imposes limits on the extent to which the basic designs themselves can be altered. The logic of designed systems, therefore, suggests precisely the kind of top-down pattern that we see in both the history of our own technological innovation[30] and in the history of life following the Cambrian explosion (compare Figs. 18.5a and 18.5b). On the other hand, competing materialistic evolutionary theories would not lead us to expect the fossil record to manifest such a "top-down" pattern, but the opposite.

Sudden Appearance and Missing Ancestors

The theory of intelligent design *can* also help to account for the abrupt appearance of complex anatomical structures and animal body plans in the fossil record. Intelligent agents sometimes produce material entities through a series of gradual modifications (as when a sculptor shapes a

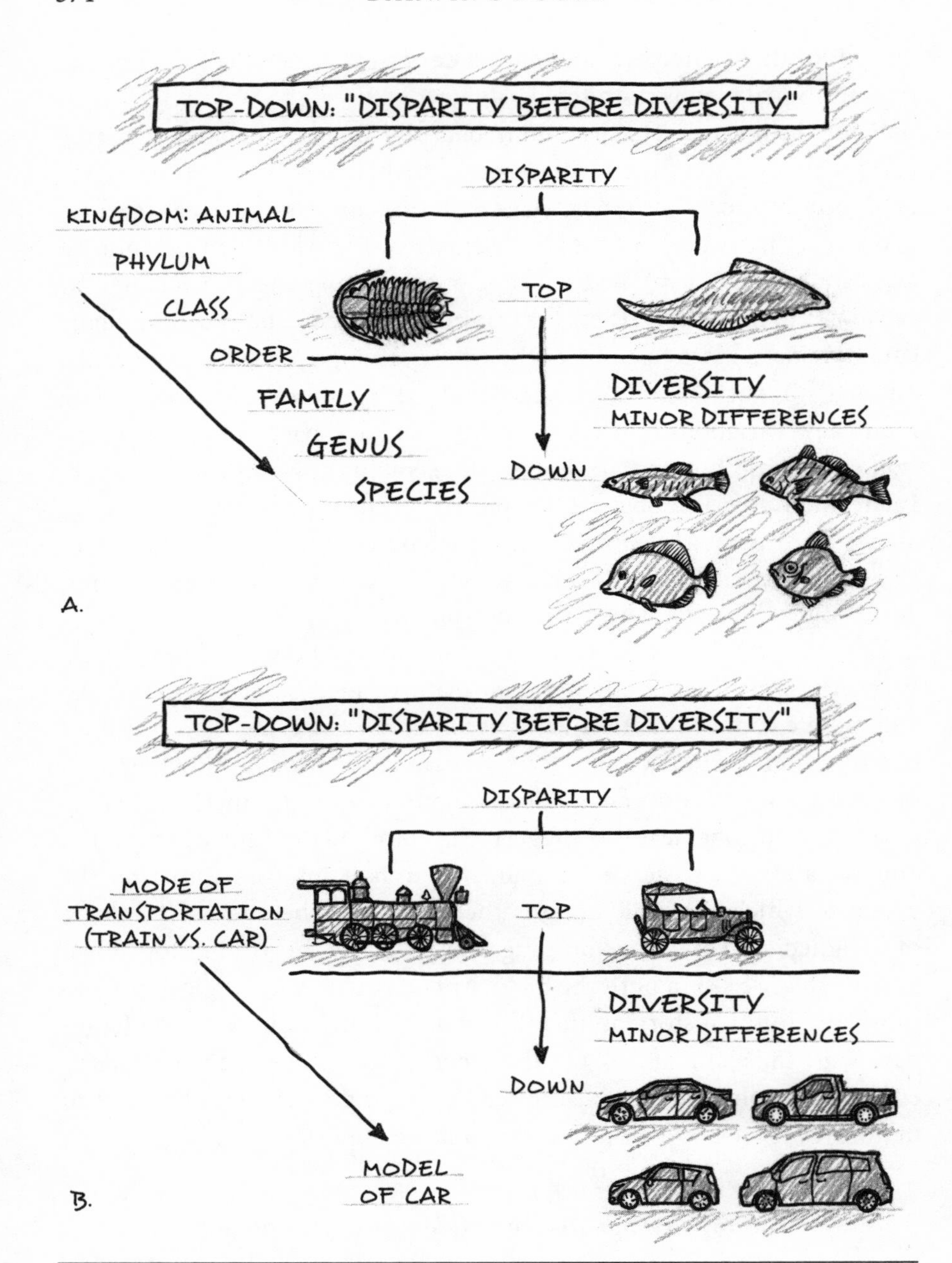

FIGURE 18.5
Figure 18.5a (top): A top-down pattern of appearance found in the history of animal life.
Figure 18.5b (bottom): A top-down pattern of appearance in human technology.

sculpture over time). Nevertheless, intelligent agents also have the capacity to introduce complex technological systems into the world fully formed. Often such systems bear no resemblance to earlier technological systems—their *invention* occurs without a *material* connection to earlier, more rudimentary technologies. When the radio was first invented, it was unlike anything that had come before, even other forms of communication technology. For this reason, although intelligent agents need not generate novel structures abruptly, they can do so. Thus, invoking the activity of a mind provides a causally adequate explanation for the pattern of abrupt appearance in the Cambrian fossil record.

On the other hand, strictly materialistic theories of evolution necessarily envision a "bottom-up" mode of causation in which material parts or materially instantiated intermediate forms of organization necessarily precede the emergence of fully developed body plans. For this reason, the sudden appearance of novel animal forms contradicts the expectations of most materialistic evolutionary theories. Neo-Darwinism, in particular, would not expect the sudden appearance of animal forms. As Darwin himself insisted: "*Natura non facit saltum*" ("Nature takes no leaps"). Yet intelligent agents can act suddenly or discretely in accord with their powers of rational choice or volition, even if they do not always do so. Thus, the sudden appearance of the Cambrian animals does suggest, at least, the possibility of a volitional act of a conscious agent—a designer.

Intelligent design likewise helps explain the absence of ancestral precursors. If body plans arose as the result of an intelligent agent actualizing an immaterial plan or idea, then an extensive series of material precursors to the first animals need not exist in the fossil record, anymore than such a series is always present in the history of technology. The radio did not evolve gradually from the telegraph. Mental plans or concepts need not leave a material trace. Thus, intelligent design can account for the dearth of material precursors in the Precambrian strata, whereas "bottom-up" materialistic evolutionary theories cannot, especially given the failure of the artifact hypothesis discussed earlier.

Stasis (or Persistent Morphological Isolation)

Finally, intelligent design also explains the observed stasis in the fossil record. As advocates of punctuated equilibrium established, Cambrian

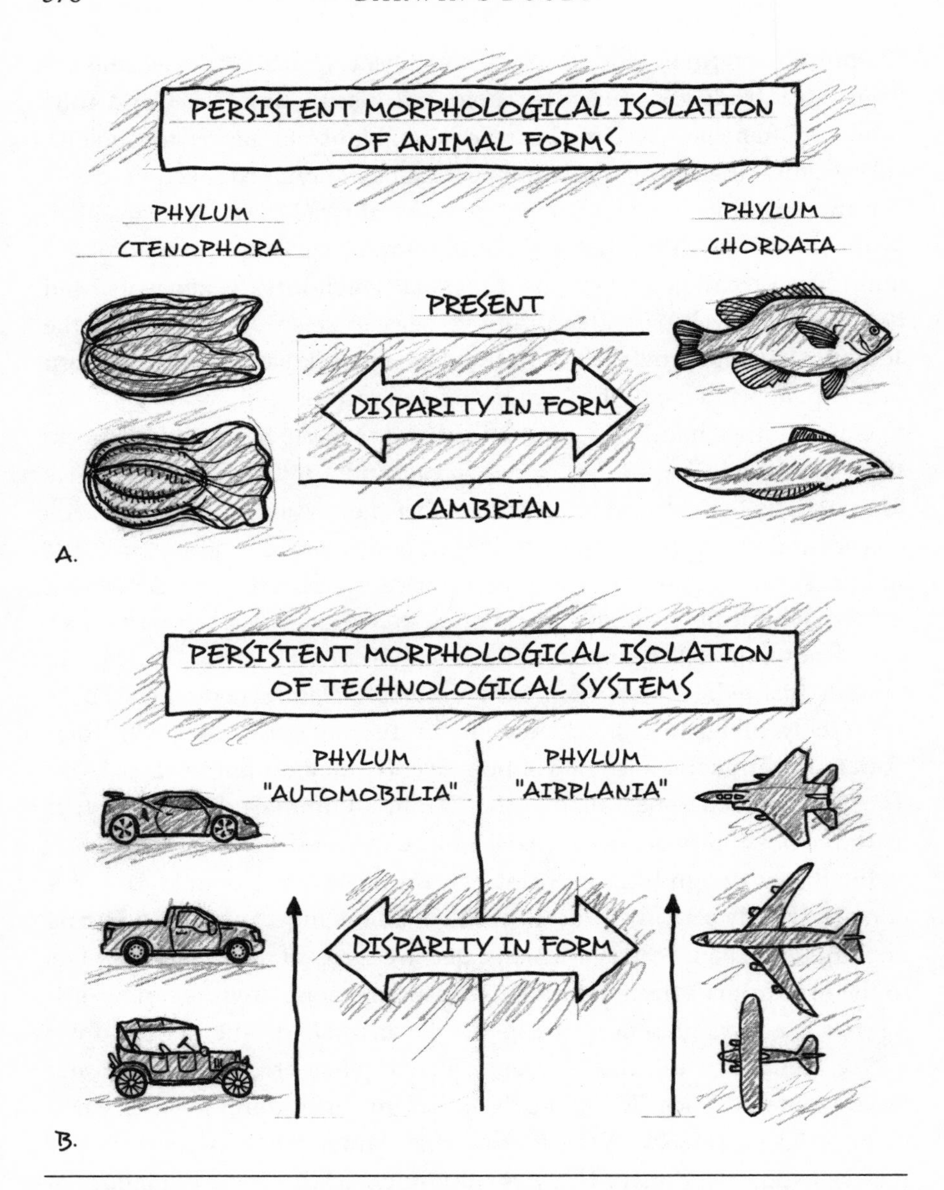

FIGURE 18.6
Comparison of persistent morphological isolation in two animal forms that first arose in the Cambrian period, and similar morphological isolation in two technological systems.

species tend to persist unchanged in their basic forms over time. Animal body plans that define the higher taxa, including classes and phyla, also remain especially stable in their basic architectural designs, showing "no directional change"[31] over geological history after their first appearance in the Cambrian. As a result, the morphological disparity between distinct animal body plans remains unbridged. Moreover, as noted in Chapter 13, developmental mechanisms constrain the degree to which organisms may vary without deleterious consequences.

The persistent morphological disparity and isolation of animal body plans is completely unexpected by neo-Darwinian, and all other gradualist, evolutionary theories, at least. Given such models of evolutionary change, theorists should expect the fossil record to exhibit forms of life that grade imperceptibly from one to another. Indeed, absent either a compelling version of the artifact hypothesis or an adequate mechanism of punctuated large-scale evolutionary change, "morphospace" should be mostly filled in. The fossil record should not display mostly morphologically disparate or separate forms of animal life.

Yet, if living systems did arise as the result of intelligent design, then such morphological isolation, and such *persistent* isolation over time, is just what we should expect to see—precisely because that is what we do see in the history of other intelligently designed systems (see Fig. 18.6). Indeed, experience suggests that designed objects have a functional integrity that makes the modification of some of their essential parts and their basic organization and architecture difficult or impossible. Though the Model-A has been replaced by everything from the Yugo to the Honda Accord, the automobile "body plan" with several essential functional and/or structural elements has remained unchanged from its first appearance in the late nineteenth century.

Further, despite the design of many innovative variations, automobiles have retained their "morphological distance" or structural disparity from other functionally distinct technological devices. Indeed, persistent morphological disparity in biological systems (manifested as stasis in the fossil record) has a direct parallel in our own technology. In biology, what we recognize as different organismal body plans are systems that differ fundamentally from each other in their overall organization. A crab and a starfish, for example, may exhibit some similarities in their low-level protein parts, but they differ fundamentally in their digestive and nervous systems and in the overall organization of their organs and body parts. In the same way, automobiles and airplanes may have many similar parts, even as they differ in the composition of their distinguishing parts and overall organization.

The presence of such structural disparities and isolation among complex functionally integrated systems represents another distinctive feature of the intelligently designed systems known from our own world of technology. For example, the basic technology of the CD-ROM (as employed in audio systems and computers) did not "evolve" incrementally from earlier technologies, such as magnetic media (e.g., digital tape or disc storage) or analog systems such as the once standard long-playing (LP) record. Indeed, it could not. In an analog recording, information is stored as three-dimensional microscopic grooves in a vinyl surface and is detected mechanically by a diamond stylus. This means of storing and detecting information differs fundamentally, *as a system,* from the digitally encoded pits storing data in the silvered surface of a CD-ROM, where information is detected optically, not mechanically, by a laser beam. The CD-ROM had to be engineered from scratch and, as a result, displays a striking structural difference and isolation from other technological devices, even those that perform roughly the same function. Although minor new features may "accrete" to its basic design architecture, a deep and impassable functional gulf separates the CD-ROM as a system from other technological systems. As biologist Michael Denton expresses it, "What is true of sentences and watches is also true of computer programs, airplane engines, and in fact of all known complex systems. Almost invariably, function is restricted to unique and fantastically improbable combinations of subsystems, tiny islands of meaning lost in an infinite sea of incoherence."[32] In fact, such structural disparity or morphological isolation constitutes a diagnostic of designed systems—that is, a feature of systems for which only one kind of cause—an intelligent cause—is known.

ACTS OF MIND

Studies in the history and philosophy of science have shown that to explain an event or a set of facts, scientists must typically cite a *cause* capable of producing that event or those facts. When scientists do not have the luxury of directly observing the cause of a particular event or effect under study, as historical scientists typically do not, they must cite a cause that is *otherwise* known to produce the facts in question. That means historical scientists must show that the event or facts of interest must in some way represent the *expected* outcome of a particular cause having acted in the past—that the event or facts should have occurred "as a matter of course."

To many scientists, especially those steeped in the materialistic assumptions of contemporary scientific culture, the idea of intelligent design seems inherently implausible or even incoherent. Science to them involves not only observing and studying material entities and phenomena, but explaining them by reference to materialistic entities. For these scientists, it makes no sense even to consider the idea of intelligent design, with its explicit reference to the activity of a designing mind.

Yet it turns out that both the Cambrian animal forms themselves and their pattern of appearance in the fossil record exhibit precisely those features that *we should expect to see* if an intelligent cause had acted to produce them (see Fig. 18.7). Further, the Cambrian animal forms and their manner of appearance contradict what we should expect to find in the fossil record and in the animal world given a purely materialistic "bottom-up" process of evolution. Thus, despite its potential for disturbing the materialistic sensibilities of many scientists, it is hard logically to avoid the conclusion that the design hypothesis actually provides a better, more casually adequate explanation for key features of the Cambrian event.

When Darwin first acknowledged the problem of the Cambrian fossil record, and the small but persistent doubt it raised for him about his theory, his nemesis Louis Agassiz not only rejected his theory of evolution, but also affirmed an alternative understanding of the nature and origin of animal life. To Agassiz, the pattern of animal classification and the fossil record reinforced the idea that living forms exemplified basic "types"—ideas that had originated in the mind of a designing intelligence. Thus, he would argue that the Cambrian fossils tell of "acts of mind."[33]

As noted in Chapter 1, Darwin himself acknowledged both Agassiz's immense paleontological knowledge and the validity of the problems that Agassiz raised. Even so, his affirmation of a positive alternative to Darwin's theory in the form a design hypothesis might well have seemed premature in the 1860s and certainly did reflect something of the preju-

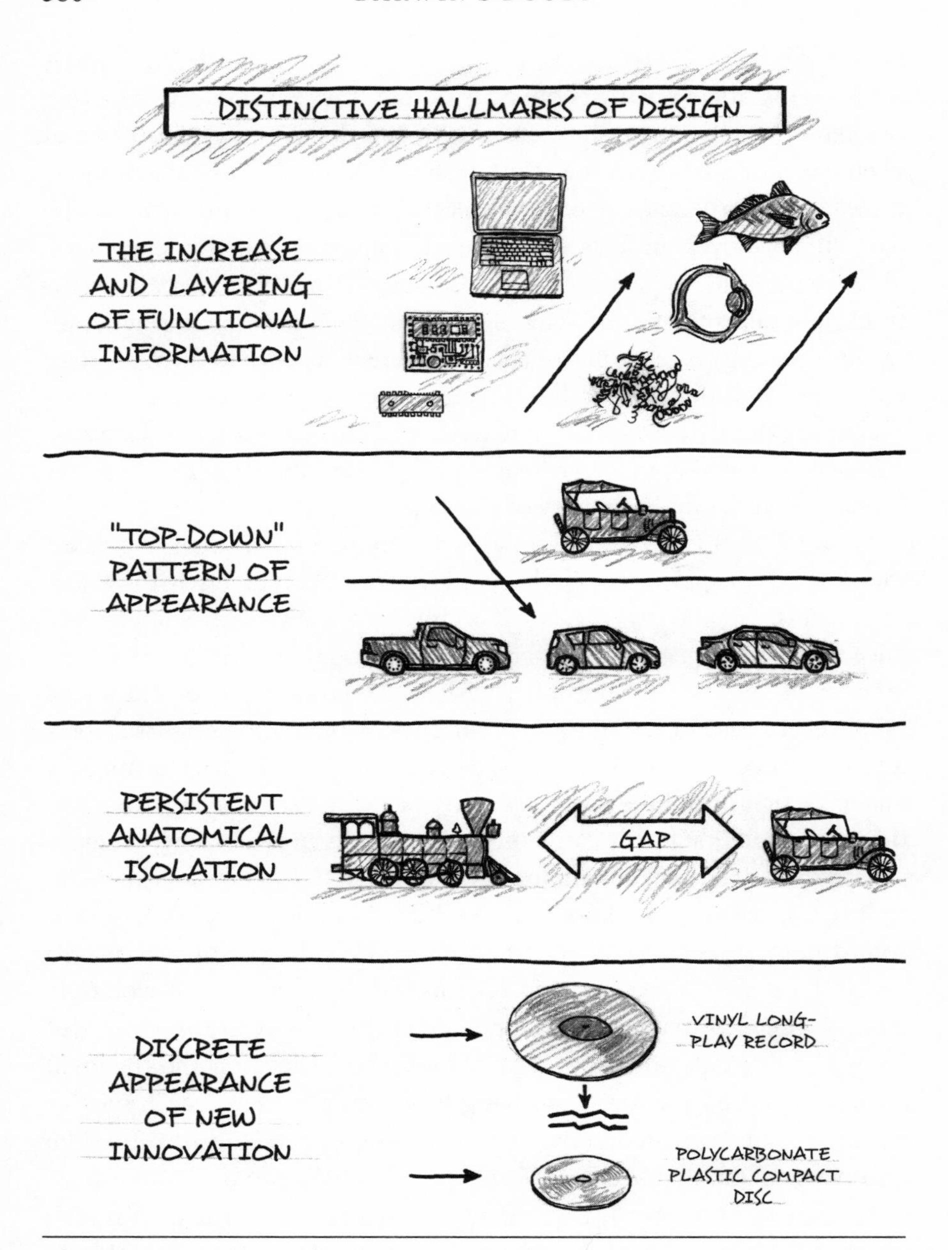

FIGURE 18.7
Both the Cambrian animal forms and their pattern of appearance in the fossil record exhibit distinctive features or hallmarks of designed systems—features that we should expect to see if an intelligence acted to produce them.

dice of the times. But more than a century and a half later, after many failed attempts to discover—and explain away—the missing fossil ancestors, and after discoveries in molecular and developmental biology have revolutionized our understanding of the complexity of animal life, continuing to regard the Cambrian explosion as merely a niggling problem for established theory—a lone question mark or negative clue—now seems not so much cautious, as simply unresponsive to the evidence.

The animal forms that arose in the Cambrian not only did so without any clear material antecedent; they came on the scene complete with digital code, dynamically expressed integrated circuitry, and multi-layered, hierarchically organized information storage and processing systems.

In light of these marvels and the persistent pattern of the fossil record, should we now continue, as Darwin did (who knew nothing of them), to regard the Cambrian explosion as just an anomaly? Or may we now consider the features of the Cambrian event as evidence supporting another view of the origin of animal life? If so, is there now a compelling logic for considering a different kind of causal history?

In fact, there is. The features of the Cambrian event point decisively in another direction—not to some as-yet-undiscovered materialistic process that merely mimics the powers of a designing mind, but instead to an actual intelligent cause. When we encounter objects that manifest any of the key features present in the Cambrian animals, or events that exhibit the patterns present in the Cambrian fossil record, and we know how these features and patterns arose, invariably we find that intelligent design played a causal role in their origin. Thus, when we encounter these same features in the Cambrian event, we may infer—based upon established cause-and-effect relationships and uniformitarian principles—that the same kind of cause operated in the history of life. In other words, intelligent design constitutes the best, most causally adequate explanation for the origin of information and circuitry necessary to build the Cambrian animals. It also provides the best explanation for the top-down, explosive, and discontinuous pattern of appearance of the Cambrian animals in the fossil record.

19

THE RULES OF SCIENCE

The argument of the previous chapter raises an obvious question. If intelligent design provides such a clear and satisfying resolution to the mystery of the Cambrian explosion, why have so many brilliant scientists missed it?

While reflecting on this question, I came across a short story by G. K. Chesterton called "The Invisible Man," which may cast some light on it. In "The Invisible Man," Chesterton tells the story of someone who is murdered in an apartment with only one entrance, an entrance watched by four honest men. These men insist that during their watch no one entered or left the building. A brilliant French detective investigates the case, along with his friend, a dusty little Catholic priest. They query the guards, each of whom insists that no one entered or exited the building. But then the unimpressive looking priest, Father Brown, all but forgotten in the background, pipes up to ask, "Has nobody been up and down stairs, then, since the snow began to fall?"

"Certainly not," they assure him.

"Then I wonder what that is?" Father Brown asks, gazing at the white snow on the outside entrance stairs. Everyone turns to find a "stringy pattern of grey footprints" there.

"God!" one of them cries, "An invisible man!"

After asking a few more questions, Father Brown quickly unravels the mystery. "When those four quite honest men said that no man had gone into the Mansions, they did not really mean that no man had gone into them," Father Brown explains to his detective friend. "They meant no man whom they could suspect of being your man. A man did go into the house and did come out of it, but they never noticed him."

"An invisible man?"

"A mentally invisible man," the priest explains.

What does a mentally invisible man look like?

"He is dressed rather handsomely in red, blue and gold," the priest explains, "and in this striking, and even showy, costume he entered Himylaya Mansions [the name of the apartment complex] under eight human eyes; he killed . . . [the murder victim] in cold blood, and came down into the street again carrying the dead body. . . . You have not noticed such a man as this."

At that moment, he reaches out and puts his hand on "an ordinary passing postman," one who had almost slipped by them unnoticed.

"Nobody ever notices postmen somehow," Father Brown muses. "Yet they have passions like other men, and even carry large bags where a small corpse can be stowed quite easily."[1]

The passing postman, of course, is the murderer. He walked up and down the stairs under the four men's noses, but because of their mental blinders telling them whom to consider and whom to ignore, they overlooked the postman entirely.

The theme is a favorite of detective-story authors: the obvious possibility missed by the experts, because their assumptions prevent them from considering what might otherwise seem to be an obvious possibility. Could something like that be at work in the investigation of the Cambrian explosion? Could evolutionary biologists and paleontologists be wearing a set of mental blinders that keeps them from considering a possible explanation of the Cambrian mystery?

Odd as it may seem, that is exactly what has been going on in the investigation of the Cambrian explosion. In this case, however, those wearing the mental blinders have elevated an unwillingness to consider certain explanations to a principle of scientific method. That principle is called "methodological naturalism" or "methodological materialism." Methodological naturalism asserts that to qualify as scientific, a theory must explain phenomena and events in nature—even events such as the origin of the universe and life or phenomena such as human consciousness—by reference to strictly material causes. According to this principle, scientists may not invoke the activity of a mind or, as one philosopher of science puts it, any "creative intelligence."[2]

To see how adherence to this principle has prevented scientists from considering a possibly true (even "causally adequate") explanation for the Cambrian explosion, let's revisit the case reported in Chapter 11 of Richard Sternberg (see Fig. 19.1), the evolutionary biologist at the Smithsonian's National Museum of Natural History. After Sternberg published my article arguing for intelligent design as the best explanation of the Cambrian information explosion in the technical journal *Proceedings of the Biological Society of Washington*,[3] he suffered professional retribution at the hands of Smithsonian administrators.[4] The Biological Society of Washington, the governing body that oversees the publication of the journal that Sternberg then edited, also issued a public statement repudiating his decision.[5] Its statement did not, however, cite any factual errors in the article or seek to rebut it. Further, the president of the society, Smithsonian zoologist Roy McDiarmid, wrote Sternberg privately and told him that he (McDiarmid) had reviewed the file containing the peer-review reports and had found everything to be in order.[6]

What, then, had Sternberg done to deserve public rebuke?

Sternberg published a paper that violated a presumed rule of science: methodological naturalism. Without saying it in so many words, the Biological Society made crystal clear that this was the crucial issue. When it distanced itself from Sternberg and the review essay, it did not invite a scientific refutation of the article, as if the problem had been a

FIGURE 19.1
Richard Sternberg.
Courtesy Laszlo Bencze.

misrepresentation or misinterpretation of the evidence. Instead, it attempted to settle the issue by releasing a *policy* statement. As a writer in the *Wall Street Journal* reported at the time, "The Biological Society of Washington released a vaguely ecclesiastical statement regretting its association with the article. It did not address its arguments but denied its orthodoxy, citing a resolution of the American Association for the Advancement of Science that defined ID [the theory of intelligent design] as, by its very nature, unscientific."[7]

The Biological Society of Washington "deemed the paper inappropriate for the pages of the *Proceedings*."[8] The Society attempted to justify this claim, first, on thin procedural grounds, claiming that a paper about the origin of animal body plans represented a "departure" from its more typical concern with issues of animal classification. Second, and more tellingly, it cited the policy statement of the American Association for the Advancement of Science (AAAS) "calling upon its members to understand the nature of science" and to recognize "the inappropriateness of 'intelligent design theory' as subject matter for science education."[9] Setting aside the obvious point that my paper was written not as a curricular manifesto but as an evidence-based scientific argument, the AAAS statement affirmed an implicitly and strictly materialistic understanding of the nature of science. It did so to disqualify intelligent design from consideration—not only in science education, but in science itself.

The Sternberg case—like numerous others in which the academic freedom of scientists advocating intelligent design has been abridged[10]—goes a long way to answering the question of why so many otherwise brilliant and knowledgeable scientists have overlooked such a seemingly obvious possible answer to the Cambrian conundrum. As in Chesterton's story about the invisible postman, they have accepted a self-imposed limitation on the hypotheses they are willing to consider. These scientists think they are doing their duty to science. Yet if researchers refuse as a matter of principle to consider the design hypothesis, they will obviously miss any evidence that happens to support it. And the cultural pressure within biology to avoid considering the intelligent design hypothesis has long been nontrivial. Francis Crick, for example, famously admonished biologists to "constantly keep in mind that what they see was not designed, but rather evolved."[11] In 1997, in an article in the *New York Review of Books*, Harvard geneticist Richard Lewontin made explicit a similar commitment to a strictly materialistic explanation—whatever the evidence might seem to indicate. As he explained in a now often quoted passage:

> *We take the side of science in spite of the patent absurdity of some of its constructs, in spite of its failure to fulfill many of its extravagant promises of health and life, in spite of the tolerance of the scientific community for unsubstantiated just-so stories, because we have a prior commitment, a commitment to materialism. It is not that the methods and institutions of science somehow compel us to accept a material explanation of the phenomenal world, but, on the contrary, that we are forced by our a priori adherence to material causes to create an apparatus of investigation and a set of concepts that produce material explanations, no matter how counter-intuitive, no matter how mystifying to the uninitiated. Moreover, that materialism is absolute, for we cannot allow a Divine Foot in the door.*[12]

The commitment to methodological naturalism that Lewontin describes, as well as the behavior of scientists in cases such as Sternberg's, leave no doubt that many in science simply will not consider the design hypothesis as an explanation for the Cambrian explosion or any other event in the history of life, whatever the evidence. To do so would be to violate the "rules of science" as they understand them.

BUT IS IT SCIENCE?

But are these scientists right? Perhaps science must limit itself to purely naturalistic or materialistic explanations. If so, are there perhaps good reasons for excluding the design hypothesis from consideration as a scientific hypothesis? Is methodological naturalism the correct policy for science?

Though scientists routinely assert methodological naturalism as a scientific norm, that principle and its exclusion of the design hypothesis have proven difficult to justify. To claim that a specific theory does not qualify as scientific requires a definition of science or a set of definitional criteria by which to make that kind of a judgment. Some philosophers and scientists have asserted that for a scientific theory to qualify as scientific, it must meet various criteria of testability, falsifiability, observability, repeatability, and the like. Philosophers of science call these "demarcation criteria," because some scientists purport to use them to define or "demarcate" science and to distinguish it from pseudo science or from other forms of inquiry such as history, religion, or metaphysics.[13]

THE GENERAL PROBLEM OF DEMARCATION

The demarcation question has long been a vexing one. Historically, scientists and philosophers of science have thought that science could be distinguished by its especially rigorous method of study. But attempts to define science by reference to a distinctive method have proven problematic because different branches and types of science use different methods.

For example, some scientific disciplines distinguish and classify natural entities, while others attempt to formulate overarching laws that apply to all entities. Some disciplines perform laboratory experiments under controlled and replicable conditions, while others attempt to reconstruct or explain singular events in the past, often based on field studies of evidence or clues rather than laboratory experiments. Some disciplines generate mathematical descriptions of natural phenomena without positing mechanisms to explain them. Others look for mechanisms or explain law-like regularities by reference to underlying mechanisms. Some scientific disciplines make predictions to test theories, while others test competing theories by comparing their explanatory power. Some disciplines use both these methods, while some conjectures (particularly in theoretical physics) may not be testable at all. And on it goes.

An episode in the history of science illustrates the problem. During the seventeenth century, a group of scientists called the "mechanical philosophers" insisted, based largely on advances in early chemistry, that scientific theories must provide mechanistic explanations. Such explanations had to involve one material entity pushing or pulling another. Yet in physics, Isaac Newton (1642–1727) formulated an important theory that provided no mechanistic explanation. His theory of universal gravitation described mathematically, but did not explain in a mechanistic way, the gravitational attraction between planetary bodies—bodies separated from each other by miles of empty space with no means of mechanical interaction with each other whatsoever.[14] Despite provocation from the German mathematician Gottfried Wilhelm Leibniz (1646–1716), who defended the mechanistic ideal, Newton expressly refused to give any explanation, mechanistic or otherwise, for the mysterious "action at a distance" that his theory described.[15]

Did that make Newton's theory unscientific? Strictly speaking, the answer depends upon *which* definition of science someone chooses to apply. Today one would be hard-pressed to find anybody who denies that Newton's famous theory qualified as scientific. Yet we could easily find scientists still willing to say scientific theories must provide mechanisms as well as others who would deny as much.

And that is the problem. If scientists and philosophers of science do not have an agreed-upon definition of science, how can they settle questions about which theories do and do not qualify as scientific? If scientists lack such a definition, it's difficult to argue that any particular theory is *unscientific* by definition. For this reason, philosophers of science, the scholars who study the nature and definition of science, now almost universally reject the use of demarcation arguments to decide the validity of theories or settle competition between them.[16] They increasingly regard demarcation as an essentially semantic question and nothing more. Is theory X scientific or not? Answer: that depends upon which definition of science is used to decide the question.

Moreover, as the philosopher of science Larry Laudan has shown in a seminal article, "The Demise of the Demarcation Problem," attempts to apply demarcation criteria to decide the scientific status of specific theories have invariably generated irreconcilable contradictions.[17] The vortex theory of gravity that Newton's theory replaced envisioned planets swirling around the sun pushed by a substance called ether.[18] It did provide a mechanistic explanation for gravitational attraction. It failed, however, to explain the evidence and was judged by Newton and physicists following him to be manifestly false. Nevertheless, because it proposed a mechanistic cause of gravitation, it qualified as "scientific"—at least given the conception of science favored by Leibniz and the mechanical philosophers.[19] Conversely, Newton's theory failed to qualify as scientific by their definition, though it much more accurately fit the evidence.

Such contradictions have long beset the whole enterprise of demarcation. Theories that scientists have rejected as false because of their inability to explain or describe the evidence often meet the very criteria or methodological features (testability, falsifiability, repeatability, observability, etc.)

that allegedly characterize true science. On the other hand, many highly esteemed or successful theories often lack allegedly necessary features of genuine science.

Thus, philosophers of science generally think it much more important to assess whether a theory is true, or whether the evidence supports it, than whether it should or should not be classified as "science." The question of whether a theory is "scientific" is really a red herring. What we really want to know is whether a theory is true or false, supported by the evidence or not, worthy of our belief or not. And we cannot decide those questions by applying a set of abstract criteria that purport to tell in advance what all good scientific theories must look like.[20]

DEFINE AND DISMISS: DEMARCATION ARGUMENTS AGAINST INTELLIGENT DESIGN

The rejection of demarcation arguments among philosophers of science has not stopped critics of intelligent design from attempting to settle debates about biological origins by the expedient of formulating such arguments against intelligent design. Some use these arguments to justify methodological naturalism (which has the same effect).

Advocates of methodological naturalism have argued that the theory of intelligent design is inherently unscientific for some, or all, of the following reasons: (a) is not testable,[21] (b) is not falsifiable,[22] (c) does not make predictions,[23] (d) does not describe repeatable phenomena, (e) does not explain by reference to natural law,[24] (f) does not cite a mechanism,[25] (g) does not make tentative claims,[26] and (h) has no problem-solving capability.[27] They have also claimed that it is not science because it (i) refers to an unobservable entity.[28] These critics also assume, imply, or assert that materialistic evolutionary theories do meet such criteria of proper scientific method.

Readers may wish to consult *Signature in the Cell* for a more detailed response to these specific arguments. There I show that many of these claims are simply false (e.g., contrary to the claims of its critics intelligent design is testable; it does make predictions; it does formulate its claims tentatively; and it does have scientific problem-solving capability). But I also show that when the claims of those making demarcation arguments are true—when intelligent design doesn't meet a specific criterion—that fact does not provide good reason for excluding intelligent design from consideration as a scientific theory. Why? Because the materialistic evolutionary theories that intelligent design challenges, theories widely regarded by convention as "scientific," *fail to meet the very same demarcation standard.* In other words, there is no defensible definition of science, and no specific demarcation criterion, that justifies both *excluding* intelligent design from science and *including* competing materialistic evolutionary theories. Instead, attempts to use demarcation criteria specifically to disqualify intelligent design as a scientific theory have repeatedly failed to *differentiate* the scientific status of intelligent design from that of competing theories. Depending upon which criteria are used to adjudicate their scientific status, and provided metaphysically neutral criteria are selected to make such assessments, intelligent design and materialistic origins theories invariably prove *equally* scientific or unscientific.

For example, some critics of intelligent design have argued that it fails to qualify as a scientific theory because it makes reference to an unseen or unobservable entity, namely, a designing mind in the remote past. Yet many accepted theories—theories assumed to be scientific—postulate *un*observable events and entities. Physicists postulate forces, fields, and quarks; biochemists infer submicroscopic structures; psychologists discuss their patients' mental states. Evolutionary biologists themselves infer unobserved past mutations and invoke the existence of extinct organisms and transitional forms for which no fossils remain. Such things, like the actions of an intelligent designer, are *inferred* from observable evidence in the present, because of the explanatory power they may offer.

If the demarcation criterion of observability is applied rigidly, then both intelligent design and materialistic theories of evolution fail to qualify as scientific. If the standard is applied more liberally (or realistically)—acknowledging the way in which historical scientific theories often infer unobservable past events, causes, or entities—then both theories qualify as scientific.

And so it goes with other such criteria as well. There is no specific (non-question-begging) demarcation criterion that succeeds in disqualifying the theory of intelligent design from consideration as a scientific theory *without also doing the same to its materialistic rivals.*

REASONS TO REGARD INTELLIGENT DESIGN AS A SCIENTIFIC THEORY

Demarcation arguments fail to justify *excluding* intelligent design from science. But it turns out that there are some good—if convention-dependent—reasons to regard intelligent design as a scientific theory.

For example, many scientists and philosophers of science regard testability as an important feature of scientific inquiry. And intelligent design is testable in three specific and interrelated ways. First, like other scientific theories concerned with explaining events in the remote past, intelligent design is testable by comparing its explanatory power with that of competing theories. Second, intelligent design, like other historical scientific theories, is tested against our knowledge of the cause-and-effect structure of the world. As we have discussed, historical scientific theories provide adequate explanations when they cite causes that are known to produce the effects in question or "causes now in operation."[29] Because of this, the plausibility of historical scientific theories, including intelligent design, can be tested by reference to independent knowledge of cause-and-effect relationships. Third, although historical scientific theories typically cannot be tested under controlled laboratory conditions, they do sometimes generate predictions that enable scientists to compare their merit to that of other theories. Intelligent design has generated a number of specific empirical predictions that distinguish it from competing evolutionary theories and that serve to confirm the design hypothesis over its competitors. (In *Signature in the Cell,* I described ten such predictions that the theory of intelligent design has generated).[30]

There is another compelling, if convention-dependent, reason to regard intelligent design as a scientific theory. The inference to intelligent design is based upon the same method of historical scientific reasoning and the same uniformitarian principles that Charles Darwin used in *On the Origin of Species.* The similarity in logical structure runs quite deep. Both the argument for intelligent design and the Darwinian argument for descent with modification were formulated as abductive inferences to the best explanation. Both theories address characteristically historical questions; both employ typically historical forms of explanation and testing; and both have metaphysical implications. Insofar as we regard Darwin's theory as a scientific theory, it seems appropriate to designate the theory of intelligent design as a scientific theory as well.

Indeed, neo-Darwinism and the theory of intelligent design are not two different kinds of inquiry, as some critics have asserted. They are two different answers—formulated using a similar logic and method of reasoning—to the same question: "What caused biological forms and the appearance of design to arise in the history of life?" It stands to reason that if we regard one theory, neo-Darwinism or intelligent design, as scientific, we should regard the other as the same. Of course, whether either theory is true or not is another matter. An idea may be scientific and incorrect. In the history of science, many theories have proven to be so. The vortex theory of gravity, to which I referred earlier, would be one of nearly countless illustrations.

For readers who would like to consider more detailed responses to arguments about whether intelligent design qualifies as "science," I recommend Chapters 18 and 19 in *Signature in the Cell.*[31] In *Signature,* I respond in detail to other philosophical objections to the case for intelligent design. These include challenges such as: (a) intelligent design is religion, not science,[32] (b) the case for intelligent design is based on flawed analogical reasoning, (c) intelligent design is a fallacious argument from ignorance, sometimes called the "God of the Gaps" objection, (d) intelligent design is a science stopper, (e) the famous zinger, popularized by Richard Dawkins, that asks "Who designed the designer?"[33] and many others.

A NEW OBJECTION TO THE SCIENTIFIC STATUS OF INTELLIGENT DESIGN

Since the publication of *Signature in the Cell,* Robert Asher, a University of Cambridge paleontologist, has offered another reason to contest my characterization of intelligent design as a scientific theory. In his book, *Evolution and Belief,* he challenges my claim to have used the uniformitarian method of Lyell and Darwin to develop the case for intelligent design. Since his objection is new, published only in 2012 by Cambridge University Press, it deserves discussion.

Asher characterizes my thinking as follows: "The processes we know and observe today are relevant to explaining the phenomena of the past, and we know that particularly complicated things we see today have an intelligence behind them."[34] He notes that I argue certain complex technologies, such as computer software, have "only one source: human ingenuity."[35] It follows, according to Asher's paraphrase of my argument that "a similarly complex device we observe in the geological past must also have arisen as a result of something like human ingenuity, i.e., intelligence."[36]

Asher doesn't seem to understand the importance of specified information, as opposed to "complicated things," as a key indicator of design. That aside, he does claim to recognize the role of uniformitarian principles of reasoning in my argument for intelligent design. In spite of this, Asher elsewhere disputes that I employ the uniformitarian method of reasoning. Why? According to Asher, the inference to intelligent design is actually "anti-uniformitarian" because it doesn't provide a "mechanism." As he puts it, "by attempting to replace a causal mechanism (natural selection) with an attribution of agency (design), ID advocates such as Meyer are decidedly anti-uniformitarian. What process of today could possibly lead to his understanding of the past?"[37]

The answer to Asher's question seems pretty obvious. The answer is: intelligence. Conscious activity. The deliberate choice of a rational agent. Indeed, we have abundant experience in the present of intelligent agents generating specified information. Our experience of the causal powers of intelligent agents—of "conscious activity" as "a cause now in operation"—provides a basis for making inferences about the best explanation of the origin of biological information in the past. In other words, our experience of the cause-and-effect structure of the world—specifically *the* cause known to produce large amounts of specified information in the present—provides a basis for understanding what likely caused large increases in specified information in living systems in the past. It is precisely my reliance on such experience that makes possible an understanding of the type of causes at work in the history of life. It also makes my argument decidedly uniformitarian—*not* "anti-uniformitarian"—in character.

Asher confuses the uniformitarian imperative in historical scientific explanations (the need to cite a presently known or adequate cause) with a demand for citing a *material* cause, or mechanism. The theory of intelligent design does cite a cause, and indeed one known to produce the effects in question, but it does not necessarily cite a mechanistic or materialistic cause. Proponents of intelligent design *may* conceive of intelligence as a strictly materialistic phenomenon, something reducible to the neurochemistry of a brain, but they may also conceive of it as part of a mental reality that is irreducible to brain chemistry or any other physical process. They may also understand and define intelligence by reference to their own introspective experience of rational consciousness and take no particular position on the mind-brain question.

Asher assumes that intelligent design denies a materialistic or "physicalist" account of the mind (as I personally do, in fact) and rejects it as unscientific on that basis. But he offers no noncircular reason for making that judgment. He cannot say that the principle of methodological naturalism requires that all genuinely scientific theories invoke only mechanistic causes, because the principle of methodological naturalism itself needs justification. And asserting that "all genuinely scientific theories must provide mechanisms" is just to restate the principle of methodological naturalism in different words. Indeed, to say that all scientific explanations must provide a mechanism is equivalent to saying that they must cite *materialistic* causes—precisely what the principle of methodological naturalism asserts. Asher seems to be assuming without justification that all scientifically acceptable causes are *mechanistic or materialistic.* His argument thus assumes a key point at issue, which is whether there are independent—that is, metaphysically *neutral*—reasons for requiring historical scientific theories to cite materialistic causes in their explanations as opposed to explanations that invoke possibly immaterial entities such as creative intelligence, mind, mental action, agency, or intelligent design.

In any case, he confuses the logical requirement of citing a *vera causa,* a true or known cause, with an arbitrary requirement to cite only *materialistic* causes. He confuses uniformitarianism with methodological naturalism.[38] He then critiques my design argument for rejecting the former, though it only rejects the latter. In so doing, he imposes an additional requirement on explanations of past events that leads him to mistake my argument as anti-uniformitarian and to miss the evidence for intelligent design. His implicit commitment to methodological naturalism makes the evidence for intelligent design—"the postman," as it were—mentally invisible to him.

Nevertheless, the concern that he raises about the theory of intelligent design not citing a mechanism still troubles people. In fact, I frequently get questions about this issue. People will ask something like this: "I can see your point about digital code providing evidence for intelligent design, but how exactly did the designing intelligence generate that information or arrange matter to form cells or animals?" Or: "How did the intelligent designer that you infer impress its ideas on matter to form animals?" As Asher puts it, "How could a biological phenomenon, even if designed, be simply willed into existence without an actual mechanism?"[39]

To help clear things up, several points need to be considered. First, the theory of intelligent design does not provide a mechanistic account of the *origin* of biological information or form, nor does it attempt to. Instead, it offers an *alternative* causal explanation involving a mental, rather than a necessarily or exclusively material, cause for the origin of that reality. It attributes the origin of information in living organisms to thought, to the rational activity of a mind, not a strictly material process or mechanism. That does not make it deficient as a materialistic or mechanistic explanation. It makes it an *alternative* to that kind of explanation. Advocates of intelligent design do not propose intelligent causes because they cannot think of a possible mechanistic explanation for the origin of form or information. They propose intelligent design because they think it provides a better, more causally adequate explanation for these realities. Given what we know from experience about the origin of information, materialistic explanations are the deficient ones.

There is a different context in which someone might want to ask about a mechanism. He or she may wish to know by what means the information, once originated, is transmitted to the world of matter. In our experience, intelligent agents, after generating information, often use material means to transmit that information. A teacher may write on a chalkboard with a piece of chalk or an ancient scribe may have chiseled an inscription in a piece of rock with a metal implement. Often, those who want to know about the mechanism of intelligent design are not necessarily challenging the idea that information ultimately originates in thought. They want to know how, or *by what material means,* the intelligent agent responsible for the information in living systems transmitted that information to a material entity such as a strand of DNA. To use a term from philosophy, they want to know about "the efficient cause" at work.

The answer is: We simply don't know. We don't have enough evidence or information about what happened, in the Cambrian explosion or other events in the history of life, to answer questions about what exactly happened, even though we *can* establish from the clues left behind that an intelligent designer played a causal role in the origin of living forms.

An illustration from archaeology helps explain how this can be so (see Fig. 19.2). Years ago explorers of a remote island in the southwestern Pacific Ocean discovered a group of enormous stone figures. The figures displayed the distinctive shape of human faces. These figures left no doubt as to their ultimate origin in thought. Nevertheless, archeologists still don't know the exact means by which they were carved or erected. The ancient head carvers might have used metallic hammers, rock chisels, or lasers for that matter. Though archaeologists lack the evidence to decide between various hypotheses about *how* the figures were constructed, they can still definitely infer *that* intelligent agents made them. In the same way, we can infer *that* an intelligence played a causal role in the origin of the Cambrian animals, even if we cannot decide what material means, if any, the designing intelligence used to transmit the information, or shape matter, or impart its design ideas to living form. Although the theory of intelligent design infers *that* an intelligent cause played a role in shaping life's history,

FIGURE 19.2
Group of carvings of giant heads, called "Moais," on Easter Island. *Courtesy iStockphoto/Think-stock.*

it does not say *how* the intelligent cause affected matter. Nor does it have to do so.

There is a logical reason we cannot without further information determine the mechanism or means by which the intelligent agent responsible for life transmitted its design to matter. We can infer an intelligent cause from certain features of the physical world, because intelligence is known to be a necessary cause, the only known cause, of those features. That allows us to infer intelligence retrospectively as a cause by observing its distinctive effects. Nevertheless, we cannot establish a unique scenario describing *how* the intelligent agent responsible for life arranged or impressed its ideas on matter, because there are many different possible means by which an idea in the mind of an intelligent agent could be transmitted or instantiated in the physical world.

There is another even more profound reason that intelligent design—indeed, science itself—may not be able to offer a completely mechanistic account of the instantiation of thought into matter. Robert Asher worries about how "a biological phenomenon, even if designed," could be "simply willed into existence without an actual mechanism." In Asher's understanding, the uniformitarian principle asks for a precedent, a known cause that not only generates information, but translates immaterial thought into material reality, impressing itself on and shaping the physical world. Asher complains that the argument for intelligent design cannot cite such a precedent and is thus "anti-uniformitarian."

Yet a precedent comes very readily to mind, an intimately familiar one for us all. At present no one has any idea how our thoughts—the decisions and choices that occur in our conscious minds—affect our material brains, nerves, and muscles, going on to instantiate our will in the material world of objects. However, we know that is exactly what our thoughts do. We have no mechanistic explanation for the mystery of consciousness, nor what is called the "mind-body problem"—the enigma of how thought affects the material state of our brains, bodies, and the world that we affect with them. Yet there is no doubt that we can—as the result of events in our conscious minds called decisions or choices—"will into existence" information-rich arrangements of matter or otherwise affect material states in the world. Professor Asher did this when he wrote the chapter in his book—representing his ideas impressed as words onto a material object, a printed page—attempting to refute intelligent design. I am doing this right now. This example, representative of countless daily experiences in life, surely satisfies the demands of uniformitarianism.

Though neuroscience can give no mechanistic explanation for consciousness or the mind-body problem, we also know that we can recognize the product of thought, the effect of intelligent design, in its distinctive information-rich manifestations. Professor Asher recognized evidence of thought when he read the text in my book; I did so when I read his; you are doing so right now. Thus, even though it remains entirely possible that we may never know how minds affect matter and, therefore, that there may always be a gap in our attempt to account for *how* a designing mind affected the material out of which living systems were formed, it does not follow that we cannot recognize evidence of the activity of mind in living systems.

WHY IT MATTERS FOR SCIENCE

But if proponents of intelligent design admit that they do not, or perhaps even cannot, answer the question of how the mind responsible for the design of animal life impressed its ideas on matter, why does it matter that we recognize the evidence for intelligent design at all? If intelligent design just replaces one mystery with another, why not limit ourselves to materialistic explanations after all, as methodological naturalism requires, and be content with accepting the mystery we already have? Wouldn't that be simpler and more intellectually economical?

Perhaps. But it puts the mystery in the wrong place. We do know of a cause that can produce the functional information necessary to build complex systems. But we do not know exactly how mind interfaces with matter. If we were to ask what caused the Rosetta Stone to arise, and then insist despite all evidence to the contrary that a purely material process is capable of producing the information-rich etchings on that stone, we would be deluding ourselves. The information etched into that black slab of igneous rock at the British Museum provides overwhelming evidence that an intelligent agent did cause those inscriptions. Any rule that prevents us from considering such an explanation diminishes the rationality of science, because it prevents scientists from considering a possibly—and in this case obviously—true explanation. And the truth matters, not least in science. For this reason, the "rules of science" should not commit us to rejecting possibly true theories before we even consider the evidence. But that is exactly what methodological naturalism does.

Moreover, adhering to methodological naturalism and refusing to consider the evidence for intelligent design in life does not just affect the explanations that we are willing to consider for the *origins* and *history* of life. They also affect the questions we ask about life as it exists, and thus the entire biological research agenda that we pursue.

An analogy to a human artifact again shows why. If we ask exactly *how* the scribe responsible for the inscriptions on the Rosetta Stone accomplished his or her task—with a metal chisel, a sharpened piece of obsidian, a diamond stylus, or some other material means—we may not have enough evidence to answer that question. Nevertheless, it will help archaeologists to know that they are looking at an artifact of intelligence, rather than a byproduct of strictly natural processes. This will lead them to ask other more relevant questions about the stone, such as: "What do the inscriptions mean?" "Who wrote them?" and "What do they tell us about the surrounding cultures at the time?" In a similar way, what we think about how animal life arose and developed will lead us to ask different questions about living forms—questions that we might never think to ask if we were assuming that they had arisen by a purely undirected mechanism such as natural selection.

Intelligent agents and natural selection do their work very differently. The mutation and selection mechanism is a blind, trial-and-error process, one that must maintain or optimize functional advantage through a series of incremental steps. Given Darwinian assumptions, we would not expect to see structures or systems in living organisms that required foresight. Nor would we expect to see structures that needed to be produced all at once in large jumps rather than by a series of function-preserving incremental steps. We would, however, expect to see evidence of a trial-and-error process in the genomes of organisms.

But what happens if we open ourselves to the possibility of detecting design in life? We know a lot about how intelligent designers do their work. Intelligent designers use many established design strategies (or "design patterns," as engineers would say). They also have foresight that allows them to reach functional goals without the need to maintain function through a series of intermediate structures. They typically engineer new systems from scratch without relying on random, incremental, trial-and-error modifications in one system to produce another.

Because these two different types of causes operate differently and often produce different types of structures and systems, scientists should expect living systems (and the history of life) to look differently depending upon which type of cause produced the organisms or structures in question. And these differing perspectives and expectations can lead scientists to ask different research questions and make different predictions about what we should find in the structure of life itself.

THE ENCODE PROJECT AND AN ID PREDICTION

In 2012, a dramatic confirmation of one such prediction made by advocates of intelligent design occurred in the field of genomics. Three leading science journals, *Nature, Genome Research,* and *Genome Biology,* published a series of groundbreaking papers reporting on the results of a massive study of the human genome called the ENCODE project (short for Encyclopedia of DNA Elements).[40] The conclusion: at least 80 percent of the genome performs significant biological functions, "dispatching the widely held view that the human genome is mostly 'junk DNA.'"[41]

The discovery challenged a long held neo-Darwinian interpretation of the genome. According to neo-Darwinism, the genome as a whole should display evidence of the random trial-and-error process that gave rise to new genetic information. The discovery in the 1970s that only a small percentage of the genome contains information for building proteins was hailed at the time as powerful confirmation of the Darwinian view of life. The noncoding regions of the genome were assumed to be nonfunctional detritus of the trial-and-error mutational process—the same process that produced the functional code in the genome. As a result, these noncoding regions were deemed "junk DNA," including by no less a scientific luminary than Francis Crick.[42]

Because intelligent design asserts that an intelligent cause produced the genome, design advocates have long predicted that most of the nonprotein-coding sequences in the genome should perform some biological function, even if they do not direct protein synthesis. Design theorists do not deny that mutational processes might have degraded some previously functional DNA, but we have predicted that the functional DNA (the signal) should dwarf the nonfunctional DNA (the noise), and not the reverse. As William Dembski, a leading design proponent, predicted in 1998, "On an evolutionary view we expect a lot of useless DNA. If, on the other hand, organisms are designed, we expect DNA, as much as possible, to exhibit function."[43]

The ENCODE project and other recent research in genomics have confirmed this prediction. As the lead article in *Nature* reported, ENCODE has "enabled us to assign biochemical functions for 80 percent of the genome in particular outside of the well-studied protein-coding regions."[44] Other research in genomics has shown that, overall, the noncoding regions of the genome function much like the operating system in a computer. Indeed, the noncoding regions of the genome direct the timing and regulate the expression of the data modules or coding regions of the genome, in addition to possessing myriad other functions.[45] Before ENCODE, neo-Darwinists would often ask: If the information in DNA provides such compelling evidence for the activity of a designing intelligence, why is over 90 percent of the genome composed of functionless nonsense sequences? The latest genomics research now provides a ready answer to this question: it isn't.

The significance of these discoveries in genomics to the debate about design has passed largely unnoticed in the media. But repeated attempts to stigmatize the ENCODE researchers as aiding and abetting "intelligent design creationists" have inadvertently highlighted what is at stake. In this effort, a biochemist at the University of Toronto, Laurence A. Moran, emerged as point man. The Moran strategy centered on tarring scientists and science journalists who publicized ENCODE and its implications with the brush of "Intelligent Design Creationism"—an all too familiar conflation of intelligent design with a very different idea, the biblical literalism of young-earth creationism. When the distinguished journal *Science* selected ENCODE as one of the top ten science news stories of 2012, reminding readers that it had detonated the notion of junk DNA by revealing overwhelming functionality in the genome,[46] Moran jeered, "Oh well, I guess I'll just have to be content to point out that many scientists are as stupid as many Intelligent Design Creationists!"[47] In the science world, as in the media, "creationist" is a dirty word; it's like calling someone a Communist used to be in the 1950s. Such attempts to stigmatize results that challenge a favored theory illustrate how an ideological monopoly in science can stifle inquiry and discussion.

The demise of the idea of junk DNA illustrates too, in a more positive way, how a competing perspective can inspire research that contributes to new discovery. Although clearly not every scientist who performed research helping to establish the functional significance of nonprotein-coding DNA was inspired by the theory of intelligent design, at least one noteworthy scientist was. During the early part of the decade, before ENCODE made the headlines, this scientist published many articles challenging the idea of junk DNA based on genomics research that he was conducting at the National Institutes of Health. After the publication of ENCODE in 2012, his coauthor on many of those articles, the prominent University of Chicago geneticist James Shapiro, wrote an article in the *Huffington Post* commending the scientist for his groundbreaking research and for anticipating the ENCODE results years before. In the article, Shapiro acknowledged that he and his coauthor had "different evolutionary philosophies"—his charitable way of referring to his coauthor's growing interest in the theory of intelligent design.

Who was that other scientist? None other than Richard Sternberg, the evolutionary biologist who was punished for his openness to intelligent design while serving at the Smithsonian Institution (and the National Institutes of Health) in 2004. Around that time, Sternberg's doubts about neo-Darwinism and his growing interest in intelligent design led him to consider the possibility that the majority of the genome could really be functional.[48] His research subsequently confirmed what was for him, an idea inspired in part by intelligent design.

In *Signature in the Cell,* I described many other discriminating predictions of the theory of intelligent design—predictions that differ from those of competing materialistic evolutionary theories—and how those predictions might help to guide new research in various subdisciplines of biology, including some in medicine. These predictions may also lead scientists to make new discoveries—discoveries that proponents of a competing perspective might not have been inclined to make—or to accept.

OPEN VISTAS

By now it should be clear why so many brilliant scientists have missed the evidence of design in the Cambrian explosion. Scott Todd, a biologist writing in *Nature,* succinctly stated the reason: "Even if all the data point to an intelligent designer, such a hypothesis is excluded from science because it is not naturalistic."[49] When scientists decide by fiat that intelligent design lies beyond the bounds of science, their decision will prevent them from considering this possibly or probably true explanation for the origin of animal form. But it will also deprive them of a new perspective that can generate new research questions and foster new avenues of discovery. Knowing this helps solve the final mystery of this book, but it also suggests a more productive way forward to address mysteries yet unsolved. Scientists committed to methodological naturalism have nothing to lose but their chains—fetters that bind them to a creaky and exhausted nineteenth-century materialism. The future lies open before them, and us. As we in the intelligent design research community like to say, let's break some rules and follow the evidence wherever it leads.

20

WHAT'S AT STAKE

In the summer of 2002, I had the opportunity to hike up to the Burgess Shale with a group of geologists, geophysicists, and marine biologists. Our group also included my then eleven-year-old son and a teenage friend of his who was interested in the Cambrian fossils and the debate about Darwinism and design.

When we got to the top of the mountain, I was unprepared for the impact the fossils would have on me. I had seen many fossils before, of course. But seeing *these* fossils—marine animals from the dawn of animal life at the top of a mountain with their beautifully preserved appendages and organs—rendered the idea of the "Cambrian explosion" a good deal less theoretical for me than it had been. These complex sea creatures, now brushed by the thin air at an elevation of 7,500 feet in the middle of the Canadian Rockies, had apparently arisen suddenly, almost from nothing by way of ancestral forms, in the sedimentary record. Everything about them cried out for a story—a big story. It set my mind and imagination racing (see Fig. 20.1).

As wonderful as the fossils were, our trip to see them was made more memorable by two things that happened en route—one on the way up the mountain and one on the way down. As we were making our ascent, crossing a large talus slope—a section of the mountain void of vegetation and covered with only fragments of sedimentary rock—I heard my son unexpectedly call out to me from up at the front of our group. His voice had a trembling quality. I looked forward to see him, normally a fearless kid blessed by energy without bound, standing locked in place, pale and wide-eyed. I stepped around several of the other hikers on the trail to catch up with him. It turned out he was experiencing a kind of vertigo, though the mountain was not dangerously steep at that point (see Fig. 20.2). As he set out across the path that cut through the rocky slope, he had made the mistake of looking down the mountain. Without trees as a reference point and with hundreds of feet of loose rock fragments above and beneath him, he became disoriented and frightened. I steadied him as we walked in step, stride for stride, with me directly behind him across that open stretch of the mountain. Before long we were back to a place on the trail where trees and other plants appeared, providing a steadying presence as a point of reference. My son's perspective quickly returned. He relaxed and soon was smiling and leaping confidently ahead of me again.

On the way down the mountain, I had a striking interaction with a member of our group, who gave voice to a different kind of disorientation. It began as a conversation between my son's friend and our official field guide, who had been assigned to us by the local Burgess Shale Geoscience Foundation. Our guide was a paleontologist and did a terrific job.

FIGURE 20.1
A trilobite fossil found at the Burgess Shale. *Courtesy Michael Melford/National Geographic Image Collection/Getty Images.*

FIGURE 20.2
Figure 20.a (left): Photograph of the author and his son Jamie at the Walcott Quarry of the Burgess Shale in British Columbia, Canada. *Figure 20.2b (right):* Pausing for a moment to reflect on the slope below the Burgess Shale outcrop.

He told many a fine story about the geological history of the surrounding formations, about the discovery of the fossils, and of course about the evolutionary history of animal life. In fact, just before we turned the final corner on the trail to ascend to a large collection of excellent fossils available for viewing at the top of the mountain, he slipped a statement of support for evolutionary orthodoxy into his description of the fossil site. Our guide was clearly unaware that many of us in the group knew the fossils we were about to see challenged the standard Darwinian story.

We hadn't made an issue of our views, of course, but nearly all the scientists on the hike were skeptical of neo-Darwinism. Paul Chien, the University of San Francisco marine biologist who had worked with J. Y. Chen in China on the sponge embryo fossils, was on the trip, and he had more than a passing acquaintance with Cambrian-era paleontology—as did several of the Canadian geologists with us. Still, not wanting to introduce any needless discord, we carefully avoided engaging the issue. We just wanted to see the fossils.

As we descended the mountain, however, my son's young friend asked our guide how he squared what we had just seen with his support for Darwinian evolution. The guide at first maintained his commitment to the Darwinian party line. He said he thought that Darwin would feel "vindicated" by the discovery of the Burgess fossils. This proved too much for the precocious, intellectual teenager, who loudly blurted out, "What?! Darwin would feel vindicated? By the sudden appearance of all those animals without any ancestors in the fossil record! Are you kidding?!"

You would have to know this endearing young man to understand how his uninhibited outburst only charmed and amused our guide. But fortunately it did. The rest of us, however, were initially mortified. This was the discussion we were trying to avoid, knowing exactly the intense emotions it often provokes. With scientists, it is generally safer to discuss religion and politics.

Nevertheless, to his credit, our guide took the challenge in stride. He explained how the Burgess fossils demonstrated evidence of change over time, how the rock column showed the great age of the earth, and how the discovery of the fossils high on a mountain revealed the evolution of the planet. Our young friend had spent too much time reading up on the subject to let the point go at that. He brushed aside the issue of the earth's age, which like our guide he reckoned in billions of years, and assured the man that he accepted evidence for change over time in the sedimentary record. He didn't question evolution in that sense. He questioned *Darwinian* evolution. "Where is the evidence of *gradual* change?" he demanded, as his teenage voice cracked with excitement as it moved into the upper register. He continued, "What mechanism could produce *so many new* animals *so quickly*?"

An odd thing happened then. The paleontologist now leading us down the trail suddenly ceased to act the part of "guide." He dropped any pretense of superior authority and said, "You know, I've wondered about that myself." I thought I could hear in his voice the candid amazement of the fourteen-year-old boy *he* once was.

"How do *you* explain it?" he asked my son's friend.

Our young spokesman confidently piped up and asserted, "Intelligent design, of course!"

At which point, our guide began to ask the probing questions. Soon my son's friend had exhausted his store of knowledge and began to look to me to join the conversation. I did so, reluctantly at first. I explained the information argument for intelligent design and how the Cambrian explosion contributed to it. Our guide asked me the hard questions: How can we detect design? Is intelligent design science? Aren't we just arguing from ignorance and giving up on science, or at least mainstream evolutionary science, too soon? He also wanted to know who I personally thought the designer was. His challenges were tough and honest. A terrific conversation ensued.

When we reached the trailhead, he surprised me, thanking me for the conversation—and thanking my young friend for starting it. He then spoke a bit more personally and revealed that he sometimes found thinking about biological origins disturbing. He said that as a scientist he was committed to the evolutionary perspective. But he also found its denial of purpose depressing. He wondered if there was some way to affirm both science and the kind of purpose and meaning in life that religion speaks about. As we parted ways, he said he would like to learn more about intelligent design. He told me he was intrigued by the perspective we were developing. I felt that we had made a genuine human connection rather than, as sometimes happens in the evolution debate, merely flinging assertions at one another.

Over the years, as I've researched and thought about biological origins, I have had numerous similar conversations with people of many persuasions and backgrounds: religious and nonreligious; scientists, engineers, medical doctors; businessmen and -women, appliance repairmen and taxi drivers. These conversations usually start innocently enough as the result of someone asking me what I do for a living. Though I often euphemize my response ("I work for a research organization") to avoid getting trapped in a heavy conversation on an airplane or over a broken dishwasher, often the conversations come whether I want them or not. People are interested in how life began and they instinctively understand that whatever theory we adopt has larger philosophical, religious, or worldview implications. People are usually energized by considering those larger implications and questions. Many would like to find a way to harmonize the evidence from science with a view of the world that addresses their deepest existential longings as human beings, their yearning for purpose and significance. But like our guide, many have been frustrated by the difficulty of arriving at a coherent synthesis.

It's not hard to see why. On the one hand, many people of faith have little real interest in what science has to say about life's origins. Indeed, many well-meaning religious believers have adopted a view of the relationship between science and faith that rejects the testimony of science as irrelevant or even dangerous and affirms that just reading the Bible will give all the insight needed to understand how life came to be. Their approach does not really attempt to harmonize faith and science, since it takes faith in the Bible, and often a particular interpretation of the Bible, as the only reliable source of information about life's origin.

On the other hand, many scientists and others who think that science has something to teach us about the big questions have started by assuming the neo-Darwinian account of biological origins, despite its many scientific difficulties—and despite its denial of any role for purposive intelligence in the history of life.

In particular, two popular ideas about how Darwinism informs worldview have come to different conclusions about the worldview it affirms—or allows. The first view, the "New Atheism," has been articulated by spokesmen such as Richard Dawkins in his book *The God Delusion* and the late Christopher Hitchens in *God Is Not Great*.[1] It purports to refute the existence of God as "a failed hypothesis,"[2] as another New Atheist book puts it. Why? Because, according to Dawkins and others, there is no evidence of design in nature. Indeed, Dawkins's argument for atheism hinges upon his claim that natural selection and random mutation can explain away all "appearances" of design in nature. And since, he asserts, the design argument always provided the strongest argument for believing in God's existence, belief in God, he concludes, is extremely improbable—tantamount to "a delusion." For the New Atheists, Darwinism makes theistic belief both implausible and unnecessary. As Dawkins has famously put it, "Darwin made it possible to be an intellectually fulfilled atheist."[3]

The New Atheists took the publishing world by storm in 2006 when *The God Delusion* first appeared. But nothing about the "New" Atheism was actually "new." Instead, it represents a popularization of a science-based philosophy, called scientific materialism, that came into currency among scientists and philosophers during the late nineteenth century in the wake of the Darwinian revolution. For many scientists and scholars at the time, a scientifically informed worldview was a materialistic worldview in which entities such as God, free will, mind, soul, and purpose played no role.

Scientific materialism, following classical Darwinism, denied evidence of any design in nature and, therefore, any ultimate purpose to human existence. As British philosopher and mathematician Bertrand Russell put it early in the twentieth century, "Man is the product of causes which had no prevision of the end they were achieving" and which predestine him "to extinction in the vast death of the solar system."[4]

An alternative and increasingly popular view is known as theistic evolution. Popularized by Christian geneticist Francis Collins in his book *The Language of God* (also published in 2006),[5] this perspective affirms the existence of God *and* the Darwinian account of biological origins. Yet it provides few details about how God might or might not influence the evolutionary process, or how to reconcile seemingly contradictory claims in the Darwinian and Judeo-Christian accounts of origins. For example, Collins has declined to say whether he thinks God in any way directed or guided the evolutionary process, though he affirms neo-Darwinism, which specifically denies that natural selection is guided in any way. Darwinism and neo-Darwinism insist that the appearance of design in living organisms is an illusion because the mechanism that produces that appearance is unguided and undirected. Does God, in Collins's view, guide the unguided process of natural selection? He, and many other theistic evolutionists, don't say. This ambiguity has made an uneasy reconciliation of science and faith possible, but it has also left many questions unanswered. In fairness, many theistic evolutionists would argue that not all such questions *can* be answered, because science and faith occupy separate, non-overlapping realms of inquiry, knowledge, and experience. But that answer itself underscores the limits of the harmonization of science and faith that Collins and others holding his view has achieved.

The argument of this book presents a scientific challenge to both of these views. In the first place, the evidence and arguments we have seen show that the scientific premise of the New Atheist argument is flawed. The mechanism of mutation and natural selection does not have the creative power attributed to it and, thus, cannot explain all "appearances" of design in life. The neo-Darwinian mechanism does not explain, for example, either the new genetic or epigenetic information necessary to produce fundamentally new animal body plans.

This book has presented four separate scientific critiques demonstrating the inadequacy of the neo-Darwinian mechanism, the mechanism that Dawkins assumes can produce the appearance of design without intelligent guidance. It has shown that the neo-Darwinian mechanism fails to account for the origin of genetic information because: (1) it has no means of efficiently searching combinatorial sequence space for functional genes and proteins and, consequently, (2) it requires unrealistically *long waiting times* to generate even a single new gene or protein. It has also shown that the mechanism cannot produce *new body plans* because: (3) early acting mutations, the only kind capable of generating large-scale changes, are also invariably deleterious, and (4) genetic mutations cannot, in any case, generate the *epi*genetic information necessary to build a body plan. Thus, despite the commercial success of *The God Delusion* and its wide cultural currency, the New Atheist philosophy lacks credibility because it has based its understanding of the metaphysical implications of modern science on a scientific theory that itself lacks credibility—as even many leading evolutionary biologists now acknowledge.[6]

Second, this book poses a strong challenge to theistic evolutionists such as Francis Collins for many of the same scientific reasons. Collins places great trust in modern Darwinism as the unifying theory of biology, but seems completely unaware of the formidable scientific problems now afflicting the theory—in particular, the challenges to the creative power of the natural selection/mutation mechanism. He makes no attempt to address or answer any of these challenges. In addition, many of his arguments for universal common descent—the defense of which was his main concern in *The Language of God*—are based upon the alleged presence of nonfunctional or "junk" elements in the genomes of different organisms. Though the theory of intelligent design, which Collins says he opposes, does not necessarily challenge this part (common descent) of Darwinian theory, the factual basis of his arguments has now also largely evaporated as the result of ENCODE and other developments in genomics.[7] Thus, this popular view of biological origins, and its conception of God's relationship to the natural world, now stands starkly at odds with the evidence. But why attempt to reconcile traditional Christian theology with Darwinian theory, as Collins tries to do, if the theory itself has begun to collapse?

The perspective of this book offers a potentially more coherent and satisfying way of addressing the big questions, of synthesizing science and metaphysics (or faith), than either of the currently popular views on offer. The Cambrian explosion, like evolutionary theory itself, raises larger worldview questions precisely because it raises questions of origins and of design, and with them, the question that all worldviews must address: What is the thing or the entity from which everything comes? But unlike strict Darwinian materialism and the New Atheism built atop it, the theory of intelligent design affirms the reality of a design*er*—a mind or personal intelligence behind life. This case for design restores to Western thought the possibility that human life in particular may have a purpose or significance beyond temporary material utility. It suggests the possibility that life may have been designed by an intelligent person, indeed, one that many would identify as God.

Unlike the theistic evolution of Francis Collins, however, the theory of intelligent design does not seek to confine the activity of such an agency to the beginning of the universe, conveying the impression of a decidedly remote and impersonal *deistic* entity. Nor does the theory of intelligent design merely assert the existence of a creative intelligence behind life. It identifies and detects activity of the designer of life, and does so at different points in the history of life, including the explosive show of creativity on display in the Cambrian event. The ability to detect design makes belief in an intelligent designer (or a creator, or God) not only a tenet of faith, but something to which the evidence of nature now bears witness. In short, it brings science and faith into real harmony.

Just as importantly, perhaps, the case for design supports us in our existential confrontation with the void and the seeming meaninglessness of physical existence—the sense of survival for survival's sake that follows inexorably from the materialist worldview. Richard Dawkins and other New Atheists may find it untroubling, even amusing and certainly profitable, to muse over the prospect of a universe without purpose. But for the vast majority of thoughtful people, that idea is tinged with terror. Modern life suspends many of us, so we feel, high over a chasm of despair. It provokes feelings of dizzying anxiety—in a word, vertigo. The evidence of a purposeful design behind life, on the other hand, offers the prospect of significance, wholeness, and hope.

As my son walked out across the mountain high above the Yoho Valley, he was surrounded by many slabs of rock containing some of the very fossils we had come to see. But as he surveyed that barren portion of landscape, he lost perspective on where he was and what he had come to do. Without landmarks or steadying points of reference, he felt as if he were lost in a sea of sensory impressions. Without his sense of balance, he feared even to take a step. He called out for his father.

It occurred to me only much later how closely his experience parallels our own as human beings trying to make sense of the world around us. To gain a true picture of the world and our place in it we need facts—empirical data. But we also need perspective, sometimes called wisdom, the reference points that a coherent view of the world provides. Historically, that wisdom was provided for many men and women by the traditions of Western monotheism—by our belief in God. The theory of intelligent design generates both excitement and loathing because, in addition to providing a compelling explanation of the scientific facts, it holds out the promise of help in integrating two things of supreme importance—science and faith—that have long been seen as at odds.[8]

The theory of intelligent design is not based upon religious belief, nor does it provide a *proof* for the existence of God. But it does have faith-affirming implications precisely because it suggests the design we observe in the natural world is real, just as a traditional theistic view of the world would lead us to expect. Of course, that by itself is not a reason to accept the theory. But having accepted it for other reasons, it may be a reason to find it important.

EPILOGUE: RESPONSES TO CRITICS OF THE FIRST EDITION OF *DARWIN'S DOUBT*

Having considered the arguments in *Darwin's Doubt*, readers will recognize the challenge it offers to traditional evolutionary thinking and perhaps wonder how stalwart defenders of evolutionary theory have responded. I would divide these critical reviews into two camps: the majority, who chose not to confront the central arguments of the book, and a distinguished minority who, more interestingly, either did grapple with those arguments or at least offered substantive points of scientific critique about factual claims made in the book. Three critiques of this latter kind stand out.

NICHOLAS MATZKE, CLADISTICS, AND MISSING ANCESTORS

Of the reviews of *Darwin's Doubt*, one in a seemingly out-of-the-way venue emerged as a touchstone for many others. Again and again, writers in journals ranging from *The New Yorker* to the ecumenical monthly *First Things* cited a review by Nicholas Matzke that appeared in *Panda's Thumb*, a popular blog dedicated to defending evolutionary theory. University of Chicago evolutionary biologist Jerry Coyne, author of the widely read website *Why Evolution Is True*, has emerged in recent years as an American equivalent of Richard Dawkins, the popular proselytizing spokesman for the neo-Darwinian viewpoint. In a telling gesture, Dr. Coyne pointed his readers to Matzke's review as a definitive response to *Darwin's Doubt*.

Currently a postdoctoral fellow at the National Institute for Mathematical and Biological Synthesis, Matzke has won renown for his tireless campaign to rebuke skeptics of evolutionary theory, a campaign going back to his days with the National Center for Science Education, an advocacy group in Oakland, California.

By his own account, Matzke is also a dizzyingly fast reader and writer. On June 19, 2013, the day after *Darwin's Doubt* was released and first made available for purchase, Matzke published a 9,400-word critical review at *Panda's Thumb*. Reading a book of this size and composing a review of that length all in little more than twenty-four hours would have to be recognized by anyone as a remarkable achievement. Challenged on how it was even possible unless the review had been largely prewritten before he saw a copy, Matzke explained in a later post how he fit in his work on the review with other responsibilities: at lunchtime, in "snippets of the afternoon,"[1] and then by pulling an all-nighter. I, for one, am content to grant him this prodigy.

But what of the content of Matzke's critique?

Matzke's main criticism of *Darwin's Doubt* is that it fails to inform readers about how evolutionary biologists have been able to establish the existence of ancestors of the Cambrian animals using a method of analysis known as cladistics. According to Matzke, cladistic analysis has established the existence of "transitional" and "intermediate" forms[2] between the animals that first arose in the Cambrian. In his view, cladistics has solved the problem of the missing ancestral fossils discussed in Part 1 (Chapters 1–7) of the book. As he asserts, "phylogenetic methods can establish, and have established, the existence of Cambrian intermediate forms, which are *collateral* ancestors of various prominent living phyla."[3] Matzke argues that my failure to inform readers of this disqualifies the book from serious consideration as an analysis of the Cambrian explosion.

Of course, in making this argument, Matzke scarcely addresses the central argument of the book: the problem of the origin of biological information. Neither does he offer any serious rebuttal to my argument in Chapter 11 showing that his 2004 article (coauthored with Gishlick and Elsberry) failed to solve that problem.[4] As I show in that chapter, Matzke and his colleagues at best described several mechanisms by which preexisting genes, rich in *preexisting* genetic information, can be shuffled and recombined.

In Chapter 11, and in the whole second part of the book (Chapters 8–14), I show that what most needs to be explained about the Cambrian explosion is essentially a question of biological engineering—in particular, what *caused* the origin of the information necessary to specify the novel animal structures and architectures that arose in the Cambrian. Cladistics, by contrast, is a method of taxonomic classification, which, like all such methods, takes these structures (or characters) as givens, without considering how they came into being. Thus cladistics bypasses the problem of greatest interest.

Even so, Matzke did challenge a key secondary argument of the book, namely, its claim that the absence of discernible ancestral forms in the Precambrian fossil record represents a mystery from the neo-Darwinian point of view. As I have noted, neo-Darwinism depicts the history of life as a gradually unfolding branching tree in which all forms of complex animal life arise by descent with modification from simpler ancestral precursors. Now, this depiction of the history of life may be true or false, but as an empirical claim, it cannot support itself. For that, evidence is required. If the evidence is not forthcoming, however—if, for instance, the fossils documenting the many morphological transformations required by this historical thesis are missing from the paleontological record—then simply restating (or presupposing) the thesis will do nothing to repair that evidential defect. The question, therefore, is exactly what *evidential* support does cladistics provide for of the Darwinian picture of the history of animal life—in particular, does it provide evidence for the existence of the presumed ancestors of the Cambrian animals that the fossil record does not document? As noted, Matzke claims that cladistic analysis can establish, and has established, the existence of various kinds of ancestors of the Cambrian animals.

But is this so?

Some Background

Darwin's Doubt makes its case for the reality of the Cambrian explosion chiefly, but not entirely, on the basis of the fossil record. Representatives of twenty-three of the roughly twenty-seven fossilized animal phyla (and of the roughly thirty-six total animal phyla) are present in the Cambrian fossil record. Twenty of these twenty-three major groups of animals make their first appearance in the Cambrian period with no discernible ancestral forms present in either earlier Cambrian or Precambrian strata. For the vast majority of the Cambrian animals, the evidence from paleontology suggests geologically abrupt appearance—an explosion (see Chapters 2–4).[5]

In his review, Matzke insists that *other* evidence nevertheless establishes the existence of the Cambrian intermediates or transitional forms. To make this claim, he does not rely on any of the most common arguments against the reality of the Cambrian explosion. He does not claim that the Ediacaran organisms represent plausible ancestral forms of the Cambrian animals (see Chapter 4); nor does he claim that these ancestral forms were not preserved because they were too small or too soft (see Chapter 3); nor does he rely on phylogenetic reconstructions based on comparative gene sequences to establish Precambrian ancestors as advocates of deep divergence, for example, have done (see Chapters 5 and 6). All these proposals *Darwin's Doubt* addresses and refutes.

Instead, Matzke invokes a more recently developed but arguably even less plausible approach to explaining away the absence of presumed ancestral forms. Matzke argues that phylogenetic reconstructions based on cladistic analysis establish the presence of intermediates and transitional forms that do not appear in the fossil record. *Darwin's Doubt* critiques this proposal only in passing (see page 60), and instead provides an extensive critique of more commonly used methods of reconstructing evolutionary history based upon comparative analyses of DNA sequences. Thus, *Darwin's Doubt* does not devote the space to cladistics that Matzke thinks it deserves. As Matzke argues:

> *Meyer never presents for his readers the point that cladistic analyses reveal the order in which the characters found in living groups were acquired, nor the fact that stem taxa are the transitional fossils the creationists are allegedly looking for. And he especially avoids giving his readers any real sense of the number of transitional forms we know about for some groups, and the detail known about their relationships and about the order in which the characters of modern groups originated.*[6]

Matzke also claims that *Darwin's Doubt* makes two significant errors regarding the classification of Cambrian animals. He claims that the book incorrectly refers to *Anomalocaris* as an arthropod, whereas, he argues, they are actually "stem-group" arthropods.[7] He also claims that the book incorrectly refers to Lobopodia as a phylum since, in his opinion, it represents a paraphyletic group (a group that contains some, though not all, descendants of the common ancestor of a group), likely encompassing the extant phyla Tardigrada and Onychophora. Matzke insists that these alleged "basic errors"[8] demonstrate my "ignorance"[9] of systematics.

A few days later, my colleague Casey Luskin replied to Matzke, pointing out that *Darwin's Doubt* includes two chapters with lengthy critiques of attempts to reconstruct phylogenetic histories using the similar technique of comparative sequence analysis as well as a discussion of the distinction in cladistics between stem and crown groups.[10] (Indeed, in endnote 5 on page 453, I explain why making the distinction between stem and crown groups does not help explain what *caused* the Cambrian explosion or origin of the biological information and anatomical characters that arose in it.) Luskin also noted that many Cambrian scientific authorities have called *Anomalocaris* (and other members of its family, the anomalocaridids) "arthropods" of one type or another,[11] while other top authorities—including J. Y. Chen, James Valentine, and Douglas Erwin—have designated Lobopodia as a phylum.[12] Moreover, he noted that in the book I acknowledge the uncertainty about the classification of anomalocaridids by describing them as "either arthropods or creatures closely related to them" (see page 53). Indeed, the very paper Matzke cites in recounting the history of phylogenetic analysis of Cambrian fossils states, "*Anomalocaris* is now recognized as an arthropod."[13] *Darwin's Doubt* does not discuss whether anomalocaridids were true arthropods or just stem group arthropods, but, as Luskin pointed out, the book does correctly note that they are generally regarded as arthropods of some type.

This set Matzke off again. That's just the point, he argued, in another lengthy response.[14] The difference between stem and crown groups is, he asserted, crucial to reconstructing evolutionary histories. According to Matzke, that *Darwin's Doubt* didn't provide a detailed discussion of this distinction showed, again, that I didn't understand how evolutionary biologists do phylogenetic reconstructions using cladistic analysis.

So what is this debate all about? What exactly is cladistics? What are stem groups and crown groups, and does the distinction between the two allow evolutionary biologists to establish the existence of arthropod ancestors? And can cladistics establish the existence of the intermediates between, and the ancestors of, the Cambrian animals?

A Short Primer on Cladistics

Cladistics generates branching patterns of relationship based upon an analysis of the number of "characters" (i.e., features, structures, or traits) shared by different types of organisms. The basic concept is simple. Systematists (experts in classification) examine a species to determine what characters it possesses. They then "score" whether the same characters are present in other presumably related taxonomic groups. After doing this for multiple characters and multiple species, they compare the number of characters each species shares with other species. Species that share more characters are deemed to be more closely related than those that share fewer characters.

For cladists, not every shared character is important in their analysis. Cladistics compares "shared derived" characters—those characters exclusively shared by all species in a group that can be traced (by inference) to the common ancestor of that group. Such characters are called synapomorphies. According to cladistics, the more shared derived characters two or more species exhibit, the closer their evolutionary relationship.

For example, let's assume that in a group of species there are five characters of interest—A, B, C, D, and E. Let's also assume a simple distribution of characters, where one species possesses only character A, another has AB, another ABC, and so on. The resulting representation of the relationships between these species, called a cladogram, would look like Figure 21.1 (see next page).[15]

By using such a diagram and interpreting it as a representation of evolutionary history, evolutionary biologists can represent where various derived characters might have arisen, as seen in the tick marks on Figure 21.2 (see next page).

Of course, reconstructing cladograms is never as simple as my idealized diagrams suggest. For each cladistic analysis, there will be many more characters than just the few in Figures 21.1 and 21.2 that could be compared. Systematists are often confronted with many characters within a group of species—the presence or absence of different anatomical structures, molecules, patterns of development, behaviors, and so forth—any of

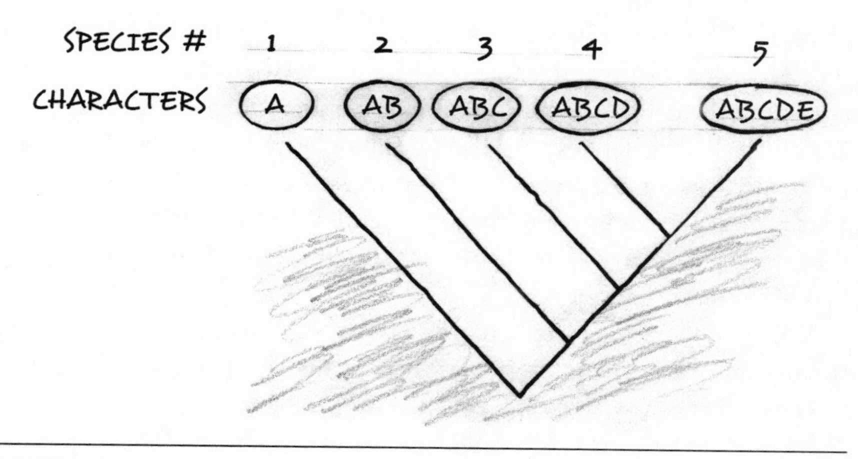

FIGURE 21.1
A simple cladogram with five species, showing how five separate characters, A, B, C, D, and E, are distributed among the species.

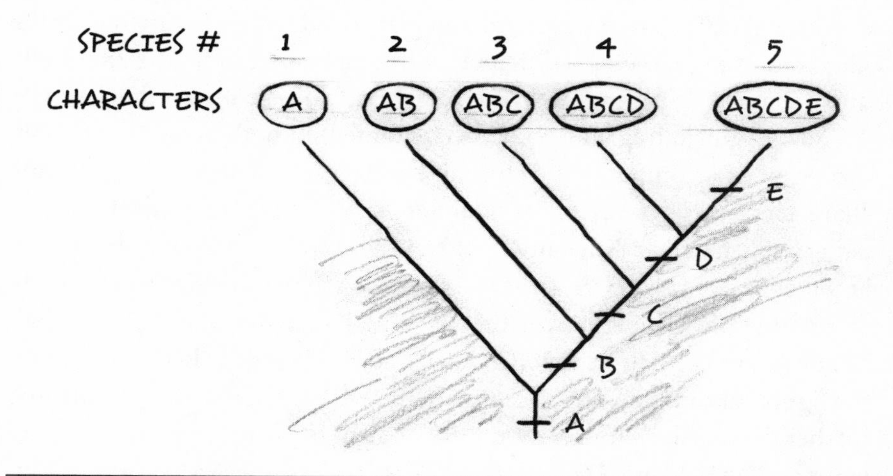

FIGURE 21.2
A simple cladogram with five species, showing how five separate characters, A, B, C, D, and E, are distributed among the species, as well as where they would have arisen on different lines of descent during the history of those groups.

which, or any combination of which, could form the basis for producing cladograms. Thus systematists face uncertainty about which characters (or combinations of characters) to include in their analyses, and further uncertainty about how to weight the characters they do include—at least, for those who practice character weighting" (see endnote for further discussion).[16] Once they have chosen those they regard as the most relevant (or "phylogenetically informative") shared derived characters, systematists feed the data about which animals possess which characters into an algorithm that generates the treelike cladograms. These algorithms perform searches for *the* tree (or a set of trees)—among a huge number of possible trees—that provides the best overall fit with the data and involves the fewest separate evolutionary events (i.e., the fewest instances of gain or loss of characters).

Yet, as systematists include more characters in their analysis, the potential increases for generating inconsistent pictures of the history of life. So too does the need to apply subjective, post hoc, or theory-laden judgments about which characters to include or how to weight the different characters—at least, that is, if the algorithms are to produce reasonably coherent trees that conform to theoretical expectations about the nature of evolutionary change. An analysis of a group of species based upon one small set of characters may produce a clear, unambiguous cladogram. An analysis of the same group emphasizing a different set of characters can render an equally unambiguous branching tree pattern that is inconsistent with the first tree. An analysis including all the characters present in both datasets, however, can generate a complicated picture of evolutionary history in which some characters emerge or disappear on different branches independently. These patterns of character distribution are typically attributed to convergent evolution or loss of characters. (Alternatively, the algorithm may identify many conflicting phylogenetic trees that are equally parsimonious.)

For example, imagine that in addition to characters A, B, C, D, and E in the figures above, a systematist also analyzes characters F and G. Imagine further that when characters F and G are included in the analysis, F occurs on branches 1, 3, and 5 (but not on 2 or 4), and G appears on branches 2, 4, and 5 (but not on 1 or 3), as shown in Figure 21.3. Explaining this pattern requires invoking multiple separate origins of the same characters (convergent evolution) and/or instances of character loss.

Since cladistics presupposes universal common descent and evolutionary biologists generally think the likelihood is low of characters appearing multiple times on separate lines of descent, this type of analysis strives to minimize the number of unexpected evolutionary events (especially separate origins of the same characters) necessary to explain the observed distribution. This attempt to generate a tree depicting the least number of

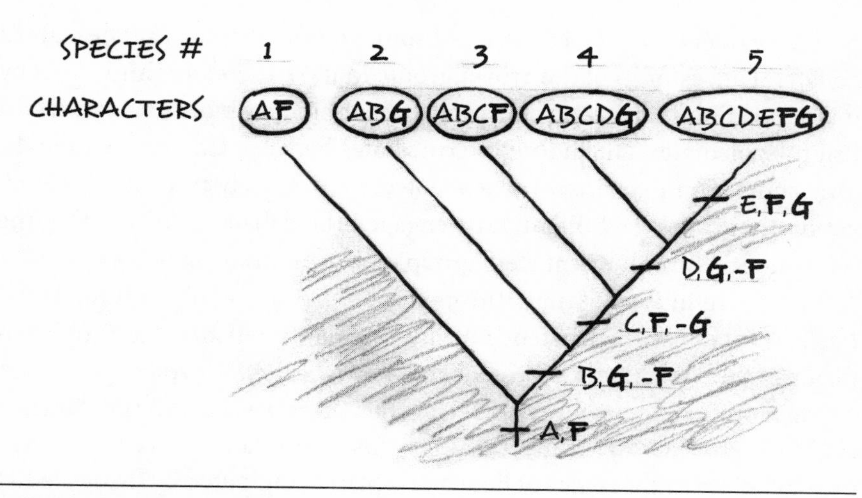

FIGURE 21.3
A cladogram with five species produced by a larger dataset of characters, A, B, C, D, E, as well as F and G. The characters shown in bold represent those that must have arisen by convergent evolution or those that were lost at some point. Minus signs in front of letters indicate evolutionary events in which characters were lost.

steps is called maximizing parsimony. However, maximizing parsimony (and minimizing the number of events involving convergence or loss) frequently becomes difficult as systematists include more characters in their analysis.

As noted, cladograms are constructed to take into account only shared derived characters. Groups that include species with all the shared derived characteristics that define a certain group (such as arthropods, for example) are called crown groups. Species that have some, though not all, of the shared derived characteristics defining the crown group are said to belong to the stem group. In Figure 21.2, for example, if traits A, B, C, D, and E are the shared derived characteristics that define the phylum ABCDE, then species with those characteristics are part of that crown group. However, species possessing characters AB, ABC, and ABCD would be said to be members of the stem group of ABCDE.

Matzke thinks cladistic methods can establish evolutionary history, including both the sequence in which the characters defining a crown group arose and the existence of various intermediates of the Cambrian animals. He claims that if paleontologists find an animal that shares some but not all the shared characters that define a crown group, then that animal can provide evidence for the existence of some intermediate—what he calls "a collateral ancestor"—of the crown group. Matzke therefore thinks that by distinguishing stem and crown groups, evolutionary biologists can establish intermediates—including intermediates between Cambrian animals. This conviction explains why he reacted so negatively to Luskin's observation that *Darwin's Doubt* hadn't engaged the debate about whether the anomalocaridids represent stem group or crown group arthropods.

Instead, in his view, "the arthropods are *instructive*" about how cladistics can establish the existence of intermediates between the Cambrian forms, since "when fossils are analyzed cladistically, we typically discover a bunch of species that morphological characters place below the crown—i.e., 'stem groups.'"[17] He thinks making this distinction helps evolutionary biologists "learn the basics of how 'body plans' originated by using cladistics (or more sophisticated methods) to estimate the order and timing of each character change found in the crown group."[18] He claims further that "cladistic analyses reveal the order in which the characters found in living groups were acquired." Specifically, he argues that because *Anomalocaris* possesses some, though not all, of the features of true arthropods, "it is one of many fossils with transitional morphology between the crown-group arthropod phylum, and the next closest living crown group, *Onychophora* (velvet worms).[19]

Matke's claims notwithstanding, there are several reasons to doubt that cladistic methods and the distinction between stem and crown groups can establish *ancestral* precursors or ancestral intermediates for the Cambrian animal groups, including putative ancestors of the arthropods. (More on whether Matzke actually claims that below.)

Ghost Lineages and Chronological Inversions

Matzke thinks that cladistic analysis of *Anomalocaris* and other fossils reveals some kind of "intermediate," "transitional," or "ancestral" arthropods, effectively solving the mystery of the missing ancestral forms of these animals. Yet using cladistics to infer such ancestral arthropods requires postulating "ghost lineages" that imply the existence of still *more* missing fossils. The need to invoke hypothetical ghost lineages commonly arises when evolutionary biologists attempt to use cladistics to infer ancestors otherwise unattested by the fossil record. The reason for this is that

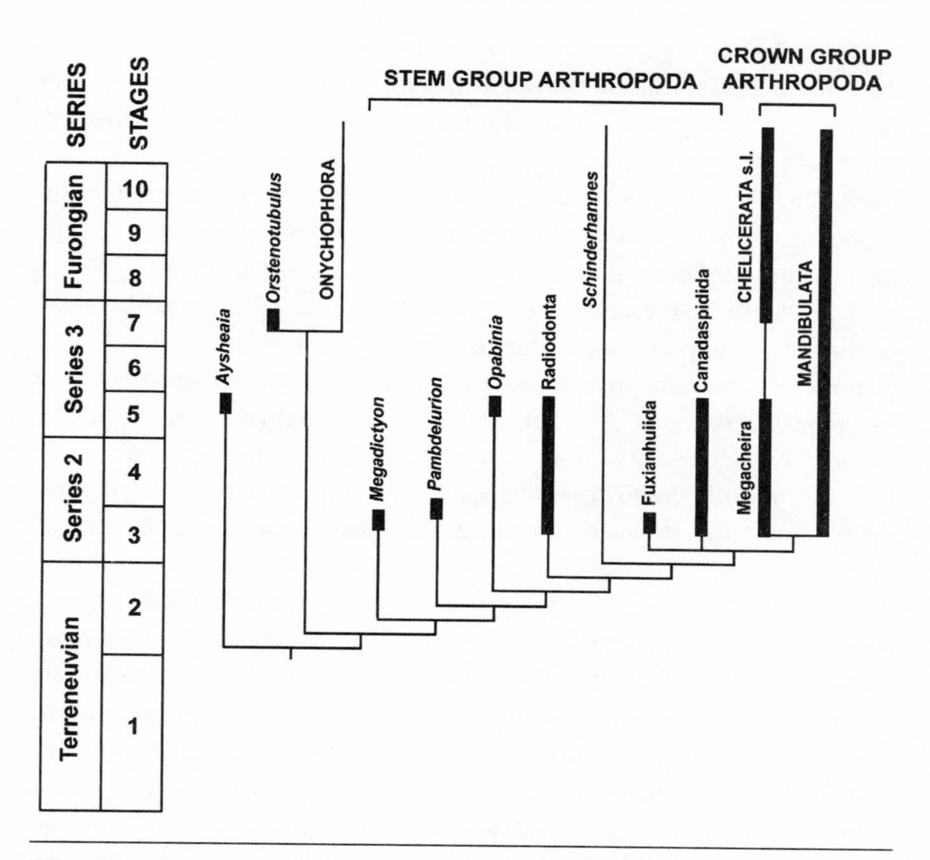

FIGURE 21.4
The fossil record of stem group and crown arthropods, and other related animals, plotted against Cambrian time scale. Thick black lines represent known fossil record. Thin black lines represent presumed evolutionary history for which there are no known fossils. *Courtesy Elsevier, Inc.*

the fossil record often reveals so-called stem groups arising contemporaneously with, or even after, crown groups. Theropod dinosaurs provide a classic example of this problem. They first appear in the fossil record millions of years after the birds that allegedly evolved from them.[20] Similarly, many supposed members of stem group arthropods appear in the Cambrian fossil record *contemporaneously* with or *after* members of the crown group arthropods that they supposedly preceded.

The anomalocaridids (and other species) that, according to Matzke, represent Cambrian intermediates illustrate this kind of chronological problem. Recall that it was my supposed misclassification of these animals to which Matzke objected. In response, he provided a cladogram from a 2012 paper lead-authored by David Legg,[21] then at the Natural History Museum of London, showing *Anomalocaris* as a stem group arthropod, which Matzke would classify as intermediate or transitional to true arthropods. Yet arthropod specialist Gregory Edgecombe reports that Radiodonta, the larger group to which *Anomalocaris* belongs, appears in the fossil record at the same time as true arthropods, *not before*.[22] Indeed, as seen in Figure 21.4, reproduced from a 2010 paper by Edgecombe, *none* of the stem group arthropods appear in the fossil record *before* the appearance of their supposed evolutionary descendants, the crown group (or "true") arthropods. As a 2013 article by Edgecomb and Legg explains, *Anomalocaris* appears at about the same time as true arthropods—*not before*—"in both the Burgess Shale in Cambrian Stage 5 in Canada (on the palaeocontinent Laurentia) and in the Chengjiang biota in Cambrian Stage 3 in China."[23]

Similarly, the 2012 paper by Legg and colleagues that Matzke cited reported an analysis of the stem group arthropod *Nereocaris*. Matzke posts a cladogram from the paper showing this animal as an intermediate between *Anomalocaris* and true arthropods with a caption purporting to show "the phylogenetic position of *Nereocaris*."[24] But such a claim implies a chronological inversion, since *Nereocaris* is known from the Tulip Beds locality of the Burgess Shale, dated at about 505 million years ago—some 15 million years *after* the first true arthropods appeared.[25]

Why are such inversions a problem? For evolutionary biologists to produce phylogenetic trees depicting evolutionary history consistent with cladistic analysis in cases involving inversions, they must draw long branches representing lineages for which they lack fossil representatives. In the case of arthropods, these ghost lineages must stretch well back in time to connect to the hypothetical ancestor of all stem and crown group arthropods. Figure 21.4 depicts this problem. In that figure the thick black lines represent time periods from which fossils of various (both stem and crown group) arthropods are known, as well as other groups that are supposedly close relatives of arthropods. The thin black lines represent animals inferred based on cladistic analysis—animals that are *not* found in the fossil record. Note that all the putative ancestors of arthropods that might link them to other groups are represented by thin black lines. Indeed, *none* of the supposed *ancestral* stem arthropods, or their evolutionary histories, or related nonarthropod ancestral groups from which arthropods supposedly evolved are documented in the fossil record.

For example, lobopods (represented in Figure 21.4 by *Orstenotubulus*, and possibly *Aysheaia*) are thought to be closely related to the ancestors of arthropods, if not directly ancestral to arthropods, but as we see in Figure 21.4, they don't appear until millions of years *after* the first true arthropods. Yet Matzke claims that "the arthropod and velvet-worm phyla [Onychophora] evolved from lobopods, and lobopods contain a whole series of transitional forms showing the basics of how this happened."[26] Do we see this "whole series" in the fossil record? Richard Fortey comments in *Science* that "Onychophora (velvet worms) were probably the most closely related group to the arthropods as a whole; this group and the arthropods must have diverged . . . in the Precambrian." Rather than finding a "whole series of transitional forms," however, Fortey calls this "earlier" evolution of arthropods "Precambrian hidden history," acknowledging that "fossils of these alleged ancestral arthropods are lacking."[27] To put some numbers on the problem, a 2011 paper in *Science* used molecular clocks to date the split of arthropods and Onychophora to over 600 million years ago,[28] but neither group appears in the fossil record until around 521 million years ago or later. That's at least 80 million years of "hidden history" of arthropods, with a group (the lobopods) representing supposed arthropod "collateral ancestors" appearing *after* arthropods.

The case of *Schinderhannes bartelsi*, an anomalocaridid known only from rocks of the lower Devonian, provides another striking example of such an inversion. Indeed, when touting the findings of cladistic analysis, Matzke might easily have cited a 2009 paper in *Science* reporting the discovery of *Schinderhannes*, which called *Schinderhannes* a "stem lineage" arthropod and included cladistic analysis making it appear intermediate (by Matzke's standard) between *Anomalocaris* and true arthropods.[29] Or he might have cited a cladogram from a 2011 paper in *Nature* showing *Schinderhannes* as one of the closest relatives (what Matzke might call a "collateral ancestor") to crown group arthropods.[30] But had he done so, it would have again highlighted the need to formulate ghost lineages to generate a coherent phylogenetic tree. Indeed, *Schinderhannes* appeared *over 100 million years after* the first true arthropods are found. This chronological inversion requires postulation of a ghost lineage of over 100 million years to place *Schinderhannes* in a correct phylogenetic relationship to other arthropods.

So does cladistic analysis—which shows that *Anomalocaris* (and some of its close relatives) lacks some characters of crown group arthropods—establish that *Anomalocaris* represents some kind of an *ancestral* intermediate between onychophorans and crown group arthropods? Matzke can make this argument only by assuming that these stem group arthropods (or their relatives) existed *before* crown group arthropods first appeared. But since the fossil record does not document the existence of *Anomalocaris* or its relatives in the earlier fossil record, those using cladistics to infer the evolutionary history of crown group arthropods must also posit ghost lineages of earlier fossil *Anomalocaris*-like ancestors reaching back long before crown group arthropods appeared. Did such a sequence really exist? Who knows? But it hardly solves the problem of missing fossil ancestors of the Cambrian animals to use cladistics to posit a phylogenetic hypothesis that requires, as a condition of its plausibility, the postulation of ghost lineages representing still more missing fossils.

Wrong or Irrelevant

There is another problem with Matzke's use of cladistics. Many cladists themselves do not think cladograms necessarily indicate anything about evolutionary history. Instead, they regard them as tools for classifying different taxa. Matzke himself acknowledges at least one important limitation on what cladograms can reveal about evolutionary history by conceding that "phylogenetic methods as they exist now can only rigorously detect sister-group relationships, not direct ancestry."[31] This weaker claim about what cladistics can tell us is much easier to defend given the paucity of ancestral forms in the Precambrian fossil record. But if that is *all* Matzke means to claim about what cladistics can establish, then the significance of his argument evaporates.[32]

Oddly, in his discussion of the use of cladistics in phylogenetic reconstruction, Matzke never defines exactly what he means by an intermediate. Does he mean an intermediate in the sense of an animal possessing only some of the features of the crown group? An anatomical intermediate? Or does he mean a true *ancestral* intermediate of the kind that *Darwin's Doubt* argues is missing? Matzke does not specify, though presumably he would insist that he does not necessarily mean to imply that the anomalocaridids were direct *ancestral* precursors to the trilobites or other crown group arthropods. Thus he uses the ambiguous word "collateral" as a modifier to "ancestor" when he states: "I claimed that phylogenetic methods can establish . . . *collateral* ancestors of various prominent living phyla."[33]

It's not entirely clear what Matzke means by "collateral ancestor," because he never defines the term. Indeed a search of Pubmed for the term reveals virtually nothing in the technical literature: only three hits were returned for "collateral ancestor" or "collateral ancestors," and in none was the term used in the way that Matzke uses it. The term does, however, have meaning in a legal context. As the online *Encyclopedia of Genealogy* explains, "Collateral ancestor is a legal term referring to a person [who is] not in the direct line of ascent, but is of an ancestral family. This is generally taken to mean a brother or sister of an ancestor (hence a 'collateral ancestor' is never an ancestor of the subject)."[34] In other words, a collateral ancestor is not a direct, actual, or common ancestor, and thus does not solve the problem of the missing fossils of the common (or direct) ancestors of the arthropods and other major Cambrian groups.

Matzke might reply that what I've said misrepresents the subtlety of *his* position. He might say that he is not claiming that the anomalocaridids are the *ancestors,* or at least the *direct ancestors,* of arthropods, or that members of stem groups are necessarily the direct ancestors of specific members of crown groups. Instead, he might say he only means to affirm that they are intermediates in the sense of possessing some but not all the characters of the crown group, and that they are *collateral* ancestors, meaning that they reside somewhere on the evolutionary tree below the crown group and thus are in some way related to the direct ancestors of the crown group.

But again, if that is *all* Matzke means, then surely his use of cladistics does not solve the problem highlighted in the first third of *Darwin's Doubt.* Recall that my first seven chapters argued that neither fossil nor genetic evidence establishes the existence of *ancestral* precursors for most of the Cambrian animals. If Matzke's intermediates are not direct *ancestral* precursors of the Cambrian animals, then they do not provide the missing intermediates highlighted by *Darwin's Doubt.* If, on the other hand, Matzke *is* claiming that cladistics resolves the mystery of missing ancestral fossils, he is simply wrong, because the temporal order of the appearance of character states in the Cambrian requires the postulation of ghost lineages representing still other missing ancestral fossils. Since Matzke never clearly defines what he means by an intermediate, it is not possible to establish upon which horn of this dilemma his position ultimately founders. Either way, cladistics-based phylogenetic hypotheses do not solve the problem of missing ancestral fossils.

In any case, there are still further difficulties with his position.

Begging the Question

In 2012, molecular biologist Michael Syvanen observed in *Annual Review of Genetics* that "one needs to be continually reminded that submitting multiple sequences (DNA, protein, or other character states) to phylogenetic analysis produces trees because that is the nature of the algorithms used."[35] The same can be said about analyses of shared derived characters and cladograms constructed on the basis of cladistic analysis (and other forms of character-based phylogenetic analysis). For those who regard cladograms as depicting real events in evolutionary history, the algorithms used during phylogenetic reconstructions and cladistic analysis presuppose, rather than demonstrate, the common ancestry of the groups they analyze. Indeed, the assumption of common ancestry is inherent in the method of cladogram construction—at least for those who regard such trees as representations of evolutionary history rather than mere classificatory devices. As University of Wisconsin philosopher of biology Elliott Sober states, when evolutionary biologists construct cladograms, "the typical question is which tree is the best one, not whether there is a tree in the first place."[36]

Furthermore, if one interprets the results of cladistic analysis as an indicator of evolutionary history, then the number of shared derived characters represents a measure of the *historical* relatedness of two or more groups. Viewed this way, cladistic analysis assumes that more shared derived characters indicate—all other things being equal—a closer evolutionary relationship and a more shallow divergence *from a common ancestor.* Conversely, it also assumes that fewer shared derived characters indicate a more distant evolutionary relationship and a deeper divergence *from a common ancestor.* Thus interpreting the results of cladistic analysis historically entails the assumption that each of the groups analyzed evolved from a common ancestor. It thus presupposes, rather than demonstrates, the existence of such ancestors.

One sees this assumption of common ancestry in nearly every phase of cladistics and similar forms of character-based phylogenetic analysis. When systematists choose which characters to include and which to exclude, they make judgments about which characters are most "phylogenetically informative."[37] In practice, this means selecting the characters that are most likely to generate congruent treelike patterns requiring the fewest evolutionary events. Judgments about how to weight different characters or which species to place at the base of a given phylogenetic tree (how to "root" the tree) are made with similar considerations in mind and are always informed by the assumption of common ancestry.

In Chapters 5 and 6, I made the same point about the use of comparative sequence analysis to generate phylogenetic trees. By presupposing that degrees of difference correspond to the amount of time elapsed since divergence from a common ancestor, those methods also presuppose, rather than demonstrate, a common ancestor. Those who cite either comparative sequence or character-based phylogenetic analyses to establish the existence of such ancestral forms elide a simple point of logic. No method that presupposes the truth of a proposition can be used to prove or establish the truth of that same proposition without begging the question. By asserting that phylogenetic reconstructions based on cladistic analysis "can establish, and have established, the existence of Cambrian intermediate forms"[38] on the animal tree of life, Matzke relies on precisely such a question-begging method.

Index of Inconsistency

There is another reason not to regard cladograms as representations of evolutionary history as opposed to classificatory devices. Characters that appear homologous do not always reflect common ancestry. Even evolutionary biologists who assume universal common descent recognize this. As I note in Chapter 6, the assumption that shared characteristics result from common ancestry commonly breaks down in the case of those characteristics thought to have arisen via "convergent" evolution. The notes in this book cite textbooks on phylogenetic methods that acknowledge the *assumption* of common ancestry, as well as the difficulty that assumption can pose for reconstructing phylogenetic trees using cladistic analysis. One source states:

> *Cladistics can run into difficulties in its application because not all character states are necessarily homologous. Certain resemblances are convergent—that is, the result of independent evolution. We cannot always detect these convergences immediately, and their presence may contradict other similarities, "true homologies" yet to be recognized. Thus, we are obliged to assume at first that, for each character, similar states are homologous, despite knowing that there may be convergence among them.*[39]

The problem of convergent evolution is rampant in cladistic studies of Cambrian arthropods. To see why, consider how often it is necessary to invoke convergent evolution (or loss) to explain the distribution of characters in cladograms. Evolutionary biologists have quantitative ways of measuring this. One method is called the consistency indices (CI). It is a statistical measure of how often the assumption of common ancestry succeeds in explaining the distribution of shared biological characters. It is calculated simply by taking the minimum number of evolutionary events required by the overall dataset (which is equivalent to the total number of characters being studied) and dividing by the number of evolutionary events implied by a given tree. A high CI (closer to 1) indicates that characters are naturally distributed in a treelike pattern without having to invoke additional evolutionary events. A lower CI (closer to 0) means that it is difficult to explain the distribution of characters without invoking many instances of convergent evolution (or loss).

In his review of my book, Matzke insists on assessing the "statistical support"[40] for an evolutionary tree as a prerequisite to challenging its veracity. Let's therefore consider the consistency index of the two cladograms he cites, each of which, according to Matzke, establishes definitive evolutionary relationships among Cambrian arthropods. One cladogram from Legg and his colleagues (2012) has a consistency indices of 0.565,[41] meaning that about 43.5 percent of the time, a given character was *not* distributed in a treelike pattern. To put it another way, around 43.5 percent of the time, the assumption of common ancestry failed to explain the distribution of a character. Even worse, the other cladogram Matzke cites has a CI of 0.384, which the original authors admitted was "rather low."[42] This is striking: roughly 61.6 percent of the time, the shared characters in these groups did *not* result from descent with modification from a common ancestor. That is, in 61.6% of the cases, the assumption of homology failed to explain the distribution of characters.

If an assumption fails more often than it holds true, is it justified? Whatever the answer, it's clear that the characters of the Cambrian arthropods often fail to fit the treelike pattern required by universal common descent. This further undermines Matzke's claim that cladistics establishes the existence of the ancestors implied by trees generated from this method.

Conflicting Trees

There is another problem with treating cladograms as depictions of evolutionary history. The same set of characters often generates many equally parsimonious cladograms. Depending upon the choices one makes about how to do a character-based phylogenetic analysis—which characters to emphasize as homologous, how strongly to weight characters, which computer programs to use in the analysis, whether to take into account paleontological data about specific taxa, how to "root"[43] trees, and so on—one can generate many conflicting trees from the same data. Officially a "pure cladist" will not weight characters in his analysis (see note 16). Nevertheless, in practice many evolutionary biologists doing character-based phylogenetic analysis, including many who consider themselves cladists, do weight characters differently—causing phylogenetic algorithms to generate different, and sometimes conflicting, trees in response to different choices about *how* to weight characters.

For example, some evolutionary biologists think that both lobopods and radiodonts (the order that includes anomalocaridids) are closely related to arthropods. Lobopods have arthropod-like legs but lack heads and eyes. Radiodonts lack legs but have arthropod-like heads and eyes. Depending on which characters are weighted more heavily, phylogenetic analysis will generate starkly different and incompatible trees, showing, in one case, lobopods arising first and radiodonts later and, in another case, just the reverse.[44] Decisions about other factors—how to root trees, whether to take paleontological data into account, and so on—can also result in multiple, conflicting trees.

Evolutionary biologists who interpret these differing trees as depictions of evolutionary history immediately face a difficult problem. Which of the conflicting treelike diagrams reflects the true evolutionary history of a given group? If the same raw data generate many trees, how can we say that cladistic data are sending a clear historical signal?

Matzke ignores this problem by affirming a single unequivocal history of the arthropods. He confidently asserts that "the arthropod and velvet-worm phyla *evolved from lobopods*, and lobopods contain a whole series of transitional forms" in which *Anomalocaris* has "transitional morphology between the crown-group arthropod phylum, and the next closest living crown group, Onychophora (velvet worms)."[45] Nevertheless, the history of the arthropods that Matzke affirms represents just one of many possible histories allowed by character-based phylogenetic analyses. In affirming a particular historical progression as *the* correct history, Matzke represents phylogenetic analysis—and indeed, cladistics, narrowly construed—as far more capable of establishing a definitive picture of the history of animals than the relevant scientific literature indicates.[46] Thus, for example, Douglas Erwin and James Valentine's 2013 book *The Cambrian Explosion* calls arthropod origins "far from settled" and even "problematic."[47] Or as Edgecombe writes, "Arthropod phylogeny is sometimes presented as an almost hopeless puzzle wherein all possible competing hypotheses have support."[48] Even Legg and his colleagues, whose article Matzke cites in support of his critique of *Darwin's Doubt,* note that "the origin of arthropods is a contentious issue. . . . [T]here is little consensus regarding the details of their origins."[49]

Conflicting Histories

Furthermore, even in an idealized case where cladistics generates one cladogram that is clearly the most parsimonious, that tree itself necessarily corresponds to many possible evolutionary histories. Oddly, a scholar Matzke cites in critique of *Darwin's Doubt* makes exactly this point. To support his claim that cladistic analysis enables evolutionary biologists to establish the "evolutionary history" of animal groups and the existence of "transitional forms,"[50] Matzke cites as authoritative a 2008 paper by historian of paleontology Keynyn Brysse. In her Figure 3 (reproduced here as Figure 21.5), Brysse shows graphically why cladistic analysis does *not*

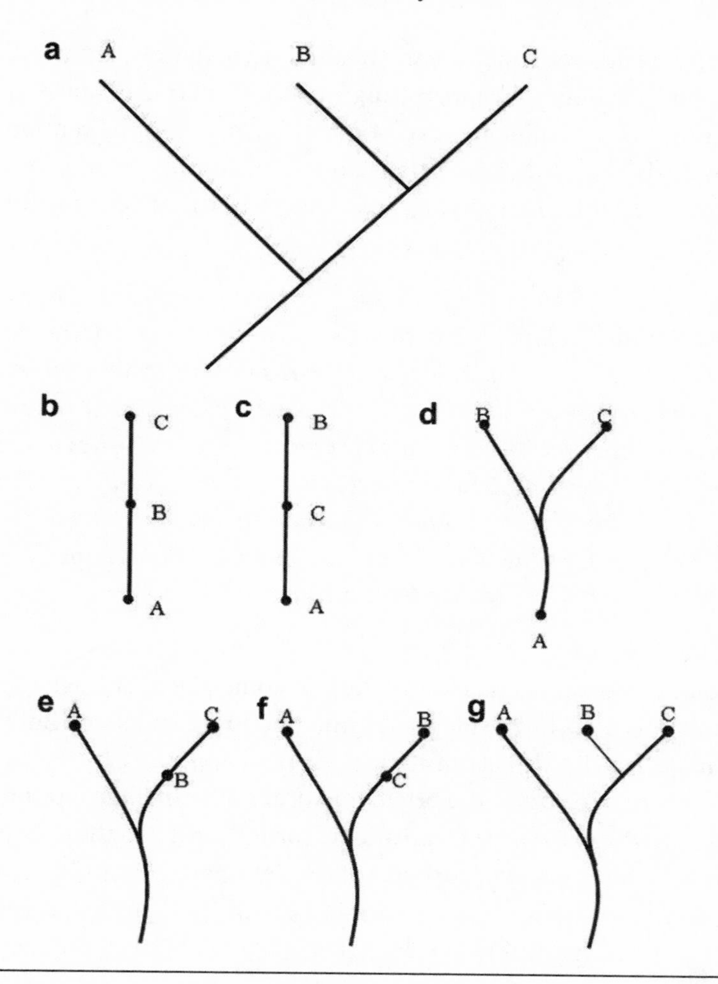

FIGURE 21.5
Brysse's Figure 3, showing that for every cladogram, there are many possible corresponding evolutionary histories. As Brysse explains: "The cladogram at the top (a) shows that among three related taxa, A, B, and C, B and C are more closely related to each other than either is to A. There is no evidence, however, to distinguish the possible ancestor-descendant relationships among the three taxa. For example, A might have evolved into B, and then into C (b); A might have evolved into C and then into B (c); B and C might both be descendants of A (d); A and B might be descendants of a common ancestor, with B evolving into C (e); A and C might be descendants of a common ancestor, with C evolving into B; or finally, B and C might have evolved from a common ancestor which itself shares a common ancestor with A. As these examples illustrate, the cladogram is not an expression of ancestor-descendant relationships; it is not a phylogenetic tree. The fact that at least six trees can be drawn from a single cladogram is evidence that the correct tree is underdetermined by the cladistic data. As cladists argue, then, there is only enough reliable information available to construct cladograms, not trees." *Courtesy Elsevier, Inc.*

establish definitive evolutionary history. That figure demonstrates how one simple cladogram representing just three character states is equally consistent with *six* different evolutionary histories. In the caption, she explains that "the cladogram is not an expression of ancestor-descendant relationships; it is not a phylogenetic tree."[51] In the article, she further explains the limitation of cladistics:

> *Cladograms depict sister groups—taxa that are thought to be each other's closest relative—but do not show ancestors and descendants. It is possible to use a cladogram as the basis for constructing an evolutionary tree, but any given cladogram* is usually consistent with multiple phylogenetic trees. *This means that according to cladistics at least, the correct ancestral-descendant relationships among fossils are always underdetermined by the available evidence. In other words, the very goal of evolutionary systematics—the determination of ancestor-descendant relationships—is on the cladistic method not just unattained but unattainable.*[52]

Of course, Matzke does acknowledge, in some places at least, that cladistics can only establish sister groups and not direct ancestors. But throughout his review, he also claims that cladistics can establish the sequence in which characters arose, the branching order of animal groups on the tree of life, and the existence of transitional forms leading to the Cambrian animals—in short, many key aspects of "evolutionary history."[53] Yet a critical source he cites shows that establishing definitive historical claims about specific animal groups using cladistics is, to say the least, problematic.

Cladistics Cannot Determine Causes

Brysse's paper is instructive in another respect. She argues that cladistics cannot establish anything about the *processes* that might have produced the characters represented, and the patterns depicted, in cladograms. After recounting the history of classification of Cambrian organisms, she argues that cladistics does nothing to solve the mystery of the Cambrian explosion. Brysse explains that cladistics allowed evolutionary scientists "to construct clearly stated hypotheses about the relationships among the organisms under examination" but not "to investigate the tempo and mode of evolution." Instead, cladistic analysis necessarily "ignores" questions about the causal processes that generate evolutionary novelty.[54] To emphasize her point, Brysse cites another authority, Henry Gee, who puts the point succinctly: "Cladistics is concerned with the pattern produced by the evolutionary process; it is not concerned with the process that created the pattern, or the swiftness or slowness with which that process acted."[55]

Cladistics describes patterns of relationships among organisms; it provides tools for classifying organisms. It *might* also suggest historical reconstructions of evolutionary history *if* its question-begging assumptions in that context are granted. But it cannot determine what caused the patterns of relationship depicted by cladograms or what *caused the origin of the complex animal features* that it analyzes. For this reason, cladistics cannot be used to rebut the central argument of *Darwin's Doubt*, which addresses precisely the question of what *caused* the Cambrian animals to arise.

And that is why Matzke's review of *Darwin's Doubt* fails to address the central argument of the book. Cladistics does not, and cannot, offer any explanation of what caused the Cambrian animals to come into existence. Nor can it account for the origin of genetic and epigenetic information necessary to produce them.

DONALD PROTHERO AND THE DURATION OF THE CAMBRIAN EXPLOSION

In his review, Matzke also challenges a factual claim I make about the duration of the Cambrian explosion. Whereas I describe the Cambrian explosion as an event encompassing roughly 10 million years, with its most explosive period occurring over 5 to 6 million years, Matzke argues that the explosion took about 30 million years. In August 2013, Donald Prothero, the geologist discussed in Chapter 4, also claimed, in a lengthy review in *Skeptic* magazine, that I exaggerated the brevity of the Cambrian explosion.[56] For the same reasons he did before, Prothero again defined the Cambrian explosion as an 80-million-year-long event.[57] Since I already rebut Prothero's version of this argument in Chapter 3, and a colleague and I published extensive rebuttals[58] to the other versions of this argument (including Matzke's) in the first months after publication, I will not reprise those arguments in detail here. Instead, I refer readers back to Chapter 3 for my original rebuttal of Prothero and to our published responses. These provide extensive documentation verifying that leading

Cambrian authorities typically do regard the Cambrian explosion as a 10-million-year event—just as I do in the book.

One point is worth reiterating and amplifying here, however. As I noted in Chapter 3, it is entirely possible to assign a different duration to the "Cambrian explosion" depending upon how many separate paleontological events scientists choose to include within that designation. Nevertheless, quibbling of that sort reduces the debate to one of semantics. The key question is not how many different events should be included within the designation "Cambrian explosion." Nor is it about the total amount of time that some arbitrarily designated series of separate paleontological events encompasses. Instead, the key question is what *caused* the discontinuous appearance of morphological novelty within specific, and measurably narrow, windows of geological time—whatever we choose to call these events. Thus, *Darwin's Doubt* focuses on the crucial Tommotian and Atdabanian stages of the Cambrian explosion—where 13–16 new animal phyla arose within a 5–6 million year window—as a key defining challenge to the efficacy of the neo-Darwinian mechanism. Defining the Cambrian explosion (or radiation) to include other separate paleontological events—the appearance of small shelly fossils or the Ediacaran fauna, for example—does nothing to explain what caused the abrupt appearance of so many (13–16) separate animal body plans within a much later and definably narrow window of geologic time.

CHARLES MARSHALL AND THE ORIGIN OF BIOLOGICAL INFORMATION

Three months after publication on September 20, 2013, following several critical, even frivolous, reviews that did not address the main arguments of the book,[59] the journal *Science* published a review of *Darwin's Doubt* that did address those arguments. The review, titled "When Prior Belief Trumps Scholarship," was written by the distinguished paleontologist and evolutionary biologist Charles Marshall of the University of California, Berkeley. His review was respectful and, in some minor ways, complimentary. Nevertheless, Marshall rendered a decidedly negative verdict overall, suggesting, as the title of his review indicates, that I had allowed prior theistic beliefs to distort my assessment of the scientific facts.

Even so, I was delighted to see the review, not only because it appeared in *Science* (which is one of the top science journals in the English-speaking world), but also because Marshall did address my main argument about the origin of biological information. Truth be told, I also appreciated the review because it demonstrated—if inadvertently—the severity of that problem and that leading Cambrian paleontologists and evolutionary biologists (such as Marshall) are nowhere close to solving it.

Darwin's Doubt argues that intelligent design provides the best explanation for the origin of the genetic (and epigenetic) information necessary to produce the novel forms of animal life that arose in the Cambrian period. In his review, Marshall did *not* attempt to show that the main evolutionary mechanisms I address in the book *can* produce the information necessary to build Cambrian animals. Instead, he sought to refute my argument by disputing the claim that significant amounts of new genetic information (and many new protein folds) would have been necessary to build these animals. Marshall claimed that "rewiring" of the developmental gene regulatory networks (dGRNs), discussed in Chapter 12, would have sufficed to produce new animals from a set of preexisting genes.

As he argues:

> *[Meyer's] case against the current scientific explanations of the relatively rapid appearance of the animal phyla rests on the claim that the origin of new animal body plans requires vast amounts of novel genetic information coupled with the unsubstantiated assertion that this new genetic information must include many new protein folds. In fact, our present understanding of morphogenesis indicates that new phyla were not made by new genes but largely emerged through the rewiring of the gene regulatory networks (GRNs) of already existing genes.*[60]

In this paragraph, Marshall claims a lot in a few words. He implies that evolutionary biologists have an adequate explanation for the process of body plan morphogenesis—one that does not require the generation of new (or at least not much new) genetic information. Essentially, Marshall attempted to circumvent rather than solve the problem of the origin of genetic information by claiming that animal body plans "emerged through the rewiring of the gene regulatory networks (GRNs) of already existing genes." Nevertheless, his proposed approach subtly presupposes at least three unexplained significant sources of genetic information: (1) the information stored in the gene regulatory networks, (2) the information required to rewire them, and (3) the genetic information stored in preexisting genes for building the proteins necessary to various anatomical structures and novelties—the necessity of which Marshall acknowledges elsewhere in his published work. Quite simply, Marshall's proposed explanation begs the question as to the origin of several necessary sources of genetic information.

Let's examine each of these sources in turn.

The Genes in Gene Regulatory Networks Contain Genetic Information

Recall the discussion in Chapters 9 and 10 regarding recent mutagenesis experiments and how these experiments have established the extreme rarity of functional genes and proteins among the many (combinatorially) possible ways of arranging nucleotide bases or amino acids within their corresponding "sequence spaces." Recall also that the rarity of functional genes and proteins within sequence space makes it overwhelmingly more likely than not that a series of random mutation searches will *fail* to generate even a single new gene or protein fold within available evolutionary time. This extreme rarity also helps to explain why mathematical biologists, using standard population genetics models, are calculating exceedingly long waiting times (well in excess of available evolutionary time) for the production of new genes and proteins, when producing such genes or proteins requires even a few coordinated mutations (see Chapters 9–12). Marshall's review did not address these mathematical[61] or experimentally based[62] challenges to the efficacy of the neo-Darwinian mechanism. Instead, his proposal presupposes but does not explain the prior existence of the genetic information necessary to produce new forms of animal life.

He did this, in the first and most obvious way, by presupposing the existence of gene regulatory networks. As noted in Chapter 12, developmental gene regulatory networks (dGRNs) are integrated networks of specific genes and gene products (proteins) that interact to control and direct cell differentiation and organization during animal development. The protein molecules that perform a signaling function within these networks are typically transcription factors that bind to specific regions on the DNA to promote (or repress) the expression of specific genes at specific times. Many of these genes in turn produce other transcription factors that regulate the expression of other genes, likewise in a tightly choreographed way. Clearly, the many genes that code for the production of these transcription factors contain a vast amount of genetic information—the origin of which Marshall does not explain. Instead, his scenario for rewiring gene regulatory networks presupposes the prior existence of the information-rich genes that constitute these networks. But how did these genes arise? Marshall doesn't say. Thus, his proposal begs the question as to the origin of at least one significant, and necessary, source of genetic information.

Yet Marshall clearly acknowledges the need for these regulatory genes, not only implicitly in his review of my book but also explicitly when writing elsewhere. In his published papers, Marshall insists that *Hox* genes, in particular, must have played a causal role in producing the Cambrian explosion. He notes that developmental considerations "point to the origin of the bilaterian developmental system, including the origin of *Hox* genes, etc., as the primary cause of the 'explosion.'"[63] *Hox* genes are regulatory elements that play important roles in many gene regulatory networks. While in these papers Marshall also emphasizes the importance of rewiring gene regulatory networks to generate new body plans, he clearly acknowledges that preexisting genes would be necessary to produce new animals—though, again, he does not explain the origin of these information-rich genes. He merely presupposes them.

Rewiring Networks Requires Informational Inputs

In any case, rewiring genetic circuitry requires reconfiguring the temporal and spatial expression of genetic information. Such reconfiguring would entail fixing some material states and excluding others. Thus it would constitute an infusion of new information (in the most general theoretical sense) into the biosphere.[64]

To see why, consider changing a wiring diagram representing a developmental gene regulatory network (see Figure 13.4). Just altering the diagram to produce a new diagram representing a new network would require changing the arrangement of nodes (representing genes) and edges (representing interactions between genes and gene products). Changing the arrangement of these elements would mean adding information to the diagram depicting the system. In the same way, changing the arrangements of genetic elements in an actual network would also require informative changes to the arrangement of the network.

Specifically, rewiring a gene regulatory network in the way Marshall envisions would require multiple coordinated mutations in the *cis*-regulatory regions adjacent to various genes in these regulatory networks. It would also require coordinated mutations in genes that code for specific regulatory proteins. Changing these genes into new functional genes requires altering sequences of nucleotide bases and thus generating new genetic information. Moreover, producing new functional genes in this manner—that is, as the result of multiple coordinated mutations—would require implausibly long waiting times of the kind discussed in Chapter 12. Thus Marshall's rewiring proposal does not eliminate the need for new information to build the Cambrian animals. Rather, it tacitly concedes additional sources of genetic information that cannot be readily explained by the standard mechanism of mutation and natural selection.

Additionally, altering the temporal and spatial expression of preexisting genetic elements in a dGRN would in all likelihood require the addition of new taxon-specific genes or gene products, hence an additional source of new *genetic* information. Indeed, experiments on dGRNs in modern representatives of the animal phyla show that different organisms use taxon-specific DNA-binding proteins to regulate the expression of genetic data files. For instance, contrary to theoretical expectations,[65] the morphogen Bicoid, essential for normal anterior-posterior body plan specification in *Drosophila,* is found only in the cyclorrhaphan flies.[66] Similarly, the body plan of the freshwater polyp *Hydra* is specified by taxonomically restricted proteins.[67] Nor are these isolated cases. The remarkable disparity of animal morphologies at the macroscopic (anatomical or body plan) level tends to correspond to differences at *other* levels (the microscopic or molecular). Moreover, studies of evo-devo model systems have repeatedly revealed that the cell and tissue specification programs that generate distinctive animal morphologies depend upon taxon-specific regulatory factors (proteins and RNAs). As Oliveri and Davidson note:

> *The specification apparatus very frequently also includes transcriptional repressors, which, within the specified spatial domain, target key regulatory genes whose expression is required for alternative regulatory states that could have been available to these cells. This is a so-called "exclu-*

> *sion effect," and numerous examples can be found across species. . . . In each developmental case,* the identity of the specific transcription factor that executes the repression is distinct, as are the specifically excluded target transcription factors. *The design is the same,* the biochemical actors diverse.[68]

Of course, building the species-specific transcription factors necessary to animal development requires genetic information. And creating these proteins in the first place would, again, have required the origin of *new* genetic information.

Anatomical Novelties Require a "Genetic Toolkit"

When Marshall said in his review that new animals "emerged through the rewiring of the gene regulatory networks (GRNs) of already existing genes," seems to be referring to already existing genes *in genetic regulatory networks,* not *other preexisting genes* such as those that are *necessary for building the anatomical structures* that characterize the Cambrian animals (the expression of which dGRNs regulate). Nevertheless, when writing elsewhere, Marshall and other evolutionary biologists have made clear that building new animal body plans during the Cambrian period would require many other preexisting genes—indeed, a preexisting, preadapted genetic "developmental toolkit"—for building many specific anatomical structures.[69]

For example, in "Explaining the Cambrian 'Explosion' of Animals" (2006), he noted, "Animals cannot evolve if the genes for making them are not yet in place. So clearly, developmental/genetic innovation must have played a central role in the radiation."[70] Later in the same paper, he argues, "It is also clear that the genetic machinery for making animals must have been in place, at least in a rudimentary way, before they could have evolved." As noted above, in other published work Marshall emphasizes that preexisting *Hox* and other regulatory genes (within dGRNS) were important for producing Cambrian animals. He also emphasizes the need for "gene novelties" for building the proteins that make up the anatomical structures and novelties of the various animals that arise in the Cambrian period.[71]

Of course, he's right about this. Building the Metazoa (multicellular animals) would not have required just new *Hox* genes or genes for building new regulatory (DNA-binding) proteins. Instead, the evolutionary process would need to produce a whole range of proteins necessary to build and service specific forms of animal life. Indeed, different forms of complex animal life exhibit unique cell types, and typically each cell type depends upon other specialized or dedicated proteins—which in turn requires genetic information. In Chapter 8, "The Cambrian Information Explosion," I offer numerous examples of this. The first arthropods would likely have required genes for building the complex protein lysyl oxidase (see p. 191). Why? Because what we know from studies of modern arthropods shows that this protein is necessary to support the stout body structure of arthropod exoskeletons.[72] Similarly, the first animals with guts (and epithelial cells lining them) would have required specialized digestive enzymes and specialized regulatory enzymes to control the secretion of those enzymes (see p. 162). Other examples abound. Tunicates require specialized proteins for building their distinctive "tunics." Mollusks require scores of specialized proteins for building their shells. Indeed, building metazoans at all requires specialized proteins (and metabolic pathways) just to produce the kind of extracellular matrices that allow developing animals to knit cells into tissues, tissues into organs, and organs and tissues into fully developed animals. Our observations of animals representing known phyla that first arose in the Cambrian show that such animals would also have needed other specialized proteins: for facilitating adhesion, for regulating development, for building specialized tissues or structural parts of specialized organs, for producing eggs and sperm as well as many other distinctive functions and structures. Obviously, these proteins must have arisen sometime in the history of life. Since most major metazoan body plans first arose in the Cambrian explosion, it is reasonable to infer that the proteins necessary to sustain those forms of life—and the genetic information necessary to synthesize them—had come into existence by that time.

In addition, recent genomic studies of animals representing phyla that first arose in the Cambrian show that these animals depend upon many *unique* genes not present in other taxa. These genes perform many functions besides just specifying body plan development. Known as taxonomically restricted or ORFan genes, they are ubiquitous in animal life and represent 10 percent or more of the genomes of each species that scientists have investigated.[73]

The presence of ORFan genes in all sequenced present-day animal genomes—indeed, in all life[74]—suggests that the genomes of Cambrian animals would likely have also contained many ORFan genes. That in turn suggests that a considerable amount of new genetic information not present in simpler Precambrian organisms would have originated before or during the Cambrian radiation. Moreover, even if some universal Precambrian genome originally contained all the genes that later became taxonomically restricted, whatever process distributed these genes to some lineages but not others necessarily involved the addition of new information into the biosphere.[75]

Begging the Central Question

In *Darwin's Doubt,* I argue that new Cambrian animals would have required new genetic information and new proteins (and therefore, new protein folds). Although Marshall characterizes my claim as "unsubstantiated," he doesn't dispute the need for genetic information to build the proteins required by each new form of metazoan life. Instead, Marshall has acknowledged the need for new genes or "gene novelties" for building specific parts of the Cambrian animals.[76] He only seems to dispute whether all that information necessarily arose *during* the Cambrian explosion. Indeed, in both his technical publications and his review of *Darwin's Doubt,* Marshall simply assumes that most of the genetic information necessary to build the Cambrian animals already existed *before* the Cambrian explosion. In fact, he seems to presuppose the existence of what Susumu Ohno called a "pananimalian genome"[77] or what others have referred to as the "ur-bilaterian genome"[78]—a nearly complete set of genes for building the anatomical structures of Cambrian animals present within some phenotypically simpler, ur-Metazoan ancestor. An article that accompanied Marshall's review in *Science,* this one by M. Paul Smith and David Harper, also presupposes such a universal gene toolkit and suggests that it might have arisen 100 million years or more before the Cambrian.[79]

Nevertheless, this question-begging assumption does not solve the central problem posed by *Darwin's Doubt*—that of *the origin* of the genetic (and epigenetic) information necessary to produce the Cambrian animals.

It merely pushes the problem back tens or hundreds of millions of years. (Marshall also makes no attempt to explain the origin of new *epigenetic* information, a problem that has led other prominent evolutionary biologists to express skepticism about the adequacy of the neo-Darwinian mechanism.[80] Indeed, it is well known that developmental gene regulatory networks are only adequate to specify animal body plans in the presence of "maternal inputs,"[81] that is, assuming the existence of a maternal egg carrying many sources of epigenetic information of the kind discussed in Chapter 13.) In any case, Marshall does not explain how the neo-Darwinian mechanism could have overcome the combinatorial search problem or the waiting times problem described in Chapters 10 and 12, both of which must be overcome to produce just the new *genetic* information necessary to build new proteins and Cambrian animals.

Marshall's review and rewiring proposal does not offer a solution to the problem of combinatiorial inflation and the rarity of genes and proteins in sequence space. Instead, it begs the question of the origin of three extensive sources of genetic information necessary for building Cambrian animals.

A Further Difficulty

There is another related issue with Marshall's proposal for circumventing the problem of where the novel biological information came from. The experimental evidence of developmental biology contradicts his claim that the evolutionary process could have somehow rewired developmental gene regulatory networks. Marshall asserts that dGRNs, which are necessary for the development of animals, could have been more labile or flexible in the past.[83] He makes this claim to challenge my contention in Chapter 13 that the observed inflexibility of these regulatory networks represents a major impediment to the evolutionary transformation of one animal body plan into another (see pp. 264–70). Not so, he argues: "Today's GRNs have been overlain with half a billion years of evolutionary innovation (which accounts for their resistance to modification), whereas GRNs at the time of the emergence of the phyla were not so encumbered."[84]

Yet, contrary to Marshall's speculation, all available observational evidence establishes that dGRNs do not tolerate random perturbations to their basic control logic. Mutagenesis experiments conducted on the genes present in dGRNs have repeatedly shown that even modest mutation-induced changes to these genes either produce no change in the developmental trajectory of animals (due to preprogrammed buffering or redundancy) or produce catastrophic (most often, lethal) effects within developing animals. Disrupt the central control nodes and the developing animal does not shift to a different, viable, stably heritable body plan. Instead, the system crashes, and the developing animal usually dies.[85] As developmental biologist Eric Davidson has noted,

> *There is always an observable consequence if a dGRN subcircuit is interrupted. Since these consequences are always catastrophically bad, flexibility is minimal, and since the subcircuits are all interconnected, the whole network partakes of the quality that there is only one way for things to work. And indeed the embryos of each species develop in only one way.*[86]

To claim that dGRNs might have been more elastic in the past contradicts what developmental biologists have learned over several decades from mutagenesis studies of many biological "model systems," including *Drosophila* (fruit flies), *Caenorhabditis* (nematodes), *Strongylocentrotus* (sea urchins), *Danio* (zebrafish), and other animals.[87] As Marshall himself has noted elsewhere, there is a good reason for this inflexibility. He explains that "many of the characters that evolved during the origin of the phyla are no longer able to change. The reason for this is that selectable variation is absent: either *the characters are invariant* or *mutants that carry this variation are sterile or lethal*."[88]

Of course, Marshall thinks that these networks could have had a more flexible character in the past. Yet given what dGRNs do—enabling cell types to organize themselves and differentiate themselves from each other in precise ways at precise times during animal development—it is hard to see how dGRNs could have functioned as regulatory networks and also exhibited the kind of flexibility Marshall envisions. Developmental gene regulatory networks are control systems. A labile dGRN would generate (uncontrolled) variable outputs, precisely the opposite of what a functional control system does. It is telling that although many evolutionary theorists (like Marshall) have speculated about early labile dGRNs, no one has ever described such a network in any functional detail—and for good reason. No developing animal that biologists have observed exhibits the kind of labile developmental gene regulatory network required by the evolution of new body plans. Indeed, when discussing hypothetical labile dGRNs, Eric Davidson acknowledges that we are speculating "where no modern dGRN provides a model" since they "must have differed in fundamental respects from those now being unraveled in our laboratories."[89]

By ignoring this evidence, Marshall and other defenders of evolutionary theory reverse the epistemological priority of the historical scientific method as pioneered by Charles Lyell, Charles Darwin, and others.[90] Rather than treating our present experimentally based knowledge as the key to evaluating the plausibility of theories about the past, Marshall uses an evolutionary assumption about what must have happened in the past (transmutation) to justify disregarding experimental observations of what does, and *does not,* occur in biological systems. The requirements of evolutionary doctrine thus trump our observations about how nature and living organisms actually behave. What we know best from observation takes a backseat to prior beliefs about how life must have arisen.

What we do know from experience, however, is that large increases in functionally specified information—especially information expressed in an alphabetic or digital form—are always produced by conscious and rational agents. So the best explanation for the explosion of information necessary to produce the Cambrian animals (whether that explosion occurred during or before the Cambrian period) remains intelligent design.[91]

ACKNOWLEDGMENTS

Though I am not a biologist, but a philosopher of biology, I have the good fortune of overseeing an interdisciplinary scientific research effort that gives me a bird's-eye view on some cutting-edge discoveries and insights coming from some exceptional scientists. With that in mind, I would like to thank my colleagues at the Discovery Institute and Biologic Institute, in particular Paul Nelson, Douglas Axe, Jonathan Wells, Michael Behe, Ann Gauger, Richard Sternberg, Paul Chien, and Casey Luskin—whose research has made the argument of this book possible. I would especially like to acknowledge Paul Nelson for his assistance writing Chapters 6 and 13, expanded versions of which we plan to publish jointly as technical articles. In addition, Casey Luskin, Discovery research coordinator, has repeatedly gone above and beyond the call of duty in his commitment to, and skillful work on, this book. I'd also like to acknowledge the two anonymous biologists and two paleontologists who gave such careful attention to improving the scientific rigor and accuracy of the manuscript during the peer-review process. I would also like to express my thanks to Paul Chien, Marcus Ross, and Paul Nelson for the research they did in support of our 2003 article "The Cambrian Explosion: Biology's Big Bang," which provided a scaffolding for the argument developed here in a much expanded form.

In addition, I deeply appreciate the work of our Discovery Institute writers and editors—Jonathan Witt, David Klinghoffer, Bruce Chapman, and Elaine Meyer—who have made this manuscript infinitely more readable. Jonathan Witt deserves special mention for helping me to launch this project and for his help developing story elements and ideas. I'd like to thank my assistant Andrew McDiarmid for his diligent work on the bibliography and managing the flow of information. And Ray Braun cannot be thanked enough for his beautiful artwork. Finally, I'd like to express my gratitude to the good people at Harper: Lisa Zuniga for her exceptional coordination and skill in moving the book through the production process; Ann Moru for her expert copyediting; and to my senior editor Roger Freet for his vision, patience, and uncommon strategic guidance.

NOTES

PROLOGUE

1. Quastler, *The Emergence of Biological Organization,* 16.

2. In fairness I should mention that a few of my critics did attempt to refute the book's actual arguments about the origin of life, and my colleagues and I addressed those in several essays collected into a book called *Signature of Controversy,* ed. David Klinghoffer (Seattle: Discovery Institute Press, 2010; http://www.discovery.org/f/6861).

3. See Francisco Ayala, "On Reading the Cell's Signature," Biologos.org, January 7, 2010, http://biologos.org/blog/on-reading-the-cells-signature.

4. Venema, "Seeking a Signature," 278.

5. Dobzhansky, "Discussion of G. Schramm's Paper," 310.

6. De Duve, *Blueprint for a Cell,* 187.

7. Gould, "Is a New and General Theory of Evolution Emerging?" 120.

8. For examples, see such books as Kauffman, *The Origins of Order;* Goodwin, *How the Leopard Changed Its Spots;* Eldredge, *Reinventing Darwin;* Raff, *The Shape of Life;* Müller and Newman, *On the Origin of Organismal Form;* Valentine, *On the Origin of Phyla;* Arthur, *The Origin of Animal Body Plans;* and Shapiro, *Evolution,* to name but a few.

9. Futuyma asserts, "There is absolutely no disagreement among professional biologists on the fact that evolution has occurred. . . . But the theory of how evolution occurs is quite another matter, and is the subject of intense dispute" ("Evolution as Fact and Theory," 8). Of course, to admit that natural selection cannot explain the appearance of design is in effect to admit that it has failed to perform the role that is claimed for it as a "designer substitute."

10. See Scott Gilbert, Stuart Newman, and Graham Budd, as quoted in Whitfield, "Biological Theory"; Mazur, *The Altenberg 16.*

11. See, e.g., Kauffman, 361; Raff, *The Shape of Life;* Miklos, "Emergence of Organizational Complexities During Metazoan Evolution."

12. Gilbert et al., "Resynthesizing Evolutionary and Developmental Biology."

13. Webster, *How the Leopard Changed Its Spots,* 33; Webster and Goodwin, *Form and Transformation,* x; Gunter Theissen, "The Proper Place of Hopeful Monsters in Evolutionary Biology," 351; Marc Kirschner and John Gerhart, *The Plausibility of Life,* 13; Schwartz, *Sudden Origins,* 3, 299–300; Erwin, "Macroevolution Is More Than Repeated Rounds of Microevolution"; Davidson, "Evolutionary Bioscience as Regulatory Systems Biology," 35; Koonin, "The Origin at 150," 473–75; Conway Morris, "Walcott, the Burgess Shale, and Rumours of a Post-Darwinian World," R928-R930; Carroll, "Towards a New Evolutionary Synthesis," 27; Wagner, "What Is the Promise of Developmental Evolution?"; Wagner and Stadler, "Quasi-independence, Homology and the Unity of Type"; Becker and Lönnig, "Transposons: Eukaryotic," 529–39; Lönnig and Saedler, "Chromosomal Rearrangements and Transposable Elements," 402; Muller and Newman, "Origination of Organismal Form," 7; Kauffman, *At Home in the Universe,* 8; Valentine and Erwin, "Interpreting Great Developmental Experiments," 96; Sermonti, *Why Is a Fly Not a Horse?*; Lynch, *The Origins of Genome Architecture,* 369; Shapiro, Evolution, 89, 128. David J. Depew and Bruce H. Weber, writing in the journal *Biological Theory,* are even more frank, "Darwinism in its current scientific incarnation has pretty much reached the end of its rope" (89–102).

14. "Statement on Teaching Evolution by the Board of Directors of the American Association for the Advancement of Science," St. Louis, Missouri, February 16, 2006; www.aaas.org/news/releases/2006/pdf/0219boardstatement.pdf (accessed October 26, 2012).

15. Dean, "Scientists Feel Miscast in Film on Life's Origin."

16. Eugenie Scott, quoted in Stutz, "State Board of Education Debates Evolution Curriculum"; also requoted in Stoddard, "Evolution Gets Added Boost in Texas Schools."

CHAPTER 1: DARWIN'S NEMESIS

1. Darwin, *On the Origin of Species,* 484. In other places in *The Origin*, Darwin hedged his bets, referring to life "having been originally breathed into a few forms or into one." Darwin, *The Origin of Species*, 490.

2. Though Darwin emphasized natural selection as the "chief agent of change," he also emphasized "sexual selection"—the preference that sexually reproducing animals have for some traits over others in prospective mates—as a mechanism responsible for some changes in evolving populations.

3. Darwin, *On the Origin of Species,* 30.

4. Darwin, *On the Origin of Species,* 21.

5. Darwin, *On the Origin of Species,* 306–7.

6. See the University of California at Berkeley Museum of Paleontology, "Brachiopoda," www.ucmp.berkeley.edu/brachiopoda/brachiopodamm.html (accessed October 23, 2012).

7. See www.geo.ucalgary.ca/~macrae/trilobite/siluria.html (accessed October 23, 2012), where Roderick Murchison's skillful trilobite drawings are reproduced.

8. Agassiz, "Evolution and the Permanence of Type," 99.

9. Ward, *On Methuselah's Trail,* 29.

10. Darwin, *On the Origin of Species,* 307. Darwin's original quotation used the "Silurian" rather than the "Cambrian," because in Darwin's time what we now label as the Cambrian period was subsumed within the concept of the lower Silurian. In a later sixth edition of the *Origin,* Darwin adopted the term "Cambrian" in place of "Silurian." See Darwin, *On the Origin of Species;* sixth edition, 286.

11. Agassiz, *Essay on Classification,* 102.

12. Agassiz, "Evolution and the Permanence of Type," 10.

13. Murchison, *Siluria,* 469.

14. Letter from Adam Sedgwick to Charles Darwin, November 24, 1859.

15. Between approximately 450 and 440 million years ago, many animal species went extinct. Known as the Ordovician extinction, this event resulted in the disappearance of a huge number of marine invertebrates. This was the second biggest extinction in the history of life, being superseded only by the great Permian extinction (about 252 million years ago). Dott and Prothero, *Evolution of the Earth,* 259.

16. Mintz, *Historical Geology,* 146, 153–54, 124–27.

17. Prothero, *Bringing Fossils to Life: An Introduction to Paleobiology,* 84–85. Dott and Prothero, *Evolution of the Earth,* 376–79.

18. Dott and Prothero, *Evolution of the Earth,* 425–26; Li et al., "An Ancestral Turtle from the Late Triassic of Southwestern China"; Gaffney, "The Comparative Osteology of the Triassic Turtle *Proganochelys.*"

19. William Smith, 1815 Geological Map of England and Wales.

20. Radiometric dating methods estimate the age of rocks based upon measurements of the ratios of unstable radioactive isotopes and their daughter products given known radioactive decay rates.

21. Gould, *Wonderful Life,* 54.

22. Darwin, *On the Origin of Species,* 302.

23. Darwin, *On the Origin of Species,* 311.

24. Gould, *Wonderful Life,* 57.

25. Agassiz, "Evolution and the Permanence of Type," 97.

26. Robinson, *Runner on the Mountain Tops,* 215.

27. Lurie, *Louis Agassiz,* vii.

28. Robinson, *Runner on the Mountain Tops,* foreword.

29. Quoted in Holder, *Louis Agassiz,* 180.

30. Dupree, *Asa Gray,* 227.

31. Dupree, *Asa Gray,* 226. Though the German idealists believed that living forms reflected divine archetypes, they did not all oppose the idea of the transmutation of species. In Schelling's view, for example, each species reflected a pre-existing archetype, but emerged in time through a gradual transition of form.

32. Dupree, *Asa Gray,* 260.

33. Lurie, *Louis Agassiz,* 284.

34. Lurie, *Louis Agassiz,* 88.

35. Lurie, *Louis Agassiz,* 282.

36. Gillespie, *Charles Darwin and the Problem of Creation.*

37. Agassiz, "Evolution and the Permanence of Type," 101.

38. Robinson, *Runner on the Mountain Tops,* 231–34.

39. Quoted in Lurie, *Nature and the American Mind,* 41.

40. Quoted in Lurie, *Nature and the American Mind,* 42.

41. Thanks go to my former student Jack Ross Harris and his probing, unpublished 1993 essay "Louis Agassiz: A Re-evaluation of the Nature of His Opposition to the Darwinian View of Natural History." His work on the subject first brought to my attention the way in which historians

writing after the widespread acceptance of Darwinism had attempted to minimize the scientific basis of Agassiz's objections to his theory.

42. Lurie, *Louis Agassiz*, 244.

43. Lurie, *Louis Agassiz*, 246.

44. Quoted in Lurie, *Louis Agassiz*, 373.

45. Gillespie, *Charles Darwin and the Problem of Creation*, 51.

46. Agassiz, "Evolution and the Permanence of Type," 92–101.

47. Gray, *Darwiniana*, 127.

48. See the parallel lines on the sides of the hills on either side of the valley in this beautiful picture: www.swisseduc.ch/glaciers/earth_icy_planet/icons-15/16.jpg (last accessed Oct. 23rd, 2012). www.uh.edu/engines/epi857.htm (last accessed Oct. 23rd, 2012). See also Tyndall, "The Parallel Roads of Glen Roy."

49. Darwin, *Autobiography*, 84. Also, Gertrude Himmelfarb notes that Darwin took more than twenty years to concede his mistake. In his autobiography, Darwin labeled it "one long gigantic blunder from beginning to end. . . . Because no other explanation was possible under our then state of knowledge, I argued in favour of sea-action; and my error has been a good lesson to me never to trust in science to the principle of exclusion." See Himmelfarb's discussion in *Darwin and the Darwinian Revolution*, 107.

50. See Oosthoek, "The Parallel Roads of Glen Roy and Forestry."

51. Darwin, *On the Origin of Species*, 307.

52. Darwin, *On the Origin of Species*, 308.

53. Agassiz, "Evolution and the Permanence of Type," 97. That Agassiz's objection was first and foremost empirically driven can be seen from his comments elsewhere as well. For instance, in "Researches on the Fossil Fishes," he writes, "More than fifteen hundred species of fossil fishes, which I have learned to know, tell me that species do not pass insensibly one into another, but that they appear and disappear unexpectedly, without direct relations with their precursors; for I think no one will seriously pretend that the numerous types of Cycloids and Ctenoids, almost all of which are contemporaneous with one another, have descended from the Placoids and Ganoids. As well might one affirm that the Mammalia, and man with them, have descended directly from fishes." Quoted in Agassiz, *Louis Agassiz and Correspondence*, 244–45.

CHAPTER 2: THE BURGESS BESTIARY

1. Gould, *Wonderful Life*, 71. A slightly different version of this story is found in Charles Schuchert, "Charles Doolittle Walcott," 124. See also Schuchert, "Charles Doolittle Walcott Paleontologist—1850–1927," 455–58.

2. Gould, *Wonderful Life*, 71.

3. Gould, *Wonderful Life*, 71–75.

4. A distant relationship between *Marrella* and chelicerates is currently the favored hypothesis. See García-Bellido, D. C., and Collins, D. H., "A New Study of Marrella Splendens (Arthropoda, Marrelomorpha) from the Middle Cambrian Burgess Shale, British Columbia, Canada," 721–42; Hou, X. G., and Bergström, J., "Arthropods of the Lower Cambrian Chengjiang Fauna, Southwest China," 109; Bergström, J., and Hou, X. G., "Arthropod Origins," 323–34.

5. Other authorities, such as Douglas Erwin at the Smithsonian, arrive at a slightly higher total. By counting groups that some paleontologists count as subphyla or classes as phyla, Erwin argues that about twenty-five phyla first appear in the Cambrian out of about thirty-three total (by his way of counting) phyla known from the fossil record. [See Erwin et al., "The Cambrian Conundrum: Early Divergence and Later Ecological Success in the Early History of Animals," 1091–97.] I've used a slightly more conservative method for estimating the total number of phyla that first appear in the Cambrian (see Fig. 2.5). I came to my count consulting the following sources:

Phylum references listed in same order as they appear on the chart.

Cnidaria: Chen et al., "Precambrian Animal Life: Probable Developmental and Adult Cnidarian Forms from Southwest China."

Mollusca: Fedonkin and Waggoner. "The Late Precambrian Fossil *Kimberella* Is a Mollusc-Like Bilaterian Organism."

Porifera: Love, G. D. "Fossil steroids record the appearance of Demospongiae during the Cryogenian period."

Annelida: Conway Morris and Peel, "The Earliest Annelids: Lower Cambrian Polychaetes from the Sirius Passet Lagerstätte, Peary Land, North Greenland."

Brachiopoda: Skovsted and Holmer, "Early Cambrian Brachiopods from North-East Greenland."

Bryozoa: Landing et al., "Cambrian Origin of All Skeletalized Metazoan Phyla—Discovery of Earth's Oldest Bryozoans (Upper Cambrian, southern Mexico)."

Chaetognatha: Szaniawski, H. "Cambrian Chaetognaths Recognized in Burgess Shale Fossils."
Chordata: Chen et al., "A Possible Early Cambrian Chordate"; Chen, "Early Crest Animals and the Insight They Provide into the Evolutionary Origin of Craniates"; Janvier, "Catching the First Fish"; Monnereau, "An Early Cambrian Craniate-Like Chordate"; Conway Morris and Caron, "*Pikaia gracilens* Walcott, a Stem-Group Chordate from the Middle Cambrian of British Columbia"; Sansom et al., "Non-Random Decay of Chordate Characters Causes Bias in Fossil Interpretation"; Shu et al., "An Early Cambrian Tunicate from China"; Shu et al. "Lower Cambrian Vertebrates from South China."
Coeloscleritophora: Bengtson and Hou, "The Integument of Cambrian Chancelloriids."
Ctenophora: Chen, J. Y. et al. "Raman Spectra of a Lower Cambrian Ctenophore Embryo from Southwestern Shaanxi, China"; Conway Morris and Collins, "Middle Cambrian Ctenophores from the Stephen Formation, British Columbia, Canada."
Echinodermata: Foote, "Paleozoic Record of Morphological Diversity in Blastozoan Echinoderms"; Shu et al., "Ancestral Echinoderms from the Chengjiang Deposits of China"; Zamora et al., "Middle Cambrian Gogiid Echinoderms from Northeast Spain: Taxonomy, Palaeoecology, and Palaeogeographic Implications."
Entoprocta: Zhang et al., "A Sclerite-Bearing Stem Group Entoproct from the Early Cambrian and Its Implications."
Euarthropoda: Cisne, J. L., "Trilobites and the Origin of Arthropods"; Daley, "The Morphology and Evolutionary Significance of the Anomalocaridids"; Grosberg, "Out on a Limb: Arthropod Origins"; Siveter, "A Phosphatocopid Crustacean with Appendages from the Lower Cambrian."
Hemichordata: Shu et al., "Reinterpretation of *Yunnanozoon* as the Earliest Known Hemichordate"; Shu et al., "A New Species of Yunnanozoan with Implications for Deuterostome Evolution."
Hyolitha: Malinky and Skovsted, "Hyoliths and Small Shelly Fossils from the Lower Cambrian of North-East Greenland"; note that some authors consider Hyolitha to belong to phylum Mollusca, whereas others consider Hyolitha to represent an independent phylum.
Lobopodia: Liu et al., "A Large Xenusiid Lobopod with Complex Appendages from the Lower Cambrian Chengjiang Lagerstätte"; Liu et al., "Origin, Diversification, and Relationships of Cambrian Lobopods"; Liu et al., "An Armoured Cambrian Lobopodian from China with Arthropod-Like Appendages"; Ou et al., "A Rare Onychophoran-Like Lobopodian from the Lower Cambrian Chengjiang Lagerstätte, Southwestern China, and Its Phylogenetic Implications."
Loricifera: Peel, "A Corset-Like Fossil from the Cambrian Sirius Passet Lagerstatte of North Greenland and Its Implications for Cycloneuralian Evolution."
Nematomorpha: Xian-guang and Wen-guo, "Discovery of Chengjiang Fauna at Meishucun, Jinning, Yunnan."
Phoronida: Erwin et al., "The Cambrian Conundrum: Early Divergence and Later Ecological Success in the Early History of Animals."
Priapulida: Wills et al., "The Disparity of Priapulid, Archaeopriapulid and Palaeoscolecid Worms in the Light of New Data"; Hu et al., "A New Priapulid Assemblage from the Early Cambrian Guanshan Fossil *Lagerstätte* of SW China."
Sipuncula: Huang et al., "Early Cambrian Sipunculan Worms from Southwest China." Some consider sipunculan worms to be a subgroup of the phylum Annelida based on phylogenomic analyses. See Struck et al., "Phylogenomic Analyses Unravel Annelid Evolution."
Tardigrada: Muller et al., "'Orsten' Type Phosphatized Soft-Integument Preservation and a New Record from the Middle Cambrian Kuonamka Formation in Siberia."
Vetulicolia: Shu, "On the Phylum Vetulicolia."
Nematoda: Erwin et al., "The Cambrian Conundrum: Early Divergence and Later Ecological Success in the Early History of Animals."
Nemertea(?): Schram, "Pseudocoelomates and a Nemertine from the Illinois Pennsylvanian." Note that the presence of Nemertea in the fossil record is contested.
Platyhelminthes: Poinar, "A Rhabdocoel Turbellarian (Platyhelminthes, Typhloplanoida) in Baltic Amber with a Review of Fossil and Sub-Fossil Platyhelminths."
Rotifera: Swadling et al., "Fossil Rotifers and the Early Colonization of an Antarctic Lake."

6. Erwin and Valentine, *The Cambrian Explosion,* 66–70.
7. Hennig, *Phylogenetic Classification.*
8. If anything, using a "rank-free" classification system may actually intensify the mystery of the Cambrian explosion. A single phylum may include many unique modes of organizing tissues, organs, and body parts, and these differences in organization may deserve to be recognized as different body plans as much as the differences that distinguish different phyla. As one proponent

of the rank-free approach put it to me, "Why shouldn't clams and squids [both of which belong to the single phylum Mollusca] be recognized as exemplifying unique body architectures every bit as much as trilobites and star fish [which belong to two different phyla, the arthropods and the echinoderms]?" In the traditional system, however, both clams and squids, and many other animals that exemplify equally pronounced differences in form within other phyla, will all fall within their respective individual phyla. For this reason, measuring the explosiveness of the Cambrian radiation solely by reference to the number of phyla that first appear in the Cambrian may actually minimize the severity of the problem, whereas dispensing with taxonomic ranking may actually tend to accentuate it.

9. In addition, in 1999 paleontologists in southern China also found fossil remains of fish in the Cambrian period. Fish are vertebrates and members of the phylum chordata. Shu et al., "Lower Cambrian Vertebrates from South China," 42–46; Shu et al., "Head and Backbone of the Early Cambrian Vertebrate *Haikouichthys*."

10. Quoted in Yochelson, *Charles Doolittle Walcott, Paleontologist*, 33.

11. Gould, *Wonderful Life*, 49.

12. Gould, *Wonderful Life*, 125–36; Budd, "The Morphology of *Opabinia Regalis* and the Reconstruction of the Arthropod Stem-Group," 1–14.

13. Darwin, *On the Origin of Species*, 307.

14. Ward, *On Methuselah's Trail*, 29–30.

15. Darwin's prediction is, of course, not a prediction in the narrow sense of forecasting a future event or process. But historical scientists regularly speak of predictions in the sense of expectations about what will be revealed about the past if and when the relevant body of evidence is uncovered.

16. Great Canadian Parks, Yoho National Park, www.greatcanadianparks.com/bcolumbia/yohonpk/page3.htm (accessed October 23, 2012).

17. Dawkins, *Unweaving the Rainbow*, 201.

18. Darwin, *On the Origin of Species*, 120.

19. Darwin, *On the Origin of Species*, 125.

20. Lewin, "A Lopsided Look at Evolution," 292.

21. Erwin, Valentine, and Sepkoski, "A Comparative Study of Diversification Events," 1183. See also Erwin et al., "The Cambrian Conundrum: Early Divergence and Later Ecological Success in the Early History of Animals," 1091–97; Bowring et al., "Calibrating Rates of Early Cambrian Evolution," 1293–98.

22. Hennig, *Phylogenetic classification*, 219. Erwin and Valentine in their 2013 book, *The Cambrian Explosion*, reaffirm the remarkable morphological disparity present from the beginning of the Cambrian period despite low species diversity. They note that since they first brought attention to the top-down pattern of Cambrian disparity preceding diversity using classical Linnaean categories in 1987, paleontologists have developed measures of disparity within the phylogenetic classification system—measures that reaffirm the same pattern within a rank-free classification system (217).

23. For a more technical description of the processes behind the Burgess Shale fossil formation, see Briggs, Erwin, and Collier, *The Fossils of the Burgess Shale*, 21–32; Conway Morris, *The Crucible of Creation*, 106–107.

24. Gould, *Wonderful Life*, 274–75.

25. Walcott, "Cambrian Geology and Paleontology II," 15.

26. Gould, *Wonderful Life*, 108.

27. Gould, *Wonderful Life*, 273.

CHAPTER 3: SOFT BODIES AND HARD FACTS

1. Nash, "When Life Exploded," 66–74.

2. Today fossils from the Burgess can be viewed in Canada at the Royal Ontario Museum in Toronto, the Tyrell Museum in Drumheller, Alberta, and at a smaller exhibition close to the Burgess Shale in Golden, British Columbia.

3. Briggs, Erwin, and Collier, *Fossils of the Burgess Shale*.

4. Desmond Collins, *Misadventures in the Burgess Shale*, 952–53.

5. Some paleontologists have also gone so far as to argue that the Cambrian explosion is nothing more than an artifact of classification, and therefore does not require explanation. Budd and Jensen, for example, argue that the problem of the Cambrian explosion resolves itself if one keeps in mind the cladistic distinction between "stem" and "crown" groups. Since crown groups arise whenever new characters are added to simpler, more ancestral stem groups during the evolutionary process, new phyla will inevitably arise once a stem group has arisen. Thus, for Budd and

Jensen, what requires explanation is not the crown groups corresponding to the new phyla, but the earlier, less derived, stem groups that presumably arose deep in the Precambrian. Yet since these earlier stem groups are by definition less derived, explaining them will be, in their view, considerably easier than explaining the origin of the Cambrian animals *de novo*. In any case, for Budd and Jensen, the explosion of new phyla in the Cambrian does not require explanation. As they put it, "given that the early branching points of major clades is an inevitable result of clade diversification, the alleged phenomenon of phyla appearing early and remaining morphologically static does not seem to require particular explanation." [Budd and Jensen, "A Critical Reappraisal of the Fossil Record of the Bilaterian Phyla," 253.] Nevertheless, Budd and Jensen's attempt to explain away the Cambrian explosion begs crucial questions. Granted, as new characters are added to existing forms, novel morphology and greater morphological disparity will likely result. But what causes new characters to arise? And how does the biological information necessary to produce new characters originate? (See Chapters 9–16.) Budd and Jensen do not specify. Nor can they say how derived the ancestral forms are likely to have been, and what processes might have been sufficient to produce them. Instead they simply assume the sufficiency of some unspecified evolutionary mechanisms. Yet, as I show in Chapters 10–16, this assumption is now problematic. In any case, Budd and Jensen do not explain what *causes* the origin of biological form and information in the Cambrian.

6. See Cloud, "The Ship That Digs Holes in the Sea," 108. See also a history of offshore drilling on the website of the National Ocean Industries Association, http://www.noia.org/website/article.asp?id=123 (accessed July 8, 2011).

7. Schuchert and Dunbar, *Textbook of Geology, Part II, Historical Geology,* 72–76, 125–30; Stokes, *Essentials of Earth History: An Introduction to Historical Geology,* 162–64; Zumberge, *Elements of Geology,* 62–67, 214–15; Dunbar, *Historical Geology,* 13–15, 129–33.

8. Müller et al., "Digital Isochrons of the World's Ocean Floor," 3212.

9. Walcott, "Cambrian Geology and Paleontology II," 2–4.

10. Gould, *Wonderful Life,* 275.

11. Some have even suggested that the transitional intermediate forms leading to the Cambrian animals only existed in the larval stage. See Davidson, Peterson, and Cameron, "Origin of Bilaterian Body Plans," 1319.

12. Wray, Levinton, and Shapiro, "Molecular Evidence for Deep Pre-Cambrian Divergences Among Metazoan Phyla." For other recent expressions of this version of the artifact hypothesis, see Simpson, *Fossils and the History of Life,* 72–74; Ward, *Out of Thin Air,* 5; Eldredge, *The Triumph of Evolution and the Failure of Creationism,* 46; and Schirber, "Skeletons in the Pre-Cambrian Closet." Though contemporary paleontologists commonly attribute the absence of Precambrian ancestral forms to their alleged lack of hard parts or appreciable size, earlier geologists and paleontologists have also employed this version of the artifact hypothesis. For example, in 1941 Charles Schuchert and Carl Dunbar stated: "We may infer, therefore, that life probably was abundant in the seas of Cryptozoic time and especially during the Proterozoic, but was of a low order and doubtless small and soft-tissued, so that there was little chance for actual preservation of fossils" (*Textbook of Geology, Part II,* 124). And as early as 1894 W. K. Brooks asserted: "the zoological features of the Lower Cambrian are of such a character as to indicate that it is a decided and unmistakable approximation to the primitive fauna of the bottom, beyond which life was represented only by minute and simple surface animals not likely to be preserved as fossils" ("The Origin of the Oldest Fossils and the Discovery of the Bottom of the Ocean," 360–61).

13. Marshall, "Explaining the Cambrian 'Explosion' of Animals," 357, 372. For an authoritative refutation of this version of the artifact hypothesis, see Conway Morris, *The Crucible of Creation,* 140–44; Conway Morris, "Darwin's Dilemma: The Realities of the Cambrian 'Explosion'," 1069–83.

14. Schopf and Packer, "Early Archean (3.3-Billion to 3.5-Billion-Year-Old) Microfossils from Warrawoona Group, Australia," 70; Schopf, "Microfossils of the Early Archean Apex Chert."

15. Schopf and Packer, "Early Archean (3.3-Billion- to 3.5-Billion-Year-Old) Microfossils from Warrawoona Group, Australia," 70; Hoffmann et al., "Origin of 3.45 Ga Coniform Stromatolites in Warrawoona Group, Western Australia."

16. Jan Bergström states: "Animals such as arthropods and brachiopods cannot exist without hard parts. The absence of remains of skeletons and shells in the Precambrian therefore proves that the phyla came into being with the Cambrian, not before, even if the lineages leading to the phyla were separate before the Cambrian" ("Ideas on Early Animal Evolution," 464).

17. Valentine and Erwin, "Interpreting Great Developmental Experiments."

18. Valentine, "Fossil Record of the Origin of *Bauplan* and Its Implications," especially 215.

19. Chen and Zhou, "Biology of the Chengjiang Fauna," 21.

20. Chen and Zhou, "Biology of the Chengjiang Fauna," 21. Or as Valentine explains, "the interpretation of the explosion as an artifact of the evolution of durable skeletons has got it backward: the skeletons are artifacts, more or less literally, of the evolutionary explosion." Valentine, *On the Origin of Phyla,* 181.

21. Ivantsov, "A New Reconstruction of *Kimberella,* a Problematic Metazoan," 3.

22. Edgecombe, "Arthropod Structure and Development," 74–75.

23. Frederick Schram, *The Crustacea.*

24. Simpson, *Fossils and the History of Life,* 73. Indeed, an exoskeleton is far more than a mere covering for the soft parts of, say, a chelicerate or crustacean, because it provides the sites for the attachment of the muscles and various other tissues. Further, the limbs (including the mouthparts and in some instances certain reproductive components) are encased in exoskeletal elements that can articulate, allowing the arthropod to move, feed, and mate. An exterior skeleton of any shrimp, for example, also has interior projections that comprise its endophragmal system, which provides support for the animal's internal musculature and organs. At the same time, the skeleton of any arthropod is a product of, and in turn regulates, its metabolism and physiology. In order for the first members of *Fuxianhuia* or *Marrella* to have grown (and possibly metamorphosed during their development), they would have had to have successively secreted a new skeleton beneath the old one; to have shed the used exoskeletons; and to have hardened each new exoskeleton. This tight functional integration suggests the implausibility of evolutionary models that envision the Arthropod exoskeleton arising late as a kind of accretion to an already integrated system of soft parts.

25. Brocks et al., "Archean Molecular Fossils and the Early Rise of Eukaryotes."

26. Conway Morris, *The Crucible of Creation,* 47–48; Gould, "The Disparity of the Burgess Shale Arthropod Fauna and the Limits of Cladistic Analysis."

27. *Wiwaxia* is considered soft-bodied, but it does have harder scales and spines. See Conway Morris, *The Crucible of Creation,* 97–98.

28. Valentine, "The Macroevolution of Phyla," sec. 3.2, "Soft-Bodied Body Fossils," 529–31.

29. Conway Morris, *The Crucible of Creation,* 82.

30. Conway Morris, *The Crucible of Creation,* 76, 99.

31. Conway Morris, *The Crucible of Creation,* 68, 73, 74; "Burgess Shale Faunas and the Cambrian Explosion."

32. Conway Morris, *The Crucible of Creation,* 107.

33. Conway Morris, *The Crucible of Creation,* 92, 184.

34. Conway Morris, "Burgess Shale Faunas and the Cambrian Explosion."

35. Conway Morris, *The Crucible of Creation,* 103; Conway Morris, "Burgess Shale Faunas and the Cambrian Explosion."

36. A recent scientific paper reinterprets *Nectocaris* as a cephalopod mollusk, though it also acknowledges the problems long associated with the definitive classification of this animal. See Smith and Caron, "Primitive Soft-Bodied Cephalopods from the Cambrian."

37. Conway Morris, *The Crucible of Creation,* 140.

38. Hou et al., *The Cambrian Fossils of Chengjiang, China,* 10.

39. Hou et al., *The Cambrian Fossils of Chengjiang, China,* 10, 12.

40. Hou et al., *The Cambrian Fossils of Chengjiang, China,* 13.

41. Hou et al., *The Cambrian Fossils of Chengjiang, China,* 10, 12.

42. Hou et al., *The Cambrian Fossils of Chengjiang, China,* 23.

43. Bergström and Hou, "Chengjiang Arthropods and Their Bearing on Early Arthropod Evolution," 152.

44. Burgess Shale fossils from the middle Cambrian (515 million years ago) confirm that many of these fully soft-bodied Cambrian organisms were long-lived and geographically widespread.

45. Chen et al., "Weng'an Biota"; Chien et al., "SEM Observation of Precambrian Sponge Embryos from Southern China."

46. Chen et al., *The Chengjiang Biota: A Unique Window of the Cambrian Explosion.* This book is currently available only in the Chinese language. The translated English version is being completed by Paul K. Chien, of the University of San Francisco.

47. Chen et al., "Weng'an Biota"; Chien et al., "SEM Observation of Precambrian Sponge Embryos from Southern China."

48. For other alternative interpretations, see Huldtgren et al., "Fossilized Nuclei and Germination Structures Identify Ediacaran 'Animal Embryos' as Encysting Protists," 1696–99; Xiao et al., "Comment on 'Fossilized Nuclei and Germination Structures Identify Ediacaran 'Animal Embryos' as Encysting Protists'," 1169; Huldtgren et al., "Response to Comment on 'Fossilized Nuclei and Germination Structures Identify Ediacaran 'Animal Embryos' as Encysting Protists'," 1169.

49. Erwin and Valentine, *The Cambrian Explosion,* 778.

50. Chien et al., "SEM Observation of Precambrian Sponge Embryos from Southern China."

Sponges are assumed by most evolutionary biologists to represent a side branch, not a node on evolutionary tree of life leading to the Cambrian phyla. Thus, sponges are not regarded as plausible transitional intermediates between Precambrian and Cambrian forms (nor are they regarded as ancestral to other Cambrian animals).

51. Some have challenged the interpretation of these Precambrian microfossils as embryos, arguing that they are, instead, large micro-organisms. For example, Therese Huldtgren and colleagues have argued that these fossils "have features incompatible with multicellular metazoan embryos" and that "the developmental pattern is [more] comparable with nonmetazoan holozoans," a group that includes one-celled protozoans. [Huldtgren et al., "Fossilized Nuclei and Germination Structures Identify Ediacaran 'Animal Embryos' as Encysting Protists," 1696–99.]

Critics of Huldtgren's proposal instead think they may well be tiny metazoan embryos, though of unknown affiliation. [Xiao et al., "Comment on 'Fossilized Nuclei and Germination Structures Identify Ediacaran 'Animal Embryos' as Encysting Protists'," 1169. Huldtgren and colleagues have defended their interpretation here: Huldtgren et al., "Response to Comment on 'Fossilized Nuclei and Germination Structures Identify Ediacaran "Animal Embryos" as Encysting Protists'," 1169.]

Another interpretation is that the fossils represent giant sulphur bacteria, since "sulphur bacteria of the genus *Thiomargarita* have sizes and morphologies similar to those of many Doushantuo microfossils, including symmetrical cell clusters that result from multiple stages of reductive division in three planes." [Bailey et al., "Evidence of giant sulphur bacteria in Neoproterozoic phosphorites," 198–201.] Critics of this hypothesis doubt that sulphur bacteria could be fossilized because they "collapse easily and have only patchy biofilms that are limited to the multi-layered envelope." [Cunningham et al., "Experimental taphonomy of giant sulphur bacteria: implications for the interpretation of the embryo-like Ediacaran Doushantuo fossils," 1857–64.]

The debate over whether the Doushantuo microfossils should be interpreted as metazoan embryos, protozoans, or giant sulphur bacteria will doubtless continue. Whatever the outcome, however, the fact remains: small, fragile, and soft-bodied organisms of some kind have been found fossilized in this Precambrian strata, raising the question of why the same layers of rock were unable to preserve the immediate precursors to the numerous metazoan phyla that emerge so abruptly in the Cambrian layers above them.

52. Similarly, paleontologists rarely find the remains of parasites that live in the soft tissues of other organisms (indeed, parasitic organisms represent several of the phyla that have no fossil record). As noted, the geological record does preserve soft tissues, but only infrequently. When it does, researchers fortunate enough to make such finds will rarely want to destroy important specimens (of soft-tissue organs) in order to examine them for traces of parasitic infection or habitation. Not surprisingly, therefore, paleontologists have not found the remains of many parasitic organisms in the fossil record.

53. *The Emergence of Animals*, 91. This point is also based on personal conversation with Professor (Mark) McMenamin.

54. Erwin and Valentine, *The Cambrian Explosion*, 8.

55. Foote, "Sampling, Taxonomic Description, and Our Evolving Knowledge of Morphological Diversity," 181. Another statistical paleontologist, Michael J. Benton, and his colleagues have reached a similar conclusion. They note that "if scaled to the . . . taxonomic level of the family [and above], the past 540 million years of the fossil record provide uniformly good documentation of the life of the past" (Benton, Wills, and Hitchin, "Quality of the Fossil Record Through Time," 534). In another article Benton also writes: "It could be argued that there are fossils out there waiting to be found. It is easy to dismiss the fossil record as seriously, and unpredictably, incomplete. For example, certain groups of organisms are almost unknown as fossils. . . . This kind of argument cannot be answered conclusively. However, an argument based on effort can be made. Paleontologists have been searching for fossils for years and, remarkably, very little has changed since 1859, when Darwin proposed that the fossil record would show us the pattern of the history of life" ("Early Origins of Modern Birds and Mammals," 1046).

56. Foote, "Sampling Taxonomic Description, and Our Evolving Knowledge of Morphological Diversity," 181. I should note that there is one way in which my analogy to colored marbles in a barrel fails to capture the nature of the challenge of Cambrian fossil discontinuity. If after pulling samples from a barrel for a while you finally came up with a green and orange ball to go along with the piles of red, blue, and yellow balls, you still wouldn't have much confidence that the barrel had a rainbow of ball colors finely grading from one to another. Yet you could at least say that the orange ball stands between the yellow and red ball, and the green ball stands between the blue and yellow balls (like the hybrid produced from two plants). But many of the new Cambrian animal forms that have been discovered since Darwin's time aren't seen as intermediates between the previously known animal forms representing known phyla. They aren't evolutionary intermediates

between one existing phylum and another. Instead, scientists consider them as existing out in morphological space all their own, standing not as intermediates but as phyla that themselves are in need of intermediate forms—almost as if, by stretching my analogy, some new primary color had been discovered.

57. See Erwin et al., "The Cambrian Conundrum: Early Divergence and Later Ecological Success in the Early History of Animals," 1091–97.

58. Bowring et al., "Calibrating Rates of Early Cambrian Evolution."

59. Bowring et al., "Calibrating Rates of Early Cambrian Evolution," 1297.

60. Lili, "Traditional Theory of Evolution Challenged," 10.

61. There are a few putative survivors of the Ediacaran fauna. For example, the enigmatic and frond-like *Thaumaptilon* found in the Burgess Shale might be a descendant of Ediacaran fronds, though this is contested. Jensen et al., "Ediacara-Type Fossils Cambrian Sediment," 567–69; Conway Morris, "Ediacaran-like Fossils in Cambrian Burgess Shale-Type Faunas of North America," 593–635.

62. Bowring et al., "Calibrating Rates of Early Cambrian Evolution," 1297. See also McMenamin, *The Emergence of Animals.*

63. Erwin et al., "The Cambrian Conundrum: Early Divergence and Later Ecological Success in the Early History of Animals," 1091–97.

64. Shu et al., "Lower Cambrian Vertebrates from South China."

65. Chen et al., "A Possible Early Cambrian Chordate"; Chen and Li, "Early Cambrian Chordate from Chengjiang, China"; Dzik, "*Yunnanozoon* and the Ancestry of Chordates." Note, however, that the assertion that *Yunnanozoon* is a chordate has been challenged. See Shu, Zhang, and Chen, "Reinterpretation of *Yunnanozoon* as the Earliest Known Hemichordate"; Shu, Morris, and Zhang, "A *Pikaia*-like Chordate from the Lower Cambrian of China." Paleontologists also have found a single specimen of a possible cephalchordate, *Cathaymyrus,* from the lower Cambrian Qiongzhusi Formation near Chengjiang. The status of *Cathaymyrus* as a valid taxon has also been challenged; some paleontologists argue that the single specimen of *Cathaymyrus* may actually be a dorsoventrally compressed *Yunnanozoon;* see Chen and Li, "Early Cambrian Chordate from Chengjiang, China."

66. Chen, Huang, and Li, "An Early Cambrian Craniate-like Chordate," 518.

67. Shu et al., "Lower Cambrian Vertebrates from South China."

68. Shu et al., "An Early Cambrian Tunicate from China."

69. The chordates discovered in the Cambrian strata near Chengjiang represent just one of many new animal body plans found there, some of them designated as new phyla, others as new subphyla, classes, or families within existing phyla. For example, paleontologists classify *Occacaris oviformis,* a round, egg-shaped animal with large forcep-like structures, as a member of the well-known phylum Arthropoda. [Hou et al., *The Cambrian Fossils of Chengjiang, China,* 130.] Nevertheless, *Occacaris,* which has been found only in the Maotianshan shale, clearly exemplifies a unique way of arranging organs and tissue—one unlike any previously known arthropod outside the Chengjiang biota. Arguably, it represents a unique body plan. In other cases, Chinese paleontologists have discovered animals with such unusual morphologies that they have been unable to classify them within any known phylum. These so-called problematica, such as the mysterious *Batofasciculus ramificans*, a hot-air-balloon-shaped cactus-like animal, has been given a species and genus name, but as yet, no specific phyletic designation, though it clearly exemplifies a unique body plan. [Hou et al., *The Cambrian Fossils of Chengjiang, China,* 196.] While such difficult-to-classify organisms do not increase the official count of novel phyla that first arose during the Cambrian, they do frequently display novel body plans.

CHAPTER 4: THE *NOT* MISSING FOSSILS?

1. Gradstein, Ogg, Schmitz, and Ogg, *The Geological Time Scale 2012.*

2. Grotzinger et al., "Biostratigraphic and Geochronologic Constraints on Early Animal Evolution." A few Ediacarans may have survived until the middle Cambrian. See Conway Morris, "Ediacaran-like Fossils in Cambrian Burgess Shale-Type Faunas of North America."

3. Monastersky, "Ancient Animal Sheds False Identity."

4. Monastersky, "Ancient Animal Sheds False Identity."

5. Monastersky, "Ancient Animal Sheds False Identity."

6. Fedonkin and Waggoner, "The Late Precambrian Fossil *Kimberella* is a Mollusc-like Bilaterian Organism," 868.

7. For example, Graham Budd, a Swedish paleontologist and Cambrian expert, has expressed skepticism about this classification. He acknowledges that "the strongest case for an Ediacaran

bilaterian body fossil has been made by Fedonkin and Waggoner (1997) for *Kimberella*," but nevertheless disputes the classification of *Kimberella* as a true mollusk. He argues that "*Kimberella* does not possess any unequivocal derived molluscan features, and its assignment to the Mollusca or even the Bilateria must be considered to be unproven" (Budd and Jensen, "A Critical Reappraisal of the Fossil Record of the Bilaterian Phyla," 270).

8. Another reason the Ediacaran body fossils cannot be assigned to the animal phyla in a decisive manner is because of the coarse grain size of the beds in which they occur. Details of body form are too vague to allow a clear decision, and until better means of analysis or new beds with finer grain texture are found, these fossils will remain as intriguing "problematica," problematic forms about which it is not possible to come to a decision. See Miklos, "Emergence of Organizational Complexities During Metazoan Evolution." See also Bergström, "Metazoan Evolution Around the Precambrian–Cambrian Transition."

9. Retallack, "Growth, Decay and Burial Compaction of *Dickinsonia*," 215. As Retallack goes on to explain, "Like fungi and lichens, *Dickinsonia* was firmly attached to its substrate, ground-hugging, moderately flexible, and very resistant to burial compaction" (236).

10. Glaessner, *The Dawn of Animal Life*, 122.

11. Birket-Smith, "A Reconstruction of the Pre-Cambrian *Spriggina*," 237–58.

12. Glaessner, *The Dawn of Animal Life*, 122.

13. Glaessner, *The Dawn of Animal Life*, 122.

14. McMenamin, "*Spriggina* Is a Trilobitoid Ecdysozoan," 105.

15. Glaessner, *The Dawn of Animal Life*, 122–23. McMenamin, *The Emergence of Animals*, 20, 24, 118.

16. In addition, a recent depiction of an Ediacaran organism closely related to *Spriggina*, called *Yorgia*, by Ediacaran expert Andrey Ivantsov, shows it exhibiting a smoother, less jagged edge than the bodies of trilobites with no protruding spines. Ivantsov's reconstruction seems to underscore a shift in opinion away from regarding *Spriggina*-like organisms as basal arthropods in possession of distinctive arthropod characteristics (such as genal spines). Ivantsov, "Giant Traces of Vendian Animals." See also Ivantsov, "Vendia and Other Precambrian 'Arthropods.'"

17. Brasier and Antcliffe, "*Dickinsonia* from Ediacara," 312.

18. Brasier and Antcliffe, "*Dickinsonia* from Ediacara," 312. Ivantsov, "Vendia and other Precambrian 'Arthropods.'"

19. "Asymmetry in the Fossil Record," 137.

20. Erwin, Valentine, and Jablonski, "The Origin of Animal Body Plans," 132.

21. Erwin, Valentine, and Jablonski, "The Origin of Animal Body Plans," 132.

22. Erwin, Valentine, and Jablonski, "The Origin of Animal Body Plans," 132.

23. Cooper and Fortey, "Evolutionary Explosions and the Phylogenetic Fuse," 151–56.

24. Fortey, "Cambrian Explosion Exploded," 438.

25. Knoll and Carroll, "Early Animal Evolution," 2129.

26. Ward, *On Methuselah's Trail*, 36.

27. "Life on Land," 153–54. Recently, Gregory Retallack has published a controversial hypothesis about the Ediacaran fauna. Retallack has studied the depositional environments of key Ediacaran fossils such as *Dickinsonia*. He has concluded that these organisms should not be classified as marine animals, because they were deposited on land. According to Retallack, the rocks that bore these Ediacaran fossils "have a variety of features that are more like the biological soil crusts of desert and tundra than the parallel wrinkled, and undulose hydrated microbial mats of intertidal flats and shallow seas." [Retallack, "Ediacaran Life on Land," 89.] Retallack's thesis has received a cool reception from other Ediacaran experts, however. They have not only questioned his analysis of ancient sediments but pointed out that the Ediacaran forms that he analyzed from Australia are also preserved in clearly marine sediments (from Newfoundland, for example) and that it is unlikely that the same organisms would live both on land and in the sea. [Callow, Brasier, Mcilroy, "Discussion: 'Were the Ediacaran siliciclastics of South Australia coastal or deep marine?' " 1–3.]

28. Erwin et al., "The Cambrian Conundrum." Others believe that these Precambrian "worm" trails could have been created by giant protists. See Matz et al., "Giant Deep-Sea Protist Produces Bilaterian-like Traces," 1849–54.

29. Valentine, Erwin, and Jablonski, "Developmental Evolution of Metazoan Body Plans"; Runnegar, "Evolution of the Earliest Animals."

30. Runnegar, "Proterozoic Eukaryotes"; Gehling, "The Case for Ediacaran Fossil Roots to the Metazoan Tree."

31. Budd and Jensen, "A Critical Reappraisal of the Fossil Record of the Bilaterian Phyla," 270.

32. See Matz et al., "Giant Deep-Sea Protist Produces Bilaterian-like Traces."

33. Erwin et al., "The Cambrian Conundrum."

34. Sperling, Pisani, and Peterson, "Poriferan paraphyly and Its Implications for Precambrian Palaeobiology"; Erwin and Valentine, *The Cambrian Explosion,* 80.

35. Conway Morris, "Evolution: Bringing Molecules into the Fold," 5.

36. McMenamin and McMenamin, *The Emergence of Animals,* 167–68.

37. Peterson et al., "The Ediacaran Emergence of Bilaterians."

38. Shen et al., "The Avalon Explosion," 81.

39. See, e.g., Cooper and Fortey, "Evolutionary Explosions and the Phylogenetic Fuse." The Cambrian period 543 mya is marked by the appearance of small shelly fossils consisting of tubes, cones, and possibly spines and scales of larger animals. These fossils, together with trace fossils, gradually become more abundant and diverse as one moves upward in the earliest Cambrian strata (the Manykaian Stage, 543–530 mya).

40. Bowring et al., "Calibrating Rates of Early Cambrian Evolution," 1293–98; Erwin et al., "The Cambrian Conundrum: Early Divergence and Later Ecological Success in the Early History of Animals," 1091–97.

41. Meyer et al., "The Cambrian Explosion: Biology's Big Bang," 323–402.

42. Valentine, "Prelude to the Cambrian Explosion," 289.

43. Animals with fivefold symmetry extending from a central body cavity are described technically as radially symmetric "pentamerous" animals.

44. Budd and Jensen, "A Critical Reappraisal of the Fossil Record of the Bilaterian Phyla," 261. As Budd and Jensen explain in more detail: "although this fossil possesses pentaradial symmetry, its small size, combined with preservation in relatively coarse sand means that other echinoderm-specific features are not readily visible. Its assignment to the Echinodermata thus largely rests on this single character, and must be at present regarded as an open question" (261).

45. Valentine, *On the Origin of Phyla,* 287, 397.

46. Bottjer, "The Early Evolution of Animals," 47.

47. Cnidarians and ctenophores, for example, are radially symmetric. (One might think that echinoderms, which as adults have pentaradial [fivefold] symmetry, are not bilaterians, but "in early development, echinoderms are bilateral" and thus classed among the Bilateria.) For more discussion, see Valentine, *On the Origin of Phyla,* 391.

48. Bengtson and Budd, "Comment on 'Small Bilaterian Fossils from 40 to 55 Million Years Before the Cambrian,'" 1291a.

49. As Bengtson and Budd explain, "The specimens presented by Chen et al. represent a common mode of preservation of microfossils in phosphatic sediments, including those of the Doushantuo." In particular, they note that "the layers have a regular banding of color and thickness that is different between the specimens but consistent within the individual specimens." They argue that "this pattern defies biological explanation but is easily explained as representing two to three generations of diagenetic overgrowth." They also note that "rather than being sinuously folded, as would be expected from deformed tissue layers," the layers in the imprint display features typical of (inorganic) diagenetic crusts. They conclude that although the imprint may have encased the remains of eukaryotic microfossils, "their reconstructed morphology as bilaterians is an artifact generated by cavities being lined by diagenetic crusts. The appearance of the fossils now has little resemblance to that of the living organisms that generated them" ("Comment on 'Small Bilaterian Fossils from 40 to 55 Million Years Before the Cambrian,'" 1291a).

50. Bengtson et al., "A Merciful Death for the 'Earliest Bilaterian,' *Vernanimalcula,*" 421.

51. Bottjer, "The Early Evolution of Animals," 47.

52. Bengtson et al., "A Merciful Death for the 'Earliest Bilaterian,' *Vernanimalcula,*" 426.

53. Bengtson et al., "A Merciful Death for the 'Earliest Bilaterian,' *Vernanimalcula,*" 426.

54. Marshall and Valentine, "The Importance of Preadapted Genomes in the Origin of the Animal Bodyplans and the Cambrian Explosion," 1190, emphasis added.

55. Budd and Jensen, "The Limitations of the Fossil Record and the Dating of the Origin of the Bilateria," 183.

56. Budd and Jensen, "The Limitations of the Fossil Record and the Dating of the Origin of the Bilateria," 168.

CHAPTER 5: THE GENES TELL THE STORY?

1. Dawkins, *The Greatest Show on Earth,* 111.

2. According to Zvelebil and Baum, "The key assumption made when constructing a phylogenetic tree from a set of sequences is that they are all derived from a single ancestral sequence, i.e., they are homologous" (*Understanding Bioinformatics,* 239). Lecointre and Le Guyader note: "Cladistics can run into difficulties in its application because not all character states are

necessarily homologous. Certain resemblances are convergent—that is, the result of independent evolution. We cannot always detect these convergences immediately, and their presence may contradict other similarities, 'true homologies' yet to be recognized. Thus, we are obliged to assume at first that, for each character, similar states are homologous, despite knowing that there may be convergence among them" (*The Tree of Life,* 16).

3. Coyne, *Why Evolution Is True,* 10.

4. Budd and Jensen, "A Critical Reappraisal of the Fossil Record of the Bilaterian Phyla," 253–95; Budd and Jensen, "The Limitations of the Fossil Record and the Dating of the Origin of the Bilateria," 166–89 ("The expected Darwinian pattern of a deep fossil history of the bilaterians, potentially showing their gradual development, stretching hundreds of millions of years into the Precambrian, has singularly failed to materialize . . . whatever the resolution of the misfit between the fossil record and molecular evidence for the origin of animals, it does not come about through a misunderstanding of the known fossil record . . . The known fossil record has not been misunderstood, and there are no convincing bilaterian candidates known from the fossil record until just before the beginning of the Cambrian (c. 543 Ma), even though there are plentiful sediments older than this that should reveal them"); Jensen et al., "Trace fossil preservation and the early evolution of animals," 19–29 ("A literal reading of the body fossil record suggests that the diversification of bilaterian animals did not significantly precede the Neoproterozoic–Cambrian boundary (ca. 545 Ma) . . . Despite reports to the contrary, there is no widely accepted trace fossil record from sediments older than about 560–555 Ma. . . . The above conclusions place serious constraints on the time of appearance of bilaterian animals. For example, assuming that key bilaterian features could only have been acquired in moderately large benthic animals, the absence of an ancient trace fossil record suggests that the Cambrian 'explosions' are a reality in terms of the relatively rapid appearance and diversification of macroscopic bilaterians"); Conway Morris, "Darwin's Dilemma: The Realities of the Cambrian 'Explosion'," 1069–83 ("The 'ancient school' argues that animals evolved long before the Cambrian and that the 'explosion' is simply an artefact, engendered by the breaching of taphonomic thresholds, such as the onset of biomineralization and/or a sudden increase in body size. The alternative 'realist school', to which I largely subscribe, proposes that while the fossil record is far from perfect and is inevitably skewed in significant ways, none is sufficient to destroy a strong historical signal"); Peterson et al., "MicroRNAs and Metazoan Macroevolution: Insights into Canalization, Complexity, and the Cambrian Explosion," 736–47; Fortey, "The Cambrian Explosion Exploded?" 438–39; Wray et al., "Molecular Evidence for Deep Precambrian Divergences Among Metazoan Phyla," 568–73 ("Darwin recognized that the sudden appearance of animal fossils in the Cambrian posed a problem for his theory of natural selection. He suggested that fossils might eventually be found documenting a protracted unfolding of Precambrian metazoan evolution. Many paleontologists today interpret the absence of Precambrian animal fossils that can be assigned to extant clades not as a preservational artifact, but as evidence of a Cambrian or late Vendian origin and divergence of metazoan phyla. This would make the Cambrian the greatest evolutionary cornucopia in the history of the earth. Definitive representatives of all readily fossilizable animal phyla (with the exception of bryozoans) have been found in Cambrian rocks, as have representatives of several soft-bodied phyla. Recent geochronological studies have reinforced the impression of a 'big bang of animal evolution' by narrowing the temporal window of apparent divergences to just a few million years") (internal citations omitted); Erwin et al., "The Cambrian Conundrum: Early Divergence and Later Ecological Success in the Early History of Animals," 1091–97 ("When Charles Darwin published *The Origin of Species* (1), the sudden appearance of animal fossils in the rock record was one of the more troubling facts he was compelled to address. He wrote: 'There is another and allied difficulty, which is much graver. I allude to the manner in which numbers of species of the same group, suddenly appear in the lowest known fossiliferous rocks' (306). Darwin argued that the incompleteness of the fossil record gives the illusion of an explosive event, but with the eventual discovery of older and better-preserved rocks, the ancestors of these Cambrian taxa would be found. Studies of Ediacaran and Cambrian fossils continue to expand the morphologic variety of clades, but the appearance of the remains and traces of bilaterian animals in the Cambrian remains abrupt.").

5. As Alan Cooper and Richard Fortey explain, "Molecular evidence indicates that prolonged periods of evolutionary innovation and cladogenesis lit the fuse long before the 'explosions' apparent in the fossil record. ("Evolutionary Explosions and the Phylogenetic Fuse," 151.) Also, according to Welch, Fontanillas, and Bromham: "However, a wide range of molecular dating studies have suggested that the major lineages of animals arose long before the Cambrian, at over 630 mya [million years ago]. This raises the possibility that there was a long cryptic period of animal evolution preceding the explosion of fossils in the Cambrian" ("Molecular Dates for the 'Cambrian Explosion,'" 672–73).

6. "The molecular clock . . . is the assumption that lineages have evolved at equal rates" (Felsenstein, *Inferring Phylogenies*, 118).

7. Smith and Peterson, "Dating the Time and Origin of Major Clades," 72.

8. Wray, Levinton, and Shapiro, "Molecular Evidence for Deep Precambrian Divergences Among Metazoan Phyla"; for another similar study of molecular sequence data that comes to the same conclusion, see Wang, Kumar, and Hedges, "Divergence Time Estimates for the Early History of Animal Phyla and the Origin of Plants, Animals and Fungi," 163; see also Vermeij, "Animal Origins"; and Fortey, Briggs, and Wills, "The Cambrian Evolutionary 'Explosion' Recalibrated."

9. These proteins were ATP-ase, cytochrome c, cytochrome oxidase I and II, alpha and beta hemoglobin, and NADH I.

10. The ribosomal RNA they used was 18S rRNA.

11. These proteins were aldolase, methionine adenosyltransferase, ATP synthase beta chain, catalase, elongation factor 1 alpha, triosephosphate isomerase, and phosphofructokinase.

12. The three RNA molecules they used were 5.8S rRNA, 18S rRNA, and 28S rRNA.

13. Erwin et al., "The Cambrian Conundrum," 1092.

14. For example, Bronham and colleagues find that mitochondrial DNA and 18S rRNA data yielded divergence dates that varied by more than 1 billion years ("Testing the Cambrian Explosion Hypothesis by Using a Molecular Dating Technique"); Xun suggests that from a total of 22 nuclear genes the divergence time between *Drosophila* and vertebrates was about 830 mya (million years ago; "Early Metazoan Divergence Was About 830 Million Years Ago"); Doolittle and colleagues date the protostome-deuterostome split at 670 million years ago ("Determining Divergence Times of the Major Kingdoms of Living Organisms with a Protein Clock"); Nikoh and colleagues date the split between eumetazoa and parazoa—animals with tissues, such as cnidarians, from those without, like sponges—at 940 mya, and the split between vertebrates and amphioxus at 700 mya ("An Estimate of Divergence Time of Parazoa and Eumetazoa and That of Cephalochordata and Vertebrata by Aldolase and Triose Phosphate Isomerase Clocks"); and Wang, Kumar, and Hedges suggest the basal animal phyla (Porifera, Cnidaria, Ctenophora) diverged between about 1200–1500 mya, and that nematodes were found to have diverged from the lineage leading to arthropods and chordates at 1177–79 mya ("Divergence Time Estimates for the Early History of Animal Phyla and the Origin of Plants, Animals and Fungi").

15. Wray, Levinton, and Shapiro, "Molecular Evidence for Deep Precambrian Divergences Among Metazoan Phyla," 568.

16. Wray, Levinton, and Shapiro, "Molecular Evidence for Deep Precambrian Divergences Among Metazoan Phyla," 569.

17. Wray, Levinton, and Shapiro, "Molecular Evidence for Deep Precambrian Divergences Among Metazoan Phyla," 568.

18. Quoted in Hotz, "Finding Turns Back Clock for Earth's First Animals," A1, A14.

19. See, e.g., Xun, "Early Metazoan Divergence Was About 830 Million Years Ago"; Aris-Brosou and Yang, "Bayesian Models of Episodic Evolution Support a Late Precambrian Explosive Diversification of the Metazoa." See also an early study by Bruce Runnegar in 1982, which measured the percent sequence difference between globin molecules in various animal phyla and, from this, postulated that "the initial radiation of the animal phyla occurred at least 900–1000 million years ago" (Runnegar, "A Molecular-Clock Date for the Origin of the Animal Phyla," 199). For other examples, see Bronham et al., "Testing the Cambrian explosion hypothesis by using a molecular dating technique," 12386–12389 (finding that mitochondrial DNA and 18S rRNA data yielded divergence dates that varied by more than 1 billion years); Xun, "Early Metazoan Divergence Was About 830 Million Years Ago," 369–71 (suggesting "From a total of 22 nuclear genes, we estimate that the divergence time between *Drosophila* and vertebrates was about 830 million years ago (mya)"); Doolittle, "Determining Divergence Times of the Major Kingdoms of Living Organisms with a Protein Clock," 470–77 (dating the protostome-deuterostome split at 670 million years ago); Nikoh et al., "An Estimate of Divergence Time of Parazoa and Eumetazoa and That of Cephalochordata and Vertebrata by Aldolase and Triose Phosphate Isomerase Clocks," 97–106 (dating the split between eumetazoa and parazoa—animals with tissues from those without, like sponges—at 940 mya, and the split between vertebrates and amphioxus at 700 mya); Wang et al., "Divergence Time Estimates for the Early History of Animal Phyla and the Origin of Plants, Animals and Fungi," 163–71 (suggesting "the basal animal phyla (Porifera, Cnidaria, Ctenophora) diverged between about 1200–1500 Ma" and "Nematodes were found to have diverged from the lineage leading to arthropods and chordates at 1177–79 Ma").

20. Valentine, Jablonski, and Erwin, "Fossils, Molecules and Embryos," 851.

21. Nikoh et al., "An Estimate of Divergence Time of Parazoa and Eumetazoa and That of Cephalochordata and Vertebrata by Aldolase and Triose Phosphate Isomerase Clocks."

22. Wang, Kumar, and Hedges, "Divergence Time Estimates for the Early History of Animal Phyla and the Origin of Plants, Animals and Fungi," 163.

23. Bronham et al., "Testing the Cambrian Explosion Hypothesis by Using a Molecular Dating Technique."

24. Xun, "Early Metazoan Divergence Was About 830 Million Years Ago."

25. Aris-Brosou and Yang, "Bayesian Models of Episodic Evolution Support a Late Precambrian Explosive Diversification of the Metazoa." Other literature surveys report that molecular clock-based estimates of the split between protostomes and deuterostomes have ranged from 588 million to 1.5 billion years ago. See Erwin, Valentine, and Jablonski, "The Origin of Animal Body Plans"; and Benton and Ayala, "Dating the Tree of Life."

26. Graur and Martin, "Reading the Entrails of Chickens: Molecular Timescales of Evolution and the Illusion of Precision."

27. Graur and Martin, "Reading the Entrails of Chickens," 85. Smith and Peterson also note: "The second area where molecules and morphology are in serious disagreement concerns the origins of the metazoan phyla. Although the difference between the molecular and morphological estimates for bird and mammal origins may be as much as 50 million years, the discord between the two for the animal phyla may be as much as 500 million years, almost the entire length of the Phanerozoic" ("Dating the Time and Origin of Major Clades," 79).

28. Ayala, Rzhetsky, and Ayala, "Origin of the Metazoan Phyla."

29. Ayala and his team eliminated the 18S rRNA, an RNA-coding gene because of problems with obtaining a reliable alignment. They also added an additional twelve protein-coding genes.

30. Ayala, Rzhetsky, and Ayala, "Origin of the Metazoan Phyla," 611.

31. Valentine, Jablonski, and Erwin, "Fossils, Molecules and Embryos," 856.

32. Behe, "Histone Deletion Mutants Challenge the Molecular Clock Hypothesis."

33. Some evolutionary biologists have attempted to explain their extreme conservation (similarity) by "strong selection" for their essential functional role: the close packing, or wrapping, of DNA in eukaryotic chromosomes. This hypothesis is hard to square, however, with experimental data showing that yeast tolerates dramatic deletions in their H4 histones. Behe, "Histone Deletion Mutants Challenge the Molecular Clock Hypothesis."

34. As Baverstock and Moritz explain in more detail: "The single most important component . . . of a phylogenetic analysis is the decision as to which method(s) or sequence(s) are appropriate to the phylogenetic question at hand. The method chosen must yield sufficient variation as to be phylogenetically informative, but not so much variation that convergence and parallelisms overwhelm informative changes" ("Project Design," 25).

35. Valentine, Jablonski, and Erwin, "Fossils, Molecules and Embryos," 856.

36. Ho et al., "Accuracy of Rate Estimation Using Relaxed-Clock Models with a Critical Focus on the Early Metazoan Radiation," 1355.

37. Smith and Peterson, "Dating the Time and Origin of Major Clades," 73.

38. Smith and Peterson, "Dating the Time and Origin of Major Clades," 73. Smith and Peterson elaborate: "All molecular clock approaches require one or more calibration points using dates derived either from the fossil record or from biogeographic constraints. There are two approaches—either calibration can rely on one or a small number of 'well documented' dates where paleontological evidence seems highly reliable, or calibration can be achieved using a large number of independent dates so that a range of estimates is arrived at. The former approach has been criticized by both Lee (1999) and Alroy (1999) for placing too much reliance on a single paleontological date without considering its error" (75). See Lee, "Molecular Clock Calibrations and Metazoan Divergence Dates"; and Alroy, "The Fossil Record of North American Mammals." Dan Graur and William Martin concur. They note that gross uncertainties often afflict assumptions about (1) the age of fossils used to calibrate the molecular clock, (2) the rate of mutations in various genes, and (3) the conclusions of comparative sequence analyses based upon the use of molecular clocks ("Reading the Entrails of Chickens").

39. Conway Morris, "Evolution," 5–6.

40. Valentine Jablonski, and Erwin, "Fossils, Molecules and Embryos," 856.

41. Zvelebil and Baum, *Understanding Bioinformatics*, 239.

42. Lecointre and Le Guyader, *The Tree of Life*, 16.

43. For a related discussion see Wagner and Stadler, "Quasi-Independence, Homology and the Unity of Type."

44. Osigus, Eitel, Schierwater, "Chasing the Urmetazoon: Striking a Blow for Quality Data?" 551–57; Conway Morris, "The Cambrian 'Explosion' and Molecular Biology," 505–506.

45. Osigus, Eitel, Schierwater, "Chasing the Urmetazoon: Striking a Blow for Quality Data?" 551–57. As Osigus and colleagues note: "The sum of molecular trees based on large numbers of gene sequences does not resolve phylogenetic relationships at the base of the Metazoa. Conflicting

scenarios have been published in short sequence and each single analysis can be criticized for one or the other reason. It is unclear to many whether the base of Metazoa can ever be resolved by means of sequence data even if whole genomes and extensive taxon sampling is used" (555).

46. Conway Morris, "Early Metazoan Evolution," 870.

47. Graur and Martin, "Reading the Entrails of Chickens"; Smith and Peterson, "Dating the Time and Origin of Major Clades"; Valentine, Jablonski, and Erwin, "Fossils, Molecules and Embryos."

CHAPTER 6: THE ANIMAL TREE OF LIFE

1. "The Darwinian Sistine Chapel," April 14, 2009, www.bbc.co.uk/darwin/?tab=21 (accessed October 31, 2012).
2. Hellström, "The Tree as Evolutionary Icon," 1.
3. Ruse, *Darwinism Defended,* 58.
4. Dobzhansky, "Nothing in Biology Makes Sense Except in the Light of Evolution," 125.
5. Dawkins, *The Greatest Show on Earth,* 315.
6. As Coyne has asserted, "both the visible traits of organisms and their DNA sequences usually give the same information about evolutionary relationships" (*Why Evolution Is True,* 10).
7. Atkins, *Galileo's Finger,* 16.
8. Coyne, *Why Evolution Is True,* 7.
9. An enormous literature exists analyzing how "similarity," which can be directly observed and measured, comes to be interpreted as "homology," a theoretical construct that cannot be directly observed. The two terms should not be equated. According to Van Valen, "For molecular biologists . . . a good touchstone is that homology is always an inference, never an observation. What we observe is similarity or identity, never homology" ("Similar, but Not Homologous," 664).
10. Prothero, *Evolution,* 140.
11. Dawkins, *A Devil's Chaplain: Reflections on Hope, Lies, Science, and Love,* 112.
12. Wiley and Lieberman, *Phylogenetics,* 6.
13. Degnan and Rosenberg, "Gene Tree Discordance, Phylogenetic Inference and the Multispecies Coalescent," 332.
14. Dávalos et al., "Understanding Phylogenetic Incongruence: Lessons from Phyllostomid Bats," 993.
15. Syvanen and Ducore, "Whole Genome Comparisons Reveals a Possible Chimeric Origin for a Major Metazoan Assemblage," 261–75.
16. Quoted in Lawton, "Why Darwin Was Wrong About the Tree of Life," 39.
17. Rokas, "Spotlight: Drawing the Tree of Life."
18. Rokas, Krüger, and Carroll. "Animal Evolution and the Molecular Signature of Radiations Compressed in Time," 1933–34.
19. Rokas and Carroll, "Bushes in the Tree of Life," 1899–1904.
20. Rokas and Carroll, "Bushes in the Tree of Life," 1899–1904 (internal citations omitted).
21. Rokas and Carroll, "Bushes in the Tree of Life," 1899–1904 (internal citations omitted).
22. Rokas and Carroll, "Bushes in the Tree of Life," 1899–1904.
23. Rokas, Krüger, and Carroll, "Animal Evolution and the Molecular Signature of Radiations Compressed in Time," 1935.
24. Zuckerkandl and Pauling, "Evolutionary Divergence and Convergence in Proteins," 101.
25. Zuckerkandl and Pauling, "Evolutionary Divergence and Convergence in Proteins," 101.
26. Theobald, "29+ Evidences for Macroevolution."
27. Hyman, *The Invertebrates, vol. 1: Protozoa Through Ctenophora.*
28. Holton and Pisani, "Deep Genomic-Scale Analyses of the Metazoa Reject Coelomata."
29. Aguinaldo et al., "Evidence for a Clade of Nematodes, Arthropods and Other Moulting Animals." See also Telford et al., "The Evolution of the Ecdysozoa"; Halanych and Passamaneck, "A Brief Review of Metazoan Phylogeny and Future Prospects in Hox-Research"; and Mallatt, Garey, and Shultz, "Ecdysozoan Phylogeny and Bayesian Inference."
30. Aguinaldo et al., "Evidence for a Clade of Nematodes, Arthropods and Other Moulting Animals." See also Halanych, "The New View of Animal Phylogeny."
31. Telford et al., "The Evolution of the Ecdysozoa."
32. Aguinaldo et al., "Evidence for a Clade of Nematodes, Arthropods and Other Moulting Animals," 492.
33. In 2004, for instance, researcher Yuri Wolf and his colleagues at the National Center for Biotechnology Information (NCBI) published a phylogeny based on molecular data (500 sets of proteins as well as insertion/deletion patterns in similar proteins) supporting the earlier Coelomata hypothesis. Wolf's team concluded, "All of these approaches supported the coelomate clade

and showed concordance between evolution of protein sequences and higher-level evolutionary events" ("Coelomata and Not Ecdysozoa," 29). Another NCBI study, by Jie Zheng and colleagues published in 2007, analyzed conserved intron positions in the genomes of various animals; it supported the Coelomata clade and rejected Ecdysozoa (introns are sections of the genome that do not encode information for building proteins and occur in the genome between the regions, called exons, that do code for proteins; "Support for the Coelomata Clade of Animals from a Rigorous Analysis of the Pattern of Intron Conservation").

34. In 2008, for example, Scott Roy (who also works at the NCBI) and Manuel Irimia (at the University of Barcelona) argued that the intron data actually supported the Ecdysozoa hypothesis ("Rare Genomic Characters Do Not Support Coelomata").

35. Holton and Pisani, "Deep Genomic-Scale Analyses of the Metazoa Reject Coelomata."

36. In this tree, now generally known as the Ecdysozoa hypothesis, the bilaterians are divided first into the protostomes and deuterostomes. the protostomes (or Protostomia) are then further subdivided into two distinct groups: (1) the Lophotrochozoa (so called because of two distinctive anatomical characters, a ciliated larva [trochophore] and a ciliated feeding structure [lophophore]) and (2) the Ecdysozoa (the molting animals).

37. Maley and Marshall, "The Coming of Age of Molecular Systematics," 505.

38. As Maley and Marshall conclude, "Different representative species, in this case brine shrimp or tarantula for the arthropods, yield wildly different inferred relationships among phyla" ("The Coming of Age of Molecular Systematics," 505).

39. According to Valentine, Jablonski, and Erwin, "Molecular evidence has produced a new view of metazoan phylogeny, prompting new analyses of morphological, ultrastructural and developmental characters" ("Fossils, Molecules and Embryos," 854).

40. See Nielsen, *Animal Evolution*, 82.

41. Rokas et al., "Conflicting Phylogenetic Signals at the Base of the Metazoan Tree"; Halanych, "The New View of Animal Phylogeny"; Borchiellini et al., "Sponge Paraphyly and the Origin of Metazoa."

42. Rokas et al., "Conflicting Phylogenetic Signals at the Base of the Metazoan Tree"; Halanych, "The New View of Animal Phylogeny."

43. Gura, "Bones, Molecules . . . or Both?" 230. A 2004 paper in the *Annual Review of Ecology and Systematics* puts it this way: "Molecular tools have profoundly rearranged our understanding of metazoan phylogeny." (Halanych, "The New View of Animal Phylogeny," 229.)

44. Dávalos et al., "Understanding Phylogenetic Incongruence: Lessons from Phyllostomid Bats," 993.

45. One might object here that building phylogenetic trees among the higher taxonomic groups, such as the animal phyla, is an inherently tricky business—but that phylogenetic trees describing lower taxonomic groups, such as those within the phyla, show consistency among different kinds of evidence. Of course, strictly speaking, even if there was evidence of a single coherent tree connecting groups within a phylum, it would do nothing to establish ancestors of the phyla themselves, but instead only members of the smaller groups within specific phyla. Nevertheless, even among lower taxa the primary literature on phylogenetic inference challenges the treelike picture of animal history.

Consider the crustacea, for instance, a large group within the phylum Arthropoda. The crustacea include such familiar creatures as shrimp and lobsters. (Darwin himself published his major technical work in biology on the classification of barnacles, a group [Cirripedia] within the crustacea.) Given the claims of Dawkins, Coyne, and Atkins, we might have expected that evolutionary biologists would have long ago established a single univocal evolutionary history for a well-studied group such as the crustacea, and that molecular data would by now only be confirming what biologists have known all along. But note, instead, that Ronald Jenner, a zoologist and expert on crustacea at the British Museum of Natural History, describes crustacean phylogeny as "essentially unresolved." As he explains the situation, "Conflict is rife, irrespective of whether one compares different morphological studies, molecular studies, or both" ("Higher-Level Crustacean Phylogeny," 143). The area of study remains "intensely contentious," he continues, and "published studies show very few points of consensus, even if one constrains the comparison to just the most comprehensive and careful analyses" (151).

Other studies of different classes of organisms within the arthropod phyla introduce further uncertainty. Insects provide another important example of such incongruity. Based on anatomical evidence, systematists have long held that insects are most closely related to the group that contains centipedes and millipedes (called the myriapod group). Nevertheless, molecular studies by F. Nardi and colleagues indicate that insects are more closely related to crustaceans. Similarly, the same molecular study suggested that some wingless insects are more closely related to crustaceans than they are to other insects, though anatomical studies indicate the opposite for obvious reasons. This led the authors of the paper to conclude that insects (hexapods) are not monophyletic—a view never anticipated by most evolutionary biologists. Because of incongruities between molecular and morphology-based trees, the Nardi team offered a puzzled observation: "Although this tree shows many interesting outcomes, it also contains some evidently untenable relationships, which nevertheless have strong statistical support" ("Hexapod Origins," 1887). But see Delsuc et al., "Comment on 'Hexapod Origins: Monophyletic or Paraphyletic?' " 1482d; Nardi et al., "Response to Comment on 'Hexapod Origins: Monophyletic or Paraphyletic?' " 1482e.

Studies of vertebrates, a subphylum of another phylum, the chordates, have revealed similar contradictory phylogenetic relationships. For example, a recent paper on bat phylogenetics, noted that "For more than a decade, evolutionary relationships among members of the New World bat family Phyllostomidae inferred from morphological and molecular data have been in conflict." The authors "ruled out paralogy, lateral gene transfer, and poor taxon sampling and outgroup choices among the processes leading to incongruent gene trees in phyllostomid bats." The authors further note that "differential rates of change and evolutionary mechanisms driving those rates produce incongruent phylogenies. Incongruence among phylogenies estimated from different sets of characters is pervasive. Phylogenetic conflict has become a more acute problem with the advent of genome-scale data sets. These large data sets have confirmed that phylogenetic conflict is common, and frequently the norm rather than the exception." Dávalos et al., "Understanding phylogenetic incongruence: lessons from phyllostomid bats," 991–1024 (internal citations omitted). See also Patterson et al., "Congruence Between Molecular and Morphological Phylogenies," 153–88.

46. Schwartz and Maresca, "Do Molecular Clocks Run at All?" 357.

47. James Valentine, for example, disputes the reality of the coelom as a shared characteristic that defines a group, as advocates of the Coelomata hypothesis contended. As Valentine notes, the "assumption of the monophyly of coelomic spaces" was one of the "major principles used to relate the phyla." In his view, however, the coelom evolved multiple times independently and thus cannot be used as a homologous character defining a monophyletic group. He argues, instead, that "coeloms are polyphyletic. Few characters are simpler than fluid-filled cavities, and it is not difficult to visualize them as evolving many times for a number of purposes" (*On the Origin of Phyla,* 500).

48. Figure 6.2 derived from: Figure 1 of Edgecombe et al., "Higher-Level Metazoan Relationships: Recent Progress and Remaining Questions."

49. For further discussion of the central position of germ cells in evolution, see Ewen-Campen, Schwager, and Extavour, "The Molecular Machinery of Germ Line Specification." As they explain: "Sexually reproducing animals must ensure that one particularly important cell type is determined: the germ cells. These cells will be the sole progenitors of eggs and sperm in the sexually mature adult, and as such, their correct specification during embryonic development is critical for reproductive success and species survival" (3).

50. As is the case with mutations affecting other major organismal features, there is a remarkable absence of examples of successful (i.e., stably transmitted) mutations that significantly modify PGC formation in any animal group. Searching the experimental literature of the model systems in developmental biology, such as fruit flies (*Drosophila*), mice (*Mus*), and nematodes (*C. elegans*), reveals instead many loss-of-function or loss-of-structure examples, including total absence of egg cells (oocytes) in fruit flies (Lehmann, "Germ-Plasm Formation and Germ-Cell Determination in *Drosophila*"), germ-cell reduction and sterility in mice (Pellas et al., "Germ-Cell Deficient [*gcd*], an Insertional Mutation Manifested as Infertility in Transgenic Mice"), and oocyte elimination with consequent sterility in hermaphroditic nematodes (Kodoyianni, Maine, and Kimble, "Molecular Basis of Loss-of-Function Mutations in the *glp-1* Gene of *Caenorhabditis elegans*"). See also Youngren, "The *Ter* mutation in the dead end gene causes germ cell loss and testicular germ cell tumours," 360–64. Such examples of deleterious or catastrophic mutations could be multiplied indefinitely from these and other model systems.

51. Andrew Johnson, associate professor and reader in genetics at the University of Nottingham, and several coauthors frame the point this way: "Germ cell development acting as a constraint on embryonic morphogenesis is at first difficult to accept. However, the retention of a pool of PGCs, which will later produce gametes, is a fundamental constraint on any sexually reproducing organism, *because the inability to pass inherited traits to subsequent generations will terminate an individual's lineage. Therefore, changes in developmental processes that endanger the maintenance of PGCs will not be retained*" ("Evolution of Predetermined Germ Cells in Vertebrate Embryos: Implications for Macroevolution," 425, emphasis added).

52. Extavour, "Evolution of the Bilaterian Germ Line," 774. See Fig. 6.4.

53. Extavour, "Gray Anatomy," 420. Instead, she postulates that "convergent evolution has resulted in many different morphological, and possibly molecular genetic, solutions to the various problems posed by sexual reproduction."

54. Willmer and Holland, "Modern Approaches to Metazoan Relationships," 691, emphasis in original.

55. Willmer and Holland, "Modern Approaches to Metazoan Relationships," 690.

56. Brusca and Brusca, *Invertebrates,* 120; 2nd ed., 115.

57. Jenner, "Evolution of Animal Body Plans," 209.

58. In addition to convergent evolution, evolutionary biologists have offered a whole host of explanations to account for the many instances where shared anatomical and molecular similarity is not explicable by reference to vertical descent from a common ancestor, including: differing rates of evolution (resulting from positive selection or purifying selection), long branch attraction, rapid evolution, whole genome-fusion, coalescence (e.g., incomplete lineage sorting), DNA contamination, and horizontal gene transfer.

Horizontal gene transfer (HGT) occurs when organisms (usually prokaryotes such as bacteria) transfer genes to neighboring individuals. This mechanism provides a plausible explanation for some phylogenetic incongruence in prokaryotes, although some mission-critical housekeeping genes are thought to resist HGT. The mechanisms by which HGT occurs are well characterized, and include transformation (incorporation of free DNA from the environment into a recipient cell), transduction (transfer of DNA from one cell to another by a bacterial virus called a bacteriophage), and conjugation (transfer of DNA by direct cell-to-cell contact via a pilus). Horizontal gene transfer is less plausible however, in eukaryotes, where it occurs much less frequently and potential mechanisms are far less well characterized, although it is thought to occur in some rare cases between eukaryotes and prokaryotes, though more rarely, if at all, between different animals. [See Doolittle, "Phylogenetic Classification and the Universal Tree," 2124–28; Hall, "Contribution of Horizontal Gene Transfer to the Evolution of *Saccharomyces cerevisiae*," 1102–15; Kondo, "Genome Fragment of *Wolbachia* Endosymbiont Transferred to X Chromosome of Host Insect," 14280–85.]

Another proposed explanation of phylogenetic conflict is called long-branch attraction, an artifact of the phylogenetic algorithms, that results in preferentially grouping together related lineages that diverged quickly, and then evolved separately over long periods of time. [Bergsten, "A Review of Long-Branch Attraction," 163–93.]

Another proposed cause of incongruity is a phenomenon called incomplete lineage sorting. This occurs when a lineage splits and then rapidly splits again to yield three daughter species. This second split takes place before the process of sorting is complete (i.e., the process by which a daughter species gradually acquires its own unique set of genetic variants). This event is followed by the loss of a random variant by genetic drift, and this may cause two species to group together which otherwise would not. This process only accounts for phylogenetic incongruity between closely related species, however.

While some of these explanations may be plausible in some cases, they remain what they are: attempts to explain how two similar genes or traits could have arisen without those genes or traits having been inherited from a common ancestor. Thus in each case they provide counter-examples to the premise upon which all phylogenetic reconstruction is based—namely that similarity is an indicator of common ancestry.

CHAPTER 7: PUNK EEK!

1. Lecture Notes, Paul Nelson, University of Pittsburgh, 9-28-83.

2. Gould and Eldredge, "Punctuated Equilibrium: The Tempo and Mode of Evolution Reconsidered," 147.

3. Gould and Eldredge, "Punctuated Equilibrium: The Tempo and Mode of Evolution Reconsidered," 115.

4. Eldredge, *The Pattern of Evolution,* 21.

5. Sepkoski, " 'Radical' or 'Conservative'? The Origin and Early Reception of Punctuated Equilibrium," 301–25. Gould and Eldredge continued to jointly offered papers elaborating and refining the theory of punctuated equilibrium until 1993. See Gould and Eldredge, "Punctuated Equilibrium Comes of Age," 223–27.

6. As the U.S. National Academy of Sciences explained, punctuated equilibrium sought to account for the absence of transitional intermediates by showing that "changes in populations might occur too rapidly to leave many transitional fossils" (*Teaching About Evolution and the Nature of Science,* 57).

7. Though Gould and Eldredge formulated punctuated equilibrium several years before studies of the molecular evidence discussed in Chapter 6, their theory could also help explain the conflicting phylogenetic histories discussed there as well. As Rokas, Krüger, and Carroll would later argue ("Animal Evolution and the Molecular Signature of Radiations Compressed in Time"),

if the evolutionary process acts quickly enough, leaving little time for differences to accumulate in key molecular markers, then biologists should expect phylogenetic studies to generate conflicting trees.

8. Sepkoski, " 'Radical' or 'Conservative'?" 304.

9. Gould and Eldredge, "Punctuated Equilibrium Comes of Age"; Theobald, "Punctuated Equilibrium" ("Punctuated equilibrium immediately lit a scientific controversy that has smoldered ever since"); Bell, "Gould's Most Cherished Concept" ("Whether or not you agree with Gould that punctuated equilibrium has become the conventional wisdom, it certainly has led to a healthy debate concerning the sufficiency of neo-Darwinian theory to explain macroevolution, the analysis of biostratigraphic sequences, and the increased incorporation of paleontological data into evolutionary theory"); Dawkins, *The Blind Watchmaker,* 240–41; Dennett, *Darwin's Dangerous Idea,* 282–99; Ridley, "The Evolution Revolution"; Gould, "Evolution: Explosion, Not Ascent"; Boffey, "100 Years after Darwin's Death, His Theory Still Evolves"; Gleick, "The Pace of Evolution"; Maynard Smith, "Darwinism Stays Unpunctured"; Levinton, "Punctuated Equilibrium"; Schopf, Hoffman, and Gould, "Punctuated Equilibrium and the Fossil Record"; Lewin, "Punctuated Equilibrium Is Now Old Hat" (noting that "the tenor of the debate" over punctuated equilibrium "at times has been strident"); Levinton, "Bryozoan Morphological and Genetic Correspondence"; Lemen and Freeman, "A Test of Macroevolutionary Problems with Neontological Data"; Charlesworth, Lande, and Slatkin, "A Neo-Darwinian Commentary on Macroevolution"; Douglas and Avise, "Speciation Rates and Morphological Divergence in Fishes."

10. Rose, ed. *The Richness of Life: The Essential Stephen Jay Gould,* 6. See also Turner, "Why We Need Evolution by Jerks"; Rée, "Evolution by Jerks."

11. Sepkoski, " 'Radical' or 'Conservative'?"

12. For an urn with 100 balls, 50 being red and 50 being blue, the probability of getting only blue balls by randomly taking 50 balls out of the urn is given by the following considerations:

First, there's only one way to select only blue balls. Second, in general for an urn with N balls, there are C(N,k) different ways of choosing k balls from among these N (with k greater than or equal to 0 but less than or equal to N). C(N,k) is equal to N! divided by the product of k! and (N-k)!, where the exclamation mark is read "factorial" and equals the product of all numbers less than or equal to the number in question down to 1. Thus "six factorial" = 6! = 6 x 5 x 4 x 3 x 2 x 1 = 720. Factorials increase very quickly, faster than exponentials. C(N,k) is read "N choose k."

So for the problem above, the total number of ways to choose 50 specific balls out of 100 balls, ignoring color, is C(100,50), which equals 100! divided by 50! times 50!, or 100!/(50! x 50!).

This number admits an exact calculation, which can be expressed in Mathematica as the following:

$$C(100,50) = 100{,}891{,}344{,}545{,}564{,}193{,}334{,}812{,}497{,}256 \text{ which is approximately } 1.00891 \times 10^{29}.$$

Thus, the probability of selecting k specific balls out of N total balls is of course the inverse of that number, and can be distilled to the following equation:

$$p = k! \times (N - k)! / N!$$

Applied to this problem, the odds of randomly selecting all 50 blue balls in a collection of 50 blue and 50 red balls is 1 divided by C(100,50), or approximately 9.91165×10^{-30}.

13. Using the same equation discussed in the above endnote, the probability of selecting 4 specific balls out of 8 total balls is given by the same equation as follows: 4! x (8 - 4)! / 8! = 1 in 70.

14. Gould and Eldredge, "Punctuated Equilibria: The Tempo and Mode of Evolution Reconsidered," 117.

15. Eldredge and Gould, "Punctuated Equilibria: An Alternative to Phyletic Gradualism," 84.

16. Lieberman and Vrba, "Stephen Jay Gould on Species Selection: 30 Years of Insight"; Gould, "The Meaning of Punctuated Equilibrium and Its Role in Validating a Hierarchical Approach to Macroevolution."

17. Gould, *The Structure of Evolutionary Theory,* 703. As Gould and Eldredge also emphasized elsewhere: "The main insight for revision [of evolutionary theory] holds that all substantial evolutionary change must be reconceived as higher-level sorting based on differential success of certain kinds of stable species, rather than as progressive transformation within lineages [i.e., species]" ("Punctuated Equilibrium Comes of Age," 224).

18. If natural selection acts upon a larger unit of selection, the species rather than the individual, it followed logically that evolution would occur in larger more discrete jumps. Nevertheless, Gould and Eldredge rarely emphasized this implication of their conception of species selection explicitly, instead highlighting allopatric speciation as the main reason for fossil discontinuity. Stanley did, however, often draw a connection between the activity of species selection as a

mechanism of evolutionary change and fossil discontinuity. As he noted, "The validity of the species as the fundamental unit of large-scale evolution depends upon the presence of discontinuities between many species in the tree of life" (*Macroevolution,* 3).

19. Schopf, Editorial Introduction to Eldredge and Gould, "Punctuated Equilibria: An Alternative to Phyletic Gradualism," 82; Stanley, *Macroevolution: Pattern and Process,* 3.

20. Valentine and Erwin, "Interpreting Great Developmental Experiments"; see diagram on p. 92.

21. Valentine and Erwin note that "transitional alliances are unknown or unconfirmed for any of the [Cambrian] phyla," and yet "the evolutionary explosion near the beginning of Cambrian time was real and produced numerous [new] body plans" ("Interpreting Great Developmental Experiments," 84, 89).

22. Valentine and Erwin, "Interpreting Great Developmental Experiments," 96.

23. Gould and Eldredge, "Punctuated Equilibrium Comes of Age."

24. Gould and Eldredge, "Punctuated Equilibrium Comes of Age."

25. Schopf, Editorial Introduction to Eldredge and Gould, "Punctuated Equilibria: An Alternative to Phyletic Gradualism," 84.

26. Foote argued that "given estimates of [a] completeness [of the fossil record], [b] median species duration, [c] the time required for evolutionary transitions, and [d] the number of . . . higher-level transitions, we could obtain an estimate of the number of major transitions we should expect to see in the fossil record." His method provided a way to evaluate, as he puts it, "whether the small number of documented major transitions provides strong evidence against evolution" ("On the Probability of Ancestors in the Fossil Record," 148).

Because variables [a], [b], and [d] are reasonably well established, [c] the time required for plausible mechanisms to produce macroevolutionary transitions stands as the crucial variable in the analysis of any specific evolutionary model, including punctuated equilibrium. If the time required to produce major evolutionary change is high, as it is for the neo-Darwinian mechanism of change, then given current estimates of [a], [b], and [d], neo-Darwinism fails to account for the data of the fossil record. Conversely, if a theory such as punctuated equilibrium can identify a fast enough acting mechanism, then it could account for the paucity of transitional intermediates.

27. Foote and Gould, "Cambrian and Recent Morphological Disparity," 1816.

28. Darwin, *On the Origin of Species,* 177.

29. As Gould and Eldredge explained: "Most evolutionary change, we argued, is concentrated in rapid (often geologically instantaneous) events of speciation in small, peripherally isolated populations (the theory of allopatric speciation)" ("Punctuated Equilibria: The Tempo and Mode of Evolution Reconsidered," 116–17). See also Lewin, "Punctuated Equilibrium Is Now Old Hat."

30. Shu et al., "Lower Cambrian Vertebrates from South China."

31. Dawkins, *The Blind Watchmaker,* 265.

32. Levinton, *Genetics, Paleontology, and Macroevolution,* 208.

33. Gould, *The Structure of Evolutionary Theory,* 710.

34. Charlesworth, Lande, and Slatkin, "A Neo-Darwinian Commentary on Macroevolution," 493. As David Jablonski concluded in 2008, "The extent and efficacy of the specific processes [of species selection] remain poorly known" ("Species Selection," 501).

35. Gould, *The Structure of Evolutionary Theory,* 1005.

36. Gould, *The Structure of Evolutionary Theory,* 55, emphasis added.

37. Gould and Eldredge, "Punctuated Equilibria: The Tempo and Mode of Evolution Reconsidered," 134.

38. Sepkoski, "'Radical' or 'Conservative'?" 307.

39. Sepkoski, "'Radical' or 'Conservative'?" 7.

40. Gould, "Is a New and General Theory of Evolution Emerging?" 120. Because his colleagues understood Gould to be offering a theory of macroevolution, many of Gould's scientific colleagues at the time thought of him, as Sepkoski notes, as an "ardent proponent of a radical (and perhaps misguided) view of evolutionary change" ("'Radical' or 'Conservative'?" 302).

41. Sepkoski, "'Radical' or 'Conservative'?" 302.

42. Valentine and Erwin, "Interpreting Great Developmental Experiments," 96.

CHAPTER 8: THE CAMBRIAN INFORMATION EXPLOSION

1. Bowler, *Theories of Human Evolution,* 44–50.
2. Vorzimmer, "Charles Darwin and Blending Inheritance," 371–90.
3. Jenkins, *Genetics,* 13–15.
4. Muller, "Artificial Transmutation of the Gene," 84–87.
5. As Mayr and Provine put it, "Various geneticists . . . demonstrated that seemingly

continuous variation is caused by discontinuous genetic factors [mutations] that obey the Mendelian rules in their mode of inheritance" (*The Evolutionary Synthesis,* 31).

6. Bowler, *Evolution: The History of an Idea,* 331–39.
7. Huxley, "The Evolutionary Vision," 249, 253.
8. Huxley, quoted in "'At Random': A Television Preview," 45.
9. Watson and Crick, "A Structure for Deoxyribose Nucleic Acids," 737–38.
10. For an animated demonstration, see the short video "Journey Inside the Cell" on my website at SignatureintheCell.com.
11. Valentine, "Late Precambrian Bilaterians."
12. Brocks et al., "Archean Molecular Fossils and the Early Rise of Eukaryotes."
13. Grotzinger et al., "Biostratigraphic and Geochronologic Constraints on Early Animal Evolution."
14. Ruppert et al, *Invertebrate Zoology,* 82.
15. Bowring et al., "Calibrating Rates of Early Cambrian Evolution."
16. Valentine, *Origin of the Phyla,* 73.
17. Koonin, "How Many Genes Can Make a Cell?"
18. Gerhart and Kirschner, *Cells, Embryos, and Evolution,* 121; Adams et al., "The Genome Sequence of *Drosophila melanogaster*"; see also www.ncbi.nlm.nih.gov/genome/?term=drosophila%20melanogaster (accessed November 1, 2012).
19. Moreover, in addition to requiring a vast amount of new genetic information, building a new animal from a single-celled organism also requires a way of arranging gene products—proteins—into higher levels of organization, including cell types, organs, and body plans. Later, in Chapter 14, I will discuss the importance of these higher-level arrangements and why they also constitute a kind of information—one that, although not stored in genes alone, nevertheless has to be explained as well.
20. Shannon, "A Mathematical Theory of Communication."
21. To determine how much Shannon information is present in any sequence of characters, information scientists use a formula that converts probability measures into informational measures using a negative logarithmic function. A simple form of that equation can be expressed as $I = -\log_2 p$, where the negative sign indicates the inverse relationship between probability and information.
22. Yockey, *Information Theory and Molecular Biology,* 110.
23. Shannon and Weaver, *The Mathematical Theory of Communication,* 8.
24. Schneider, "Information Content of Individual Genetic Sequences"; Yockey, *Information Theory and Molecular Biology,* 58–177.
25. DNA clearly does not convey meaningful information in the sense of "knowledge" conveyed to, and comprehended by, a conscious agent, although the precise sequences of bases could be said to be meaningful in the sense that they are 'significant' to the function DNA performs. Clearly, however, the cellular machinery that uses and "reads" the information in DNA to build proteins is not conscious. Nevertheless, semantically meaningful information—a message, the meaning of which is understood by a conscious agent—represents only a special kind of *functional* information. And all sequences of characters containing functional information can be distinguished from mere Shannon information in that the precise arrangement of characters or symbols in such sequences matters to the function that they perform.
26. Crick, "On Protein Synthesis," 144, 153. See also Sarkar, "Biological Information," 191.

CHAPTER 9: COMBINATORIAL INFLATION

1. Eden, "Inadequacies of Neo-Darwinian Evolution as a Scientific Theory," 11.
2. The quotation and the historical material about the Geneva gathering are drawn from G. R. Taylor, *Great Evolution Mystery,* 4.
3. Schützenberger, "Algorithms and the Neo-Darwinian Theory of Evolution," 73–75.
4. Schützenberger, "Algorithms and the Neo-Darwinian Theory of Evolution," 74–75.
5. Commenting on the symposium thirty years later in a now infamous article in *Commentary* magazine, mathematician David Berlinski amplified Eden's argument. As he explains, "However it may operate in life, randomness in language is the enemy of order, a way of annihilating meaning. And not only in language, but in any language-like system" ("The Deniable Darwin").
6. King and Jukes, "Non-Darwinian Evolution," 788.
7. Eden, "Inadequacies of Neo-Darwinian Evolution as a Scientific Theory," 110.
8. Schützenberger, "Algorithms and the Neo-Darwinian Theory of Evolution," 74.
9. Ulam, "How to Formulate Mathematically Problems of Rate of Evolution," 21.
10. Eden, "Inadequacies of Neo-Darwinian Evolution as a Scientific Theory," 7.

11. Denton, *Evolution: A Theory in Crisis,* 309–11.
12. Maynard Smith, "Natural Selection and the Concept of a Protein Space."
13. Denton, *Evolution,* 324.
14. Reidhaar-Olson and Sauer, "Functionally Acceptable Substitutions in Two Alpha-Helical Regions of Lambda Repressor."
15. Yockey, "On the Information Content of Cytochrome C."
16. Yockey, "On the Information Content of Cytochrome C."
17. Behe, "Experimental Support for Regarding Functional Classes of Proteins," 66.
18. Lau and Dill, "Theory for Protein Mutability and Biogenesis."
19. Behe, "Experimental Support for Regarding Functional Classes of Proteins."

CHAPTER 10: THE ORIGIN OF GENES AND PROTEINS

1. Dawkins, *The Blind Watchmaker: Why the Evidence of Evolution Reveals a Universe Without Design,* 46–47.
2. For a critique of Dawkins' simulation see Chapter 13 of *Signature in the Cell.* See also Ewert, et al., "Efficient Per Query Information Extraction from a Hamming Oracle," 290–97; Dembski, *No Free Lunch,* 181–216. See also Weasel Ware–Evolutionary Simulation at http://evoinfo.org/weasel.
3. Reidhaar-Olson and Sauer, "Functionally Acceptable Substitutions in Two Alpha-Helical Regions of Lambda Repressor," 315.
4. Protein scientists recognize an additional level of structure called quaternary structure. Quaternary structures are formed from multiple protein folds, or multiple whole proteins.
5. As Reidhaar-Olson and Sauer note: "At [amino-acid] positions that are buried in the structure, there are severe limitations on the number and type of residues allowed. At most surface positions, many different [amino-acid] residues and residue types are tolerated" ("Functionally Acceptable Substitutions in Two Alpha-Helical Regions of Lambda Repressor," 306).
6. Axe, "Extreme Functional Sensitivity to Conservative Amino Acid Changes on Enzyme Exteriors."
7. Dawkins, *Climbing Mount Improbable.*
8. Jensen, "Enzyme Recruitment in Evolution of New Function," 409–25.
9. Axe, "Extreme Functional Sensitivity to Conservative Amino Acid Changes on Enzyme Exteriors."
10. Axe, "Extreme Functional Sensitivity to Conservative Amino Acid Changes on Enzyme Exteriors," 585–96. Experimental work that Axe performed with colleague Ann Gauger, published in 2011, also confirmed this result. See Gauger and Axe, "The Evolutionary Accessibility of New Enzyme Functions: A Case Study from the Biotin Pathway."
11. Axe, "Extreme Functional Sensitivity to Conservative Amino Acid Changes on Enzyme Exteriors."
12. Unfolded proteins also stick to other molecular entities within the cell or form what are called *inclusion bodies,* in both cases impeding proper protein function. Moreover, even slight elevations in temperature will accelerate the unfolding of already destabilized protein folds.
13. Axe's experiments using a more sensitive screen for function had shown him that even most *single* amino-acid changes will diminish the function of a protein enough to diminish its fitness, even in cases where such changes do not eliminate function altogether.
14. Blanco, Angrand, and Serrano, "Exploring the Conformational Properties of the Sequence Space Between Two Proteins with Different Folds: An Experimental Study." As they explain, "Both the hydrophobic core residues and the surface residues are important in determining the structure of the proteins" (741).
15. Neutral evolution in this context refers to a process alleged to explain the origin of new functional genes and proteins from gene duplicates unhinged from selection pressure. A neutral model of gene evolution is part of a more expansive and general neutral theory of evolution proposed by Motoo Kimura in 1968. As Long et al. explain, Kimura's model helped "describe how gene duplicates could acquire new functions and ultimately be preserved in a lineage" ("The Origin of New Genes," 868). However, Kimura's model of neutral evolution attempted to explain facts and phenomena beyond just the origin of new genes. Thus, not everyone who accepts a neutral model of gene origins subscribes to the whole of Kimura's theory. Kimura, *The Neutral Theory of Molecular Evolution.*
16. Matthew Hahn notes that, "There appear to be 4 major mechanisms by which DNA is duplicated: (1) unequal crossing-over, (2) duplicative (DNA) transposition, (3) retrotransposition, and (4) polyploidization" ("Distinguishing Among Evolutionary Models for the Maintenance of Gene Duplicates," 606).

17. Axe, "Estimating the Prevalence of Protein Sequences Adopting Functional Enzyme Folds."
18. Dembski, *The Design Inference,* 175.
19. Michael Behe made this calculation in *The Edge of Evolution* based on a paper in *Proceedings of the National Academy of Sciences U.S.A.* that observed that approximately 10^{30} prokaryotes are formed on earth each year. [Whitman, "Prokaryotes: The unseen majority," 6578–83.] Since prokaryotes make up the overwhelming majority of organisms, he multiplied that number by 10^{10}, which is about twice the number of years the age of the earth. This allowed him to estimate the total number of organisms that have lived on earth as "slightly fewer than 10^{40} cells." Behe, *The Edge of Evolution,* 64.
20. Bowring et al., "Calibrating Rates of Early Cambrian Evolution"; "A New Look at Evolutionary Rates in Deep Time"; "Geochronology Comes of Age"; Kerr, "Evolution's Big Bang Gets Even More Explosive"; Monastersky, "Siberian Rocks Clock Biological Big Bang."
21. Ohno, "The Notion of the Cambrian Pananimalia Genome."
22. Dembski, *The Design Inference,* 175–223. Dembski often uses the figure 1 in 10^{150} as his universal probability bound, but this figure derives from Dembski rounding up the exponent in the figure that he actually calculates. See my discussion of the derivation of Dembski's universal probability bound in *Signature in the Cell,* Chapter 10.
23. Dawkins, *The Blind Watchmaker,* 139.

CHAPTER 11: ASSUME A GENE

1. Stephen C. Meyer, "The Origin of Biological Information and the Higher Taxonomic Categories."
2. For detailed discussions of facts of the Sternberg case, see "Smithsonian Controversy," www.richardsternberg.com/smithsonian.php; U.S. Office of Special Counsel Letter (2005) at www.discovery.org/f/1488; United States House of Representatives Committee on Government Reform, Subcommittee Staff Report, "Intolerance and the Politicization of Science at the Smithsonian" (December 2006), at www.discovery.org/f/1489; Appendix, United States House of Representatives Committee on Government Reform, Subcommittee Staff Report (December 2006) at www.discovery.org/f/1490.
3. See Holden, "Defying Darwin"; Giles, "Peer-Reviewed Paper Defends Theory of Intelligent Design," 114; Agres, "Smithsonian 'Discriminated 'Against Scientist"; Stokes, ". . . And Smithsonian Has ID Troubles"; Monastersky, "Society Disowns Paper Attacking Darwinism."
4. Powell, "Controversial Editor Backed"; Klinghoffer, "The Branding of a Heretic."
5. Hagerty, "Intelligent Design and Academic Freedom.
6. See www.talkreason.org/AboutUs.cfm.
7. Gishlick, Matzke, and Elsberry, "Meyer's Hopeless Monster," www.talkreason.org/About Us.cfm.
8. Jones, *Kitzmiller et al. v. Dover Area School District.*
9. Matzke and Gross, "Analyzing Critical Analysis," 42.
10. Matzke and Gross, "Analyzing Critical Analysis," 42.
11. Matzke and Gross, "Analyzing Critical Analysis," 42.
12. See Chapter 10, n. 15, for the four major mechanisms by which DNA is duplicated.
13. For example, see Zhen et al., "Parallel Molecular Evolution in an Herbivore Community"; Li et al., "The Hearing Gene Prestin Unites Echolocating Bats and Whales"; Jones, "Molecular Evolution"; Christin, Weinreich, and Bresnard, "Causes and Evolutionary Significance of Genetic Convergence"; Rokas and Carroll, "Frequent and Widespread Parallel Evolution of Protein Sequences." According to Dávalos and colleagues, "In-depth analyses of specific genes in the context of multi-locus phylogenies have also shown that adaptive evolution leading to convergence, once thought to be extremely rare, is as much a source of conflict among gene trees as it is between morphological and molecular phylogenies" ("Understanding Phylogenetic Incongruence," 993).
14. Shen et al., "Parallel Evolution of Auditory Genes for Echolocation in Bats and Toothed Whales"; Li et al., "The Hearing Gene Prestin Unites Echolocating Bats and Whales"; Jones, "Molecular Evolution."
15. Khalturin et al., "More Than Just Orphans"; Merhej and Raoult, "Rhizome of Life, Catastrophes, Sequence Exchanges, Gene Creations, and Giant Viruses"; Beiko, "Telling the Whole Story in a 10,000-Genome World."
16. Suen et al., "The Genome Sequence of the Leaf-Cutter Ant *Atta cephalotes* Reveals Insights into Its Obligate Symbiotic Lifestyle" ("We also found 9,361 proteins that are unique to *A. cephalotes,* representing over half of its predicted proteome," 5). See also Smith et al., "Draft Genome of the Globally Widespread and Invasive Argentine Ant (*Linepithema humile*)" ("A total of 7,184 genes (45%) were unique to *L. humile* relative to these three other species," 2).

17. Tautz and Domazet-Lošo, "The Evolutionary Origin of Orphan Genes"; Beiko, "Telling the Whole Story in a 10,000-Genome World"; Merhej and Raoult, "Rhizome of Life, Catastrophes, Sequence Exchanges, Gene Creations, and Giant Viruses."

18. Lyell, *Principles of Geology*.

19. See Pray and Zhaurova, "Barbara McClintock and the Discovery of Jumping Genes (Transposons)."

20. Long et al., "The Origin of New Genes," 867.

21. Nurminsky et al., "Selective Sweep of a Newly Evolved Sperm-Specific Gene in *Drosophila*," 574.

22. Chen, DeVries, and Cheng, "Evolution of Antifreeze Glycoprotein Gene from a Trypsinogen Gene in Antarctic Notothenioid Fish," 3816.

23. Courseaux and Nahon, "Birth of Two Chimeric Genes in the *Hominidae* Lineage."

24. Knowles and McLysaght, "Recent de Novo Origin of Human Protein-Coding Genes."

25. Wu, Irwin, and Zhang, "De Novo Origin of Human Protein-Coding Genes."

26. Guerzoni and McLysaght, "De Novo Origins of Human Genes"; see also Wu, Irwin, and Zhang, "De Novo Origin of Human Protein-Coding Genes."

27. Siepel, "Darwinian Alchemy."

28. Siepel, "Darwinian Alchemy."

29. Siepel, "Darwinian Alchemy."

30. As Siepel notes, "These apparent *de novo* gene origins raise the question of how evolution by natural selection can produce functional genes from noncoding DNA. While a single gene is not as complex as a complete organ, such as an eye or even a feather, it still has a series of nontrivial requirements for functionality, for instance, an ORF [an open reading frame], an encoded protein that serves some useful purpose, a promoter capable of initiating transcription, and presence in a region of open chromatin structure that permits transcription to occur. How could all of these pieces fall into place through the random processes of mutation, recombination, and neutral drift . . . ?" ("Darwinian Alchemy").

31. Of course, one could argue that the mutational processes that Long invokes to explain the origin of new genes from preexisting cassettes of genetic information *themselves* explain the origin of the information in those cassettes in the first place. This view would suggest that the scenarios that Long cites do not so much beg the question as much as generate a regress of explanation terminating with the ultimate origin of biological information at the point of the origin of the first life. This view implies that, although the ultimate origin of biological information and the closely associated question of life's first origin may remain a mystery, the processes that Long cites account for all subsequent informational increases during the course of biological evolution. But this view still does not explain how shuffling preexisting cassettes of information generates the specific arrangements of the characters that make up those cassettes.

32. Zhang, Zhang, and Rosenberg, "Adaptive Evolution of a Duplicated Pancreatic Ribonuclease Gene in a Leaf-Eating Monkey." The genes that code for these two proteins differ by 12 nucleotides in their coding sequences. Those nucleotide differences produce two proteins that differ from each other in their overall electronegativity. This difference, in turn, allows the protein *RNASE1B* to operate at a slightly lower pH than the other protein, *RNASE1*. Since other primates have only the *RNASE1* protein, Zhang, Zhang, and Rosenberg hypothesize that the evolution of this second gene and protein gave individuals within the species of monkey a selective advantage. To explain the origin of the second gene, they posit a common ancestral gene, a gene duplication event, and the accumulation of different mutations on the duplicate copy (the *RNASE1B*) over time.

33. In this study ("Adaptive Evolution of Cid, a Centromere-Specific Histone in Drosophila"), Malik and Henikoff infer that "adaptive evolution has occurred on both the *D. melanogaster* and *D. simulans* lineages since their split from a common ancestor." They base this inference on an analysis of the ratio of "synonymous" and "nonsynonymous" mutations in the genomes of these organisms. The study found that many of the differences/mutations in nucleotide base sequences changed amino-acid sequence (called "nonsynonymous" mutations), while some did not (called synonymous, or "silent," mutations). A higher percentage of these differences changed the amino-acid sequence than would be expected from neutral evolution alone, leading the authors to infer that "adaptive evolution has occurred on both the *D. melanogaster* and *D. simulans* lineages since their split from a common ancestor." Since some of these differences exist in the region of the protein that binds to the chromosome, they may have affected the protein's functional binding ability. But the authors of the paper identify no specific functional effects of these amino-acid differences and base their claim of "strong evidence for the adaptive evolution of Cid" solely upon comparisons of the relative numbers of a handful of synonymous and nonsynonymous differences between the genes.

34. Enard et al., "Molecular Evolution of *FOXP2,* a Gene Involved in Speech and Language"; Zhang, Webb, and Podlaha, "Accelerated Protein Evolution and Origins of Human-Specific Features."

35. Enard et al., "Molecular Evolution of *FOXP2,* a Gene Involved in Speech and Language."

36. Long et al., "The Origin of New Genes," 866.

37. Darnell and Doolittle, "Speculations on the Early Course of Evolution"; Hall, Liu, and Shub, "Exon Shuffling by Recombination Between Self-Splicing Introns of Bacteriophage T4"; Rogers, "Split-Gene Evolution"; Gilbert, "The Exon Theory of Genes"; Doolittle et al., "Relationships of Human Protein Sequences to Those of Other Organisms."

38. For example, Arli A. Parikesit and colleagues note that "Although there is a statistically significant correlation between protein domain boundaries and exon boundaries, about two thirds of the annotated protein domains are interrupted by at least one intron, and on average a domain contains 3 or 4 introns" ("Quantitative Comparison of Genomic-Wide Protein Domain Distributions," 96–97; internal citations omitted).

39. Gauger, "Why Proteins Aren't Easily Recombined."

40. Axe, "The Limits of Complex Adaptation." See also Voigt et al., "Protein Building Blocks Preserved by Recombination."

41. Experimental shuffling of genes has proven fruitful only when the parent genes are highly similar. See He, Friedman, and Bailey-Kellogg, "Algorithms for Optimizing Cross-Overs in DNA Shuffling."

42. In 2012, a research group at the University of Washington reported success in designing a few stable protein folds using a few rules, a lot of computational analysis, and trial and error (only 10 percent of the designed proteins folded as predicted). Though these proteins did form stable folds, they did not perform any actual biological function. The researchers acknowledge that there is likely a tradeoff between stability and functionality in natural proteins. It remains to be seen whether these methods of sequence engineering can create new, stable folds capable of enzymatic activity. What this research highlights is the extreme difficulty of intelligently designing a stable *and* functional protein from scratch, even with the best minds and computational resources working on the problem. There is, therefore, little reason to think that unguided process of exon shuffling could generate both a stable and functional protein. See Nobuyasu et al., "Principles for Designing Ideal Protein Structures"; Marshall, "Proteins Made to Order."

43. Altamirano et al., "Directed Evolution of New Catalytic Activity Using the Alpha/Beta-Barrel Scaffold." See also Altamirano et al., "Retraction: Directed Evolution of New Catalytic Activity Using the Alpha/Beta-Barrel Scaffold."

44. Gauger, "Why Proteins Aren't Easily Recombined."

45. Axe, "The Case Against Darwinian Origin of Protein Folds."

46. See, e.g., Long and Langley, "Natural Selection and the Origin of *Jingwei,* a Chimeric Processed Functional Gene in *Drosophila*"; Wang et al., "Origin of *Sphinx,* a Young Chimeric RNA Gene in *Drosophila melanogaster*"; Begun, "Origin and Evolution of a New Gene Descended from Alcohol Dehydrogenase in *Drosophila.*"

47. Long et al., "Exon Shuffling and the Origin of the Mitochondrial Targeting Function in Plant Cytochrome cl Precursor."

48. Long et al., "The Origin of New Genes. See also Begun, "Origin and Evolution of a New Gene Descended from Alcohol Dehydrogenase in *Drosophila.*"

49. Nurminsky et al., "Selective Sweep of a Newly Evolved Sperm-Specific Gene in *Drosophila.*"

50. Nurminsky et al., "Selective Sweep of a Newly Evolved Sperm-Specific Gene in *Drosophila.*"

51. Brosius, "The Contribution of RNAs and Retroposition to Evolutionary Novelties."

52. Begun, "Origin and Evolution of a New Gene Descended from Alcohol Dehydrogenase in *Drosophila.*"

53. Begun, "Origin and Evolution of a New Gene Descended from Alcohol Dehydrogenase in *Drosophila.*"

54. Papers cited by Long where natural selection was invoked even though the function of the gene and thus the function being selected for was unknown include Begun, "Origin and Evolution of a New Gene Descended from Alcohol Dehydrogenase in *Drosophila*"; Long and Langley, "Natural Selection and the Origin of Jingwei, a Chimeric Processed Functional Gene in *Drosophila*"; and Johnson et al., "Positive Selection of a Gene Family During the Emergence of Humans and African Apes."

55. Logsdon and Doolittle, "Origin of Antifreeze Protein Genes."

56. Courseaux and Nahon, "Birth of Two Chimeric Genes in the *Hominidae* Lineage."

57. Paulding, Ruvolo, and Haber, "The *Tre2* (*USP6*) Oncogene Is a Hominoid-Specific Gene."

58. Chen, DeVries, and Cheng, "Convergent Evolution of Antifreeze Glycoproteins in Antarctic Notothenioid Fish and Arctic Cod."

59. Logsdon and Doolittle, "Origin of Antifreeze Protein Genes."

60. Johnson et al., "Positive Selection of a Gene Family During the Emergence of Humans and African Apes."

61. See Nurminsky et al., "Selective Sweep of a Newly Evolved Sperm-Specific Gene in *Drosophila*"; Chen, DeVries, and Cheng, "Evolution of Antifreeze Glycoprotein Gene from a Trypsinogen Gene in Antarctic Notothenioid Fish"; Courseaux and Nahon, "Birth of Two Chimeric Genes in the *Hominidae* Lineage"; Knowles and McLysaght, "Recent de Novo Origin of Human Protein-Coding Genes"; Wu, Irwin, and Zhang, "De Novo Origin of Human Protein-Coding Genes"; Siepel, "Darwinian Alchemy."

CHAPTER 12: COMPLEX ADAPTATIONS AND THE NEO-DARWINIAN MATH

1. Frazzetta, "From Hopeful Monsters to Bolyerine Snakes?" 62–63.
2. Frazzetta, "From Hopeful Monsters to Bolyerine Snakes?" 63.
3. Gould, "Return of the Hopeful Monsters," 28.
4. Frazzetta, "From Hopeful Monsters to Bolyerine Snakes?" 63.
5. Frazzetta, *Complex Adaptations in Evolving Populations,* 20.
6. As Darwin wrote in the *Origin,* "If we must compare the eye to an optical instrument, we ought in imagination to take a thick layer of transparent tissue, with a nerve sensitive to light beneath, and then suppose every part of this layer to be continually changing slowly in density, so as to separate into layers of different densities and thicknesses, placed at different distances from each other, and with the surfaces of each layer slowly changing in form" (188–89).
7. Frazzetta, *Complex Adaptations in Evolving Populations,* 21.
8. Frazzetta, "Modeling Complex Morphological Change in Evolution," 129.
9. Frazzetta, "Modeling Complex Morphological Change in Evolution," 130.
10. Ehrlich and Holm, *The Process of Evolution,* 157.
11. Until a famous experiment performed by Oswald Avery of the Rockefeller Institute in 1944, many biologists still suspected that proteins might actually be the repositories of genetic information. Meyer, *Signature in the Cell,* 66. Avery, MacCleod, and McCarty, "Induction of Transformation by a Deoxyribonucleic Acid Fraction Isolated from Pneumococcus Type III."
12. Bateson, "Heredity and Variation in Modern Lights," 83–84.
13. Withgott, "John Maynard Smith Dies."
14. Salisbury, "Natural Selection and the Complexity of the Gene," 342–43.
15. Maynard Smith, "Natural Selection and the Concept of a Protein Space," 564.
16. As Maynard Smith writes in *Nature:* "If evolution by natural selection is to occur, functional proteins must form a continuous network which can be traversed by unit mutational steps without passing through nonfunctional intermediates" ("Natural Selection and the Concept of a Protein Space," 564).
17. Maynard Smith, "Natural Selection and the Concept of a Protein Space," 564, emphasis added.
18. Maynard Smith, "Natural Selection and the Concept of a Protein Space," 564.
19. Orr, "The Genetic Theory of Adaptation," 123.
20. Behe and Snoke, "Simulating Evolution by Gene Duplication of Protein Features That Require Multiple Amino Acid Residues."
21. Wen-Hsiung, *Molecular Evolution,* 427
22. Wen-Hsiung, *Molecular Evolution,* 427.
23. Behe, *The Edge of Evolution,* 54.
24. "Powerball—Prizes and Odds," Powerball, http://www.powerball.com/powerball/pb_prizes.asp.
25. "Powerball—Prizes and Odds," Powerball, http://www.powerball.com/powerball/pb_prizes.asp.
26. Behe and Snoke, "Simulating Evolution by Gene Duplication of Protein Features That Require Multiple Amino Acid Residues," 2661.
27. Lynch and Conery, "The Origins of Genome Complexity," 1401–02.
28. Behe and Snoke, "Simulating Evolution by Gene Duplication of Protein Features That Require Multiple Amino Acid Residues," 2661.
29. For the purpose of Behe's argument it did not matter whether the mutations involved in a *single* CCC trait arose in a stepwise, or coordinated fashion. Behe was not *calculating* how long it would take for a single CCC trait to arise. That chloroquine resistance only arises once in every 10^{20} malarial cells was an observed empirical fact based upon public health studies, not a

calculation of waiting times based upon a population genetics model. [See White, "Antimalarial Drug Resistance," 1085.] What Behe was *calculating* was how long it would take for a *hypothetical* trait to arise that *did* require two coordinated mutations, each of the complexity of a single CCC, in order to function—what he called a "double CCC." Whether or not generating a single CCC itself required coordinated mutations was irrelevant. This aspect of his argument has been largely misunderstood by his critics. [See Miller, "Falling Over the Edge," 1055–56; Gross, "Design for Living," 73; Coyne, "The Great Mutator (Review of *The Edge of Evolution*, by Michael J. Behe)," 40–42; Nicholas J. Matzke, "The Edge of Creationism," 566–67.] It's worth noting that although Behe's calculation was for a hypothetical trait, he went on to argue on independent biological grounds that "life is bursting" (63) with features that would require a double CCC to arise.

30. Behe, *The Edge of Evolution*, 135.

31. This assumes that every one of these organisms would be under selection for that particular trait—a completely unrealistic assumption.

32. Durrett and Schmidt, "Waiting for Two Mutations," 1507.

33. Behe, *The Edge of Evolution*, 84–102.

34. Gauger and Axe, "The Evolutionary Accessibility of New Enzyme Functions: A Case Study from the Biotin Pathway."

35. Gauger et al., "Reductive Evolution Can Prevent Populations from Taking Simple Adaptive Paths to High Fitness"; Durrett and Schmidt, "Waiting for Two Mutations."

CHAPTER 13: THE ORIGIN OF BODY PLANS

1. Nüsslein-Volhard and Wieschaus, "Mutations Affecting Segment Number and Polarity"; Wieschaus, "From Molecular Patterns to Morphogenesis." As he comments: "If transcription of a gene was essential for embryonic development, homozygous embryos [i.e., those missing both copies of the gene] should develop abnormally when that gene was eliminated. . . . Based on these defects, it should be possible to reconstruct the normal role of each gene" (316).

2. St. Johnston, "The Art and Design of Genetic Screens," 177.

3. Quotes recorded in contemporaneous notes taken by philosopher of biology Paul Nelson, who was in attendance at this lecture.

4. Quotes recorded in contemporaneous notes taken by philosopher of biology Paul Nelson, who was in attendance at this lecture.

5. Arthur, *The Origin of Animal Body Plans*, 21; Cameron et al., "Evolution of the Chordate Body Plan: New Insights from Phylogenetic Analyses of Deuterostome Phyla"; Michael, "Arthropods: Developmental Diversity Within a (Super) Phylum"; Peterson and Davidson, "Regulatory Evolution and the Origin of the Bilaterians"; Carroll, "Endless Forms: The Evolution of Gene Regulation and Morphological Diversity"; Halder et al., "Induction of Ectopic Eyes by Targeted Expression of the *Eyeless* Gene in *Drosophila*."

6. Arthur, *The Origin of Animal Body Plans*, 21.

7. Van Valen, "How Do Major Evolutionary Changes Occur?" 173.

8. Thomson, "Macroevolution," 111.

9. John and Miklos, *The Eukaryote Genome in Development and Evolution*, 309.

10. Thomson, "Macroevolution."

11. See, e.g., the special issue of *Development* (December 1996) dedicated to the large-scale mutagenesis of the model vertebrate *Danio rerio* (the zebrafish), especially Haffter et al., "The Identification of Genes with Unique and Essential Functions in the Development of the Zebrafish, *Danio rerio*"; or the many fruit-fly mutagenesis experiments summarized in Bate and Arias, eds., *The Development of Drosophila melanogaster*. Summarizing the evidence from a wide range of animal systems, Wallace Arthur writes, "Those genes that control key early developmental processes are involved in the establishment of the basic body plan. Mutations in these genes will usually be extremely disadvantageous, and it is conceivable that they are *always so*" (*The Origin of Animal Body Plans*, 14, emphasis in original). Arthur goes on to speculate that because developmental regulatory genes often differ between phyla, perhaps "mutations of these genes are sometimes advantageous" (15). He offers no evidence for such mutations, however, other than as a deduction from his prior assumption of common descent.

12. Fisher, *The Genetical Theory of Natural Selection*, 44.

13. Wallace, "Adaptation, Neo-Darwinian Tautology, and Population Fitness," 70.

14. Nüsslein-Volhard and Wieschaus, "Mutations Affecting Segment Number and Polarity in *Drosophila*"; Lawrence and Struhl, "Morphogens, Compartments and Pattern."

15. Van Valen, "How Do Major Evolutionary Changes Occur?" 173.

16. McDonald, "The Molecular Basis of Adaptation," 93.

17. McDonald, "The Molecular Basis of Adaptation," 93.

18. Wimsatt, "Generativity, Entrenchment, Evolution, and Innateness: Philosophy, Evolutionary Biology, and Conceptual Foundations of Science"; Wimsatt and Schank, "Generative Entrenchment, Modularity and Evolvability: When Genic Selection Meets the Whole Organism."

19. Nelson notes that there is one noteworthy exception to this generalization: the *loss* of structures. A wide range of well-documented cases—including cave animals, island birds and insects, and marine and freshwater fishes—show that many animals will tolerate, or actually thrive, after losing traits to mutation—as long as those traits are not essential for survival in some specialized environment. For example, macromutations resulting in loss of vision have had no deleterious effects on some species of now blind cave fish that no longer have a need to see. Similarly, macromutations that disrupt wing formation in an insect—ordinarily devastating in an environment where functional wings are essential equipment—might well be tolerated in an island setting where that species faces no need to fly. The processes that generate these exceptions, however, do not help to explain the *origin* of form such as occurs in the Cambrian explosion. Clearly, processes that result in a loss of form and structure cannot be credibly invoked to explain the origin of form and structure in the first place.

20. Darwin, *On the Origin of Species*, 108.

21. Løvtrup, "Semantics, Logic and Vulgate Neo-Darwinism," 162.

22. Britten and Davidson, "Gene Regulation for Higher Cells," 57.

23. Britten and Davidson, "Gene Regulation for Higher Cells," 57. One exception to this rule are cells called "erythrocytes" in humans.

24. Britten and Davidson, "Gene Regulation for Higher Cells," 57.

25. Britten and Davidson, "Gene Regulation for Higher Cells," 353.

26. Cameron et al., "A Sea Urchin Genome Project," 9514.

27. Oliveri, Tu, and Davidson, "Global Regulatory Logic for Specification of an Embryonic Cell Lineage."

28. Davidson, *The Regulatory Genome*, 16.

29. Davidson, *The Regulatory Genome*, 16.

30. Davidson, "Evolutionary Bioscience as Regulatory Systems Biology," 38.

31. Davidson, "Evolutionary Bioscience as Regulatory Systems Biology," 40, emphasis added.

32. As Davidson explains, "Interference with expression of any [multiply linked dGRNs] by mutation or experimental manipulation has severe effects on the phase of development that they initiate. This accentuates the selective conservation of the whole subcircuit, *on pain of developmental catastrophe*" (Davidson and Erwin, "An Integrated View of Precambrian Eumetazoan Evolution," 8).

33. Davidson, *The Regulatory Genome*, 195.

34. Davidson, "Evolutionary Bioscience as Regulatory Systems Biology," 35–36.

CHAPTER 14: THE EPIGENETIC REVOLUTION

1. Spemann and Mangold, "Induction of Embryonic Primordia by Implantation of Organizers from a Different Species."

2. Harvey, "Parthenogenetic Merogony or Cleavage Without Nuclei in *Arbacia punctulata*"; "A Comparison of the Development of Nucleate and Non-nucleate Eggs of *Arbacia punctulata*."

3. Brachet, Denis, and De Vitry, "The Effects of Actinomycin D and Puromycin on Morphogenesis in Amphibian Eggs and *Acetabularia mediterranea*."

4. Masui, Forer, and Zimmerman, "Induction of Cleavage in Nucleated and Enucleated Frog Eggs by Injection of Isolated Sea-Urchin Mitotic Apparatus."

5. Müller and Newman, "Origination of Organismal Form," 8.

6. Müller and Newman, "Origination of Organismal Form," 8.

7. Müller and Newman, "Origination of Organismal Form," 7.

8. Müller and Newman, "Origination of Organismal Form," 7. Or as Müller also explains, the question of how "individualized constructional elements" are organized during "the evolution of organismal form" is "not satisfactorily answered by current evolutionary theories"; Müller, "Homology," 57–58.

9. Levinton, *Genetics, Paleontology, and Macroevolution*, 485.

10. In 1942 Conrad Waddington coined the word "epigenetics" to refer to the study of "the processes involved in the mechanism by which the genes of the genotype bring about phenotypic effects" ("The Epigenotype," 1). Some more recent biologists have used it to refer to information in chromosomal structures that do not depend on the underlying DNA sequence. I will use it to refer to any biological information that is not encoded in a DNA sequence.

11. Goodwin, "What Are the Causes of Morphogenesis?"; Nijhout, "Metaphors and the Role of

Genes in Development"; Sapp, *Beyond the Gene;* Müller and Newman, "Origination of Organismal Form"; Brenner, "The Genetics of Behaviour"; Harold, *The Way of the Cell.*

12. Harold, *The Way of the Cell,* 125.

13. Harold, "From Morphogenes to Morphogenesis," 2774; Moss, *What Genes Can't Do.* Of course, many proteins bind chemically with each other to form complexes and structures within cells. Nevertheless, these "self-organizational" properties do not fully account for higher levels of organization in cells, organs, or body plans. Or, as Moss has explained "Neither DNA nor any other aperiodic crystal constitutes a unique repository of heritable stability in the cell; in addition, the chemistry of the solid state does not constitute either a unique or even an ontologically or causally privileged basis for explaining the existence and continuity of order in the living world . . ." Moss, *What Genes Can't Do,* 76.

14. Wells, "Making Sense of Biology," 121.

15. Harold, *The Way of the Cell,* 125.

16. Ally et al., "Opposite-Polarity Motors Activate One Another to Trigger Cargo Transport in Live Cells"; Gagnon and Mowry, "Molecular Motors."

17. Marshall and Rosenbaum, "Are There Nucleic Acids in the Centrosome?"

18. Poyton, "Memory and Membranes"; Edidin, "Patches, Posts and Fences."

19. Frohnhöfer and Nüsslein-Volhard, "Organization of Anterior Pattern in the *Drosophila* Embryo by the Maternal Gene *Bicoid*"; Lehmann and Nüsslein-Volhard, "The Maternal Gene *Nanos* Has a Central Role in Posterior Pattern Formation of the *Drosophila* Embryo."

20. Roth and Lynch, "Symmetry Breaking During *Drosophila* Oogenesis."

21. Skou, "The Identification of the Sodium-Pump as the Membrane-Bound Na^+/K^+-ATPase."

22. Levin, "Bioelectromagnetics in Morphogenesis."

23. Shi and Borgens, "Three-Dimensional Gradients of Voltage During Development."

24. Schnaar, "The Membrane Is the Message," 34–40.

25. Schnaar, "The Membrane Is the Message," 34–40; Gabius et al., "Chemical Biology of the Sugar Code," 740-764; Gabius, "Biological Information Transfer Beyond the Genetic Code: The Sugar Code," 108–121.

26. Gabius et al., "Chemical Biology of the Sugar Code," 741. See also Gabius, "Biological Information Transfer Beyond the Genetic Code: The Sugar Code," 108–21.

27. Gabius, "Biological Information Transfer Beyond the Genetic Code," 109; Gabius et al., "Chemical Biology of the Sugar Code," 741.

28. Spiro, "Protein Glycosylation."

29. Palade, "Membrane Biogenesis."

30. Babu, Kriwacki, and Pappu, "Versatility from Protein Disorder"; Uversky and Dunker, "Understanding Protein Non-Folding"; Fuxreiter and Tompa, "Fuzzy Complexes."

31. Wells, "Making Sense of Biology: The Evidence for Development by Design," 121.

32. McNiven and Porter, "The Centrosome."

33. Lange et al., "Centriole Duplication and Maturation in Animal Cells"; Marshall and Rosenbaum, "Are There Nucleic Acids in the Centrosome?"

34. Sonneborn, "Determination, Development, and Inheritance of the Structure of the Cell Cortex," 1–13; Frankel, "Propagation of cortical differences in *tetrahymena*," 607–623; Nanney, "The ciliates and the cytoplasm," 163–170.

35. Moss, *What Genes Can't Do.*

36. Harold, "From Morphogenes to Morphogenesis," 2767.

37. See Müller and Newman, "The Origination of Organismal Form," 7.

38. Thomson, "Macroevolution," 107.

39. Miklos, "Emergence of Organizational Complexities During Metazoan Evolution."

40. Gilbert, Opitz, and Raff, "Resynthesizing Evolutionary and Developmental Biology," 361. The Brian Goodwin quotation is from *How the Leopard Changed Its Spots.*

41. Gilbert, Opitz, and Raff, "Resynthesizing Evolutionary and Developmental Biology." Specifically, they argue that changes in morphogenetic fields might produce large-scale changes in the developmental programs and, ultimately, in the body plans of organisms. However, they offer no evidence that such fields—if indeed they exist—can be altered to produce advantageous variations in body plan, though such a condition is necessary to any successful causal theory of macroevolution.

42. Webster, *How the Leopard Changed Its Spots,* 33; Webster and Goodwin, *Form and Transformation,* x; Gunter Theißen, "The Proper Place of Hopeful Monsters in Evolutionary Biology," 351; Marc Kirschner and John Gerhart, *The Plausibility of Life,* 13; Schwartz, *Sudden Origins,* 3, 299–300; Erwin, "Macroevolution Is More Than Repeated Rounds of Microevolution"; Davidson, "Evolutionary Bioscience as Regulatory Systems Biology," 35; Koonin, "The Origin at 150," 473–5; Conway Morris, "Walcott, the Burgess Shale, and Rumours of a Post-Darwinian World," R928–R930; Carroll, "Towards a New Evolutionary Synthesis," 27; Wagner, "What Is the Promise of

Developmental Evolution?"; Wagner and Stadler, "Quasi-independence, Homology and the Unity-of Type"; Becker and Lönnig, "Transposons: Eukaryotic," 529–39; Lönnig and Saedler, "Chromosomal Rearrangements and Transposable Elements," 402; Müller and Newman, "Origination of Organismal Form," 7; Kauffman, *At Home in the Universe*, 8; Valentine and Erwin, "Interpreting Great Developmental Experiments," 96; Sermonti, *Why Is a Fly Not a Horse?*; Lynch, *The Origins of Genome Architecture*, 369; Shapiro, *Evolution*, 89, 128.

The perspective of Eugene Koonin, a biologist at the National Center for Biotechnology Information at the National Institutes of Health, provides just one good example of this skepticism. He argues: "The edifice of the modern synthesis has crumbled, apparently, beyond repair . . . The summary of the state of affairs on the 150th anniversary of the *Origin* is somewhat shocking. In the postgenomic era, all major tenets of the modern synthesis have been, if not outright overturned, replaced by a new and incomparably more complex vision of the key aspects of evolution. So, not to mince words, the modern synthesis is gone. What comes next? The answer suggested by the Darwinian discourse of 2009 is a postmodern state, not so far a postmodern synthesis. Above all, such a state is characterized by the pluralism of processes and patterns in evolution that defies any straightforward generalization." Koonin, "The Origin at 150," 473–75. David J. Depew and Bruce H. Weber, writing in the journal *Biological Theory*, are even more frank: "Darwinism in its current scientific incarnation has pretty much reached the end of its rope" (89–102).

CHAPTER 15: THE POST-DARWINIAN WORLD AND SELF-ORGANIZATION

1. Conway Morris, "Walcott, the Burgess Shale and Rumours of a Post-Darwinian World," R928.
2. Conway Morris, "Walcott, the Burgess Shale and Rumours of a Post-Darwinian World," R930.
3. Mazur, *The Altenberg 16: An Exposé of the Evolution Industry*. See also Whitfield, "Biological Theory."
4. Budd, quoted in Whitfield, "Biological Theory," 282.
5. Valentine and Erwin, "Interpreting Great Developmental Experiments," 97.
6. Endler, *Natural Selection in the Wild*, 46, 248; Lewontin, "Adaptation," 212–30.
7. Gerhart and Kirschner, *The Plausibility of Life: Resolving Darwin's Dilemma*, 10.
8. Kauffman, *The Origins of Order: Self-Organization and Selection in Evolution*.
9. In *The Origins of Order*, Kauffman seeks to show that self-organizational processes could help account for both the origin of the first life and the origin of subsequent forms of life, including new animal body plans. In *Signature in the Cell*, I examined Kauffman's specific proposal for explaining the origin of the first life. Here I'll examine his proposal for explaining the origin of animal form.
10. Kauffman, *The Origins of Order*, 443.
11. Kauffman, *The Origins of Order*, 443.
12. Kauffman, *The Origins of Order*, 443.
13. Kauffman, *The Origins of Order*, 537.
14. Kauffman, *The Origins of Order*, 539.
15. Kauffman, *The Origins of Order*, 549–66.
16. Kauffman, *The Origins of Order*, 590.
17. Kauffman, *The Origins of Order*, 298.
18. Kauffman, *The Origins of Order*, 537, emphasis added.
19. Kauffman, *At Home in the Universe*, 68.
20. Kauffman, *At Home in the Universe*, 47–92.
21. Kauffman, *At Home in the Universe*, 71.
22. Kauffman, *At Home in the Universe*, 75–92.
23. Kauffman, *At Home in the Universe*, 86–88.
24. Kauffman, *At Home in the Universe*, 85, emphasis added.
25. Kauffman, *At Home in the Universe*, 53, 89, 102.
26. Kauffman, *At Home in the Universe*, 200.
27. Indeed, Kauffmann explicitly notes, "Mutants affecting early stages of development disrupt development more than do mutants affecting late stages of development. A mutation disrupting formation of the spinal column and cord is more likely to be lethal than one affecting the number of fingers that form" (*At Home in the Universe*, 200).
28. Newman, "Dynamical Patterning Modules."
29. Newman, "Dynamical Patterning Modules," 296; see also "The Developmental Genetic Toolkit and the Molecular Homology-Analogy Paradox."

30. Newman, "Dynamical Patterning Modules," 284.
31. Newman, "Dynamical Patterning Modules," 284.
32. Newman, "Dynamical Patterning Modules," 285.
33. Newman, "Dynamical Patterning Modules," 285.
34. Newman, "Animal Egg as Evolutionary Innovation," 570.
35. Newman, "Animal Egg as Evolutionary Innovation"; "Dynamical Patterning Modules."
36. Newman, "Animal Egg as Evolutionary Innovation," 470–71; see also Newman and Bhat, "Dynamical Patterning Modules: Physico-Genetic Determinants."
37. In a 2011 paper, Newman attempts to account for the origin of the crucial egg stage in animal development, a long-unsolved puzzle for neo-Darwinism. He does this by proposing, again, an interplay between the developmental-genetic regulatory toolkit and a host of self-organizational and epigenetic processes. He envisions the egg stage arising in three steps. First, he sees the interaction between "the metazoan developmental-genetic toolkit and certain physical processes" organizing "primitive animal body plans independently of an egg stage." Second, he envisions the emergence of a "proto-egg" arising as the result of the "adaptive specialization of cells" as they are released from aggregates of other cells. Third, he notes that since self-organizing processes (such as "egg-patterning processes") are known to reorganize the contents of the cellular cytoplasm during development, these same processes could have done so in the past, further sculpting the proto-egg into something more like the egg stage in animal development observed today. He insists, moreover, that the structures induced by these egg-patterning processes are not adaptive "in the sense of having been gradually arrived at through multiple cycles of selection," but instead that they have resulted from self-organizational physical and chemical processes. See Newman, "Animal Egg as Evolutionary Innovation: A Solution to the 'Embryonic Hourglass' Puzzle," 467–83.
38. Khalturin et al., "More Than Just Orphans: Are Taxonomically-Restricted Genes Important in Evolution?" 404–413; Tautz and Domazet-Lošo, "The evolutionary origin of orphan genes," 692–70; Beiko, "Telling the Whole Story in a 10,000-Genome World," 34.
39. See the discussion in Chapter 14, pages 277–81.
40. See the discussion in Chapter 14, pages 277–81.
41. Pivar, *Lifecode*; *On the Origin of Form;* Prigogine, Nicolis, and Babloyantz, "Thermodynamics of Evolution"; Wolfram, "A New Kind of Science," 398.
42. Meyer, *Signature in the Cell*, 254–55.
43. Yockey, "A Calculation of the Probability of Spontaneous Biogenesis by Information Theory," 380.
44. For example, see the popular computer game among amateur programmers, the "Game of Life," created by the British mathematician John Horton Conway.
45. Kauffman, "The End of a Physics Worldview: Heraclitus and the Watershed of Life."
46. Kauffman, "The End of a Physics Worldview: Heraclitus and the Watershed of Life."

CHAPTER 16: OTHER POST-NEO-DARWINIAN MODELS

1. See Gould, "Is a New and General Theory of Evolution Emerging?" 120.
2. Gould does not use the term "macromutation" anywhere in his famous 1980 article ("Is a New and General Theory of Evolution Emerging?"). He does, however, use the term "micromutation" (see 120) and challenges the sufficiency of accumulated micromutations to explain "macroevolution" (a term used throughout the article).
3. Goldschmidt, *The Material Basis of Evolution*, 395.
4. The rejection of large-scale mutations affecting morphology and function as adaptive nonstarters emerged early and persisted as one of the defining aspects of the neo-Darwinian synthesis. Neo-Darwinian paleontologist and macroevolution theorist Jeffrey Levinton, for instance, gives expression to the widely held skepticism about the evolutionary plausibility of such mutants in his major textbook dealing with macroevolution:

> *As a general rule, major developmental mutants give a picture of hopeless monsters, rather than hopeful change. Epigenetic and genetic pleiotropy [i.e., side effects] both impart great burden to any major developmental perturbation. Thus it is unlikely that mutants affecting any fundamental prepattern in development are likely to produce a functional organism. Genes that activate switches in prepatterns are not sufficiently isolated in effect on other parts of the phenotype to expect major saltations. The cyclops mutant of the [brine shrimp] Artermia is lethal. The homeotic mutants of Drosophila melanogaster suffer similar fates. . . . Disruptions, i.e., mutants, have drastic effects on other parts of the phenotype. . . . Thus, the accumulated evidence suggests that major developmental mutants are of minor significance in evolution. The side effects are drastic. (Genetics, Paleontology, and Macroevolution, 252–54)*

5. The central difficulty with relying on developmental macro-mutations to generate innovations in form, many neo-Darwinians noted, arises from the consequence of rapidly changing a system of genetic and developmental switches directed toward producing one "target" (stable adult form) to another system of such switches directed toward producing another form. Geneticist Bruce Wallace, trained by Theodosius Dobzhansky at Columbia University, explains: "The *Bauplan* [body plan] of an organism . . . can be thought of as the arrangement of genetic switches that control the course of the embryonic and subsequent development of the individual; such control must operate properly both in time generally and sequentially in the separately differentiated tissues. Selection, both natural and artificial, that leads to morphological change and other developmental modification does so by altering the settings and triggerings of these switches. . . . The extreme difficulty encountered when attempting to transform one organism into another but still functioning one lies in the difficulty in resetting a number of the many controlling switches in a manner that still allows for the individual's orderly (somatic) development" ("Adaptation, Neo-Darwinian Tautology, and Population Fitness," 70). Our discussion in Chapter 13 suggests that the need to alter these functionally integrated switches also presents an obstacle to the efficacy of the neo-Darwinian mechanism.

6. Gould, "Is a New and General Theory of Evolution Emerging?" 127.

7. Of course, the range of post-neo-Darwinian theories is not exhausted by this chapter's survey of four prominent contenders. In a recent review paper, evolutionary biologist Armin Moczek, of Indiana University, examined three additional ideas attempting to move beyond what Moczek calls the "unrealistic and unproductive" assumptions of gene-centered neo-Darwinian theory. Those ideas are, respectively: (1) the theory of "facilitated variation" (Gerhart and Kirschner, "The Theory of Facilitated Variation"), the theory of "genetic accommodation" and "niche construction theory." See Moczek, "The Nature of Nurture and the Future of Evodevo." Short explanations and critiques of these models are posted on the website for this book, www.darwinsdoubt.com.

8. Gould, "Is a New and General Theory of Evolution Emerging?"

9. Gilbert, Opitz, and Raff, "Resynthesizing Evolutionary and Developmental Biology," 362.

10. The assumption that evolutionary biologists could ignore the role of developmental biology derived in part from the need for population geneticists to make simplifying assumptions in order "to keep the mathematics tractable," note developmental biologists Michael Palopoli and Nipam Patel. As they explain, "It was assumed that evolutionary changes in genotype are translated into phenotypic changes by an undefined set of epigenetic laws; in other words, developmental evolution was ignored [by neo-Darwinism] in order to focus on the dynamics of allele frequency changes in populations" ("Neo-Darwinian Developmental Evolution, 502).

11. Raff, *The Shape of Life.*

12. Schwartz, "Homeobox Genes, Fossils, and the Origin of Species"; Schwartz, *Sudden Origins;* Goodwin, *How the Leopard Changed Its Spots;* Carroll, *Endless Forms Most Beautiful.*

13. Experimental mutagenesis of fruit flies (*Drosophila melanogaster*) began in earnest at Columbia University, in the breeding laboratories of Thomas H. Morgan and others, during the first decade of the twentieth century.

14. Meyer et al., *Explore Evolution,* 108.

15. Body Plans, http://ncse.com/book/export/html/2585 (accessed November 6, 2012).

16. Normal Protein, http://ncse.com/book/export/html/2580 (accessed November 6, 2012).

17. Body Plans, http://ncse.com/book/export/html/2585 (accessed November 6, 2012).

18. Prud'homme, Gompel, and Carroll, "Emerging Principles of Regulatory Evolution."

19. Prud'homme, Gompel, and Carroll, "Emerging Principles of Regulatory Evolution," 8605.

20. Hoekstra and Coyne note explicitly that the best examples of alleged CRE-induced mutations showed "losses of traits rather than the origin of new traits" ("The Locus of Evolution," 1006).

21. Hoekstra and Coyne, "The Locus of Evolution," 996.

22. Schwartz, *Sudden Origins,* 3.

23. Schwartz, *Sudden Origins,* 13. According to Schwartz, "At the genetic level, major morphological novelty can indeed be accomplished in the twinkling of an eye. All that is necessary is that homeobox genes are either turned on or they are not" (362).

24. McGinnis and Kurziora, "The Molecular Architects of Body Design," 58.

25. Lindsey and Grell, *Guide to Genetic Variations of Drosophila melanogaster.*

26. Lewis, "A Gene Complex Controlling Segmentation in *Drosophila*"; Peifer and Bender, "The Anterobithorax and Bithorax Mutations of the Bithorax Complex"; Fernandes et al., "Muscle Development in the Four-Winged *Drosophila* and the Role of the Ultrabithorax Gene."

27. Szathmáry, "When the Means Do Not Justify the End," 745.

28. Scott and Carroll, "The Segmentation and Homeotic Gene Network in Early Drosophila Development."

29. Davidson and Erwin, "An Integrated View of Precambrian Eumetazoan Evolution."
30. Schwartz, *Sudden Origins,* 362.
31. Szathmáry, "When the Means Do Not Justify the End," 745.
32. Schwartz, *Sudden Origins,* 362.
33. Szathmáry, "When the Means Do Not Justify the End," 745.
34. Panganiban et al., "The Origin and Evolution of Animal Appendages."
35. Instead, epigenetic information and structures actually determine the function of many *Hox* genes. This can be seen dramatically when the same *Hox* gene (as determined by nucleotide sequence homology) regulates the development of the strikingly different (i.e., classically nonhomologous) anatomical features found in different phyla. For instance, in arthropods, the *Hox* gene *Distal-less* is required for the normal development of limbs, but homologous genes are found in vertebrates (e.g., the *Dlx* gene in mice), where the gene also plays a key role in limb development—albeit a vertebrate (internal skeleton), not arthropod (external skeleton) limb. *Distal-less* homologues in yet other phyla, such as echinoderms, regulate the development of tube feet and spines—again, anatomical features classically not homologous to arthropod or vertebrates limbs.

In each case, the roles of the *Hox* genes are governed "top-down" by the higher-level organismal contexts in which they occur. Panganiban et al., "The Origin and Evolution of Animal Appendages."

36. Despite all this, some evolutionary theorists have argued that the emergence of *Hox* genes in Precambrian organisms may have triggered the Cambrian explosion by providing the raw materials for the diversification of body plans (Carroll, *Patterns and Processes of Vertebrate Evolution*). Yet, in addition to those difficulties already noted, recent studies have highlighted another problem with attributing the origin of body plans to *Hox* genes. *Hox* genes first emerged long before the diversification of the various bilaterian phyla, suggesting—because of the length of the time lag—that something else must have responsible for the Cambrian explosion. As a paper in the journal *Science* explains, "The temporal lag between the initial construction of these networks and the eventual appearance of bilaterian fossils suggests that the solution to the dilemma of the Cambrian explosion lies not solely with this genomic and developmental potential, but instead must also be found in the ecology of the Cambrian radiation itself" (Erwin et al., "The Cambrian Conundrum," 1095). See also de Rosa et al., "Hox Genes in Brachiopods and Priapulids and Protostome Evolution."
37. Lynch, "The Origins of Eukaryotic Gene Structure," 454.
38. Lynch, "The Origins of Eukaryotic Gene Structure," 450. He actually goes even further, contending that "random genetic drift can impose a strong barrier to the advancement of molecular refinements by adaptive processes." Lynch, "Evolutionary Layering and the Limits to Cellular Perfection."
39. Jurica, "Detailed Closeups and the Big Picture of Spliceosomes," 315. See also Butcher, "The Spliceosome as Ribozyme Hypothesis Takes a Second Step," 12211–12; Nilsen, "The Spliceosome: The Most Complex Macromolecular Machine in the Cell?" 1147–49.
40. Quoted in Azar, "Profile of Michael Lynch," 16015.
41. On my website for this book, www.darwinsdoubt.com, I explain why the origin of the eukaryotic cell presents such a formidable challenge to all theories of unguided evolution.
42. Lynch, "The Origins of Eukaryotic Gene Structure," 450–68 (emphasis added).
43. See Lynch, "The Frailty of Adaptive Hypotheses for the Origin of Organismal Complexity," 8597–604.
44. Gauger points out that Lynch, "offers no explanation of how non-adaptive forces can produce the functional genomic and organismal complexity we observe in modern species." [Ann Gauger, "The Frailty of the Darwinian Hypothesis, Part 2."] Jerry Coyne observes much the same: "Both drift and natural selection produce genetic change that we recognize as evolution. But there's an important difference. Drift is a random process, while selection is the anti-thesis of randomness. . . . As a purely random process, genetic drift can't cause the evolution of adaptations. It could never build a wing or an eye. That takes nonrandom natural selection." [Coyne, *Why Evolution Is True,* 123.]
45. Lynch and Abegg, "The Rate of Establishment of Complex Adaptations," 1404.
46. Lynch and Abegg, "The Rate of Establishment of Complex Adaptations," 1414.
47. Axe, "The Limits of Complex Adaptation: An Analysis Based on a Simple Model of Structured Bacterial Populations," 3.
48. Axe, "The Limits of Complex Adaptation: An Analysis Based on a Simple Model of Structured Bacterial Populations," 3.
49. Darwin, *On the Origin of Species,* 134–38. Darwin's own theory of transmission genetics, dubbed "pangenesis," postulated that a host of minute heredity particles, which he called "gemmules," accumulated in the reproductive organs of organisms, carrying information about the life

history and environmental circumstances of the parent. This information would then be transmitted at reproduction to offspring, allowing the "inheritance" of "acquired" characteristics.

50. Darwin. *The Illustrated Origin of Species* (6th edition), 95, writes, "I think there can be no doubt that use in our domestic animals has strengthened and enlarged certain parts, and disuse diminished them; and such modifications are inherited."

51. Some question exists about the historical accuracy of calling these twenty-first-century ideas "neo-Lamarckian" in light of the actual content of Jean-Baptiste de Lamarck's views when compared with the enormous growth of knowledge about heredity over the past two hundred years. Given that Jablonka adapts the term "Lamarckism" to her own position, however, I follow that practice, with the caveats about differences in content noted.

52. Jablonka and Raz, "Transgenerational Epigenetic Inheritance," 168.

53. Jablonka and Raz, "Transgenerational Epigenetic Inheritance," 138, emphasis added.

54. Jablonka and Raz, "Transgenerational Epigenetic Inheritance,"162.

55. Shapiro, *Evolution,* 2.

56. Shapiro contends that the neo-Darwinian insistence on fundamental randomness arose for philosophical, not empirical (or observational) reasons, having to do with the exclusion of "supernatural intervention" in the origin of organisms.

57. Shapiro, *Evolution,* 12.

58. Shapiro, *Evolution,* 14.

59. Shapiro, *Evolution,* 14.

60. Shapiro, *Evolution,* 16.

61. Shapiro, *Evolution,* 16.

62. Shapiro, *Evolution,* 14.

63. As biologist Bénédicte Michel observes, "Clearly, it is important for bacteria to keep all levels of the SOS response under tight control. There is no utility to the organism of using error-prone polymerases longer than absolutely necessary" ("After 30 Years, the Bacterial SOS Response Still Surprises Us," 1175).

64. Shapiro, "Darwin's Black Box: The Biochemical Challenge to Evolution-Book Reviews."

CHAPTER 17: THE POSSIBILITY OF INTELLIGENT DESIGN

1. As Dawkins explains: "Natural selection, the blind, unconscious, automatic process which Darwin discovered and which we now know is the explanation for the existence and apparently purposeful form of all life, has no purpose in mind. It has no mind and no mind's eye" (*The Blind Watchmaker,* 5).

2. Mayr, Foreword, in Ruse, ed., *Darwinism Defended,* xi–xii.

3. Ayala, "Darwin's Greatest Discovery," 8572.

4. As Dawkins notes: "Biology is the study of complicated things that give the appearance of having been designed for a purpose" (*The Blind Watchmaker,* 1). Crick likewise explains: "Organisms appear as if they had been designed to perform in an astonishingly efficient way, and the human mind therefore finds it hard to accept that there need be no Designer to achieve this" (*What Mad Pursuit,* 30). Lewontin also observes that living organisms "appear to have been carefully and artfully designed" ("Adaptation").

5. Dawkins, *River Out of Eden,* 17.

6. Dawkins, *The Blind Watchmaker,* 1.

7. Gillespie, *Charles Darwin and the Problem of Creation,* 83–108.

8. Thaxton, Bradley, and Olsen, *The Mystery of Life's Origin,* 211.

9. We have observational evidence in the present that intelligent investigators can (and do) build contrivances to channel energy down nonrandom chemical pathways to bring about some complex chemical synthesis, even gene building. May not the principle of uniformity then be used in a broader frame of consideration to suggest that DNA had an intelligent cause at the beginning? (Thaxton, Bradley, and Olsen, *The Mystery of Life's Origin,* 211).

10. Peirce, *Collected Papers,* 2:372–88; Meyer, "Of Clues and Causes," 25. See also Whewell, "Lyell's Principles of Geology"; *The Philosophy of the Inductive Sciences.*

11. Peirce, *Collected Papers,* 2:372–88. See also Fann, *Peirce's Theory of Abduction,* 28–34; Whewell, "Lyell's Principles of Geology"; *The Philosophy of the Inductive Sciences.*

12. Gould, "Evolution and the Triumph of Homology," 61. See also Whewell, "Lyell's Principles of Geology"; *The Philosophy of the Inductive Sciences.*

13. Peirce, "Abduction and Induction," 150–54.

14. Chamberlain, "The Method of Multiple Working Hypotheses."

15. Conniff, "When Continental Drift Was Considered Pseudoscience."

16. Conniff, "When Continental Drift Was Considered Pseudoscience."

17. Oreskes, "From Continental Drift to Plate Tectonics," 12.
18. Heirtzler, "Sea-Floor Spreading"; Hurley, "The Confirmation of Continental Drift"; Vine, "Reversals of Fortune."
19. For example, Xavier Le Pichon recalls: "I was progressively forced by the convincing power of the magnetic [data]" ("My Conversion to Plate Tectonics," 212).
20. Lipton, *Inference to the Best Explanation,* 1.
21. Lipton, *Inference to the Best Explanation,* 1.
22. Lyell, *Principles of Geology,* 75–91.
23. Kavalovski, "The *Vera Causa* Principle," 78–103.
24. Scriven, "Explanation and Prediction in Evolutionary Theory," 480; Gallie, "Explanations in History and the Genetic Sciences"; Sober, *Reconstructing the Past,* 1–5.
25. Meyer, "Of Clues and Causes," 96–108.

CHAPTER 18: SIGNS OF DESIGN IN THE CAMBRIAN EXPLOSION

1. Davidson, "Evolutionary Bioscience as Regulatory Systems Biology," 35.
2. Erwin and Davidson, "The Evolution of Hierarchical Gene Regulatory Networks," 141.
3. Davidson, "Evolutionary Bioscience as Regulatory Systems Biology," 6.
4. Erwin, "Early Introduction of Major Morphological Innovations," 288.
5. Erwin, "Early Introduction of Major Morphological Innovations," 288, emphasis added.
6. As Erwin puts it: "*There is every indication that the range of morphological innovation possible in the Cambrian is simply not possible today*" ("The Origin of Body Plans," 626, emphasis added).
7. Conway Morris, *Life's Solution: Inevitable Humans in a Lonely Universe,* 327; see also "Evolution: Bringing Molecules into the Fold"; "The Cambrian Explosion of Metazoans."
8. Conway Morris, *Life's Solution,* 327. See also Conway Morris, "Bringing Molecules into the Fold," 8.
9. Shapiro, *Evolution.*
10. Dawkins, *River Out of Eden,* 17.
11. Hood and Galas, "The Digital Code of DNA."
12. Gates, *The Road Ahead,* 188.
13. Quastler, *The Emergence of Biological Organization,* 16.
14. Of course, the phrase "large amounts of specified information" begs a quantitative question, namely, "How much specified information would the minimally complex cell have to have before it implied design?" In *Signature in the Cell,* I give and justify a precise quantitative answer to this question. I show that the *de novo* emergence of 500 or more bits of specified information reliably indicates design. Meyer, *Signature in the Cell,* 294.
15. Dawkins, *The Blind Watchmaker,* 47–49; Küppers, "On the Prior Probability of the Existence of Life"; Scheider, "The Evolution of Biological Information"; Lenski, "The Evolutionary Origin of Complex Features." For a critique of these genetic algorithms and claims that they simulate the ability of natural selection and random mutation to generate new biological information apart from intelligent activity, see Meyer, *Signature in the Cell,* 281–95.
16. Berlinski, "On Assessing Genetic Algorithms."
17. Rodin, Szathmáry, and Rodin, "On the Origin of the Genetic Code and tRNA Before Translation," 2.
18. Denton, *Evolution: A Theory in Crisis,* 309–11.
19. Polanyi, "Life Transcending Physics and Chemistry"; "Life's Irreducible Structure."
20. Oliveri and Davidson, "Built to Run, Not Fail." All animal embryos, notes Davidson, must "*turn on the right regulatory genes in the right place.* These genes must also be dynamically locked on; the regulatory state of cells in a given spatial domain must further be made dependent on signaling among them all; the expression of these same regulatory genes must be specifically forbidden anywhere else; and then, on top of all that, specific alternative states must be excluded. These components are of course interlinked. . . . In the sea urchin embryo, where all of the above are to be found, *disarming any one of these subcircuits produces some abnormality in expression*" (511; emphasis added).
21. Davidson, *Genomic Regulatory Systems,* 54.
22. "Sameness" (i.e., homology) is determined by nucleotide-sequence conservation within a reading frame.
23. Newman, "The Developmental Genetic Toolkit and the Molecular Homology-Analogy Paradox," 12.
24. Within evolutionary theory, several proposals have attempted to resolve the paradox. Perhaps the most popular modifies the neo-Darwinian definition of the concept of "homology" by

preceding it with the adjective "deep," a verbal slipcover that—if one is cynically inclined—ought to elicit sardonic commentary befitting a late-night comedy sketch. Strictly speaking, we should remember that the neo-Darwinian prediction about genes, phenotypes, and homology was found to be *wrong*. Nothing wrong ever turned out right by putting the adjective "deep" in front of it.

25. Carroll, *Endless Forms Most Beautiful*, 72.

26. Gould, *The Structure of Evolutionary Theory*, 1065. Stuart Newman observes: "It came as a big surprise to workers in the fields of evolutionary and developmental biology . . . that the *Drosophila eyeless* (*ey*) gene has extensive DNA similarity to the mouse and human *Pax-6* genes" because "evolutionary theory has traditionally held that very long periods of time were needed for natural selection to generate extreme differences in morphological organization." Yet, he notes, if over long periods of time, natural selection and mutation had produced extensive changes in morphology, it ought to have produced extensive changes in genes as well ("The Developmental Genetic Toolkit and the Molecular Homology-Analogy Paradox," 12).

27. Nelson and Wells, "Homology in Biology," 316.

28. Nelson and Wells, "Homology in Biology," 316.

29. Erwin, "Disparity." Erwin observes: "*The distribution of organic forms is clumpy at any scale* from populations to the highest taxonomic categories, and whether considered within clades or within eco-systems. *The fossil record provides little support for expectations that the morphological gaps between species or groups of species have increased through time as it might if the gaps were created by extinction of a more homogeneous distribution of morphologies.* As the quantitative assessments of morphology have replaced counts of higher taxa as a metric of morphological disparity, numerous studies have demonstrated *the rapid construction of morphospace early in evolutionary radiations*" (57, emphasis added).

30. Dembski, "Intelligent Design as a Theory of Technological Evolution"; Savransky, *Engineering of Creativity*, 8, 24; Bracht, "Inventions, Algorithms, and Biological Design."

31. Valentine, "Why No New Phyla After the Cambrian?"; see also Bergström, "Ideas on Early Animal Evolution." Bergström comments: "There is absolutely no sign of convergence between phyla as we follow them backwards to the Early Cambrian. They were as widely apart from the beginning as they are today. Hierarchical levels apparently include a biological reality, not only classificatory convention. In fact, the overwhelming taxonomic difficulty is to recognize relationships between phyla, not to distinguish between them" (464).

32. Denton, *Evolution*, 313.

33. Agassiz, "Evolution and the Permanence of Type," 101.

CHAPTER 19: THE RULES OF SCIENCE

1. Chesterton, "The Invisible Man."

2. Murphy, "Phillip Johnson on Trial," 33. Nancey Murphy is a philosopher and seminary professor who strongly affirms methodological naturalism. Here's what she says in full: "Science *qua* science seeks naturalistic explanations for all natural processes. Christians and atheists alike must pursue scientific questions in our era without invoking a Creator. . . . Anyone who attributes the characteristics of living things to creative intelligence has by definition stepped into the arena of either metaphysics or theology." See also Willey, "Darwin's Place in the History of Thought," 15 ("Science must be provisionally atheistic, or cease to be itself"); Grizzle, "Some Comments on the 'Godless' Nature of Darwinian Evolution," 176.

3. Meyer, "The Origin of Biological Information and the Higher Taxonomic Categories."

4. For detailed discussions of facts of the Sternberg case, see "Smithsonian Controversy," www.richardsternberg.com/smithsonian.php; U.S. Office of Special Counsel Letter (2005) at www.discovery.org/f/1488; United States House of Representatives Committee on Government Reform, Subcommittee Staff Report, "Intolerance and the Politicization of Science at the Smithsonian" (December 2006), at www.discovery.org/f/1489; Appendix, United States House of Representatives Committee on Government Reform, Subcommittee Staff Report (December, 2006) at www.discovery.org/f/1490.

5. See "Statement from the Council of the Biological Society of Washington," originally at http://www.biolsocwash.org/id_statement.html; now at http://ncse.com/news/2004/10/bsw-strengthens-statement-repudiating-meyer-paper-00528 or http://web.archive.org/web/20070926214521/http://www.biolsocwash.org/id_statement.html.

6. E-mail from Roy McDiarmid to Hans Sues, "Re: Request for information" (January 28, 2005, 2:25 p.m.), cited in United States House of Representatives Committee on Government Reform, "Intolerance and the Politicization of Science at the Smithsonian," available at http://www.discovery.org/f/1489 ("I have seen the review file and comments from 3 reviewers on the Meyer paper. All three with some differences among the comments recommended or suggested

publication. I was surprised but concluded that there was not inappropriate behavior vs a vis [sic] the review process").

7. Klinghoffer, "The Branding of a Heretic."

8. See "Statement from the Council of the Biological Society of Washington," originally at http://www.biolsocwash.org/id_statement.html; now at http://ncse.com/news/2004/10/bsw-strengthens-statement-repudiating-meyer-paper-00528 or http://web.archive.org/web/20070926214521/http://www.biolsocwash.org/id_statement.html.

9. "AAAS Board Resolution on Intelligent Design Theory," AAAS News Archives, October 18, 2002, at http://www.aaas.org/news/releases/2002/1106id2.shtml.

10. See Siegal, "Riled by Intelligent Design"; *World Net Daily,* "Intelligent Design Torpedoes Tenure"; Brumfiel, "Darwin Sceptic Says Views Cost Tenure"; Dillon, "Regents Deny Gonzalez's Tenure Appeal"; Meyer, "A Scopes Trial for the '90s"; West, *Darwin Day in America,* 234–38; "Background to the Guillermo Gonzalez Story," at http://www.evolutionnews.org/gg-bckgrndr.final.pdf; "Intelligent Design Was the Issue After All (Updated)," at http://www.evolutionnews.org/ID_was_the_Issue_Gonzalez_Tenure.pdf; *LeVake v. Independent School District,* 625 N.W.2d 502, 506 (Minn. Ct. App. 2001), *cert. denied,* 534 U.S. 1081 (2002); Mims, "The Scientific American Affair"; Crocker, *Free to Think;* Vedantam, "Eden and Evolution"; Luskin, "Darwin's Dilemma"; Anderson, "Cancellation of Darwin Film Creates Uproar"; Reardon, "California Science Center to Pay $110,000 Settlement Over Intelligent Design Film"; Boehm, "California Science Center Is Sued for Canceling a Film Promoting Intelligent Design"; Crowther, "Academic Freedom Expelled from Baylor University"; Briggs and Maaluf, "BU Had Role in Dembski Return"; Luskin, "Credibility Gap"; Black, "Intelligent Design Proponent Fired from NASA Lab"; Pitts, "Design Flaw?"; Associated Press, "Former NASA Specialist Claims He Was Fired over Intelligent Design"; Associated Press, "JPL Worker Sues over Intelligent Design Demotion"; Gallegos, "Intelligent Design Proponent Who Works at JPL Says He Experienced Religious Discrimination"; Luskin, "Intelligent Design Demoted."

11. Crick, *What Mad Pursuit,* 138.

12. Lewontin, "Billions and Billions of Demons," 28. The role of methodological materialism in artificially buttressing Darwin's theory began with Darwin himself. According to historian of science Neal Gillespie: "The uneasy reservations about natural selection among Darwin's contemporaries and the widespread rejection of it from the 1890s to the 1930s suggest that . . . it was more Darwin's insistence on totally natural explanations than on natural selection that won their adherence. . . . The primary change had not been in speciation theory but in beliefs about the nature of science." In short, Darwin's new definition of science excluded "both direct and indirect design" in living things and thereby protected it from competition (*Charles Darwin and the Problem of Creation,* 123, 147, 152).

13. Laudan, "The Demise of the Demarcation Problem"; Meyer, "The Demarcation of Science and Religion."

14. Newton, *Newton's Principia,* 543–44; Alexander, *The Leibniz-Clarke Correspondence,* 92; Hutchison, "What Happened to Occult Qualities in the Scientific Revolution?"; Leibniz, *New Essays on Human Understanding,* 61, 65–67.

15. As Newton wrote in his famous letter to Bishop Bentley: "The cause of gravity is what I do not pretend to know" (Cohen, *Isaac Newton's Papers and Letters on Natural Philosophy,* 302).

16. Laudan says, "There is no demarcation line between science and nonscience, or between science and pseudo-science, which would win assent from a majority of philosophers" (*Beyond Positivism and Relativism,* 210).

17. Laudan, "The Demise of the Demarcation Problem.

18. Newton, *Newton's Principia,* 543–44; Alexander, *The Leibniz-Clarke Correspondence,* 92; Hutchison, "What Happened to Occult Qualities in the Scientific Revolution?"; Leibniz, *New Essays on Human Understanding,* 61, 65–67.

19. Hutchison, "What Happened to Occult Qualities in the Scientific Revolution?" Leibniz, *New Essays on Human Understanding,* 61, 65–67.

20. Laudan, "The Demise of the Demarcation Problem"; Eger, "A Tale of Two Controversies."

21. Gould, "Creationism"; Ruse, "Witness Testimony Sheet"; Ebert et al., *Science and Creationism,* 8–10.

22. Kline, "Theories, Facts and Gods," 42; Gould, "Evolution as Fact and Theory," 120; Root-Bernstein, "On Defining a Scientific Theory," 72.

23. Root-Bernstein, "On Defining a Scientific Theory," 73; Ruse, "A Philosopher's Day in Court," 28; Ebert et al., *Science and Creationism,* 8–10.

24. Ruse, "Witness Testimony Sheet," 301; "A Philosopher's Day in Court," 26; "Darwinism: Philosophical Preference, Scientific Inference, and a Good Research Strategy," 1–6.

25. Ruse, *Darwinism Defended*, 59; "Witness Testimony Sheet," 305; Gould, "Evolution as Fact and Theory," 121; Root-Bernstein, "On Defining a Scientific Theory," 74.

26. Kehoe, "Modern Anti-Evolutionism"; Ruse, "Witness Testimony Sheet," 305; "A Philosopher's Day in Court," 28; Ebert et al., *Science and Creationism*, 8.

27. Kitcher, *Abusing Science*, 126–27, 176–77.

28. Skoog, "A View from the Past"; Root-Bernstein, "On Defining a Scientific Theory," 74; Scott, "Keep Science Free from Creationism," 30.

29. Lyell, *Principles of Geology*.

30. Meyer, *Signature in the Cell*, 481–97.

31. See Meyer, *Signature in the Cell*, 416–38; see also Meyer, "The Scientific Status of Intelligent Design: The Methodological Equivalence of Naturalistic and Non-Naturalistic Origins Theories," 151–212; Meyer, "The Nature of Historical Science and the Demarcation of Design and Descent" 91–130; Meyer, "Laws, Causes and Facts: A Response to Professor Ruse," 29–40; Meyer, "Sauce for the Goose: Intelligent Design, Scientific Methodology, and the Demarcation Problem," 95–131; Meyer, "The Methodological Equivalence of Design and Descent," 67–112.

32. See Meyer, *Signature in the Cell*, 441–42.

33. Meyer, *Signature in the Cell*, 373–95.

34. Asher, *Evolution and Belief*, 32.

35. Asher, *Evolution and Belief*, 32.

36. Asher, *Evolution and Belief*, 32.

37. Asher, *Evolution and Belief*, 32.

38. Oddly, also Asher observes that overall "Meyer professes a low regard for naturalism, but high regard for uniformitarianism" (*Evolution and Belief*, 32). If by naturalism, he means treating the principle of methodological naturalism as normative for all scientific inquiry, his characterization of my position is, in this respect, accurate.

39. Asher, *Evolution and Belief*, 32.

40. Ecker et al., "Genomics."

41. Ecker et al., "Genomics," 52.

42. Orgel and Crick, "Selfish DNA."

43. Dembski, "Science and Design," 26.

44. The ENCODE Project Consortium, "An Integrated Encyclopedia of DNA Elements in the Human Genome," 57.

45. Shapiro and Von Sternberg, "Why Repetitive DNA is Essential to Genome Function"; Von Sternberg and Shapiro, "How Repeated Retroelements Format Genome Function"; Han, Szak, and Boeke, "Transcriptional Disruption by the L1 Retrotransposon"; Janowski et al., "Inhibiting Gene Expression at Transcription Start Sites in Chromosomal DNA with Antigene RNAs"; Goodrich and Kugel, "Non-coding-RNA Regulators of RNA Polymerase II Transcription"; Li et al., "Small dsRNAs Induce Transcriptional Activation in Human Cells"; Pagano et al., "New Small Nuclear RNA Gene-like Transcriptional Units"; Van de Lagemaat et al., "Transposable Elements in Mammals"; Donnelly, Hawkins, and Moss, "A Conserved Nuclear Element"; Dunn, Medstrand, and Mager, "An Endogenous Retroviral Long Terminal Repeat"; Burgess-Beusse et al., "The Insulation of Genes from External Enhancers and Silencing Chromatin"; Medstrand, Landry, and Mager, "Long Terminal Repeats Are Used as Alternative Promoters"; Mariño-Ramírez et al., "Transposable Elements Donate Lineage-Specific Regulatory Sequences to Host Genomes"; Green, "The Role of Translocation and Selection"; Figueiredo et al., "A Central Role for *Plasmodium Falciparum* Subtelomeric Regions"; Henikoff, Ahmad, and Malik, "The Centromere Paradox"; Bell, West, and Felsenfeld, "Insulators and Boundaries"; Pardue and Debaryshe, "*Drosophila* Telomeres"; Henikoff, "Heterochromatin Function in Complex Genomes"; Schueler et al., "Genomic and Genetic Definition of a Functional Human Centromere"; Jordan et al., "Origin of a Substantial Fraction of Human Regulatory Sequences from Transposable Elements"; Chen, DeCerbo, and Carmichael, "*Alu* Element-Mediated Gene Silencing"; Jurka, "Evolutionary Impact of Human Alu Repetitive Elements"; Lev-Maor et al., "The Birth of an Alternatively Spliced Exon"; Kondo-Iida et al., "Novel Mutations and Genotype–Phenotype Relationships in 107 Families"; Mattick and Makunin, "Non-coding RNA"; McKenzie and Brennan, "The Two Small Introns of the *Drosophila affinidisjuncta Adh* Gene"; Arnaud et al., "SINE Retroposons Can Be Used In Vivo"; Rubin, Kimura, and Schmid, "Selective Stimulation of Translational Expression by Alu RNA"; Bartel, "MicroRNAs"; Mattick and Makunin, "Small Regulatory RNAs in Mammals"; Dunlap et al., "Endogenous Retroviruses Regulate Periimplantation Placental Growth and Differentiation"; Hyslop et al., "Downregulation of NANOG Induces Differentiation"; Peaston et al., "Retrotransposons Regulate Host Genes in Mouse Oocytes and Preimplantation Embryos"; Morrish et al., "DNA Repair Mediated by Endonuclease-Independent LINE-1 Retrotransposition"; Tremblay, Jasin, and Chartrand, "A

Double-Strand Break in a Chromosomal LINE Element Can Be Repaired"; Grawunder et al., "Activity of DNA Ligase IV Stimulated by Complex Formation with XRCC4 Protein in Mammalian Cells"; Wilson, Grawunder, and Liebe, "Yeast DNA Ligase IV Mediates Non-Homologous DNA End Joining"; Mura et al., "Late Viral Interference Induced by Transdominant Gag of an Endogenous Retrovirus"; Goh et al., "A Newly Discovered Human Alpha-Globin Gene"; Kandouz et al., "Connexin43 Pseudogene Is Expressed"; Tam et al., "Pseudogene- Derived Small Interfering RNAs Regulate Gene Expression in Mouse Oocytes"; Watanabe et al., "Endogenous siRNAs from Naturally Formed dsRNAs Regulate Transcripts in Mouse Oocytes"; Piehler et al., "The Human ABC Transporter Pseudogene Family"; Mattick and Gagen, "The Evolution of Controlled Multitasked Gene Networks"; Pandey and Mukerji, "From 'JUNK' to Just Unexplored Noncoding Knowledge"; Balakirev and Ayala, "Pseudogenes"; Pink et al., "Pseudogenes"; Wen et al., "Pseudogenes Are Not Pseudo Any More"; Franco-Zorrilla et al., "Target Mimicry Provides a New Mechanism for Regulation of MicroRNA Activity"; Colas et al., "Whole-Genome MicroRNA Screening Identifies let-7 and miR-18 as Regulators"; Carrier et al., "Long Non-Coding Antisense RNA Controls Uchl1 Translation"; Kelley and Rinn, "Transposable Elements Reveal a Stem Cell Specific Class of Long Noncoding RNAs"; Wang et al., "Alternative Isoform Regulation in Human Tissue Transcriptomes"; Louro et al., "Conserved Tissue Expression Signatures of Intronic Noncoding RNAs"; Hoeppner et al., "Evolutionarily Stable Association of Intronic SnoRNAs and MicroRNAs with Their Host Genes"; Monteys et al., "Structure and Activity of Putative Intronic miRNA Promoters"; Mondal et al., "Characterization of the RNA Content of Chromatin"; Rodriguez-Campos and Azorin, "RNA Is an Integral Component of Chromatin That Contributes to Its Structural Organization"; George et al., "Evolution of Diverse Mechanisms for Protecting Chromosome Ends"; Von Sternberg, "On the Roles of Repetitive DNA Elements in the Context of a Unified Genomic-Epigenetic System."

46. "Breakthrough of the Year Newsfocus: Genomics Beyond Genes," *Science* 338 (December 21, 2012): 1528.

47. Moran, "Intelligent Design Creationists Chose ENCODE as the #1 Evolution Story of 2012."

48. Shapiro and von Sternberg, "Why Repetitive DNA Is Essential to Genome Function"; Von Sternberg and Shapiro, "How Repeated Retroelements Format Genome Function." See also Von Sternberg, "On the Roles of Repetitive DNA Elements in the Context of a Unified Genomic-Epigenetic System."

49. Todd, "A View from Kansas on That Evolution Debate."

CHAPTER 20: WHAT'S AT STAKE

1. Dawkins, *The God Delusion;* Hitchens, *God Is Not Great;* Stenger, *God: The Failed Hypothesis.*

2. Stenger, *God: The Failed Hypothesis.*

3. Dawkins, *The Blind Watchmaker,* 6.

4. Russell, quoted in Conant, *Modern Science and Modern Man,* 139–40.

5. Collins, *The Language of God*. See also Giberson, *Saving Darwin;* Miller, *Finding Darwin's God.*

6. For prominent critics of the neo-Darwinian consensus, see Chapter 14, n. 42.

7. For some of the latest studies, see Chapter 19, n. 45.

8. As Alfred North Whitehead said, "When we consider what religion is for mankind and what science is, it is no exaggeration to say that the future course of history depends upon the decision of this generation as to the relations between them" (*Science and the Modern World,* 260).

EPILOGUE: RESPONSES TO CRITICS OF THE FIRST EDITION OF *DARWIN'S DOUBT*

1. Matzke, "Luskin's Hopeless Monster."

2. Matzke, "Meyer's Hopeless Monster, Part II"; Matzke, "Luskin's Hopeless Monster."

3. Matzke, "Luskin's Hopeless Monster" (emphasis added). See also Matzke, "Meyer's Hopeless Monster, Part II."

4. See Luskin, "Does Natural Selection Leave 'Detectable Statistical Evidence in the Genome'?"

5. See Casey Luskin's response to Nick Matzke reinforcing this claim. Luskin, "How 'Sudden' Was the Cambrian Explosion?"

6. Matzke, "Meyer's Hopeless Monster, Part II."

7. Matzke, "Meyer's Hopeless Monster, Part II"; Matzke, "Luskin's Hopeless Monster."

8. Matzke, "Incompetent and Wrong."
9. Matzke, "Meyer's Hopeless Monster, Part II."
10. See Luskin, "Rush to Judgment."
11. The following sources identify *Anomalocaris* as a type of arthropod: Paterson et al., "Acute Vision in the Giant Cambrian Predator *Anomalocaris* and the Origin of Compound Eyes"; Yong, "The Sharp Eyes of Anomalocaris, a Top Predator That Lived Half a Billion Years Ago"; Waggoner, "Phylogenetic Hypotheses of the Relationships of Arthropods to Precambrian and Cambrian Problematic Fossil Taxa"; Liu et al., "A Large Xenusiid Lobopod with Complex Appendages from the Lower Cambrian Chengjiang Lagerstätte"; Marshall and Valentine, "The Importance of Preadapted Genomes in the Origin of the Animal Bodyplans and the Cambrian Explosion"; Budd and Jensen, "A Critical Reappraisal of the Fossil Record of the Bilaterian Phyla"; Butterfield, "Exceptional Fossil Preservation and the Cambrian Explosion"; Hou et al., *The Cambrian Fossils of Chengjiang, China*, 94.
12. The following sources identify Lobopodia as a phylum: Erwin et al., "The Cambrian Conundrum," Table S1; Cavalier-Smith, "A Revised Six-Kingdom System of Life"; M. Alan Kazlev, "Panarthropoda: Lobopodia," *Palaeos.com* ("The Lobopodia or Lobopoda are an evolutionary grade or phylum"); Erwin and Valentine, *The Cambrian Explosion*, 350; Hou et al., *The Cambrian Fossils of Chengjiang*, 82.
13. Brysse, "From Weird Wonders to Stem Lineages," 303.
14. See Matzke, "Luskin's Hopeless Monster."
15. For a classic example of such cladistic analysis, see Eldredge and Cracraft, *Phylogenetic Patterns and the Evolutionary Process*, 28. See also Funk and Wheeler, "Symposium: Character Weighting, Cladistics and Classification."
16. "Weighting" refers to the practice of determining evolutionary relationships based on the characters that systematists decide are most important for defining a particular group. Many systematists do not regard character weighting as a part of cladistics, at least strictly speaking. See Brysse, "From Weird Wonders to Stem Lineages," 305. One distinctive feature of cladistics as a method of evolutionary systematics has been its attempt to eliminate subjectivity from its analytical methods. Accordingly, strict cladists reject differential weighting of characters as a means of producing the most parsimonious cladograms. Nevertheless, many evolutionary systematists who do character-based phylogenetic analysis choose to weight some characters more heavily than others. Whether such weighting represents part of cladistics specifically or just part of evolutionary systematics more generally is partly a semantic issue and one that is controversial among practitioners. In any case, character-based phylogenetic analysis often does involve weighting characters, and decisions about how to weight characters affect which of the many possible trees are generated by the phylogenetic algorithms.
17. Matzke, "Meyer's Hopeless Monster, Part II."
18. Matzke, "Meyer's Hopeless Monster, Part II."
19. Matzke, "Meyer's Hopeless Monster, Part II."
20. See Martin, *Missing Links*, 153 (stating theropods "all occur in the fossil record after *Archaeopteryx* and so cannot be directly ancestral"); Swisher et al., "Cretaceous Age for the Feathered Dinosaurs of Liaoning."
21. Legg et al., "Cambrian Bivalved Arthropod Reveals Origin of Arthrodization."
22. Edgecombe, "Arthropod Phylogeny."
23. Edgecombe and Legg, "The Arthropod Fossil Record," 394.
24. Legg et al., "Cambrian Bivalved Arthropod Reveals Origin of Arthrodization."
25. Legg et al., "Cambrian Bivalved Arthropod Reveals Origin of Arthrodization."
26. Matzke, "Meyer's Hopeless Monster, Part II" (emphasis removed).
27. Fortey, "Evolution."
28. Erwin et al., "The Cambrian Conundrum, 1094.
29. Kühl et al., "A Great-Appendage Arthropod with a Radial Mouth from the Lower Devonian Hunsrück Slate, Germany."
30. See Paterson et al., "Acute Vision in the Giant Cambrian Predator *Anomalocaris* and the Origin of Compound Eyes," 239, Figure 3.
31. Matzke, "Meyer's Hopeless Monster, Part II."
32. Arguably, organisms like *Anomalocaris* also make poor anatomical intermediates to arthropods, since they bear unique features—entirely ignored by cladistics—like strange flexible side-lobes or a mouth with ringed teeth, which are foreign to crown group arthropods. See Whittington and Briggs, "The Largest Cambrian Animal, *Anomalocaris*," 604 ("*Anomalocaris* . . . is unlike any known arthropod, particularly in the nature of the jaw apparatus and the close-spaced, strongly overlapping lateral lobes"); Brysse, "From Weird Wonders to Stem Lineages," 307 ("Cla-

distics, by contrast, gives no weight to unique characters"); Briggs and Fortey, "The Early Radiation and Relationships of the Major Arthropod Groups," 243 ("The cladistic approach, on the other hand, focuses on shared characters. Unique attributes, autapomorphies, are of no use in assessing relationship and are consequently accorded little significance.").

33. Matzke, "Luskin's Hopeless Monster" (emphasis added).

34. Robert Shaw, "Collateral Ancestor," *Encyclopedia of Genealogy,* www.eogen.com/collateral ancestor.

35. Syvanen, "Evolutionary Implications of Horizontal Gene Transfer."

36. Sober and Steele, "Testing the Hypothesis of Common Ancestry," 395.

37. This telltale phrase is remarkably widespread in the cladistics literature. See, for instance, Edgecombe et al., "Higher-Level Metazoan Relationships," 152: ". . . the complexity and number of potential phylogenetically informative characters often forced scientists to base their hypotheses on a few selected characters—e.g. larval ciliary bands, excretory systems, or embryology in the case of deep metazoan relationships—leaving out much other important information and carrying the risk of producing biased results based on homoplasies." See also Wheat and Wahlberg, "Phylogenomic Insights into the Cambrian Explosion."

38. Matzke, "Luskin's Hopeless Monster" (emphasis removed).

39. Lecointre and Le Guyader, *The Tree of Life*, 16.

40. Matzke, "Meyer's Hopeless Monster, Part II."

41. Legg et al., "Cambrian Bivalved Arthropod Reveals Origin of Arthrodization."

42. Briggs and Fortey, "The Early Radiation and Relationships of the Major Arthropod Groups."

43. Rooting refers to the practice by which systematists decide which taxon among a group of taxa under analysis is most likely to have been the last common ancestor of the group and force a phylogenetic algorithm to generate only trees that reflect that decision.

44. See Liu et al., "An Armoured Cambrian Lobopodian from China with Arthropod-like Appendages"; Mounce and Wills, "Phylogenetic Position of *Diania* Challenged"; Legg et al., "Lobopodian Phylogeny Reanalyzed"; Liu et al., "Liu et al. Reply."

45. Matzke, "Meyer's Hopeless Monster, Part II" (emphasis in original).

46. For example, Matzke obscures the doubts about whether the legs of lobopods, or the eyes and head of anomalocaridids, are in fact homologous to (and therefore help explain the evolution of) the legs and eyes of arthropods. See Legg et al., "Cambrian Bivalved Arthropod Reveals Origin of Arthrodization"; Edgecombe and Legg, "The Arthropod Fossil Record"; Liu et al., "Liu et al. Reply"; Legg et al., "Lobopodian Phylogeny Reanalyzed," E2; Erwin and Valentine, *The Cambrian Explosion,* 202; Edgecombe, "Arthropod Phylogeny"; Richter et al., "The Arthropod Head." An organism cannot be transitional or intermediate to true arthropods if its features aren't homologous to arthropods. But if the legless anomalocaridids are more closely related to arthropods, and the arthropod head and eyes evolved first, then the legs of lobopods cannot be homologous to arthropod legs. Conversely, if the eyeless and headless lobopods are more closely related to arthropods, and arthropod legs evolved first, then the eyes and head of anomalocaridids cannot be homologous to arthropod eyes or heads. Pick either option, and you're faced with a situation that requires evolutionary loss of key arthropod traits, and at least one of Matzke's so-called transitional forms can't really be transitional. As one recent *Nature* paper puts it, one of the "puzzles of stem-group arthropod evolution" is "the absence of [arthropod-like] trunk limbs in dinocaridids (*Anomalocaris*, etc.)," even though *Anomalocaris* has "a more arthropod-like head region" than lobopods, which lack the eyes or head of arthropods but have arthropod-like "jointed trunk appendages" (Liu et al., "Liu et al. Reply," e3–e4). In other words, Matzke's clean evolutionary grade of intermediates leading to arthropods does not exist.

47. Erwin and Valentine, *The Cambrian Explosion*, 195, 202.

48. Edgecombe, "Arthropod Phylogeny."

49. Legg et al., "Cambrian Bivalved Arthropod Reveals Origin of Arthrodization."

50. Matzke, "Meyer's Hopeless Monster, Part II."

51. Brysse, "From Weird Wonders to Stem Lineages," 306.

52. Brysse, "From Weird Wonders to Stem Lineages," 306 (emphasis added).

53. Matzke, "Meyer's Hopeless Monster, Part II."

54. Brysse, "From Weird Wonders to Stem Lineages," 311–12.

55. Gee, *In Search of Deep Time*, 151.

56. Prothero, "Stephen Meyer's Fumbling Bumbling Amateur Cambrian Follies."

57. Prothero, "Stephen Meyer's Fumbling Bumbling Amateur Cambrian Follies."

58. Luskin, "How 'Sudden' Was the Cambrian Explosion?"; Luskin, "Darwin Defenders Love Donald Prothero's Ranting Review of *Darwin's Doubt*." See also Meyer, "More on Small Shelly

Fossils and the Length of the Cambrian Explosion"; Luskin, "Small Shelly Fossils, and the Length of the Cambrian Explosion."

59. John Farrell, writing in *National Review*, attempted to discredit the book largely on the basis of an allegedly misplaced ellipsis. See Farrell, "How Nature Works"; and my response in the letters section of the September 30, 2013, issue of *National Review*. See also more comprehensive responses at Luskin, "In *National Review*, John Farrell's Predictable and Misleading Review of *Darwin's Doubt*"; Klinghoffer, "In the Matter of John Farrell's Silly Review of *Darwin's Doubt*, *National Review* Sets Things Straight."

60. Marshall, "When Prior Belief Trumps Scholarship," 1344.

61. Eden, "Inadequacies of Neo-Darwinian Evolution as a Scientific Theory," 109–11; Schützenberger, "Algorithms and the Neo-Darwinian Theory of Evolution," 73–80.

62. Reidhaar-Olson and Sauer, "Functionally Acceptable Solutions in Two Alpha-Helical Regions of Lambda Repressor"; Axe, "Estimating the Prevalence of Protein Sequences Adopting Functional Enzyme Folds."

63. Marshall, "Nomothetism and Understanding the Cambrian 'Explosion.'"

64. See Shannon, "A Mathematical Theory of Communication."

65. Rasmussen, "A New Model of Developmental Constraints as Applied to the *Drosophila* System."

66. "Phylogenetic evidence suggests that *bcd* may be a new innovation in the anterior positional information gene network during the evolution of Dipterans. Despite repeated attempts, it has not been possible to clone *bcd* homologues outside of the Cyclorraphan flies (Stauber et al., 1999). Additionally, *bcd* is not present in the *Antennapedia* complex of the flour beetle *Tribolium castaneum* (Brown et al., 2002). This has caused speculation that *bcd* may have evolved late in the evolution of the Dipterans" (Rudel and Sommer, "The Evolution of Developmental Mechanisms," 25).

67. Khalturin et al., "A Novel Gene Family Controls Species-Specific Morphological Traits in Hydra."

68. Oliveri and Davidson, "Built to Run, Not to Fail," 1510–11 (emphasis added).

69. Marshall and Valentine, "The Importance of Preadapted Genomes in the Origin of the Animal Bodyplans and the Cambrian Explosion."

70. Marshall, "Explaining the Cambrian 'Explosion' of Animals," 366.

71. His words are: "Accompanied by some additional gene novelties." Marshall and Valentine, "The Importance of Preadapted Genomes in the Origin of the Animal Bodyplans and the Cambrian Explosion," 1189.

72. Ohno, "The Notion of the Cambrian Pananimalia Genome."

73. Gibson et al., "Why So Many Unknown Genes?"

74. Khalturin et al., "More Than Just Orphans."

75. Marshall seems to have an idiosyncratic view of animal evolution, depicting the evolution of animals as a reductive process in which preexisting genetic information from a universal Precambrian gene set is selectively lost to some lineages but not to others. This contrasts markedly with a more standard neo-Darwinian view in which form and information gradually accumulate over time.

76. See Marshall and Valentine, "The Importance of Preadapted Genomes in the Origin of the Animal Bodyplans and the Cambrian Explosion," 1189.

77. Ohno, "The Notion of the Cambrian Pananimalia Genome."

78. Miller and Ball, "The Gene Complement of the Ancestral Bilaterian—Was Urbilateria a Monster?"

79. Smith and Harper, "Causes of the Cambrian Explosion." Smith and Harper propose "an apparent >100-million-year gap between the evolutionary innovation and its consequences."

80. Müller and Newman, "Origination of Organismal Form."

81. Davidson, "Emerging Properties of Animal Gene Regulatory Networks." Davidson notes that "important initial inputs in this system are the transcription factors . . . encoded by maternal messenger RNAs" (913).

82. For this reason, defining the Cambrian explosion as a 25-million-year event, as Marshall does in his review, instead of a 10-million-year event, as many other Cambrian experts do—and as I do (and as Marshall himself has often done [Marshall and Valentine, "The Importance of Preadapted Genomes in the Origin of the Animal Bodyplans and the Cambrian Explosion," 1193])—makes no appreciable difference in solving the problem of the origin of genetic information. Such is the extreme rarity of functional bio-macromolecules within their relevant sequence spaces. Nor, for that matter, does positing the origin of a complete set of genes for building the Cambrian animals 100 million years before the Cambrian explosion. That strategy merely pushes the problem back and raises another unanswered question, namely exactly what selective advantages did

all these genes for building new animals have for over 100 million years before they were actually used to build the diverse animals that arose in the Cambrian? In any case, the experimentally based calculations in *Darwin's Doubt* show that neither 10 million nor several hundred million years would afford enough opportunities to produce the genetic information necessary to build even a single novel gene or protein, let alone all the new genes and proteins needed to produce new animal forms. Neither stretch of time is sufficient to allow the mutation/selection process to search more than a tiny fraction of the relevant sequence spaces of interest.

83. Marshall, "When Prior Belief Trumps Scholarship."

84. Marshall, "When Prior Belief Trumps Scholarship."

85. See Davidson, "Evolutionary Bioscience as Regulatory Systems Biology," 35–40. See also the discussion in *Darwin's Doubt*, pp. 264–70.

86. See Davidson, "Evolutionary Bioscience as Regulatory Systems Biology," 40.

87. See, for example, discussion of the model system *Strongylocentrotus purpuratus* in Oliveri et al., "Global Regulatory Logic for Specification of an Embryonic Cell Lineage." For a diagrammed schematic of the dGRN network of *Strongylocentrotus purpuratus,* see Figure 1D.

88. Shubin and Marshall, "Fossils, Genes, and the Origin of Novelty," 335 (emphasis added).

89. Davidson, "Evolutionary Bioscience as Regulatory Systems Biology," 40.

90. See Lyell, *Principles of Geology,* 75–91.

91. In his review, Marshall made two other critiques of the book. He claimed that I overlooked the importance of the small shelly fossils and thus exaggerated the brevity of the Cambrian explosion. He also claimed that my positive argument for intelligent design actually constituted a fallacious argument from ignorance or a "God-of-the-gaps" argument. For a detailed response to both these criticisms, see Meyer, "Does *Darwin's Doubt* Commit the God-of-the-Gaps Fallacy?"; Meyer, "More on Small Shelly Fossils and the Length of the Cambrian Explosion"; Luskin, "Small Shelly Fossils, and the Length of the Cambrian Explosion."

BIBLIOGRAPHY

Adams, Melissa D., et al. "The Genome Sequence of *Drosophila melanogaster*." *Science* 287 (2000): 2185–95.

Agassiz, Louis. "Evolution and the Permanence of Type." *Atlantic Monthly* 33 (1874): 92–101.

———. *Louis Agassiz: His Life and Correspondence*. Vol. 1, edited by E. C. Agassiz. Boston: Houghton, Mifflin; Cambridge, MA: Riverside, 1890.

———. *Essay on Classification*. Edited by Edward Lurie. Cambridge, MA: Harvard Univ. Press, Belknap Press, 1962.

Agres, Ted. "Smithsonian 'Discriminated 'Against Scientist." *The Scientist*, December 22, 2006.

Aguinaldo, Anna Marie, James M. Turbeville, Lawrence S. Linford, Maria C. Rivera, James R. Garey, Rudolf A. Raff, and James A. Lake. "Evidence for a Clade of Nematodes, Arthropods and Other Moulting Animals." *Nature* 387 (1997): 489–93.

Akam, Michael. "Arthropods: Developmental Diversity Within a (Super) Phylum." *Proceedings of the National Academy of Sciences USA* 97 (2000): 4438–41.

Alean, Jürg, and Michael Hambrey. "Glaciers Online." http://www.swisseduc.ch/glaciers/earth_icy_planet/icons-15/16.jpg.

Alexander, H. G., ed. *The Leibniz-Clarke Correspondence*. New York: Manchester Univ. Press, 1956.

Ally, Shabeen, A. G. Larson, K. Barlan, S. E. Rice, and V. Gelfand. "Opposite-Polarity Motors Activate One Another to Trigger Cargo Transport in Live Cells." *Journal of Cell Biology* 187 (2009): 1071–82. http://jcb.rupress.org/content/187/7/1071.full.pdf+html (accessed October 30, 2012).

Alroy, John. "The Fossil Record of North American Mammals: Evidence for a Paleocene Evolutionary Radiation." *Systematic Biology* 48 (1999): 107–18.

Altamirano, Myriam M., Jonathan M. Blackburn, Cristina Aguayo, and Alan R. Fersht. "Directed Evolution of New Catalytic Activity Using the Alpha/Beta-Barrel Scaffold." *Nature* 403 (2000): 617–22.

American Association for the Advancement of Science Board of Directors. "Statement on Teaching Evolution." St. Louis, Missouri, February 16, 2006. http://www.aaas.org/news/releases/2006/pdf/0219boardstatement.pdf.

Anderson, Troy. "Cancellation of Darwin Film Creates Uproar." *Los Angeles Daily News*, October 8, 2009.

Aris-Brosou, Stéphane, and Ziheng Yang. "Bayesian Models of Episodic Evolution Support a Late Precambrian Explosive Diversification of the Metazoa." *Molecular Biology and Evolution* 20 (2003): 1947–54.

Arnaud, Phillipe, Chantal Goubely, Thierry Pe'Lissier, and Jean-Marc Deragon. "SINE Retroposons Can Be Used In Vivo as Nucleation Centers for De Novo Methylation." *Molecular and Cellular Biology* 20, no. 10 (2000): 3434–41.

Arthur, Wallace. *The Origin of Animal Body Plans: A Study in Evolutionary Developmental Biology*. Cambridge: Cambridge Univ. Press, 1997.

Asher, Robert J. *Evolution and Belief: Confessions of a Religious Paleontologist*. Cambridge: Cambridge Univ. Press, 2012.

Associated Press. "JPL Worker Sues over Intelligent Design Demotion." *USA Today*, April 19, 2010.

———. "Former NASA Specialist Claims He Was Fired over Intelligent Design." FoxNews.com, March 11, 2012.

Avery, Oswald T., Colin M. MacCleod, and Maclyn McCarty. "Induction of Transformation by a Deoxyribonucleic Acid Fraction Isolated from Pneumococcus Type III." *Journal of Experimental Medicine* 79 (1944): 137–58.

Axe, Douglas D. "Extreme Functional Sensitivity to Conservative Amino Acid Changes on Enzyme Exteriors." *Journal of Molecular Biology* 301 (2000): 585–95.

———. "Estimating the Prevalence of Protein Sequences Adopting Functional Enzyme Folds." *Journal of Molecular Biology* 341 (2004): 1295–1315.

———. "The Case Against Darwinian Origin of Protein Folds." *BIO-Complexity* 2010, no. 1 (2010): 1–12.
———. "The Limits of Complex Adaptation: An Analysis Based on a Simple Model of Structured Bacterial Populations." *BIO-Complexity* 2010, no. 4 (2010): 1–10.
Ayala, Francisco J. "Design Without Designer: Darwin's Greatest Discovery." In *Debating Design: From Darwin to DNA,* edited by M. Ruse and W. Dembski, 55–80. Cambridge: Cambridge Univ. Press, 2004.
———. "Darwin's Greatest Discovery: Design Without Designer." *Proceedings of the National Academy of Sciences USA* 104 (2007): 8567–73.
Ayala, Francisco, A. Rzhetsky, and F. J. Ayala. "Origin of the Metazoan Phyla: Molecular Clocks Confirm Paleontological Estimates." *Proceedings of the National Academy of Sciences USA* 95 (1998): 606–11.
Azar, Beth. "Profile of Michael Lynch." *Proceedings of the National Academy of Sciences USA* 107 (2010): 16013–15.
Babcock, Loren E. "Asymmetry in the Fossil Record." *European Review* 13 (2005): 135–43.
Babu, M. Madan, Richard W. Kriwacki, and Rohit V. Pappu. "Versatility from Protein Disorder." *Science* 337 (2012): 1460–61.
Bailey, J. V., S. B. Joye, K. M. Kalanetra, B. E. Flood, and F. A. Corsetti. "Evidence of Giant Sulphur Bacteria in Neoproterozoic Phosphorites." *Nature* 445 (2007): 198–201.
Balakirev, E. S., and F. J. Ayala. "Pseudogenes: Are They 'Junk' or Functional DNA?" *Annual Review of Genetics* 37 (2003): 123–51.
Bartel, David P. "MicroRNAs: Genomics, Biogenesis, Mechanism, and Function." *Cell* 116 (2004): 281–97.
Bate, Michael, and Alfonso Martinez Arias, eds. *The Development of Drosophila melanogaster.* 2 vols. Plainview, NY: Cold Spring Harbor Laboratory Press, 1993.
Bateson, William. "Heredity and Variation in Modern Lights." In *Darwin and Modern Science,* edited by A. C. Seward, 85–101. Cambridge: Cambridge Univ. Press, 1909.
Baverstock, Peter R., and Craig Moritz. "Project Design." In *Molecular Systematics,* 2nd ed., edited by D. M. Hillis, C. Moritz, and B. K. Mable, 17–27. Sunderland, MA: Sinauer, 1996.
BBC News, "The Darwinian Sistine Chapel," April 14, 2009, http://www.bbc.co.uk/darwin/?tab=21 (accessed October 31, 2012).
Becker, H., and W. Lönnig. "Transposons: Eukaryotic." In *Nature's Encyclopedia of Life Sciences,* 18: 529–39. London: Nature Publishing, 2001.
Begun, David J. "Origin and Evolution of a New Gene Descended from Alcohol Dehydrogenase in *Drosophila.*" *Genetics* 145 (1997): 375–82.
Behe, Michael. "Histone Deletion Mutants Challenge the Molecular Clock Hypothesis." *Trends in Ecology and Evolution* 15 (1990): 374–76.
———. "Experimental Support for Regarding Functional Classes of Proteins to Be Highly Isolated from Each Other." In *Darwinism: Science or Philosophy?* edited by J. Buell and V. Hearn, 60–71. Richardson, TX: Foundation for Thought and Ethics, 1994.
———. *Darwin's Black Box: The Biochemical Challenge to Evolution.* New York: Free Press, 1996.
———. *The Edge of Evolution: The Search for the Limits of Darwinism.* New York: Free Press, 2007.
Behe, Michael J., and David W. Snoke. "Simulating Evolution by Gene Duplication of Protein Features That Require Multiple Amino Acid Residues." *Protein Science* 13 (2004): 2651–64.
Beiko, Robert G. "Telling the Whole Story in a 10,000-Genome World." *Biology Direct* 6 (2011): 34.
Bell, C., A. G. West, and G. Felsenfeld. "Insulators and Boundaries: Versatile Regulatory Elements in the Eukaryotic Genome." *Science* 291 (2001): 447–50.
Bell, Michael A. "Gould's Most Cherished Concept." *Trends in Ecology and Evolution* 23, no. 3 (2008): 121–22.
Bengtson, Stefan, and Graham E. Budd. "Comment on 'Small Bilaterian Fossils from 40 to 55 Million Years Before the Cambrian.'" *Science* 306 (2004): 1291–92.
Bengtson, Stefan, John A. Cunningham, Chongyu Yin, and Philip C. J. Donoghueb. "A Merciful Death for the 'Earliest Bilaterian' *Vernanimalcula.*" *Evolution and Development* 14, no. 5 (2012): 421–27.
Bengtson, Stefan, and Xian-guang Hou. "The Integument of Cambrian Chancelloriids." *Acta Paleontological Polonica* 46 (2001): 1–22.
Benton, Michael J. "Early Origins of Modern Birds and Mammals: Molecules vs. Morphology." *BioEssays* 21 (1999): 1043–51.
Benton, Michael, and Francisco J. Ayala. "Dating the Tree of Life." *Science* 300 (2003): 1698–1700.

Benton, Michael J., M. A. Wills, and R. Hitchin. "Quality of the Fossil Record Through Time." *Nature* 402 (2000): 534–37.
Bergsten, J. "A Review of Long-Branch Attraction." *Cladistics* 21 (2005): 163–93.
Bergström, Jan. "Metazoan Evolution Around the Precambrian–Cambrian Transition." In *The Early Evolution of Metazoa and the Significance of Problematic Taxa,* edited by A. M. Simonetta and S. Conway Morris, 25–34. Cambridge: Cambridge Univ. Press, 1991.
———. "Ideas on Early Animal Evolution." In *Early Life on Earth,* Nobel Symposium No. 84, edited by S. Bengtson, 460–66. New York: Columbia Univ. Press, 1994.
Bergström, Jan, and Xian-Guang Hou. "Chengjiang Arthropods and Their Bearing on Early Arthropod Evolution." In *Arthropod Fossils and Phylogeny,* edited by G. D. Edgecombe, 151–84. New York: Columbia Univ. Press, 1998.
———. "Arthropod Origins." *Bulletin of Geosciences* 78 (2003): 323–34.
Berlinski, David. "The Deniable Darwin." *Commentary* 101 (1996): 19–29. http://www.discovery.org/a/130.
———. "On Assessing Genetic Algorithms." Public lecture, "Science and Evidence of Design in the Universe" Conference, Yale University, November 4, 2000.
Birket-Smith, S. J. R. "A Reconstruction of the Precambrian *Spriggina.*" *Zoologische Jahrbücher Anatomie und Ontogenie der Tiere* 105 (1981): 237–58.
Black, Nathan. "Intelligent Design Proponent Fired from NASA Lab." *Christian Post,* January 26, 2011.
Blanco, F., I. Angrand, and L. Serrano. "Exploring the Conformational Properties of the Sequence Space Between Two Proteins with Different Folds: An Experimental Study." *Journal of Molecular Biology* 285 (1999): 741–53.
Boehm, Mike. "California Science Center Is Sued for Canceling a Film Promoting Intelligent Design." *Los Angeles Times,* December 29, 2009.
Boffey, Philip M. "100 Years After Darwin's Death, His Theory Still Evolves." *New York Times,* April 20, 1982.
Borchiellini, C., M. Manuel, E. Alivon, N. Boury-Esnault, J. Vacelet, and Y. Le Parco. "Sponge Paraphyly and the Origin of Metazoa." *Journal of Evolutionary Biology* 14 (2001): 171–79.
Borel, Emile. *Probability and Certainty.* Translated by D. Scott. New York: Walker, 1963.
Bottjer, David. "The Early Evolution of Animals." *Scientific American* 293 (2005): 42–47.
Bowler, P. J. *Theories of Human Evolution: A Century of Debate, 1844–1944.* Baltimore: Johns Hopkins Univ. Press, 1986.
———. *Evolution: The History of an Idea.* 3rd ed. Berkeley: Univ. of California Press, 2003.
Bowring, Samuel A., J. P. Grotzinger, C. E. Isachsen, A. H. Knoll, S. M. Pelechaty, and P. Kolosov. "Calibrating Rates of Early Cambrian Evolution." *Science* 261 (1993): 1293–98.
———. "A New Look at Evolutionary Rates in Deep Time: Uniting Paleontology and High-Precision Geochronology." *GSA Today* 8 (1998): 1–8.
———. "Geochronology Comes of Age." *Geotimes* 43 (1998): 36–40.
Brachet, J., H. Denis, and F. De Vitry. "The Effects of Actinomycin D and Puromycin on Morphogenesis in Amphibian Eggs and *Acetabularia mediterranea.*" *Developmental Biology* 9 (1964): 398–434.
Bracht, John. "Inventions, Algorithms, and Biological Design." *Progress in Complexity, Information, and Design* 1.2 (2002). Available at http://www.iscid.org.
Brasier, Martin D., and Jonathan B. Antcliffe. "*Dickinsonia* from Ediacara: A New Look at Morphology and Body Construction." *Palaeogeography, Palaeoclimatology, Palaeoecology* 270 (2008): 311–23.
Brenner, Sidney. "The Genetics of Behaviour." *British Medical Bulletin* 29 (1973): 269–71.
Briggs, Brad, and Grace Maaluf. "BU Had Role in Dembski Return." *The Lariat,* November 16, 2007, http://www.baylor.edu/lariat/news.php?action=story&story=48260.
Briggs, Derek, Douglas Erwin, and Frederick Collier. *The Fossils of the Burgess Shale.* Washington, DC: Smithsonian Institution Press, 1994.
Britten, Roy J., and Eric H. Davidson. "Gene Regulation for Higher Cells: A Theory." *Science* 165 (1969): 349–57.
Brocks, Jochen J., Graham A. Logan, Roger Buick, and Roger E. Summons. "Archean Molecular Fossils and the Early Rise of Eukaryotes." *Science* 285 (1999): 1033–36.
Bronham, Lindell, Andrew Rambaut, Richard Fortey, Alan Cooper, and David Penny. "Testing the Cambrian Explosion Hypothesis by Using a Molecular Dating Technique." *Proceedings of the National Academy of Sciences USA* 95 (1998): 12386–89.

Brooks, W. K. "The Origin of the Oldest Fossils and the Discovery of the Bottom of the Ocean." *Journal of Geology* 2 (1894): 359–76.

Brosius, Jürgen. "The Contribution of RNAs and Retroposition to Evolutionary Novelties." *Genetica* 118 (2003): 99–116.

Brumfiel, Geoff. "Darwin Sceptic Says Views Cost Tenure." *Nature* 447 (2007): 364.

Brusca, Richard C., and Gary J. Brusca. *Invertebrates.* Sunderland, MA: Sinauer, 1990.

Budd, Graham E. "The Morphology of *Opabinia regalis* and the Reconstruction of the Arthropod Stem-Group." *Lethaia* 29 (1996): 1–14.

Budd, Graham E., and Sören Jensen. "A Critical Reappraisal of the Fossil Record of the Bilaterian Phyla." *Biological Reviews of the Cambridge Philosophical Society* 75 (2000): 253–95.

———. "The Limitations of the Fossil Record and the Dating of the Origin of the Bilateria." In *Telling the Evolutionary Time: Molecular Clocks and the Fossil Record,* edited by P. C. J. Donoghue and M. P. Smith, 166–89. London: Taylor & Francis, 2003.

Burgess-Beusse, B., C. Farrell, M. Gaszner, M. Litt, V. Mutskov, F. Recillas-Targa, M. Simpson, A. West, and G. Felsenfeld. "The Insulation of Genes from External Enhancers and Silencing Chromatin." *Proceedings of the National Academy of Sciences USA* 99 (2002): 16433–37.

Butcher, Samuel E. "The Spliceosome as Ribozyme Hypothesis Takes a Second Step." *Proceedings of the National Academy of Sciences USA* 106 (2009): 12211–12.

Cameron, Chris B., James R. Garey, and Billie J. Swalla. "Evolution of the Chordate Body Plan: New Insights from Phylogenetic Analyses of Deuterostome Phyla." *Proceedings of the National Academy of Sciences USA* 97 (2000): 4469–74.

Cameron, R. A., et al. "A Sea Urchin Genome Project: Sequence Scan, Virtual Map, and Additional Resources." *Proceedings of the National Academy of Sciences USA* 97 (2000): 9514–18.

Canoe, Inc. "Yoho National Park." http://www.canadianparks.com/bcolumbia/yohonpk/index.htm (accessed March 22, 2013).

Carrieri, C., et al. "Long Non-Coding Antisense RNA Controls Uchl1 Translation Through An Embedded SINEB2 Repeat." *Nature* 49 (2012): 454–57.

Carroll, Robert L. *Patterns and Processes of Vertebrate Evolution.* Cambridge: Cambridge Univ. Press, 1997.

———. "Towards a New Evolutionary Synthesis." *Trends in Ecology and Evolution* 15 (2000): 27–32.

Carroll, Sean B. *Endless Forms Most Beautiful: The New Science of Evo Devo.* New York: Norton, 2006.

Chamberlain, Thomas C. "The Method of Multiple Working Hypotheses." *Science* (old series) 15 (1890): 92–96. Reprinted in *Science* 148 (1965): 754–59. Also reprinted in *Journal of Geology* (1931): 155–65.

Chargaff, Erwin. "Chemical Specificity of Nucleic Acids and Mechanism of Their Enzymic Degradation." In *Essays on Nucleic Acids,* 1–24. New York: Elsevier, 1963.

Charlesworth, Brian, Russell Lande, and Montgomery Slatkin. "A Neo-Darwinian Commentary on Macroevolution." *Evolution* 36, no. 3 (1982): 474–98.

Chen, J. Y. "Early Crest Animals and the Insight They Provide into the Evolutionary Origin of Craniates." *Genesis* 46 (2008): 623–39.

Chen, J. Y., J. Dzik, G. D. Edgecombe, L. Ramsköld, and G.-Q. Zhou. "A Possible Early Cambrian Chordate." *Nature* 377 (1995): 720–22.

Chen, J. Y., D. Y. Huang, and C. W. Li. "An Early Cambrian Craniate-like Chordate." *Nature* 402 (1999): 518–22.

Chen, J. Y., and C. W. Li. "Early Cambrian Chordate from Chengjiang, China." *Bulletin of the National Museum of Natural Science of Taiwan* 10 (1997): 257–73.

Chen, J. Y., C. W. Li, Paul Chien, G.-Q. Zhou, and Feng Gao. "Weng'an Biota: A Light Casting on the Precambrian World." Paper presented to "The Origin of Animal Body Plans and Their Fossil Records" Conference, Kunming, China, June 20–26, 1999, sponsored by the Early Life Research Center and the Chinese Academy of Sciences.

Chen, J. Y., P. Oliveri, F. Gao, S. Q. Dornbos, C. W. Li, D. J. Bottjer, and E. H. Davidson. "Precambrian Animal Life: Probable Developmental and Adult Cnidarian Forms from Southwest China." *Developmental Biology* 248 (2002): 182–96.

Chen, J. Y., J. W. Schopf, et al. "Raman Spectra of a Lower Cambrian Ctenophore Embryo from Southwestern Shaanxi, China." *Proceedings of the National Academy of Sciences USA* 104 (2007): 6289–92.

Chen, J. Y., and G.-Q. Zhou. "Biology of the Chengjiang Fauna." In *The Cambrian Explosion and the Fossil Record,* edited by Chen, J. Y., Y. Cheng, and H. V. Iten, 11–106. Taiwan: National Museum of Natural Science, 1997.

Chen, J. Y., G.-Q. Zhou, M. Y. Zhu, and K. Y. Yeh. *The Chengjiang Biota: A Unique Window of the Cambrian Explosion.* Taichung, Taiwan: National Museum of Natural Science, 1996.

Chen, Liangbiao, Arthur L. DeVries, and Chi-Hing C. Cheng. "Evolution of Antifreeze Glycoprotein Gene from a Trypsinogen Gene in Antarctic Notothenioid Fish." *Proceedings of the National Academy of Sciences USA* 94 (1997): 3811–16.

———. "Convergent Evolution of Antifreeze Glycoproteins in Antarctic Notothenioid Fish and Arctic Cod." *Proceedings of the National Academy of Sciences USA* 94 (1997): 3817–22.

Chen, Ling-Ling, Joshua N. DeCerbo, and Gordon G. Carmichael. "*Alu* Element-Mediated Gene Silencing." *EMBO Journal* (2008): 1–12.

Chesterton, G. K. "The Invisible Man." In *The Complete "Father Brown,"* at ebooks.adelaide.edu.auu/c/Chesterton/gk/c52fb/chapter5.html. First published in *The Saturday Evening Post*, January 28, 1911, 5–7, 30.

Chien, Paul, J. Y. Chen, C. W. Li, and Frederick Leung. "SEM Observation of Precambrian Sponge Embryos from Southern China, Revealing Ultrastructures Including Yolk Granules, Secretion Granules, Cytoskeleton, and Nuclei." Paper presented to the North American Paleontological Convention, University of California, Berkeley, June 26–July 1, 2001.

Christin, P. A., D. M. Weinreich, and G. Bresnard. "Causes and Evolutionary Significance of Genetic Convergence." *Trends in Genetics* 26 (2010): 400–405.

Cisne, J. L. "Trilobites and the Origin of Arthropods." *Science* 186 (1974): 13–18.

Cloud, Wallace. "The Ship That Digs Holes in the Sea." *Popular Mechanics* 131 (1969): 108–11, 236.

Cohen, Bernard I., ed. *Isaac Newton's Papers and Letters on Natural Philosophy.* Cambridge: Cambridge Univ. Press, 1958.

Colas, A. R., et al. "Whole-Genome MicroRNA Screening Identifies let-7 and miR-18 as Regulators of Germ Layer Formation During Early Embryogenesis." *Genes and Development* 26 (2012): 2567–79.

Collins, Desmond. "Misadventures in the Burgess Shale." *Nature* 460 (2009): 952–53.

Collins, Francis. *The Language of God.* New York: Free Press, 2006.

Conant, James B. *Modern Science and Modern Man.* New York: Doubleday Anchor, 1953.

Conniff, Richard. "When Continental Drift Was Considered Pseudoscience." *Smithsonian Magazine,* June 2012. http://www.smithsonianmag.com/science-nature/When-Continental-Drift-Was-Considered-Pseudoscience.html (accessed December 30, 2012).

Conway Morris, Simon. "Burgess Shale Faunas and the Cambrian Explosion." *Science* 246 (1989): 339–46.

———. "Ediacaran-like Fossils in Cambrian Burgess Shale-type Faunas of North America." *Palеontology* 36 (1993): 593–635.

———. "Early Metazoan Evolution: Reconciling Paleontology and Molecular Biology." *American Zoologist* 38 (1998): 867–77.

———. "Nipping the Cambrian "Explosion" in the Bud?" *BioEssays* 22 (2000): 1053–56.

———. *The Crucible of Creation: The Burgess Shale and the Rise of Animals.* Oxford: Oxford Univ. Press, 2000.

———. "Evolution: Bringing Molecules into the Fold." *Cell* 100 (2000): 1–11.

———. "The Cambrian 'Explosion' of Metazoans and Molecular Biology: Would Darwin Be Satisfied?" *International Journal of Developmental Biology* 47 (2003): 505–15.

———. "The Cambrian 'Explosion' of Metazoans." In *Origination of Organismal Form: Beyond the Gene in Developmental and Evolutionary Biology,* edited by G. B. Müller and S. A. Newman, 13–32. Cambridge, MA: MIT Press, 2003.

———. *Life's Solution: Inevitable Humans in a Lonely Universe.* Cambridge: Cambridge Univ. Press, 2003.

———. "Darwin's Dilemma: The Realities of the Cambrian 'Explosion.'" *Philosophical Transactions of the Royal Society B* 361 (2006): 1069–83.

———. "Walcott, the Burgess Shale and Rumours of a Post-Darwinian World." *Current Biology* 19 (2009): R927–31.

Conway Morris, S., and J.-B. Caron. "*Pikaia gracilens* Walcott, a Stem-Group Chordate from the Middle Cambrian of British Columbia." *Biological Reviews of the Cambridge Philosophical Society* 87 (2012): 480–512.

Conway Morris, S., and D. H. Collins. "Middle Cambrian Ctenophores from the Stephen Formation, British Columbia, Canada." *Philosophical Transactions of the Royal Society B: Biological Sciences* 351 (1996): 279–308.
Conway Morris, S., and J. S. Peel. "Articulated Halkieriids from the Lower Cambrian of North Greenland and Their Role in Early Protostome Evolution." *Philosophical Transactions of the Royal Society B: Biological Sciences* 347 (1995): 305–58.
———. "The Earliest Annelids: Lower Cambrian Polychaetes from the Sirius Passet Lagerstätte, Peary Land, North Greenland." *Acta Palaeontologica Polonica* 53 (2008): 137–48.
Cooper, Alan, and Richard Fortey. "Evolutionary Explosions and the Phylogenetic Fuse." *Trends in Ecology and Evolution* 13 (1998):151–56.
Courseaux, Anouk, and Jean-Louis Nahon. "Birth of Two Chimeric Genes in the *Hominidae* Lineage." *Science* 291 (2001): 1293–97.
Coyne, Jerry. "The Great Mutator: Review of *The Edge of Evolution,* by Michael J. Behe." *New Republic,* June 18, 2007, 38–44.
———. *Why Evolution Is True*. New York: Viking, 2009.
Crick, Francis. "On Protein Synthesis." *Symposium for the Society of Experimental Biology* 12 (1958): 138–63.
———. *What Mad Pursuit: A Personal View of Scientific Discovery.* New York: Basic Books, 1988.
Crick, Francis, and James Watson. "A Structure for Deoxyribose Nucleic Acid." *Nature* 171 (1953): 737–38.
Crocker, Caroline. *Free to Think: Why Scientific Integrity Matters.* Port Orchard, WA: Leafcutter Press, 2010.
Crowther, Robert. "Academic Freedom Expelled from Baylor University." *Evolution News and Views,* September 5, 2007. http://www.evolutionnews.org/2007/09/academic_freedom_expelled_from004189.html.
Cunningham, J. A., et al. "Experimental Taphonomy of Giant Sulphur Bacteria: Implications for the Interpretation of the Embryo-Like Ediacaran Doushantuo Fossils." *Proceedings of the Royal Society B* 279 (2012): 1857–64.
Daley, A. C. "The Morphology and Evolutionary Significance of the Anomalocaridids." *Digital Comprehensive Summaries of Uppsala Dissertations from the Faculty of Science and Technology* 714 (2010): 9–34.
Darnell, J. E., and W. F. Doolittle. "Speculations on the Early Course of Evolution." *Proceedings of the National Academy of Sciences USA* 83 (1986): 1271–75.
Darwin, Charles. *On the Origin of Species by Means of Natural Selection.* A facsimile of the first edition, published by John Murray, London, 1859. Reprint, Cambridge, MA: Harvard Univ. Press, 1964.
———. *On the Origin of Species by Means of Natural Selection.* 6th ed. London: John Murray, 1872.
———. *The Autobiography of Charles Darwin, 1809–1882.* Edited by Nora Barlow. New York: Norton, 1958.
———. *The Illustrated Origin of Species.* Abridged and introduced by Richard Leakey. 6th ed. London: Faber and Faber, 1979.
Dávalos, Liliana M., Andrea L. Cirranello, Jonathan H. Geisler, and Nancy B. Simmons. "Understanding Phylogenetic Incongruence: Lessons from Phyllostomid Bats." *Biological Reviews* 87 (2012): 991–1024.
Davidson, Eric H. *Genomic Regulatory Systems: Development and Evolution.* New York: Academic, 2001.
———. *The Regulatory Genome: Gene Regulatory Networks in Development and Evolution.* Burlington: Elsevier, 2006.
———. "Evolutionary Bioscience as Regulatory Systems Biology." *Developmental Biology* 357 (2011): 35–40.
Davidson, Eric H., and Douglas Erwin. "An Integrated View of Precambrian Eumetazoan Evolution." *Cold Spring Harbor Symposia on Quantitative Biology* 74 (2010): 1–16.
Davidson, Eric H., Kevin J. Peterson, and R. Andrew Cameron. "Origin of Bilaterian Body Plans: Evolution of Developmental Regulatory Mechanisms." *Science* 270 (1995): 1319–24.
Dawkins, Richard. *The Blind Watchmaker: Why the Evidence Reveals a Universe Without Design.* New York: Norton, 1986.
———. *River Out of Eden: A Darwinian View of Life.* New York: Basic Books, 1995.
———. *Climbing Mount Improbable.* New York: Norton, 1996.

———. *Unweaving the Rainbow: Science, Delusion, and the Appetite for Wonder.* Boston: Houghton Mifflin, 1998.

———. *A Devil's Chaplain: Reflections on Hope, Lies, Science, and Love.* Boston: Houghton Mifflin, 2003.

———. *The God Delusion.* Boston: Houghton Mifflin, 2006.

———. *The Greatest Show on Earth: The Evidence for Evolution.* New York: Free Press, 2009.

Dean, Cornelia. "Scientists Feel Miscast in Film on Life's Origin." *New York Times,* September 27, 2007.

De Duve, C. *Blueprint for a Cell: The Nature and Origin of Life.* Burlington, NC: Patterson, 1991.

Degnan, James H., and Noah A. Rosenberg. "Gene Tree Discordance, Phylogenetic Inference and the Multispecies Coalescent." *Trends in Ecology and Evolution* 24 (2009): 332–40.

Delsuc, Frederic, Matthew J. Phillips, and David Penny. "Comment on 'Hexapod Origins: Monophyletic or Paraphyletic?'" *Science* 301 (2003): 1482.

Dembski, William A. "Intelligent Science and Design." *First Things* 86 (1998): 21–27.

———. *The Design Inference: Eliminating Chance Through Small Probabilities.* Cambridge: Cambridge Univ. Press, 1998.

———. "Intelligent Design as a Theory of Technological Evolution." *Progress in Complexity, Information, and Design* 1.2 (2002). Available at http://www.iscid.org.

———. *No Free Lunch: Why Specified Complexity Cannot Be Purchased Without Intelligence.* Boston: Rowman & Littlefield, 2002.

Demuth, J. P., T. De Bie, J. E. Stajich, N. Cristianini, and M. W. Hahn. "The Evolution of Mammalian Gene Families." *PLoS One* 1 (2006): e85.

Dennett, Daniel C. *Darwin's Dangerous Idea: Evolution and the Meanings of Life.* New York: Simon & Schuster, 1995.

Denton, Michael. *Evolution: A Theory in Crisis.* London: Adler and Adler, 1985.

De Rosa, R., J. K. Grenier, T. Andreeva, C. E. Cook, A. Adoutte, M. Akam, S. B. Carroll, and G. Balavoine. "Hox Genes in Brachiopods and Priapulids and Protostome Evolution." *Nature* 399 (1999): 772–76.

Dillon, William. "Regents Deny Gonzalez's Tenure Appeal." *Ames Tribune,* February 7, 2008.

Dobzhansky, Theodosius. "Discussion of G. Schramm's Paper." In *The Origins of Prebiological Systems and of Their Molecular Matrices,* edited by S. W. Fox, 309–15. New York: Academic, 1965.

———. "Nothing in Biology Makes Sense Except in the Light of Evolution." *American Biology Teacher* 35 (1973): 125–29.

Donnelly, S. R., T. E. Hawkins, and S. E. Moss, "A Conserved Nuclear Element with a Role in Mammalian Gene Regulation." *Human Molecular Genetics* 8 (1999): 1723–28.

Doolittle, Russell F., D. F. Feng, S. Tsang, G. Cho, and E. Little. "Determining Divergence Times of the Major Kingdoms of Living Organisms with a Protein Clock." *Science* 271 (1996): 470–77.

Doolittle, W. F. "Phylogenetic Classification and the Universal Tree." *Science* 284 (1999): 2124–28.

Dott, Robert H., Jr., and Donald R. Prothero. *Evolution of the Earth.* 5th ed. New York: McGraw-Hill, 1994.

Douglas, Michael Edward, and John C. Avise. "Speciation Rates and Morphological Divergence in Fishes: Tests of Gradual Versus Rectangular Modes of Evolutionary Change." *Evolution* 36 (1982): 224–32.

Dunbar, Carl O. *Historical Geology.* New York: Wiley, 1949.

Dunlap, K.A., M. Palmarini, M. Varela, R. C. Burghardt, K. Hayashi, J. L. Farmer, and T. E. Spencer. "Endogenous Retroviruses Regulate Periimplantation Placental Growth and Differentiation." *Proceedings of the National Academy of Sciences USA* 103 (2006): 14390–95.

Dunn, C. A., P. Medstrand, and D. L. Mager. "An Endogenous Retroviral Long Terminal Repeat Is the Dominant Promoter for Human B1, 3-galactosyltransferase 5 in the Colon." *Proceedings of the National Academy of Sciences USA* 100 (2003): 12841–46.

Dupree, A. Hunter. *Asa Gray: American Botanist, Friend of Darwin.* Cambridge, MA: Harvard Univ. Press, Belknap Press, 1959.

Durrett, Rick, and Deena Schmidt. "Waiting for Two Mutations: With Applications to Regulatory Sequence Evolution and the Limits of Darwinian Evolution." *Genetics* 180 (2008): 1501–9.

Dzik, J. "*Yunnanozoon* and the Ancestry of Chordates." *Acta Palaeontologica Polanica* 40 (1995): 341–60.

Ebert, James, et al. *Science and Creationism: A View from the National Academy of Science.* Washington, DC: National Academy Press, 1984.

Ecker, J. R., W. A. Bickmore, I. Barroso, J. K. Pritchard, Y. Gilad, and E. Segal. "Genomics: ENCODE Explained." *Nature* 489 (2012): 52–55.

Eden, Murray. "Inadequacies of Neo-Darwinian Evolution as a Scientific Theory." In *Mathematical Challenges to the Neo-Darwinian Interpretation of Evolution,* edited by P. S. Moorhead and M. M. Kaplan, 109–11. Wistar Institute Symposium Monograph No. 5. New York: Liss, 1967.

Edgecombe, G. D., G. Giribet, C. W. Dunn, A. Hejnol, R. M. Kristensen, R. C. Neves, G. W. Rouse, K. Worsaae, and M. V. Sørensen. "Higher-Level Metazoan Relationships: Recent Progress and Remaining Questions." *Organisms, Diversification, and Evolution* 11 (2011): 151–72.

Edidin, Michael. "Patches, Posts and Fences: Proteins and Plasma Membrane Domains." *Trends in Cell Biology* 2 (1992): 376–80.

Eger, Martin. "A Tale of Two Controversies: Dissonance in the Theory and Practice of Rationality." *Zygon* 23 (1988): 291–326.

Ehrlich, Paul, and Richard Holm. *The Processes of Evolution*. New York: McGraw-Hill, 1963.

Eldredge, Niles. *Reinventing Darwin: The Great Debate at the High Table of Evolutionary Theory.* New York: Wiley, 1995.

———. *The Pattern of Evolution*. New York: Freeman, 1999.

———. *The Triumph of Evolution and the Failure of Creationism*. New York: Freeman, 2000.

Eldredge, Niles, and Stephen Jay Gould. "Punctuated Equilibria: An Alternative to Phyletic Gradualism." In *Models in Paleobiology,* edited by T. J. M. Schopf. San Francisco: Freeman, Cooper, 1972.

Enard, Wolfgang et al. "Molecular Evolution of *FOXP2,* a Gene Involved in Speech and Language." *Nature* 418 (2002): 869–72.

ENCODE Project Consortium. "Identification and Analysis of Functional Elements in 1% of the Human Genome by the ENCODE Pilot Project." *Nature* 447 (2007): 799–816.

Endler, John. *Natural Selection in the Wild.* Princeton, NJ: Princeton Univ. Press, 1986.

Erwin, Douglas H. "Early Introduction of Major Morphological Innovations." *Acta Palaeontologica Polonica* 38 (1994): 281–94.

———. "The Origin of Body Plans." *American Zoologist* 39 (1999): 617–29.

———. "Macroevolution Is More Than Repeated Rounds of Microevolution." *Evolution and Development* 2 (2000): 78–84.

———. "Disparity: Morphological Pattern and Developmental Context." *Palaeontology* 50 (2007): 57–73.

———. "Evolutionary Uniformatarianism." *Developmental Biology* 357 (2011): 27–34.

Erwin, Douglas H., and Eric Davidson. "The Last Common Bilaterian Ancestor." *Development* 129 (2002): 3021–32.

———. "The Evolution of Hierarchical Gene Regulatory Networks." *Nature Reviews Genetics* 10 (2009): 141–48.

Erwin, Douglas H., Marc Laflamme, Sarah M. Tweedt, Erik A. Sperling, Davide Pisani, and Kevin J. Peterson. "The Cambrian Conundrum: Early Divergence and Later Ecological Success in the Early History of Animals." *Science* 334 (2011): 1091–97.

Erwin, Douglas, James Valentine, and David Jablonski. "The Origin of Animal Body Plans." *American Scientist* 85 (1997): 126–37.

Erwin, Douglas, James Valentine, and J. J. Sepkoski. "A Comparative Study of Diversification Events: The Early Paleozoic Versus the Mesozoic." *Evolution* 41 (1987): 1177–86.

Ewen-Campen, Ben, E. E. Schwager, and C. G. Extavour. "The Molecular Machinery of Germ Line Specification." *Molecular Reproduction and Development* 77 (2010): 3–18.

Ewert, Winston, George Montañez, William Dembski, and Robert J. Marks II. "Efficient Per Query Information Extraction from a Hamming Oracle." *42nd South Eastern Symposium on System Theory* (2010): 290–97.

Extavour, Cassandra G. M. "Evolution of the Bilaterian Germ Line: Lineage Origin and Modulation of Specific Mechanisms." *Integrative and Comparative Biology* 47 (2007): 770–85.

———. "Gray Anatomy: Phylogenetic Patterns of Somatic Gonad Structures and Reproductive Strategies Across the Bilateria." *Integrative and Comparative Biology* 47 (2007): 420–26.

Fedonkin, Mikhail A., and Benjamin M. Waggoner. "The Late Precambrian Fossil *Kimberella* is a Mollusc-like Bilaterian Organism." *Nature* 388 (1997): 868–71.

Felsenstein, Joseph. *Inferring Phylogenies*. Sunderland, MA: Sinauer, 2004.

Fernandes, J., S. E. Celniker, E. B. Lewis, and K. VijayRaghavan. "Muscle Development in the Four-Winged *Drosophila* and the Role of the Ultrabithorax Gene." *Current Biology* 4 (1994): 957–64.

Figueiredo, L. M., L. H. Freitas-Junior, E. Bottius, J.-C. Olivo-Marin, and A. Scherf. "A Central Role for *Plasmodium Falciparum* Subtelomeric Regions in Spatial Positioning and Telomere Length Regulation." *EMBO Journal* 21 (2002): L815–24.

Fisher, Ronald A. *The Genetical Theory of Natural Selection*. New York: Dover, 1958.

Foote, Michael. "Paleozoic Record of Morphological Diversity in Blastozoan Echinoderms." *Proceedings of the National Academy of Sciences USA* 89 (1992): 7325–29.

———. "On the Probability of Ancestors in the Fossil Record." *Paleobiology* 22 (1996): 141–51.

———. "Sampling, Taxonomic Description, and Our Evolving Knowledge of Morphological Diversity." *Paleobiology* 23 (1997): 181–206.

Foote, Michael, and Stephen Jay Gould. "Cambrian and Recent Morphological Disparity." *Science* 258 (1992): 1816–17.

Fortey, Richard. "The Cambrian Explosion Exploded?" *Science* 293 (2001): 438–39.

Fortey, Richard A., Derek E. G. Briggs, and Matthew A. Wills. "The Cambrian Evolutionary 'Explosion' Recalibrated." *BioEssays* 19 (1997): 429–34.

Franco-Zorrilla, J. M., et al. "Target Mimicry Provides a New Mechanism for Regulation of MicroRNA Activity." *Nature Genetics* 39 (2007): 1033–37.

Frankel, J. "Propagation of Cortical Differences in *Tetrahymena*." *Genetics* 94 (1980): 607–23.

Frazzetta, Thomas H. "From Hopeful Monsters to Bolyerine Snakes?" *American Naturalist* 104 (1970): 55–72.

———. *Complex Adaptations in Evolving Populations*. Sunderland, MA: Sinauer, 1975.

———. "Modeling Complex Morphological Change in Evolution, and A Possible Ecological Analogy." *Evolutionary Theory* 6 (1982): 127–41.

Frohnhöfer, Hans Georg, and Christiane Nüsslein-Volhard. "Organization of Anterior Pattern in the *Drosophila* Embryo by the Maternal Gene *Bicoid*." *Nature* 324 (1986): 120–25.

Futuyma, Douglas J. "Evolution as Fact and Theory." *BIOS* 56 (1985): 8.

Fuxreiter, Monika, and Peter Tompa. "Fuzzy Complexes: A More Stochastic View of Protein Function." In *Fuzziness: Structural Disorder in Protein Complexes,* edited by M. Fuxreiter and P. Tompa, 1–14. Advances in Experimental Medicine and Biology 725. Austin, TX: Landes Bioscience, Springer Science, 2012.

Gabius, Hans-Joachim. "Biological Information Transfer Beyond the Genetic Code: The Sugar Code." *Naturwissenschaften* 87 (2000): 108–21.

Gabius, Hans-Joachim, H. C. Siebert, S. André, J. Jiménez-Barbero, and H. Rüdinger. "Chemical Biology of the Sugar Code." *Chembiochem* 5 (2004): 740–64.

Gagnon, James A., and Kimberly L. Mowry. "Molecular Motors: Directing Traffic During RNA Localization." *Critical Reviews of Biochemistry and Molecular Biology* 46 (2011): 229–39. http://www.ncbi.nlm.nih.gov/pmc/articles/PMC3181154/?tool=pubmed (accessed October 30, 2012).

Gaffney, Eugene S. "The Comparative Osteology of the Triassic Turtle Proganochelys." *Bulletin of the American Museum of Natural History* 194 (1990): 1–263.

Gallegos, Emma. "Intelligent Design Proponent Who Works at JPL Says He Experienced Religious Discrimination." *San Gabriel Valley Tribune,* April 18, 2010.

Gallie, Walter Bryce. "Explanations in History and the Genetic Sciences." In *Theories of History,* edited by P. Gardiner, 386–402. Glencoe, IL: Free Press, 1959.

García-Bellido, D. C., and D. H. Collins. "A New Study of *Marrella splendens* (Arthropoda, Marrelomorpha) from the Middle Cambrian Burgess Shale, British Columbia, Canada." *Canadian Journal of Earth Sciences* 43 (2006): 721–42.

Gauger, Ann. "The Frailty of the Darwinian Hypothesis, Part 2." *Evolution News & Views,* July 14, 2009. http://www.evolutionnews.org/2009/07/the_frailty_of_the_darwinian_h_1022911.html.

———. "Why Proteins Aren't Easily Recombined." May 7, 2012. http://www.biologicinstitute.org/post/22595615671/why-proteins-arent-easily-recombined.

Gauger, Ann K., and Douglas D. Axe. "The Evolutionary Accessibility of New Enzyme Functions: A Case Study from the Biotin Pathway." *BIO-Complexity* 2011, no. 1 (2011): 1–17.

Gauger, Ann K., Stephanie Ebnet, Pamela F. Fahey, and Ralph Seelke. "Reductive Evolution Can Prevent Populations from Taking Simple Adaptive Paths to High Fitness." *BIO-Complexity* 2010, no. 2 (2010): 1–9.

Gehling, J. G. "The Case for Ediacaran Fossil Roots to the Metazoan Tree." In *The World of Martin F. Glaessner,* Memoir No. 20, edited by B. P. Radhakrishna, 181–223. Bangalore: Geological Society of India, 1991.

George, J. A., K. L. Traverse, P. G. DeBaryshe, K. J. Kelley, and M. L. Pardue. "Evolution of Diverse Mechanisms for Protecting Chromosome Ends by *Drosophila* TART Telomere Retrotransposons." *Proceedings of the National Academy of Sciences USA* 107 (2010): 21052–57.
Gerhart, John, and Marc Kirschner. *Cells, Embryos, and Evolution*. London: Blackwell Science, 1997.
———. "The Theory of Facilitated Variation." *Proceedings of the National Academy of Sciences USA* 104 (2007): 8582–89.
Gilbert, S. F., J. M. Opitz, and R. A. Raff. "Resynthesizing Evolutionary and Developmental Biology." *Developmental Biology* 173 (1996): 357–72.
Gilbert, W. "The Exon Theory of Genes." *Cold Spring Harbor Symposium on Quantitative Biology* 52 (1987): 901–5.
Giles, Jim. "Peer-Reviewed Paper Defends Theory of Intelligent Design." *Nature* 431 (2004): 114.
Gillespie, Neal C. *Charles Darwin and the Problem of Creation*. Chicago: Univ. of Chicago Press, 1979.
Gishlick, Alan, Nicholas Matzke, and Wesley R. Elsberry. "Meyer's Hopeless Monster." Talk Reason.org, September 12, 2004. http://www.talkreason.org/articles/meyer.cfm.
Glaessner, M. F. "A New Genus of Late Precambrian Polychaete Worms from South Australia." *Transactions of the Royal Society of South Australia* 100 (1976): 169–70.
Gleick, James. "The Pace of Evolution: A Fossil Creature Moves to Center of Debate." *New York Times*, December 22, 1987.
Goh, Sung-Ho, Y. T. Lee, N. V. Bhanu, M. C. Cam, R. Desper, B. M Martin, R. Moharram, R. B. Gherman, and J. L. Miller. "A Newly Discovered Human Alpha-Globin Gene." *Blood* 106 (2005): 1466–72.
Goldschmidt, Richard. *The Material Basis of Evolution*. New Haven, CT: Yale Univ. Press, 1940.
Gon III, S. M. "Origins of Trilobites." http://www.trilobites.info/origins.htm.
Goodrich, J. A., and J. F. Kugel. "Non-coding RNA Regulators of RNA Polymerase II Transcription." *Nature Reviews Molecular and Cell Biology* 7 (2006): 612–16.
Goodwin, Brian C. "What Are the Causes of Morphogenesis?" *BioEssays* 3 (1985): 32–36.
———. "Structuralism in Biology." *Science Progress* 74 (1990): 227–44.
———. *How the Leopard Changed Its Spots: The Evolution of Complexity*. New York: Scribner, 1994.
Gould, Stephen Jay. "The Return of Hopeful Monsters." *Natural History* 86 (1977): 22–30.
———. "Evolution: Explosion, Not Ascent; When Change Was Slow and Safe; No Evolutionary Ladder; Everyone Has Prejudices." *New York Times*, January 22, 1978.
———. "Is a New and General Theory of Evolution Emerging?" *Paleobiology* 6 (1980): 119–30.
———. "The Meaning of Punctuated Equilibrium and Its Role in Validating a Hierarchical Approach to Macroevolution." In *Perspectives on Evolution*, edited by R. Milkman, 83–104. Sunderland, MA: Sinauer, 1982.
———. "Creationism: Genesis Versus Geology." In *Science and Creationism*, edited by A. Montagu, 126–35. New York: Oxford Univ. Press, 1984.
———. "Evolution as Fact and Theory." In *Science and Creationism*, edited by A. Montagu, 118–24. New York: Oxford Univ. Press, 1984.
———. "Evolution and the Triumph of Homology: Or, Why History Matters." *American Scientist* 74 (1986): 60–69.
———. *Wonderful Life: The Burgess Shale and the Nature of History*. New York: Norton, 1990.
———. "The Disparity of the Burgess Shale Arthropod Fauna and the Limits of Cladistic Analysis: Why We Must Strive to Quantify Morphospace." *Paleobiology* 17 (1991): 411–23.
———. *The Structure of Evolutionary Theory*. Cambridge, MA: Harvard Univ. Press, 2002.
Gould, Steven Jay, and Niles Eldredge. "Punctuated Equilibria: The Tempo and Mode of Evolution Reconsidered." *Paleobiology* 3 (1977): 115–51.
———. "Punctuated Equilibrium Comes of Age." *Nature* 366 (1993): 223–27.
Graur, Dan, and William Martin. "Reading the Entrails of Chickens: Molecular Timescales of Evolution and the Illusion of Precision." *Trends in Genetics* 20, no. 2 (2004): 80–86.
Grawunder, U., M. Wilm, X. Wu, P. Kulesza, T. E. Wilson, M. Mann, and M. R. Lieber. "Activity of DNA Ligase IV Stimulated by Complex Formation with XRCC4 Protein in Mammalian Cells." *Nature* 388 (1997): 492–95.
Gray, Asa. *Darwiniana*. Edited by A. Hunter Dupree. Cambridge, MA: Belknap Press, 1963.
Green, David G. "The Role of Translocation and Selection in the Emergence of Genetic Clusters and Modules." *Artificial Life* 13 (2007): 249–58.

Grizzle, Raymond. "Some Comments on the 'Godless' Nature of Darwinian Evolution, and a Plea to the Philosophers Among Us." *Perspectives on Science and Christian Faith* 44 (1993): 175–77.
Grosberg, R. K. "Out on a Limb: Arthropod Origins." *Science* 250 (1990): 632–33.
Gross, Paul. "Design for Living." *New Criterion* 26 (October 2007): 70–73.
Grotzinger, John P., Samuel A. Bowring, Beverly Z. Saylor, and Alan J. Kaufman. "Biostratigraphic and Geochronologic Constraints on Early Animal Evolution." *Science* 270 (1995): 598–604.
Guerzoni, Daniele, and Aoife McLysaght. "De Novo Origins of Human Genes," *PLoS Genetics* 7 (2011): e1002381.
Gura, Trisha. "Bones, Molecules, or Both?" *Nature* 406 (2000): 230–33.
Haffter, et al. "The Identification of Genes with Unique and Essential Functions in the Development of the Zebrafish, *Danio rerio*." *Development* 123 (1996): 1–36.
Hagerty, Barbara. "Intelligent Design and Academic Freedom." NPR, November 17, 2005.
Hahn, Matthew W. "Distinguishing Among Evolutionary Models for the Maintenance of Gene Duplicates." *Journal of Heredity* 100 (2009): 605–17.
Halanych, Kenneth M. "The New View of Animal Phylogeny." *Annual Review of Ecology and Systematics* 35 (2004): 229–56.
Halanych, Kenneth M., and Yale Passamaneck. "A Brief Review of Metazoan Phylogeny and Future Prospects in Hox-Research." *American Zoologist* 41 (2001): 629–39.
Halder, Georg, Patrick Callaerts, and Walter J. Gehring. "Induction of Ectopic Eyes by Targeted Expression of the *eyeless* Gene in *Drosophila*." *Science* 267 (1995): 1788–92.
Hall, C., S. Brachat, and F. S. Dietrich. "Contribution of Horizontal Gene Transfer to the Evolution of *Saccharomyces cerevisiae*." *Eukaryotic Cell* 4 (2005): 1102–15.
Hall, Dwight H., Ying Liu, and David A. Shub. "Exon Shuffling by Recombination Between Self-Splicing Introns of Bacteriophage T4." *Nature* 340 (1989): 574–76.
Han, Jeffrey S., Suzanne T. Szak, and Jeff D. Boeke. "Transcriptional Disruption by the L1 Retrotransposon and Implications for Mammalian Transcriptomes." *Nature* 429 (2004): 268–74.
Harold, Franklin M. "From Morphogenes to Morphogenesis." *Microbiology* 141 (1995): 2765–78.
———. *The Way of the Cell: Molecules, Organisms, and the Order of Life*. New York: Oxford Univ. Press, 2001.
Harris, Jack Ross. "Louis Agassiz: A Reevaluation of the Nature of His Opposition to the Darwinian View of Natural History." Senior thesis, Whitworth College, 1993.
Harvey, Ethel Browne. "Parthenogenetic Merogony or Cleavage Without Nuclei in *Arbacia punctulata*." *Biological Bulletin* 71 (1936): 101–21. http://www.biolbull.org/content/71/1/101.full.pdf+html (accessed October 30, 2012).
———. "A Comparison of the Development of Nucleate and Non-nucleate Eggs of *Arbacia punctulata*." *Biological Bulletin* 79 (1940): 166–87. http://www.biolbull.org/content/79/1/166.full.pdf+html (accessed October 30, 2012).
He, Lu, Alan M. Friedman, and Chris Bailey-Kellogg. "Algorithms for Optimizing Cross-Overs in DNA Shuffling." ACM Conference on Bioinformatics, Computational Biology and Biomedicine 2011. *BMC Bioinformatics* 2012, 13 (Suppl 3): S3.
Heirtzler, J. R. "Sea-Floor Spreading." In *Continents Adrift: Readings from Scientific American*, edited by J. T. Wilson, 68–78. San Francisco: Freeman, 1970.
Hellström, Nils Petter. "The Tree as Evolutionary Icon: TREE in the Natural History Museum, London." *Archives of Natural History* 38.1 (2011): 1–17.
Henikoff, Steven. "Heterochromatin Function in Complex Genomes." *Biochimica et Biophysica Acta* 1470 (2000): O1–O8.
Henikoff, Steven, Kami Ahmad, and Harmit S. Malik. "The Centromere Paradox: Stable Inheritance with Rapidly Evolving DNA." *Science* 293 (2001): 1098–1102.
Himmelfarb, Gertrude. *Darwin and the Darwinian Revolution*. 1959. Reprint, Chicago: Ivan R. Dee, 1996.
Ho, Simon Y. W., Matthew J. Phillips, Alexei J. Drummond, and Alan Cooper. "Accuracy of Rate Estimation Using Relaxed-Clock Models with a Critical Focus on the Early Metazoan Radiation." *Molecular Biology and Evolution* 22, no. 5 (2005): 1355–63.
Hoekstra, Hopi E., and Jerry A. Coyne. "The Locus of Evolution: Evo Devo and the Genetics of Adaptation." *Evolution* 61 (2007): 995–1016.
Hoeppner, M. P., S. White, D. C. Jeffares, and A. M. Poole. "Evolutionarily Stable Association of Intronic SnoRNAs and MicroRNAs with Their Host Genes." *Genome Biology and Evolution* 1 (2009): 420–28.

Hoffman, H. J., K. Grey, A. H. Hickman, and R. I. Thorpe. "Origin of 3.45 Ga Coniform Stromatolites in Warrawoona Group, Western Australia." *Geological Society of America Bulletin* 111 (1999): 1256–62.
Holden, Constance. "Defying Darwin." *Science* 305 (2004): 1709.
Holder, Charles F. *Louis Agassiz: His Life and Works.* Leaders in Science Series. New York: Putnam, 1893.
Holton, Theìrèse A., and Davide Pisani. "Deep Genomic-Scale Analyses of the Metazoa Reject Coelomata: Evidence from Single- and Multigene Families Analyzed Under a Supertree and Supermatrix Paradigm." *Genome Biology and Evolution* 2 (2010): 310–24.
Hood, Leroy, and David Galas. "The Digital Code of DNA." *Nature* 421 (2003): 444–48.
Hotz, R. L. "Finding Turns Back Clock for Earth's First Animals." *Los Angeles Times,* October 25, 1996, A1, A14.
Hou, Xian-guang, Richard J. Aldridge, Jan Bergström, David J. Siveter, Derek J. Siveter, and Xianhong Feng. *The Cambrian Fossils of Chengjiang, China: The Flowering of Early Animal Life.* Oxford: Blackwell, 2004.
Hou, X.-G., and J. Bergström. "Arthropods of the Lower Cambrian Chengjiang Fauna, Southwest China." *Fossils and Strata* 45 (1997): 1–116.
Hou, Xian-guang, and Wen-guo Sun. "Discovery of Chengjiang Fauna at Meishucun, Jinning, Yunnan." *Acta Paleontologica Sinica* 27 (1988): 1–12.
Hu, S., M. Steiner, M. Zhu, H. Luo, A. Forchielli, H. Keupp, F. Zhao, and Q. Liu. "A New Priapulid Assemblage From the Early Cambrian Guanshan Fossil *Lagerstätte* of SW China." *Bulletin of Geosciences* 87 (2012): 93–106.
Huang, D.-Y., J.-Y. Chen, J. Vannier, and J. I. Saiz Salinas. "Early Cambrian Sipunculan Worms from Southwest China." *Proceedings of the Royal Society B* 271 (2004): 1671–76.
Huldtgren, T., J. A. Cunningham, C. Yin, M. Stampanoni, F. Marone, P. C. J. Donoghue, and S. Bengtson. "Fossilized Nuclei and Germination Structures Identify Ediacaran 'Animal Embryos' as Encysting Protists." *Science* 334 (2011): 1696–99.
———. "Response to 'Comment on "Fossilized Nuclei and Germination Structures Identify Ediacaran 'Animal Embryos' as Encysting Protists."'" *Science* 335 (2012): 1169.
Hurley, Patrick M. "The Confirmation of Continental Drift." In *Continents Adrift: Readings from Scientific American,* edited by J. T. Wilson, 57–67. San Francisco: Freeman, 1970.
Hutchison, Keith. "What Happened to Occult Qualities in the Scientific Revolution?" *Isis* 73 (1982): 253.
Huxley, Julian. "'At Random': A Television Preview." In *Evolution After Darwin: The University of Chicago Centennial.* Vol. 3, *Issues in Evolution,* edited by S. Tax and C. Callendar, 41–65. Chicago: Univ. of Chicago Press, 1960.
———. "The Evolutionary Vision." In *Evolution After Darwin: The University of Chicago Centennial.* Vol. 3, *Issues in Evolution,* edited by S. Tax and C. Callendar, 249–61. Chicago: Univ. of Chicago Press, 1960.
Hyman, Libbie H. *The Invertebrates.* Vol. 1, *Protozoa Through Ctenophora.* New York: McGraw-Hill, 1940.
Hyslop, L., et al. "Downregulation of NANOG Induces Differentiation of Human Embryonic Stem Cells to Extraembryonic Lineages." *Stem Cells* 23 (2005): 1035–43.
Ivantsov, A. Yu. "A New Dickinsonid from the Upper Vendian of the White Sea Winter Coast (Russia, Arkhangelsk Region)." *Paleontological Journal* 33 (1999): 211–21.
———. "Vendia and Other Precambrian 'Arthropods.'" *Paleontological Journal* 35 (2001): 335–43.
———. "A New Reconstruction of *Kimberella,* a Problematic Vendian Metazoan." *Paleontological Journal* 43 (2009): 601–11.
Ivantsov, A. Yu, and Ya E. Malakhovskaya. "Giant Traces of Vendian Animals." *Doklady Earth Sciences* 385A (2002): 618–22.
Jablonka, Eva, and Marion J. Lamb. "Transgenerational Epigenetic Inheritance." In *Evolution: The Extended Synthesis,* edited by M. Pigliucci and G. B. Müller, 137–74. Cambridge, MA: MIT Press, 2010.
Janowski, Bethany, K. E. Huffman, J. C. Schwartz, R. Ram, D. Hardy, D. S. Shames, J. D. Minna, and D. R. Corey. "Inhibiting Gene Expression at Transcription Start Sites in Chromosomal DNA with Antigene RNAs." *Nature Chemical Biology* 1 (2005): 216–22.
Janvier, P. "Catching the First Fish." *Nature* 402 (1999): 21–22.
Jenkins, John B. *Genetics.* Boston: Houghton Mifflin Harcourt, 1975.

Jenner, Ronald A. "Evolution of Animal Body Plans: The Role of Metazoan Phylogeny at the Interface Between Pattern And Process." *Evolution and Development* 2 (2000): 208–21.

———. "Higher-Level Crustacean Phylogeny: Consensus and Conflicting Hypotheses." *Arthropod Structure and Development* 39 (2010): 143–53.

Jensen, Roy A. "Enzyme Recruitment in Evolution of New Function." *Annual Review of Microbiology* 30 (1976): 409–25.

Jensen, S., M. L. Droser, and J. G. Gehling. "Trace Fossil Preservation and the Early Evolution of Animals." *Palaeogeography, Palaeoclimatology, Palaeoecology* 220 (2005): 19–29.

Jensen, S., J. G. Gehling, and M. L. Droser. "Ediacara-type Fossils in Cambrian Sediments." *Nature* 393 (1998): 567–69.

Johnson, Andrew D., M. Drum, R. F. Bachvarova, T. Masi, M. E. White, and B. I. Crother. "Evolution of Predetermined Germ Cells in Vertebrate Embryos: Implications for Macroevolution." *Evolution and Development* 5 (2003): 414–31.

Johnson, Matthew E., et al. "Positive Selection of a Gene Family During the Emergence of Humans and African Apes." *Nature* 413 (2001): 514–19.

Jones, Gareth. "Molecular Evolution: Gene Convergence in Echolocating Mammals." *Current Biology* 20 (2010): R62–R64.

Jones, Judge John E., III. Decision in *Kitzmiller et al. v. Dover Area School Board.* No.04cv2688, 2005 WL 2465563, *66 (M.D.Pa. Dec. 20, 2005). http://www.pamd.uscourts.gov/kitzmiller/kitzmiller_342.pdf.

Jordan, I. K, I. B. Rogozin, G. V. Glazko, and E. V. Koonin. "Origin of a Substantial Fraction of Human Regulatory Sequences from Transposable Elements." *Trends in Genetics* 19 (2003): 68–72.

Jurica, Melissa. "Detailed Closeups and the Big Picture of Spliceosomes." *Current Opinion in Structural Biology* 18 (2008): 315–20.

Jurka, Jerzy. "Evolutionary Impact of Human *Alu* Repetitive Elements." *Current Opinion in Genetics and Development* 14 (2004): 603–8.

Kandouz, M., A. Bier, G. D Carystinos, M. A. Alaoui-Jamali, and G. Batist. "Connexin43 Pseudogene Is Expressed in Tumor Cells and Inhibits Growth." *Oncogene* 23 (2004): 4763–70.

Kauffman, Stuart A. *The Origins of Order: Self-Organization and Selection in Evolution.* Oxford: Oxford Univ. Press, 1993.

———. *At Home in the Universe: The Search for the Laws of Self-Organization and Complexity.* Oxford: Oxford Univ. Press, 1995.

———. "The End of a Physics Worldview: Heraclitus and the Watershed of Life." http://www.npr.org/blogs/13.7/2011/08/08/139006531/the-end-of-a-physics-worldview-heraclitus-and-the-watershed-of-life (accessed October 25, 2012).

Kavalovski, V. "The *Vera Causa* Principle: A Historico-Philosophical Study of a Metatheoretical Concept from Newton Through Darwin." Ph.D. dissertation, University of Chicago, 1974.

Kehoe, A. "Modern Anti-Evolutionism: The Scientific Creationists." In *What Darwin Began,* edited by L. R. Godfrey, 173–80. Boston: Allyn and Bacon, 1985.

Kelley, D. R., and J. L. Rinn. "Transposable Elements Reveal a Stem Cell Specific Class of Long Noncoding RNAs." *Genome Biology* 13 (2012): R107.

Kerr, Richard A. "Evolution's Big Bang Gets Even More Explosive." *Science* 261 (1993): 1274–75.

———. "Did Darwin Get It All Right?" *Science* 267 (1995): 1421–22.

Khalturin, K., G. Hemmrich, S. Fraune, R. Augustin, and T. C. Bosch. "More Than Just Orphans: Are Taxonomically Restricted Genes Important in Evolution?" *Trends in Genetics* 25 (2009): 404–13.

Kimura, Motoo. *The Neutral Theory of Molecular Evolution.* Cambridge: Cambridge Univ. Press, 1983.

King, Jack L., and Thomas H. Jukes. "Non-Darwinian Evolution." *Science* 164, no. 3881 (1969): 788–98.

Kirschner, Marc W., and John C. Gerhart. *The Plausibility of Life: Resolving Darwin's Dilemma.* New Haven, CT: Yale Univ. Press, 2005.

Kitcher, Philip. *Abusing Science: The Case Against Creationism.* Cambridge, MA: MIT Press, 1982.

Kline, A. David. "Theories, Facts and Gods: Philosophical Aspects of the Creation-Evolution Controversy." In *Did the Devil Make Darwin Do It?* edited by D. Wilson, 37–44. Ames: Iowa State Univ. Press, 1983.

Klinghoffer, David. "The Branding of a Heretic." *Wall Street Journal,* January 28, 2005, national edition, W11.

———. ed. *Signature of Controversy: Responses to Critics of Signature in the Cell.* Seattle: Discovery Institute Press, 2010.

Knoll, Andrew H., and Sean B. Carroll. "Early Animal Evolution: Emerging Views from Comparative Biology and Geology." *Science* 84 (1999): 2129–37.

Knowles, David G., and Aoife McLysaght. "Recent de Novo Origin of Human Protein-Coding Genes." *Genome Research* 19 (2009): 1752–59.

Kodoyianni, Voula, E. M. Maine, and J. Kimble. "Molecular Basis of Loss-of-Function Mutations in the *glp-1* Gene of *Caenorhabditis elegans*." *Molecular Biology of the Cell* 3 (1992): 1199–1213.

Koga, Nobuyasu, Rie Tatsumi-Koga, Gaohua Liu, Rong Xiao, Thomas B. Acton, Gaetano T. Montelione, and David Baker. "Principles for Designing Ideal Protein Structures." *Nature* 491 (2012): 222–27.

Kondo, N., N. Nikoh, N. Ijichi, M. Shimada, and T. Fukatsu. "Genome Fragment of *Wolbachia* Endosymbiont Transferred to X Chromosome of Host Insect." *Proceedings of the National Academy of Sciences* 99 (2002): 14280–85.

Kondo-Iida, E., et al. "Novel Mutations and Genotype–Phenotype Relationships in 107 Families with Fukuyama-Type Congenital Muscular Dystrophy (FCMD)." *Human Molecular Genetics* 8 (1999): 2303–9.

Koonin, Eugene V. "How Many Genes Can Make a Cell? The Minimal Genome Concept." *Annual Review of Genomics and Human Genetics* 1 (2002): 99–116.

———. "The *Origin* at 150: Is a New Evolutionary Synthesis in Sight?" *Trends in Genetics* 25 (2009): 473–75.

Küppers, Bernd-Olaf. "On the Prior Probability of the Existence of Life." In *The Probabilistic Revolution,* vol. 1, edited by L. Krüger, L. Daston, and M. Heidelberger, 355–69. Cambridge, MA: MIT Press, 1987.

Landing, E., A. English, and J. D. Keppie. "Cambrian Origin of All Skeletalized Metazoan Phyla—Discovery of Earth's Oldest Bryozoans (Upper Cambrian, Southern Mexico)." *Geology* 38 (2010): 547–50.

Lange, B. M. H., A. J. Faragher, P. March, and K. Gull. "Centriole Duplication and Maturation in Animal Cells." In *The Centrosome in Cell Replication and Early Development,* edited by R. E. Palazzo and G. P. Schatten, 235–49. Current Topics in Developmental Biology 49. San Diego, CA: Academic, 2000.

Lau, K. F., and K. A. Dill. "Theory for Protein Mutability and Biogenesis." *Proceedings of the National Academy of Sciences USA* 87 (1990): 638–42.

Laudan, Larry. *Beyond Positivism and Relativism: Theory, Method, and Evidence.* Boulder, CO: Westview, 1996.

Laudan, Larry. "The Demise of the Demarcation Problem." In *But Is It Science?* edited by M. Ruse, 337–50. Buffalo, NY: Prometheus, 1988.

Lawrence, P. A., and G. Struhl. "Morphogens, Compartments and Pattern: Lessons from *Drosophila*?" *Cell* 85 (1996): 951–61.

Lawton, Graham. "Why Darwin Was Wrong About the Tree of Life." *New Scientist,* January 21, 2009, 34–39.

Lecointre, Guillaume, and Hervé Le Guyader. *The Tree of Life: A Phylogenetic Classification.* Cambridge, MA: Harvard Univ. Press, 2006.

Lee, Michael S. Y. "Molecular Clock Calibrations and Metazoan Divergence Dates." *Journal of Molecular Evolution* 49 (1999): 385–91.

Lehmann, Ruth. "Germ-Plasm Formation and Germ-Cell Determination in *Drosophila*." *Current Opinion in Genetics and Development* 2 (1992): 543–49.

Lehmann, Ruth, and Christiane Nüsslein-Volhard. "The Maternal Gene *Nanos* Has a Central Role in Posterior Pattern Formation of the *Drosophila* Embryo." *Development* 112 (1991): 679–91. http://dev.biologists.org/content/112/3/679.long (accessed October 30, 2012).

Leibniz, Gottfried. *New Essays on Human Understanding.* Translated and edited by P. Remnant and J. Bennett. Cambridge: Cambridge Univ. Press, 1981.

Lemen, Cliff A., and Patricia W. Freeman. "A Test of Macroevolutionary Problems with Neontological Data." *Paleobiology* 7 (1981): 316–31.

Lenski, Richard, Charles Ofria, Robert T. Pennock, and Christopher Adami. "The Evolutionary Origin of Complex Features." *Nature* 423 (2003): 139–44.

Le Pichon, Xavier. "My Conversion to Plate Tectonics." In *Plate Tectonics: An Insider's History of the Modern Theory of the Earth,* edited by N. Oreskes, 201–26. Boulder, CO: Westview, 2003.

Levin, Michael. "Bioelectromagnetics in Morphogenesis." *Bioelectromagnetics* 24 (2003): 295–315. http://ase.tufts.edu/biology/labs/levin/publications/documents/2003BEMS.pdf (accessed October 30, 2012).

Levinton, Jeffrey S. "Punctuated Equilibrium." *Science* 231 (1986): 1490.

———. *Genetics, Paleontology, and Macroevolution*. Cambridge: Cambridge Univ. Press, 1988.

———. "Bryozoan Morphological and Genetic Correspondence: What Does It Prove?" *Science* 51 (1991): 318–19.

Lev-Maor, G., R. Sorek, N. Shomron, and G. Ast. "The Birth of an Alternatively Spliced Exon: 3' Splice-Site Selection in Alu Exons." *Science* 300 (2003): 1288–91.

Lewin, Roger. "Punctuated Equilibrium Is Now Old Hat." *Science* 231 (1986): 672–73.

———. "A Lopsided Look at Evolution." *Science* 241 (1988): 292.

Lewis, Edward B. "A Gene Complex Controlling Segmentation in *Drosophila*." *Nature* 276 (1978): 565–70.

Lewontin, Richard. "Adaptation," *Scientific American* 239 (1978): 212–30.

———. "Billions and Billions of Demons." Review of *The Demon-Haunted World: Science as a Candle in the Dark,* by Carl Sagan. *New York Review of Books,* January 9, 1997, 28.

Li, Chun, X. C. Wu, O. Rieppel, L. T. Wang, and L. J. Zhao. "An Ancestral Turtle from the Late Triassic of Southwestern China." *Nature* 456 (2008): 497–501.

Li, Long-Cheng, S.T. Okino, H. Zhao, D. Pookot, R. F. Place, S. Urakami, H. Enokida, and R. Dahiya. "Small dsRNAs Induce Transcriptional Activation in Human Cells." *Proceedings of the National Academy of Sciences USA* 103 (2006): 17337–42.

Li, Wen-Hsiung. *Molecular Evolution*. Sunderland, MA: Sinauer, 1997.

Li, Ying, Z. Liu, P. Shi, and J. Zhang. "The Hearing Gene Prestin Unites Echolocating Bats and Whales." *Current Biology* 20 (2010): R55–R56.

Lieberman, Bruce S., and Elisabeth S. Vrba. "Stephen Jay Gould on Species Selection: 30 Years of Insight." *Paleobiology* 31 (2005): 113–21.

Lienhard, John H. "No. 857: Tyndall on Parallel Roads." In *Engines of Our Ingenuity* [audio podcast], http://www.uh.edu/engines/epi857.htm (accessed March 22, 2013).

Lili, Cui. "Traditional Theory of Evolution Challenged." *Beijing Review,* March 31–April 6, 1997, 10.

Lindsey, Dan L., and E. H. Grell. *Guide to Genetic Variations of Drosophila melanogaster.* Washington, DC: Carnegie Institution of Washington Publication No. 627, 1968. http://www.carnegiescience.edu/publications_online/genetic_variations.pdf.

Lipton, Peter. *Inference to the Best Explanation*. London and New York: Routledge, 1991.

Liu, J., D. Shu, J. Han, Z. Zhang, and X. Zhang. "A Large Xenusiid Lobopod with Complex Appendages from the Lower Cambrian Chengjiang Lagerstätte." *Acta Palaeontologica Polonica* 51 (2006): 215–22.

———. "Origin, Diversification, and Relationships of Cambrian Lobopods." *Gondwana Research* 14 (2008): 277–83.

Liu, J., M. Steiner, J. A. Dunlop, H. Keupp, D. Shu, Q. Ou, J. Han, and Z. Zhang. "An Armoured Cambrian Lobopodian from China with Arthropod-Like Appendages." *Nature* 470 (2011): 526–30.

Logsdon, John M., Jr., and W. Ford Doolittle. "Origin of Antifreeze Protein Genes: A Cool Tale in Molecular Evolution." *Proceedings of the National Academy of Sciences USA* 94 (1997): 3485–87.

Long, Manyuan, Ester Betrán, Kevin Thornton, and Wen Wang. "The Origin of New Genes: Glimpses from the Young and Old." *Nature Reviews Genetics* 4 (2003): 865–75.

Long, Manyuan, and Charles H. Langley. "Natural Selection and the Origin of *Jingwei,* a Chimeric Processed Functional Gene in *Drosophila*." *Science* 260 (1993): 91–95.

Long, Manyuan, S. J. de Souza, C. Rosenberg, and W. Gilbert. "Exon Shuffling and the Origin of the Mitochondrial Targeting Function in Plant Cytochrome c1 Precursor." *Proceedings of the National Academy of Sciences USA* 93 (1996): 7727–31.

Lönnig, Wolf-Ekkehard, and Heinz Saedler. "Chromosome Rearrangements and Transposable Elements." *Annual Review of Genetics* 36 (2002): 389–410.

Love, G. D., et al. "Fossil Steroids Record the Appearance of Demospongiae During the Cryogenian Period." *Nature* 457 (2009): 718–21.

Louro, R., T. El-Jundi, H. I. Nakaya, E. M. Reis, and S. Verjovski-Almeida. "Conserved Tissue Expression Signatures of Intronic Noncoding RNAs Transcribed from Human and Mouse Loci." *Genomics* 92 (2008): 18–25.

Løvtrup, Søren. "Semantics, Logic and Vulgate Neo-Darwinism." *Evolutionary Theory* 4 (1979): 157–72.

Lurie, Edward. *Nature and the American Mind: Louis Agassiz and the Culture of Science*. New York: Science History Publications, 1974.

Luskin, Casey. "Credibility Gap: Baylor Denies Robert Marks' Situation Has Anything to Do with ID." *Evolution and News and Views,* October 1, 2007. http://www.evolutionnews.org/2007/10/credibility_gap_baylor_denies004290.html.

———. "Darwin's Dilemma: Evolutionary Elite Choose Censorship over Scientific Debate." *CNS News,* October 14, 2009.

———. "Intelligent Design Demoted." *Liberty Legal Journal,* September 7, 2010.

Lyell, Charles. *Principles of Geology: Being an Attempt to Explain the Former Changes of the Earth's Surface, by Reference to Causes Now in Operation.* 3 vols. London: Murray, 1830–33.

Lynch, Michael. "The Origins of Eukaryotic Gene Structure." *Molecular Biology and Evolution* 23 (2006): 450–68.

———. "The Frailty of Adaptive Hypotheses for the Origins of Organismal Complexity." *Proceedings of the National Academy of Sciences USA* 104 (2007): 8597–604.

———. "Evolutionary Layering and the Limits to Cellular Perfection." *Proceedings of the National Academy of Sciences USA* 109 (2012): 18851–56.

Lynch, Michael, and Adam Abegg. "The Rate of Establishment of Complex Adaptations." *Molecular Biology and Evolution* 27 (2010): 1404–14.

Ma, Xiaoya, Xianguang Hou, Gregory D. Edgecombe, and Nicholas J. Strausfeld. "Complex Brain and Optic Lobes in an Early Cambrian Arthropod." *Nature* 490 (2012): 258–62.

MacRae, Andrew. "Trilobites in Murchison's *Siluria*." http://www.talkorigins.org/faqs/trilobite/siluria.html.

Maley, Laura E., and Charles R. Marshall. "The Coming of Age of Molecular Systematics." *Science* 279 (1998): 505–6.

Malik, Harmit S., and Steven Henikoff. "Adaptive Evolution of Cid, a Centromere-Specific Histone in *Drosophila*." *Genetics* 157 (2001): 1293–98.

Malinky, J. M., and C. B. Skovsted. "Hyoliths and Small Shelly Fossils from the Lower Cambrian of Northeast Greenland." *Acta Palaeontologica Polonica* 49 (2004): 551–78.

Mallatt, Jon M., James R. Garey, and Jeffrey W. Shultz. "Ecdysozoan Phylogeny and Bayesian Inference: First Use of Nearly Complete 28S and 18S rRNA Gene Sequences to Classify the Arthropods and Their Kin." *Molecular Phylogenetics and Evolution* 31, no. 1 (2004): 178–91.

Margulis, Lynn, and Dorion Sagan. *Acquiring Genomes: A Theory of the Origins of the Species*. New York: Basic Books, 2002.

Mariño-Ramírez, L., K. C. Lewis, D. Landsmana, and I. K. Jordan. "Transposable Elements Donate Lineage-Specific Regulatory Sequences to Host Genomes." *Cytogenetic and Genome Research* 110 (2005): 333–41.

Marshall, Charles R. "Explaining the Cambrian 'Explosion' of Animals." *Annual Reviews of Earth and Planetary Sciences* 34 (2006): 355–84.

Marshall, Jessica. "Proteins Made to Order." *Nature News,* November 7, 2012. http://www.nature.com/news/proteins-made-to-order–1.11767.

Marshall, Wallace F., and Joel L. Rosenbaum. "Are There Nucleic Acids in the Centrosome?" *Current Topics in Developmental Biology* 49 (2000): 187–205.

Masui, Y., A. Forer, and A. M. Zimmerman. "Induction of Cleavage in Nucleated and Enucleated Frog Eggs by Injection of Isolated Sea-Urchin Mitotic Apparatus." *Journal of Cell Science* 31 (1978): 117–35. http://jcs.biologists.org/content/31/1/117.long (accessed October 30, 2012).

Mattick, John S., and Michael J. Gagen. "The Evolution of Controlled Multitasked Gene Networks: The Role of Introns and Other Noncoding RNAs in the Development of Complex Organisms." *Molecular Biology and Evolution* 18 (2001): 1611–30.

Mattick, J. S., and I. V. Makunin. "Small Regulatory RNAs in Mammals." *Human Molecular Genetics* 14 (2005): R121–32.

———. "Non-coding RNA." *Human Molecular Genetics* 15 (2006): R17–R29.

Matz, Mikhail V., Tamara M. Frank, N. Justin Marshall, Edith A. Widder, and Sönke Johnsen. "Giant Deep-Sea Protist Produces Bilaterian-like Traces." *Current Biology* 18 (December 9, 2008): 1849–54.

Matzke, Nicholas J. "The Edge of Creationism." *Trends in Ecology and Evolution* 22 (2007): 566–67.

Matzke, Nicholas J., and Paul R. Gross. "Analyzing Critical Analysis: The Fallback Antievolutionist Strategy." In *Not in Our Classrooms: Why Intelligent Design Is Wrong for Our Schools,* edited by E. C. Scott and G. Branch, 28–56. Boston: Beacon, 2006.

Maynard Smith, John. "Natural Selection and the Concept of a Protein Space." *Nature* 225 (1970): 563–64.

———. "Darwinism Stays Unpunctured." *Nature* 330 (1987): 516.

Mayr, Ernst. Foreword. In *Darwinism Defended: A Guide to the Evolution Controversies,* edited by M. Ruse, xi–xii. Reading, MA: Addison-Wesley, 1982.

Mayr, Ernst, and William B. Provine. *The Evolutionary Synthesis: Perspectives on the Unification of Biology.* Cambridge, MA: Harvard Univ. Press, 1998.

Mazur, Suzan. *The Altenberg 16: An Exposé of the Evolution Industry.* Berkeley, CA; North Atlantic Books, 2010.

McCall, G. J. H. "The Vendian (Ediacaran) in the Geological Record: Enigmas in Geology's Prelude to the Cambrian Explosion." *Earth-Science Reviews* 77 (2006): 1–229.

McDonald, John F. "The Molecular Basis of Adaptation: A Critical Review of Relevant Ideas and Observations." *Annual Review of Ecology and Systematics* 14 (1983): 77–102.

McGinnis, William, and Michael Kurziora. "The Molecular Architects of Body Design." *Scientific American* 270 (1994): 58–66.

McKenzie, Richard W., and Mark D. Brennan. "The Two Small Introns of the *Drosophila affinidisjuncta Adh* Gene Are Required for Normal Transcription." *Nucleic Acids Research* 24 (1996): 3635–42.

McMenamin, M. A. S. "Ediacaran Biota from Sonora, Mexico." *Proceedings of the National Academy of Sciences USA* 93 (1996): 4990–93.

———. *The Garden of Ediacara: Discovering the First Complex Life.* New York: Columbia Univ. Press, 1998.

———. *The Evolution of the Noösphere.* New York: American Teilhard Association for the Future of Man, 2001.

———. "*Spriggina* Is a Trilobitoid Ecdysozoan." *Geological Society of America Abstracts with Programs* 35 (2003): 105–6.

———. "Harry Blackmore Whittington, 1916–2010." *Geoscientist* 20 (2010): 5.

———. "Fossil Chitons and *Monomorphichnus* from the Ediacaran Clemente Formation, Sonora, Mexico." *Geological Society of America Abstracts with Programs* 43 (2011): 87.

———. "Teilhard de Chardin's Legacy in Science." In *The Legacy of Teilhard de Chardin,* edited by J. Salmon and J. Farina, 33–45. Mahwah, NJ: Paulist Press, 2011.

McMenamin, M. A. S., and D. L. S. McMenamin. *The Emergence of Animals: The Cambrian Breakthrough.* New York: Columbia Univ. Press, 1990.

Medstrand, P., J.-R. Landry, and D. L. Mager, "Long Terminal Repeats Are Used as Alternative Promoters for the Endothelin B Receptor and Apolipoprotein C-I Genes in Humans." *Journal of Biological Chemistry* 276 (2001): 1896–903.

Merhej, Vicky, and Didier Raoult. "Rhizome of Life, Catastrophes, Sequence Exchanges, Gene Creations, and Giant Viruses: How Microbial Genomics Challenges Darwin." *Frontiers in Cellular and Infectious Microbiology* 2, no. 113 (2012).

Meyer, Stephen C. "Of Clues and Causes: A Methodological Interpretation of Origin of Life Studies." Ph.D. dissertation, Cambridge University, 1990.

———. "A Scopes Trial for the '90s." *Wall Street Journal,* December 6, 1993, A14.

———. "Laws, Causes and Facts: A Response to Professor Ruse." In *Darwinism: Science or Philosophy?* edited by J. Buell and V. Hearn, 29–40. Dallas: Foundation for Thought and Ethics, 1994.

———. "The Methodological Equivalence of Design and Descent." In *The Creation Hypothesis: Scientific Evidence for Intelligent Design,* edited by J. P. Moreland, 67–112. Downer's Grove, IL: InterVarsity Press, 1994.

———. "The Nature of Historical Science and the Demarcation of Design and Descent." In *Facets of Faith and Science.* Vol. 4, *Interpreting God's Action in the World,* 91–130. Washington, DC: Univ. Press of America, 1996.

———. "The Demarcation of Science and Religion." In *The History of Science and Religion in the Western Tradition: An Encyclopedia,* edited by G. B. Ferngren, 17–23. New York: Garland, 2000.

———. "The Scientific Status of Intelligent Design: The Methodological Equivalence of Naturalistic and Non-Naturalistic Origins Theories." In *Science and Evidence for Design in the Universe, The Proceedings of the Wethersfield Institute,* 151–212. San Francisco: Ignatius, 2000.

———. "The Origin of Biological Information and the Higher Taxonomic Categories," *Proceedings of the Biological Society of Washington* 117 (2004): 213–39.

———. "The Origin of Biological Information and the Higher Taxonomic Categories." In *Darwin's Nemesis: Phillip Johnson and the Intelligent Design Movement,* edited by W. A. Dembski, 174–213. Downers Grove, IL: InterVarsity, 2006.

———. *Signature in the Cell: DNA and the Evidence for Intelligent Design.* San Francisco: HarperOne, 2009.

———. "Sauce for the Goose: Intelligent Design, Scientific Methodology, and the Demarcation Problem." In *Nature of Nature: Examining the Role of Naturalism in Science,* edited by B. L. Gordon and W. A. Dembski, 95–131. Wilmington, DE: ISI Books, 2011.

Meyer, Stephen C., Scott Minnich, Jonathan Moneymaker, Paul A. Nelson, and Ralph Seelke. *Explore Evolution: The Arguments for and Against Neo-Darwinism.* Melbourne and London: Hill House, 2007.

Meyer, Stephen C., Marcus Ross, Paul Nelson, and Paul Chien. "The Cambrian Explosion: Biology's Big Bang." In *Darwinism, Design and Public Education,* edited by J. A. Campbell and S. C. Meyer, 323–402. East Lansing: Michigan State Univ. Press, 2003.

Michel, Bénédicte. "After 30 Years, the Bacterial SOS Response Still Surprises Us." *PLoS Biology* 3 (2005): 1174–76.

Miklos, George L. G. "Emergence of Organizational Complexities During Metazoan Evolution: Perspectives from Molecular Biology, Palaeontology and Neo-Darwinism." *Memoirs of the Association of Australasian Palaeontologists* 15 (1993): 7–41.

Miller, Kenneth R. "Falling over the Edge: Review of *The Edge of Evolution,* by Michael Behe." *Nature* 447 (2007): 1055–56.

Mintz, Leigh W. *Historical Geology: The Science of a Dynamic Earth.* 2nd ed. Columbus, OH: Merrill, 1977.

Moczek, Armin P. "On the Origins of Novelty in Development and Evolution." *BioEssays* 30 (2008): 432–47.

———. "The Nature of Nurture and the Future of Evodevo: Toward a Theory of Developmental Evolution." *Integrative and Comparative Biology* 52 (2012): 108–19.

Monastersky, Richard. "Siberian Rocks Clock Biological Big Bang." *Science News* 144 (1993): 148.

———. "Ancient Animal Sheds False Identity." *Science News* 152 (1997): 32.

———. "Society Disowns Paper Attacking Darwinism." *Chronicle of Higher Education* 51, no. 5 (2004): A16.

Mondal, T., M. Rasmussen, G. K. Pandey, A. Isaksson, and C. Kanduri. "Characterization of the RNA Content of Chromatin." *Genome Research* 20 (2010): 899–907.

Monteys, A. M., R. M. Spengler, J. Wan, L. Tecedor, K. A. Lennox, Y. Xing, and B. L. Davidson. "Structure and Activity of Putative Intronic miRNA Promoters." *RNA* 16 (2010): 495–505.

Morrish, T. A., N. Gilbert, J. S. Myers, B. J. Vincent, T. D. Stamato, G. E. Taccioli, M. A. Batzer, and J. V. Moran. "DNA Repair Mediated by Endonuclease-Independent LINE-1 Retrotransposition." *Nature Genetics* 31 (2002): 159–65.

Moss, Lenny. *What Genes Can't Do.* Cambridge, MA: MIT Press, 2004.

Müller, Dietmar R., Walter R. Roest, Jean-Yves Royer, Lisa M. Gahagan, and John G. Sclater. "Digital Isochrons of the World's Ocean Floor." *Journal of Geophysical Research* 102 (1997): 3211–14.

Müller, Gerd B. "Homology: The Evolution of Morphological Organization." In *Origination of Organismal Form: Beyond the Gene in Developmental and Evolutionary Biology,* edited by G. B. Müller and S. A. Newman, 51–69. Cambridge, MA: MIT Press, 2003.

Müller, Gerd B., and Stuart A. Newman. "Origination of Organismal Form: The Forgotten Cause in Evolutionary Theory." In *Origination of Organismal Form: Beyond the Gene in Developmental and Evolutionary Biology,* edited by G. B. Müller and S. A. Newman, 3–10. Cambridge, MA: MIT Press, 2003.

Muller, H. J. "Artificial Transmutation of the Gene." *Science* 66 (1927): 84–87.

Muller, K. J., D. Bonn, and A. Zakharov. "'Orsten' Type Phosphatized Soft-Integument Preservation and a New Record from the Middle Cambrian Kuonamka Formation in Siberia." *Neues Jahrbuch für Geologie und Paläontologie, Monatshefte* 197 (1995): 101–18.

Mura, M., P. Murcia, M. Caporale, T. E. Spencer, K. Nagashima, A. Rein, and M. Palmarini. "Late Viral Interference Induced by Transdominant Gag of an Endogenous Retrovirus." *Proceedings of the National Academy of Sciences USA* 101 (2004): 11117–22.

Murchison, Roderick Impey. *Siluria: The History of the Oldest Known Rocks Containing Organic Remains.* London: John Murray, 1854.

Murphy, Nancey. "Phillip Johnson on Trial: A Critique of His Critique of Darwin." *Perspectives on Science and Christian Faith* 45 (1993): 26–36.
Nanney, D. L. "The Ciliates and the Cytoplasm." *Journal of Heredity* 74 (1983): 163–70.
Nardi, F., G. Spinsanti, J. L. Boore, A. Carapelli, R. Dallai, and F. Frati. "Hexapod Origins: Monophyletic or Paraphyletic?" *Science* 299 (2003): 1887–89.
———. "Response to Comment on 'Hexapod Origins: Monophyletic or Paraphyletic?'" *Science* 301 (2003): 1482.
Nash, J. Madeleine. "When Life Exploded." *Time* 146 (1995): 66–74.
National Academy of Sciences. *Teaching About Evolution and the Nature of Science.* Washington DC: National Academy Press, 1998.
National Ocean Industries Association. "About NOIA." http://www.noia.org/website/article.asp?id-51 (accessed March 29, 2013).
Nature editors. "Life on Land." *Nature* 492 (2012): 153–54.
Nelson, Paul, and Jonathan Wells. "Homology in Biology: Problem for Naturalistic Science and Prospect for Intelligent Design." In *Darwinism, Design and Public Education,* edited by J. A. Campbell and S. C. Meyer, 303–22. East Lansing: Michigan State Univ. Press, 2003.
Newman, Stuart. "The Developmental Genetic Toolkit and the Molecular Homology-Analogy Paradox." *Biological Theory* 1 (2006): 12–16.
———. "Dynamical Patterning Modules." In *Evolution: The Extended Synthesis,* edited by M. Pigliucci and G. B. Müller, 281–306. Cambridge, MA: MIT Press, 2010.
———. "Animal Egg as Evolutionary Innovation: A Solution to the 'Embryonic Hourglass' Puzzle." *Journal of Experimental Zoology B: Molecular and Developmental Evolution* 314 (2011): 467–83.
Newman, Stuart, and Ramray Bhat. "Dynamical Patterning Modules: Physico-Genetic Determinants of Morphological Development and Evolution." *Physical Biology* 5 (2008): 015008.
Newton, Isaac. *Newton's Principia.* Translated by Andrew Motte (1686). Translation revised by Florian Cajori. Berkeley, CA: Univ. of California Press, 1934.
Nielsen, Claus. *Animal Evolution: Interrelationships of the Living Phyla.* Oxford: Oxford Univ. Press, 2001.
Nijhout, H. F. "Metaphors and the Role of Genes in Development." *BioEssays* 12 (1990): 441–46.
Nikoh, Naruo, et al. "An Estimate of Divergence Time of Parazoa and Eumetazoa and That of Cephalochordata and Vertebrata by Aldolase and Triose Phosphate Isomerase Clocks." *Journal of Molecular Evolution* 45 (1997): 97–106.
Nilsen, Timothy W. "The Spliceosome: The Most Complex Macromolecular Machine in the Cell?" *BioEssays* 25 (2003): 1147–49.
Nurminsky, D. I., M. V. Nurminskaya, D. De Aguiar, and D. L. Hartl. "Selective Sweep of a Newly Evolved Sperm-Specific Gene in *Drosophila.*" *Nature* 396 (1998): 572–75.
Nüsslein-Volhard, C., and E. Wieschaus. "Mutations Affecting Segment Number and Polarity in *Drosophila.*" *Nature* 287 (1980): 795–801.
O'Brien, L. J., and J.-B. Caron. "A New Stalked Filter-Feeder from the Middle Cambrian Burgess Shale, British Columbia, Canada." *PLoS One* 7 (2012): 1–21.
Ohno, S. "The Notion of the Cambrian Pananimalia Genome." *Proceedings of the National Academy of Sciences USA* 93 (1996): 8475–78.
Oliveri, Paola, and Eric H. Davidson. "Built to Run, Not Fail." *Science* 315 (2007): 1510–11.
Oliveri, Paola, Qiang Tu, and Eric H. Davidson. "Global Regulatory Logic for Specification of an Embryonic Cell Lineage." *Proceedings of the National Academy of Sciences USA* 105 (2008): 5955–62.
Oosthoek, Jan. "The Parallel Roads of Glen Roy and Forestry." *Environmental History Resources.* http://www.eh-resources.org/roy.html.
Oreskes, Naomi. "From Continental Drift to Plate Tectonics." In *Plate Tectonics: An Insider's History of the Modern Theory of the Earth,* edited by N. Oreskes, 3–30. Boulder, CO: Westview, 2003.
Orgel, L. E., and F. H. Crick. "Selfish DNA: The Ultimate Parasite." *Nature* 284 (1980): 604–7.
Orr, H. Allen. "The Genetic Theory of Adaptation: A Brief History." *Nature Reviews Genetics* 6 (2005): 119–27.
Osigus, Hans-Jürgen, Michael Eitel, and Bernd Schierwater. "Chasing the Urmetazoon: Striking a Blow for Quality Data?" *Molecular Phylogenetics and Evolution* 66 (2013): 551–57.
Ou, Q., J. Liu, D. Shu, J. Han, Z. Zhang, X. Wan, and Q. Lei. "A Rare Onychophoran-Like Lobopodian from the Lower Cambrian Chengjiang Lagerstätte, Southwestern China, and Its Phylogenetic Implications." *Journal of Paleontology* 85 (2011): 587–94.

Pagano, A., M. Castelnuovo, F. Tortelli, R. Ferrari, G. Dieci, and R. Cancedda. "New Small Nuclear RNA Gene-like Transcriptional Units as Sources of Regulatory Transcripts." *PLoS Genetics* 3 (2007): e1.

Palade, George E. "Membrane Biogenesis: An Overview." *Methods in Enzymology* 96 (1983): xxix–lv.

Paley, William. *Natural Theology: Or Evidences of the Existence and Attributes of the Deity Collected from the Appearances of Nature.* 1802. Reprint, Boston: Gould and Lincoln, 1852.

Palopoli, Michael, and Nipam Patel. "Neo-Darwinian Developmental Evolution: Can We Bridge the Gap Between Pattern and Process?" *Current Opinion in Genetics and Development* 6 (1996): 502–8.

Pandey, R., and M. Mukerji. "From 'JUNK' to Just Unexplored Noncoding Knowledge: The Case of Transcribed Alus." *Briefings in Functional Genomics* 10 (2011): 294–311.

Panganiban, G., et al. "The Origin and Evolution of Animal Appendages." *Proceedings of the National Academy of Sciences USA* 94 (1997): 5162–66.

Pardue, M. L., and P. G. DeBaryshe. "*Drosophila* Telomeres: Two Transposable Elements with Important Roles in Chromosomes." *Genetica* 107 (1999): 189–96.

Parikesit, A. A., P. F. Stadler, and S. J. Prohaska. "Quantitative Comparison of Genomic-Wide Protein Domain Distributions." In *Lecture Notes in Informatics* P–173, edited by D. Schomburg and A. Grote, 93–102. Bonn: Gesellschaft für Informatik, 2010.

Patterson, Colin, David M. Williams, and Christopher J. Humphries. "Congruence Between Molecular and Morphological Phylogenies." *Annual Review of Ecology and Systematics* 24 (1993): 153–88.

Paulding, Charles A., Maryellen Ruvolo, and Daniel A. Haber. "The *Tre2* (*USP6*) Oncogene Is a Hominoid-Specific Gene." *Proceedings of the National Academy of Sciences USA* 100 (2003): 2507–11.

Peaston, E., A. V. Evsikov, J. H. Graber, W. N. de Vries, A. E. Holbrook, D. Solter, and B. B. Knowles. "Retrotransposons Regulate Host Genes in Mouse Oocytes and Preimplantation Embryos." *Developmental Cell* 7 (2004): 597–606.

Peel, J. S. "A Corset-Like Fossil from the Cambrian Sirius Passet Lagerstätte of North Greenland and Its Implications for Cycloneuralian Evolution." *Journal of Paleontology* 84 (2010): 332–40.

Peifer, Mark, and Welcome Bender. "The Anterobithorax and Bithorax Mutations of the Bithorax Complex." *EMBO Journal* 5 (1986): 2293–303. http://www.ncbi.nlm.nih.gov/pmc/articles/PMC1167113/pdf/emboj00172-0253.pdf.

Peirce, Charles S. *Collected Papers.* Vols. 1–6. Edited by C. Hartshorne and P. Weiss. Cambridge, MA: Harvard Univ. Press, 1931–35.

———. "Abduction and Induction." In *The Philosophy of Peirce,* edited by J. Buchler, 150–54. London: Routledge, 1956.

———. *Collected Papers,* Vols. 7–8. Edited by A. Burks. Cambridge, MA: Harvard Univ. Press, 1958.

Pellas, Theodore C., B. Ramachandran, M. Duncan, S. S. Pan, M. Marone, and K. Chada. "Germ-Cell Deficient (*gcd*), an Insertional Mutation Manifested as Infertility in Transgenic Mice." *Proceedings of the National Academy of Sciences USA* 88 (1991): 8787–91.

Peterson, Kevin J., James A. Cotton, James G. Gehling, and Davide Pisani. "The Ediacaran Emergence of Bilaterians: Congruence Between the Genetic and the Geological Fossil Records." *Philosophical Transactions of the Royal Society B* (2008): 1435–43.

Peterson, Kevin J., and Eric H. Davidson. "Regulatory Evolution and the Origin of the Bilaterians." *Proceedings of the National Academy of Sciences USA* 97 (2000): 4430–33.

Peterson, Kevin J., Michael R. Dietrich, and Mark A. McPeek. "MicroRNAs and Metazoan Macroevolution: Insights into Canalization, Complexity, and the Cambrian Explosion." *BioEssays* 31 (2009): 736–47.

Piehler, Armin P., M. Hellum, J. J. Wenzel, E. Kaminski, K. B. F. Haug, P. Kierulf, and W. E. Kaminski. "The Human ABC Transporter Pseudogene Family: Evidence for Transcription and Gene-Pseudogene Interference." *BMC Genomics* 9 (2008): 165.

Pink, R. C., K. Wicks, D. P. Caley, E. K. Punch, L. Jacobs, and D. R. F. Carter. "Pseudogenes: Pseudo-Functional or Key Regulators in Health and Disease?" *RNA* 17 (2011): 792–98.

Pitts, Edward Lee. "Design Flaw?" *World Magazine,* February 11, 2011.

Pivar, Stuart. *Lifecode: The Theory of Biological Self-Organization.* New York: Ryland, 2004.

———. *On the Origin of Form: Evolution by Self-Organization.* Berkeley, CA: North Atlantic Books, 2009.

Poinar, G. "A Rhabdocoel Turbellarian (Platyhelminthes, Typhloplanoida) in Baltic Amber with a Review of Fossil and Sub-Fossil Platyhelminths." *Invertebrate Biology* 122 (2003): 308–12.
Polanyi, Michael. "Life Transcending Physics and Chemistry." *Chemical and Engineering News* 45 (1967): 54–66.
———. "Life's Irreducible Structure." *Science* 160 (1968): 1308–12.
Powell, Michael. "Controversial Editor Backed." *Washington Post,* August 19, 2005.
Poyton, Robert O. "Memory and Membranes: The Expression of Genetic and Spatial Memory During the Assembly of Organelle Macrocompartments." *Modern Cell Biology* 2 (1983): 15–72.
Pray, Leslie, and Kira Zhaurova. "Barbara McClintock and the Discovery of Jumping Genes (Transposons)." *Nature Education* 1 (2008). http://www.nature.com/scitable/topicpage barbara-mcclintock-and-the-discovery-of-jumping-34083.
Prigogine, Ilya, Gregoire Nicolis, and Agnessa Babloyantz. "Thermodynamics of Evolution." *Physics Today* 25 (1972): 23–31.
Prothero, Donald R. *Bringing Fossils to Life: An Introduction to Paleobiology.* Boston: McGraw-Hill, 1998.
———. *Evolution: What the Fossils Say and Why It Matters.* New York: Columbia Univ. Press, 2007.
Prud'homme, Benjamin, Nicolas Gompel, and Sean B. Carroll. "Emerging Principles of Regulatory Evolution." *Proceedings of the National Academy of Sciences USA* 104 (2007): 8605–12.
Quastler, Henry. *The Emergence of Biological Organization.* New Haven, CT: Yale Univ. Press, 1964.
Raff, Rudolf A. *The Shape of Life: Genes, Development, and the Evolution of Animal Form.* Chicago: Univ. of Chicago Press, 1996.
Reardon, Sara. "California Science Center to Pay $110,000 Settlement over Intelligent Design Film." *Science Insider,* August 31, 2011.
Reidhaar-Olson, John, and Robert Sauer. "Functionally Acceptable Solutions in Two Alpha-Helical Regions of Lambda Repressor." *Proteins: Structure, Function, and Genetics* 7 (1990): 306–16.
Retallack, Gregory J. "Growth, Decay and Burial Compaction of *Dickinsonia,* an Iconic Ediacaran Fossil." *Alcheringa: An Australasian Journal of Palaeontology* 31 (2007): 215–40.
———. "Ediacaran Life on Land." *Nature* 493 (2013): 89–92.
———. "Reply to the Discussion by Callow et al. on 'Were the Ediacaran Siliciclastics of South Australia Coastal or Deep Marine?'" *Sedimentology* 60 (2013): 628–30.
Ridley, Mark. "The Evolution Revolution." *New York Times,* March 17, 2002.
Robinson, Mabel L. *Runner on the Mountain Tops.* New York: Random House, 1939.
Rodin, Andrei S., Eörs Szathmáry, and Sergei N. Rodin. "On the Origin of the Genetic Code and tRNA Before Translation." *Biology Direct* 6 (2011).
Rodríguez-Campos, A., and F. Azorín. "RNA Is an Integral Component of Chromatin That Contributes to Its Structural Organization." *PloS One* 2 (2007): e1182.
Rogers, John. "Split-Gene Evolution: Exon Shuffling and Intron Insertion in Serine Protease Genes." *Nature* 315 (1985): 458–59.
Rokas, Antonis. "Spotlight: Drawing the Tree of Life." Broad Institute, November 15, 2006. https://www.broadinstitute.org/news/168.
Rokas, Antonis, and Sean B. Carroll. "Bushes in the Tree of Life." *PLoS Biology* 4, no. 11 (2006): 1899–1904.
———. "Frequent and Widespread Parallel Evolution of Protein Sequences." *Molecular Biology and Evolution* 25 (2008): 1943–53.
Rokas, Antonis, Nicole King, John Finnerty, and Sean B. Carroll. "Conflicting Phylogenetic Signals at the Base of the Metazoan Tree." *Evolution and Development* 5 (2003): 346–59.
Rokas, Antonis, Dirk Krüger, and Sean B. Carroll. "Animal Evolution and the Molecular Signature of Radiations Compressed in Time." *Science* 310 (2005): 1933–38.
Rose, Steven, ed. *The Richness of Life: The Essential Stephen Jay Gould.* New York: Norton, 2006.
Roth, Siegfried, and Jeremy A. Lynch. "Symmetry Breaking During *Drosophila* Oogenesis." *Cold Spring Harbor Perspectives in Biology* 1 (2009): a001891. http://cshperspectives.cshlp.org/content/1/2/a001891.full.pdf+html.
Roy, Scott William, and Manuel Irimia. "Rare Genomic Characters Do Not Support Coelomata: Intron Loss/Gain." *Molecular Biology and Evolution* 25 (2008): 620–23.
Rubin, C. M., R. H. Kimura, and C. W. Schmid. "Selective Stimulation of Translational Expression by Alu RNA." *Nucleic Acids Research* 30 (2002): 3253–61.
Runnegar, Bruce. "A Molecular-Clock Date for the Origin of the Animal Phyla." *Lethaia* 15, no. 3 (1982): 199–205.

———. "Evolution of the Earliest Animals." In *Major Events in the History of Life,* edited by J. W. Schopf, 65–93. Boston: Jones & Bartlett, 1992.
Ruppert, E. E., R. S. Fox, and R. D. Barnes. *Invertebrate Zoology.* 7th ed. Belmont, CA: Brooks/Cole, 2004.
Ruse, Michael. *Darwinism Defended: A Guide to the Evolution Controversies.* London: Addison-Wesley, 1982.
———. "A Philosopher's Day in Court." In *But Is It Science?* edited by M. Ruse, 31–36. Buffalo, NY: Prometheus, 1988.
———. "Witness Testimony Sheet: *McLean v. Arkansas.*" In *But Is It Science?* edited by M. Ruse, 287–306. Buffalo, NY: Prometheus, 1988.
———. "Darwinism: Philosophical Preference, Scientific Inference and Good Research Strategy." In *Darwinism: Science or Philosophy?* edited by J. Buell and V. Hearn, 21–28. Richardson, TX: Foundation for Thought and Ethics, 1994.
Salisbury, Frank B. "Natural Selection and the Complexity of the Gene." *Nature* 224 (1969): 342–43.
Sansom, R. S., S. E. Gabbott, and M. A. Purnell. "Non-Random Decay of Chordate Characters Causes Bias in Fossil Interpretation." *Nature* 463 (2010): 797–800.
Sapp, Jan. *Beyond The Gene.* New York: Oxford Univ. Press, 1987.
———. "The Structure of Microbial Evolutionary Theory." *Studies in History and Philosophy of Biology and Biomedical Sciences* 38 (2007): 780–95.
Savransky, Semyon. *Engineering of Creativity: Introduction to TRIZ Methodology of Inventive Problem Solving.* Boca Raton, FL: CRC Press, 2000.
Schirber, Michael. "Skeletons in the Pre-Cambrian Closet." PhysOrg.com, January 7, 2011.
Schnaar, Ronald L. "The Membrane Is the Message." *The Sciences,* May–June, 1985, 34–40.
Schneider, Thomas D. "Information Content of Individual Genetic Sequences." *Journal of Theoretical Biology* 189 (1997): 427–41.
———. "Evolution of Biological Information." *Nucleic Acids Research* 28 (2000): 2794–99.
Schopf, Thomas J. M., Antoni Hoffman, and Stephen Jay Gould. "Punctuated Equilibrium and the Fossil Record." *Science* 219 (1983): 438–39.
Schopf, J. William, and Bonnie M. Packer. "Early Archean (3.3-Billion to 3.5-Billion-Year-Old) Microfossils from Warrawoona Group, Australia." *Science* 237 (1987): 70–73.
Schram, F. R. "Pseudocoelomates and a Nemertine from the Illinois Pennsylvanian." *Journal of Paleontology* 47 (1973): 985–89.
Schuchert, Charles. "Charles Doolittle Walcott." In *Annual Report of National Academy of Sciences Fiscal Year 1924–1925.* Washington DC: Government Printing Office, 1926.
———. "Charles Doolittle Walcott, Paleontologist, 1850–1927." *Science* 65 (1927): 455–58.
Schuchert, Charles, and Carl O. Dunbar. *A Textbook of Geology, Part II: Historical Geology.* 4th ed. New York: Wiley, 1941.
Schueler, Mary G., A. W. Higgins, M. K. Rudd, K. Gustashaw, and H. F. Willard. "Genomic and Genetic Definition of a Functional Human Centromere." *Science* 294 (2001): 109–15.
Schützenberger, M. "Algorithms and the Neo-Darwinian Theory of Evolution." In *Mathematical Challenges to the Darwinian Interpretation of Evolution,* edited by P. S. Morehead and M. M. Kaplan, 73–80. Wistar Institute Symposium Monograph No. 5. Philadelphia: Wistar Institute Press, 1967.
Schwartz, Jeffrey H. "Homeobox Genes, Fossils, and the Origin of Species." *The Anatomical Record* 257 (1999): 15–31.
———. *Sudden Origins: Fossils, Genes, and the Emergence of Species.* New York: Wiley, 1999.
Schwartz, Jeffrey H., and Bruno Maresca. "Do Molecular Clocks Run at All? A Critique of Molecular Systematics." *Biological Theory* 1 (2006): 357–71.
Science Daily. "Australian Multicellular Fossils Point to Life on Land, Not at Sea, Geologist Proposes." December 12, 2012. http://www.sciencedaily.com/releases/2012/12/121212134050.htm (accessed March 22, 2013).
Scott, Eugenie. "Keep Science Free from Creationism." *Insight,* February 21, 1994.
Scott, Matthew P., and Sean B. Carroll. "The Segmentation and Homeotic Gene Network in Early *Drosophila* Development." *Cell* 51 (1987): 689–98.
Scriven, Michael. "Explanation and Prediction in Evolutionary Theory." *Science* 130 (1959): 477–82.
Sepkoski, David. "'Radical' or 'Conservative'? The Origin and Early Reception of Punctuated Equilibrium." In *The Paleobiological Revolution: Essays on the Growth of Modern Paleontology,* edited by D. Sepkoski and M. Ruse, 301–25. Chicago: Univ. of Chicago Press, 2009.

Sermonti, Giuseppe. *Why Is a Fly Not a Horse?* Seattle: Discovery Institute Press, 2005. Translated from the original, *Dimenticare Darwin*. Milan: Rusconi, 1999.

Shannon, Claude E. "A Mathematical Theory of Communication." *Bell System Technical Journal* 27 (1948): 379–423, 623–56.

Shannon, Claude E., and Warren Weaver. *The Mathematical Theory of Communication*. Urbana: Univ. of Illinois Press, 1949.

Shapiro, James A. "A 21st Century View of Evolution: Genome System Architecture, Repetitive DNA, and Natural Genetic Engineering." *Gene* 345 (2005): 91–100.

———. *Evolution: A View from the 21st Century*. Upper Saddle River, NJ: FT Press Science, 2011.

Shapiro, James, and Richard von Sternberg. "Why Repetitive DNA Is Essential to Genome Function." *Biological Reviews of the Cambridge Philosophical Society* 80 (2005): 227–50.

Shen, Bing, Lin Dong, Shuhai Xiao, and Michal Kowalewski. "The Avalon Explosion: Evolution of Ediacara Morphospace." *Science* 319 (2008): 81–84.

Shen, Y. Y., L. Liang, G. S. Li, R. W. Murphy, and Y. P. Zhang. "Parallel Evolution of Auditory Genes for Echolocation in Bats and Toothed Whales." *PLoS Genetics* 8 (2012): e1002788.

Shi, Riyi, and Richard B. Borgens. "Three-Dimensional Gradients of Voltage During Development of the Nervous System as Invisible Coordinates for the Establishment of Embryonic Pattern." *Developmental Dynamics* 202 (1995): 101–14. http://onlinelibrary.wiley.com/doi/10.1002/aja.1002020202/pdf (accessed October 30, 2012).

Schierwater, Bernd, and Rob DeSalle. "Can We Ever Identify the Urmetazoan?" *Integrative and Comparative Biology* 47 (2007): 670–76.

Shu, D. G. "On the Phylum Vetulicolia." *Chinese Science Bulletin* 50 (2005): 2342–54.

Shu, D. G., L. Chen, J. Han, and X. L. Zhang. "An Early Cambrian Tunicate from China." *Nature* 411 (2001): 472–73.

Shu, D. G., S. Conway Morris, J. Han, Z. F. Zhang, and J. N. Liu. "Ancestral Echinoderms from the Chengjiang Deposits of China." *Nature* 430 (2004): 422–28.

Shu, D. G., S. Conway Morris, J. Han, Z. F. Zhang, K. Yasui, P. Janvierk, L. Chen, X. L. Zhang, J. N. Liu, Y. Li, and H. Q. Liu. "Head and Backbone of the Early Cambrian Vertebrate *Haikouichthys*." *Nature* 421 (2003): 526–29.

Shu, D. G., S. Conway Morris, and X. L. Zhang. "A *Pikaia*-like Chordate from the Lower Cambrian of China." *Nature* 384 (1996): 157–58.

Shu, D. G., S. Conway Morris, Z. F. Zhang, J. N. Liu, J. Han, L. Chen, X. L. Zhang, K. Yasui, and Y. Li. "A New Species of Yunnanozoan with Implications for Deuterostome Evolution." *Science* 299 (2003): 1380–84.

Shu, D. G., H. L. Lou, S. Conway Morris, X. L. Zhang, S. X. Hu, L. Chen, J. Han, M. Zhu, Y. Li, and L. Z. Chen. "Lower Cambrian Vertebrates from South China." *Nature* 402 (1999): 42–46.

Shu, D. G., X. Zhang, and L. Chen, "Reinterpretation of *Yunnanozoon* as the Earliest Known Hemichordate." *Nature* 380 (1996): 428–30.

Shubin, N. H., and C. R. Marshall. "Fossils, Genes, and the Origin of Novelty." *Paleobiology* 26, no. 4 (2000): 324–40.

Siegal, Nina. "Riled by Intelligent Design." *New York Times,* November 6, 2005.

Siepel, Adam. "Darwinian Alchemy: Human Genes from Noncoding DNA." *Genome Research* 19 (2009): 1693–95.

Simons, Andrew M. "The Continuity of Microevolution and Macroevolution." *Journal of Evolutionary Biology* 15 (2002): 688–701.

Simpson, George Gaylord. *Fossils and the History of Life*. New York: Scientific American Books, 1983.

Siveter, D. J., M. Williams, and D. Waloszek. "A Phosphatocopid Crustacean with Appendages from the Lower Cambrian." *Science* 293 (2001): 479–81.

Skoog, Gerald. "A View from the Past." *Bookwatch Reviews* 2 (1989): 1–2.

Skou, Jens C. "The Identification of the Sodium-Pump as the Membrane-Bound Na^+/K^+-ATPase: a Commentary." *Biochimica et Biophysica Acta* 1000 (1989): 435–38.

Skovsted, C. B., and L. E. Holmer. "Early Cambrian Brachiopods From Northeast Greenland." *Palaeontology* 48 (2005): 325–45.

Smith, Andrew B., and Kevin J. Peterson. "Dating the Time and Origin of Major Clades." *Annual Review of Earth and Planetary Sciences* 30 (2002): 65–88.

Smith, C. R., et al. "Draft Genome of the Globally Widespread and Invasive Argentine Ant (*Linepithema humile*)." *Proceedings of the National Academy of Sciences USA* 108 (2011): 5667–72.

Smith, William S. "A Delineation of the Strata of England and Wales with Part of Scotland Exhibiting the Collieries and Mines the Marshes and Fen Lands Originally Overflowed by the Sea and the Varieties of Soil According to the Variations in the Substrata." Reproduced by the British Geological Survey. August 1, 1815.

Sonneborn, T. M. "Determination, Development, and Inheritance of the Structure of the Cell Cortex." In *Control Mechanisms in the Expression of Cellular Phenotypes,* edited by H. A. Padykula, 1–13. London: Academic Press, 1970.

Spemann, Hans, and Hilde Mangold. "Über Induktion von Embryonalanlagen durch Implantation artfremder Organisatoren." *Archiv für Mikroskopische Anatomie und Entwicklungsmechanik* 100 (1924): 599–638. Translated by Viktor Hamburger and edited by Klaus Sander as "Induction of Embryonic Primordia by Implantation of Organizers from a Different Species." *International Journal of Developmental Biology* 45 (2001): 13–38. http://www.ijdb.ehu.es/web/paper.php?doi=11291841.

Sperling, Erik A., and Jakob Vinther. "A Placozoan Affinity for *Dickinsonia* and the Evolution of Late Proterozoic Metazoan Feeding Modes." *Evolution and Development* 12 (2010): 201–9.

Spiro, Robert G. "Protein Glycosylation: Nature, Distribution, Enzymatic Formation, and Disease Implications of Glycopeptide Bonds." *Glycobiology* 12 (2002): 43R–56R.

Stadler, B. M. R., P. F. Stadler, G. P. Wagner, and W. Fontana. "The Topology of the Possible: Formal Spaces Underlying Patterns of Evolutionary Change." *Journal of Theoretical Biology* 213 (2001): 241–74.

Stanley, Steven M. *Macroevolution: Pattern and Process.* Baltimore: Johns Hopkins Univ. Press, 1998.

Stenger, V. J. *God: The Failed Hypothesis—How Science Shows That God Does Not Exist.* Amherst, NY: Prometheus, 2007.

St. Johnston, Daniel. "The Art and Design of Genetic Screens: *Drosophila melanogaster.*" *Nature* 3 (2002): 176–88.

Stoddard, Ed. "Evolution Gets Added Boost in Texas Schools." Reuters, http://blogs.reuters.com/faithworld/2009/01/23/evolution-gets-added-boost-in-texas-schools/ (accessed October 26, 2012).

Stokes, Trevor. " . . . And Smithsonian Has ID Troubles." *The Scientist,* July 4, 2005.

Stoltzfus, Arlin. "Mutationism and the Dual Causation of Evolutionary Change." *Evolution and Development* 8 (2006): 304–17.

Struck, T. H., et al. "Phylogenomic Analyses Unravel Annelid Evolution." *Nature* 471 (2011): 95–98.

Stutz, Terrence. "State Board of Education Debates Evolution Curriculum." *Dallas Morning News,* January 22, 2009.

Suen, Garret, et al. "The Genome Sequence of the Leaf-Cutter Ant *Atta cephalotes* Reveals Insights into Its Obligate Symbiotic Lifestyle." *PLoS Genetics* 7 (2011): e1002007.

Swadling, K. M., H. J. G. Dartnall, J. A. E. Gibson, É. Saulnier-Talbot, and W. F. Vincent. "Fossil Rotifers and the Early Colonization of an Antarctic Lake." *Quaternary Research* 55 (2001): 380–84.

Syvanen, Michael, and Jonathan Ducore. "Whole Genome Comparisons Reveals a Possible Chimeric Origin for a Major Metazoan Assemblage." *Journal of Biological Systems* 18 (2010): 261–75.

Szaniawski, H. "Cambrian Chaetognaths Recognized in Burgess Shale Fossils." *Acta Palaeontologica Polonica* 50 (2005): 1–8.

Szathmáry, Eörs. "When the Means Do Not Justify the End." *Nature* 399 (1999): 745–46.

Tam, O. H., et al. "Pseudogene-Derived Small Interfering RNAs Regulate Gene Expression in Mouse Oocytes." *Nature* 453 (2008): 534–38.

Tautz, Diethard, and Tomislav Domazet-Lošo. "The Evolutionary Origin of Orphan Genes." *Nature Reviews Genetics* 12 (2011): 692–702.

Taylor, Gordon Rattray. *The Great Evolution Mystery.* New York: Harper & Row, 1983.

Telford, M. J., S. J. Bourlat, A. Economou, D. Papillon, and O. Rota-Stabelli. "The Evolution of the Ecdysozoa." *Philosophical Transactions of the Royal Society B* 363 (2008): 1529–37.

Thaxton, Charles, Walter L. Bradley, and Roger L. Olsen. *The Mystery of Life's Origin: Reassessing Current Theories.* New York: Philosophical Library, 1984.

Theissen, Günter. "The Proper Place of Hopeful Monsters in Evolutionary Biology," *Theory in Biosciences* 124 (2006): 349–69.

———. "Saltational Evolution: Hopeful Monsters Are Here to Stay," *Theory in Biosciences* 128 (2009): 43–51.

Theobald, Douglas. "Punctuated Equilibrium." In *International Encyclopedia of the Social Sciences,* 2nd ed., 6: 629–31. Detroit: Macmillan Library Reference, 2008.

———. "29+ Evidences for Macroevolution." http://www.talkorigins.org/faqs/comdesc/section1.html (accessed October 31, 2012).

Thomson, K. S. "Macroevolution: The Morphological Problem." *American Zoologist* 32 (1992): 106–12.

Todd, Scott C. "A View from Kansas on That Evolution Debate." *Nature* 401 (1999): 423.

Tremblay, A., M. Jasin, and P. Chartrand. "A Double-Strand Break in a Chromosomal LINE Element Can Be Repaired by Gene Conversion with Various Endogenous LINE Elements in Mouse Cells." *Molecular and Cellular Biology* 20 (2000): 54–60.

Tyndall, John. "The Parallel Roads of Glen Roy." In *Fragments of Science: A Series of Detached Essays, Addresses, and Reviews.* Vol. 1. New York: Appleton, 1915.

Ulam, Stanislaw M. "How to Formulate Mathematically Problems of Rate of Evolution." In *Mathematical Challenges to the Neo-Darwinian Interpretation of Evolution,* edited by P. S. Moorhead and M. M. Kaplan, 21–33. Wistar Institute Symposium Monograph No. 5. New York: Liss, 1967.

University of California at Berkeley Museum of Paleontology. "Brachiopoda." http://www.ucmp.berkeley.edu/brachiopoda/brachiopodamm.html.

Uversky, Vladimir N., and Keith Dunker. "Understanding Protein Non-Folding." *Biochimica at Biophysica Acta* 1804 (2010): 1231–64.

Valentine, James W. "Fossil Record of the Origin of *Bauplan* and Its Implications." In *Patterns and Processes in the History of Life,* edited by D. M. Raup and D. Jablonski, 209–22. Berlin: Springer-Verlag, 1986.

———. "Late Precambrian Bilaterians: Grades and Clades." In *Tempo and Mode in Evolution: Genetics and Paleontology 50 Years After Simpson,* edited by W. M. Fitch and F. J. Ayala, 87–107. Washington, DC: National Academy Press, 1995.

———. "Why No New Phyla After the Cambrian? Genome and Ecospace Hypotheses Revisited." *Palaios* 10 (1995): 190–94.

———. "Prelude to the Cambrian Explosion." *Annual Review of Earth and Planetary Sciences* 30 (2002): 285–306.

———. *On the Origin of Phyla.* Chicago: Univ. of Chicago Press, 2004.

Valentine, James W., and Douglas H. Erwin. "Interpreting Great Developmental Experiments: The Fossil Record." In *Development as an Evolutionary Process,* edited by R. A. Raff and E. C. Raff, 71–107. New York: Liss, 1987.

Valentine, James W., Douglas H. Erwin, and David Jablonski. "Developmental Evolution of Metazoan Body Plans: The Fossil Evidence." *Developmental Biology* 173 (1996): 373–81.

Valentine, James W., David Jablonski, and Douglas H. Erwin. "Fossils, Molecules and Embryos: New Perspectives on the Cambrian Explosion." *Development* 126 (1999): 851–59.

Van de Lagemaat, L. N., J. R. Landry, D. L. Mager, and P. Medstrand. "Transposable Elements in Mammals Promote Regulatory Variation and Diversification of Genes with Specialized Functions." *Trends in Genetics* 19 (2003): 530–36.

Van Valen, Leigh. "Similar, but Not Homologous." *Nature* 305 (1983): 664.

———. "How Do Major Evolutionary Changes Occur?" *Evolutionary Theory* 8 (1988): 173–76.

Vedantam, Shankar. "Eden and Evolution." *Washington Post,* February 5, 2006.

Venema, D. R. "Seeking a Signature." *Perspectives on Science and Christian Faith* 62 (2010): 276–83.

Vermeij, Geerat J. "Animal Origins." *Science* 25 (1996): 525–26.

Von Sternberg, Richard. "On the Roles of Repetitive DNA Elements in the Context of a Unified Genomic-Epigenetic System." *Annals of the New York Academy of Sciences* 981 (2002): 154–88.

Von Sternberg, Richard, and James A. Shapiro. "How Repeated Retroelements Format Genome Function." *Cytogenetic and Genome Research* 110 (2005): 108–16.

Vorzimmer, P. "Charles Darwin and Blending Inheritance." *Isis* 54 (1963): 371–90.

Waddington, Conrad. "The Epigenotype." *Endeavour* 1 (1942): 18–20.

Waggoner, B. M. "Phylogenetic Hypotheses of the Relationships of Arthropods to Precambrian and Cambrian Problematic Fossil Taxa." *Systematic Biology* 45 (1996): 190–222.

Wagner, Andreas. "The Molecular Origins of Evolutionary Innovations." *Trends in Genetics* 27 (2011): 397–410.

Wagner, G. P. "What Is the Promise of Developmental Evolution? Part II: A Causal Explanation of Evolutionary Innovations May Be Impossible." *Journal of Experimental Zoology* (*Molecular and Developmental Evolution*) 291 (2001): 305–9.

Wagner, G. P., and P. F. Stadler. "Quasi-Independence, Homology and the Unity of Type: A Topological Theory of Characters." *Journal of Theoretical Biology* 220 (2003): 505–27.
Walcott, Charles Doolittle. "Cambrian Geology and Paleontology II: Abrupt Appearance of the Cambrian Fauna on the North American Continent." *Smithsonian Miscellaneous Collections* 57 (1910): 1–16.
Wallace, Bruce. "Adaptation, Neo-Darwinian Tautology, and Population Fitness: A Reply." *Evolutionary Biology* 17 (1984): 59–71.
Wang, Daniel Y. C., Sudhir Kumar, and S. Blair Hedges. "Divergence Time Estimates for the Early History of Animal Phyla and the Origin of Plants, Animals and Fungi." *Proceedings of the Royal Society of London B* 266 (1999): 163–71.
Wang, E. T., R. Sandberg, S. Luo, I. Khrebtukova, L. Zhang, C. Mayr, S. F. Kingsmore, G. P. Schroth, and C. B. Burge. "Alternative Isoform Regulation in Human Tissue Transcriptomes." *Nature* 456 (2008): 470–76.
Wang, Wen, Frédéric Brunet, Eviatar Nevo, and Manyuan Long. "Origin of *Sphinx*, a Young Chimeric RNA Gene in *Drosophila melanogaster*." *Proceedings of the National Academy of Sciences USA* 99 (2002): 4448–53.
Ward, Peter. *On Methuselah's Trail: Living Fossils and the Great Extinctions*. New York: Freeman, 1992.
———. *Out of Thin Air: Dinosaurs, Birds, and Earth's Ancient Atmosphere*. Washington, DC: Joseph Henry Press, 2006.
Watanabe T., et al. "Endogenous siRNAs from Naturally Formed dsRNAs Regulate Transcripts in Mouse Oocytes." *Nature* 453 (2008): 539–43.
Webster, Gerry, and Brian Goodwin. *Form and Transformation: Generative and Relational Principles in Biology*. Cambridge: Cambridge Univ. Press, 1996.
Welch, John J., Eric Fontanillas, and Lindell Bromham. "Molecular Dates for the 'Cambrian Explosion': The Influence of Prior Assumptions." *Systematic Biology* 54 (2005): 672–78.
Wells, Jonathan. "Making Sense of Biology: The Evidence for Development by Design." In *Signs of Intelligence: Understanding Intelligent Design*. edited by J. Kushiner and W. A. Dembski, 118–27. Grand Rapids, MI: Brazos, 2001.
Wen, Y., L. Zheng, L. Qu, F. J. Ayala, and Z. Lun. "Pseudogenes Are Not Pseudo Any More." *RNA Biology* 9 (2012): 27–32.
West, John G. *Darwin Day in America: How Our Politics and Culture Have Been Dehumanized in the Name of Science*. Wilmington, DE: ISI Books, 2007.
Whewell, William. "Lyell's Principles of Geology." *British Critic* 9 (1830): 180–206.
Whitehead, Alfred North. *Science and the Modern World*. New York: Macmillan, 1926.
Whitfield, John. "Biological Theory: Postmodern Evolution?" *Nature* 455 (2008): 281–84.
Whitman, W. B., D. C. Coleman, and W. J. Wiebe. "Prokaryotes: The Unseen Majority." *Proceedings of the National Academy of Sciences USA* 95 (1998): 6578–83.
Wieschaus, Eric. "From Molecular Patterns to Morphogenesis: The Lessons from *Drosophila*." Nobel lecture, December 8, 1995.
Wiley, E. O., and Bruce S. Lieberman. *Phylogenetics: Theory and Practice of Phylogenetic Systematics*. New York: Wiley-Blackwell, 2011.
Willey, Basil. "Darwin's Place in the History of Thought." In *Darwinism and the Study of Society*, edited by M. Banton, 15. Chicago: Quadrangle, 1961.
Willmer, Pat G., and Peter W. H. Holland. "Modern Approaches to Metazoan Relationships." *Journal of Zoology (London)* 224 (1991): 689–94.
Wills, M. A., S. Gerber, M. Ruta, and M. Hughes. "The Disparity of Priapulid, Archaeopriapulid and Palaeoscolecid Worms in the Light of New Data." *Journal of Evolutionary Biology* 25 (2012): 2056–76.
Wilson, T. E., U. Grawunder, and M. R. Liebe. "Yeast DNA Ligase IV Mediates Non-Homologous DNA End Joining." *Nature* 388 (1997): 495–98.
Withgott, Jay. "John Maynard Smith Dies." *Science*, April 20, 2004. http://news.sciencemag.org/sciencenow/2004/04/20-01.html.
Wimsatt, William C. "Generativity, Entrenchment, Evolution, and Innateness: Philosophy, Evolutionary Biology, and Conceptual Foundations of Science." In *Where Biology Meets Psychology: Philosophical Essays*, edited by V. G. Hardcastle, 139–79. Cabridge, MA: MIT Press, 1999.
Wimsatt, William C., and J. C. Schank. "Generative Entrenchment, Modularity and Evolvability: When Genic Selection Meets the Whole Organism." In *Modularity in Development and Evolution*, edited by G. Schlosser and G. Wagner, 359–94. Chicago: Univ. of Chicago Press, 2004.

Wolf, Yuri I., Igor B. Rogozin, and Eugene V. Koonin. "Coelomata and Not Ecdysozoa: Evidence from Genome-wide Phylogenetic Analysis." *Genome Research* 14 (2004): 29–36.
Wolfe, Stephen L. *Molecular and Cellular Biology*. Belmont, CA: Wadsworth, 1993.
Wolfram, Stephen. *A New Kind of Science*. Champaign, IL: Wolfram Media, 2002.
World Net Daily. "Intelligent Design Torpedoes Tenure." *World Net Daily*, May 19, 2007.
Wray, Gregory A., Jeffrey S. Levinton, and Leo H. Shapiro. "Molecular Evidence for Deep Precambrian Divergences Among Metazoan Phyla." *Science* 274 (1996): 568–73.
Wu, Dong-Dong, David M. Irwin, and Ya-Ping Zhang. "De Novo Origin of Human Protein-Coding Genes." *PLoS Genetics* 7 (2011): e1002379.
Xiao, S., A. H. Knoll, J. D. Schiffbauer, C. Zhou, and X. Yuan. "Comment on 'Fossilized Nuclei and Germination Structures Identify Ediacaran "Animal Embryos" as Encysting Protists.'" *Science* 335 (2012): 1169.
Xun, Gu. "Early Metazoan Divergence Was About 830 Million Years Ago," *Journal of Molecular Evolution* 47 (1998): 369–71.
Yochelson, Ellis L. *Charles Doolittle Walcott, Paleontologist*. Kent, OH: Kent State Univ. Press, 1998.
Yockey, Hubert P. "On the Information Content of Cytochrome C." *Journal of Theoretical Biology* 67, no. 3 (1977): 345–76.
———. "A Calculation of the Probability of Spontaneous Biogenesis by Information Theory." *Journal of Theoretical Biology* 67 (1977): 377–98.
———. *Information Theory and Molecular Biology*. Cambridge: Cambridge Univ. Press, 1992.
Youngren, K. K., et al. "The *Ter* mutation in the Dead End Gene Causes Germ Cell Loss and Testicular Germ Cell Tumours." *Nature* 435 (2005): 360–64.
Zamora, S., R. Gozalo, and E. Linñán. "Middle Cambrian Gogiid Echinoderms from Northeast Spain: Taxonomy, Palaeoecology, and Palaeogeographic Implications." *Acta Palaeontologica Polonica* 54 (2009): 253–65.
Zhang, Jianzhi, David M. Webb, and Ondrej Podlaha. "Accelerated Protein Evolution and Origins of Human-Specific Features: FOXP2 as an Example." *Genetics* 162 (2002): 1825–35.
Zhang, Jianzhi, Y. P. Zhang, and H. F. Rosenberg. "Adaptive Evolution of a Duplicated Pancreatic Ribonuclease Gene in a Leaf-Eating Monkey." *Nature Genetics* 30, no. 4 (2002): 411–15.
Zhang, Z., et al. "A Sclerite-Bearing Stem Group Entoproct from the Early Cambrian and Its Implications." *Scientific Reports* 3 (2013): 1066.
Zhen, Ying, M. Aardema, E. M. Medina, M. Schumer, and P. Andolfatto. "Parallel Molecular Evolution in an Herbivore Community." *Science* 337 (2012): 1634–37.
Zheng, Jie, Igor B. Rogozin, Eugene V. Koonin, and Teresa M. Przytycka. "Support for the Coelomata Clade of Animals from a Rigorous Analysis of the Pattern of Intron Conservation." *Molecular Biology and Evolution* 24 (2007): 2583–92.
Zvelebil, Marketa, and Jeremy O. Baum. *Understanding Bioinformatics*. New York: Garland Science, 2008.
Zuckerkandl, Emile, and Linus Pauling. "Evolutionary Divergence and Convergence in Proteins." In *Evolving Genes and Proteins*, edited by B. Bryson and H. Vogel, 97–166. New York: Academic, 1965.

ADDITIONAL RESOURCES FOR THE EPILOGUE

Axe, Douglas D. "Estimating the Prevalence of Protein Sequences Adopting Functional Enzyme Folds." *Journal of Molecular Biology* 341 (2004): 1295–1315.
Briggs, Derek, and Richard Fortey. "The Early Radiation and Relationships of the Major Arthropod Groups." *Science* 246 (1989): 241–43.
Brysse, Keynyn. "From Weird Wonders to Stem Lineages: The Second Reclassification of the Burgess Shale Fauna." *Studies in History and Philosophy of Science Part C: Studies in History and Philosophy of Biological and Biomedical Sciences* 39, no. 3 (2008): 298–313.
Butterfield, Nicholas J. "Exceptional Fossil Preservation and the Cambrian Explosion." *Integrative & Comparative Biology* 43, no. 1 (2003): 166–77.
Cavalier-Smith, Thomas. "A Revised Six-Kingdom System of Life." *Biological Reviews of the Cambridge Philosophical Society* 73, no. 3 (1998): 203–66.
Davidson, Eric. "Emerging Properties of Animal Gene Regulatory Networks." *Nature* 468 (2010): 911–20.

Davidson, Eric H. "Evolutionary Bioscience as Regulatory Systems Biology." *Developmental Biology* 357 (2011): 35–40.

Eden, Murray. "Inadequacies of Neo-Darwinian Evolution as a Scientific Theory." In *Mathematical Challenges to the Neo-Darwinian Interpretation of Evolution,* edited by P. S. Moorhead and M. M. Kaplan, 9–11. Wistar Institute Symposium Monograph. New York: Liss, 1967.

Edgecombe, Gregory D. "Arthropod Phylogeny: An Overview from the Perspective of Morphology, Molecular Data, and the Fossil Record." *Arthropod Structure & Development* 39 (2010): 74–87.

Edgecombe, Gregory D., Gonzalo Giribet, Casey W. Dunn, Andreas Hejnol, Reinhardt M. Kristensen, Ricardo C. Neves, Greg W. Rouse, Katrine Worsaae, and Martin V. Sørensen. "Higher-Level Metazoan Relationships: Recent Progress and Remaining Questions." *Organisms, Diversity, & Evolution* 11 (2011): 151–172.

Edgecombe, Gregory D., and David A. Legg. "The Arthropod Fossil Record." In *Arthropod Biology and Evolution: Molecules, Development, Morphology,* edited by A. Minelli et al., 393–415. Berlin: Springer-Verlag, 2013.

Eldredge, Niles, and Joel Cracraft. *Phylogenetic Patterns and the Evolutionary Process.* New York: Columbia University Press, 1980.

Erwin, Douglas, and James Valentine. *The Cambrian Explosion: The Construction of Animal Biodiversity.* Greenwood Village, CO: Roberts, 2013.

Farrell, John. "How Nature Works." *National Review,* September 2, 2013.

Fortey, Richard. "Evolution: The Cambrian Explosion Exploded?" *Science* 293 (2001): 438–39.

Funk, Vicki A., and Quentin D. Wheeler. "Symposium: Character Weighting, Cladistics and Classification." *Systematic Zoology* 35, no. 1 (1986): 100–101.

Gee, Henry. *In Search of Deep Time: Beyond the Fossil Record to a New History of Life.* New York: Free Press, 2000.

Gibson, Amanda K., Zach Smith, Clay Fuqua, Keith Clay, and John K. Colbourne. "Why So Many Unknown Genes? Partitioning Orphans from a Representative Transcriptome of the Lone Star Tick *Amblyomma americanum.*" *BMC Genomics* 14 (2013): 135.

Kazlev, M. Alan. "Panarthropoda: Lobopodia." *Palaeos.com,* www.palaeos.com/metazoa/ecdysozoa/panarthropoda/lobopodia.html.

Khalturin, Konstantin, Friederike Anton-Erxleben, Sylvia Sassmann, Jörg Wittlieb, Georg Hemmrich, and Thomas C. G. Bosch. "A Novel Gene Family Controls Species-Specific Morphological Traits in Hydra." *PLoS Biology* 6 (2008): e278.

Khalturin, Konstantin, Georg Hemmrich, Sebastian Fraune, René Augustin, and Thomas C. G. Bosch. "More Than Just Orphans: Are Taxonomically-Restricted Genes Important in Evolution?" *Trends in Genetics* 25 (2009): 404–13.

Kühl, Gabriele, Derek E. G. Briggs, and Jes Rust. "A Great-Appendage Arthropod with a Radial Mouth from the Lower Devonian Hunsrück Slate, Germany." *Science* 323 (2009): 771–73.

Legg, David A., Xiaoya Ma, Joanna M. Wolfe, Javier Ortega-Hernández, Gregory D. Edgecombe, and Mark D. Sutton. "Lobopodian Phylogeny Reanalyzed." *Nature* 476 (2011): E2.

Legg, David A., Mark D. Sutton, Gregory D. Edgecombe, and Jean-Bernard Caron. "Cambrian Bivalved Arthropod Reveals Origin of Arthrodization." *Proceedings of the Royal Society B* 279 (2012): 4699–704.

Liu, Jianni, Degan Shu, Jian Han, Zhifei Zhang, and Xingliang Zhang. "A Large Xenusiid Lobopod with Complex Appendages from the Lower Cambrian Chengjiang Lagerstätte." *Acta Palaeontologica Polonica* 51, no. 2 (2006): 215–22.

Liu, Jianni, Michael Steiner, Jason A. Dunlop, Helmut Keupp, Degan Shu, Qiang Ou, JianHan, Zhifei Zhang, and Xingliang Zhang. "Liu et al. Reply." *Nature* 476 (2011): e3–e4.

Luskin, Casey. "Darwin Defenders Love Donald Prothero's Ranting Review of *Darwin's Doubt.*" *Evolution News & Views,* July 23, 2013, www.evolutionnews.org/2013/07/darwin_defender_1074791.html.

Luskin, Casey. "Does Natural Selection Leave 'Detectable Statistical Evidence in the Genome'?" *Evolution News & Views,* August 7, 2013, www.evolutionnews.org/2013/08/does_natural_se075171.html.

Luskin, Casey. "How 'Sudden' Was the Cambrian Explosion? Nick Matzke Misreads Stephen Meyer and the Paleontological Literature; *New Yorker* Recycles Misrepresentation." *Evolution News & Views,* July 16, 2013, www.evolutionnews.org/2013/07/how_sudden_was_074511.html.

Luskin, Casey. "Rush to Judgment: Nick Matzke's Hasty Review of *Darwin's Doubt* Makes Bogus Charges of Errors and Ignorance." *Evolution News & Views,* June 25, 2013, www.evolutionnews.org/2013/06/rush_to_judgmen073791.html.
Luskin, Casey. "Small Shelly Fossils, and the Length of the Cambrian Explosion." *Evolution News & Views,* October 23, 2013, www.evolutionnews.org/2013/08/does_natural_se075171.html.
Lyell, Charles. *Principles of Geology: Being an Attempt to Explain the Former Changes of the Earth's Surface, by Reference to Causes Now in Operation.* 3 vols. London: Murray, 1830–33.
Marshall, Charles R. "Explaining the Cambrian 'Explosion' of Animals." *Annual Reviews of Earth and Planetary Sciences* 34 (2006): 355–84.
Marshall, Charles R. "Nomothetism and Understanding the Cambrian 'Explosion.'" *Palaois* 18 (2003): 195–96.
Marshall, Charles R. "When Prior Belief Trumps Scholarship." *Science* 341 (2013): 1344.
Marshall, Charles R., and James W. Valentine. "The Importance of Preadapted Genomes in the Origin of the Animal Bodyplans and the Cambrian Explosion." *Evolution* 64 (2010): 1189–201.
Martin, Robert A. *Missing Links: Evolutionary Concepts and Transitions Through Time.* Boston: Jones and Bartlett, 2004.
Matzke, Nick. "Incompetent and Wrong." Amazon.com, June 20, 2013, www.amazon.com/review/R2ZTHZ38V22QG8?cdPage=16.
Matzke, Nick. "Luskin's Hopeless Monster." *Panda's Thumb,* June 27, 2013, http://pandasthumb.org/archives/2013/06/luskins-hopeles.html.
Matzke, Nick. "Meyer's Hopeless Monster, Part II." *Panda's Thumb,* June 19, 2013, http://pandasthumb.org/archives/2013/06/meyers-hopeless-2.html.
Meyer, Stephen C. "More on Small Shelly Fossils and the Length of the Cambrian Explosion: A Concluding Response to Charles Marshall." *Evolution News & Views,* October 23, 2013, www.evolutionnews.org/2013/10/more_on_small_s078251.html.
Miller, David J., and Eldon E. Ball. "The Gene Complement of the Ancestral Bilaterian—Was Urbilateria a Monster?" *Journal of Biology* 8 (2009): 89.
Mounce, Ross C. P., and Matthew A. Wills. "Phylogenetic Position of *Diania* Challenged." *Nature* 476 (August 11, 2011): E1.
Müller, Gerd B., and Stuart A. Newman. "Origination of Organismal Form: The Forgotten Cause in Evolutionary Theory." In *Origination of Organismal Form: Beyond the Gene in Developmental and Evolutionary Biology,* edited by Gerd B. Müller and Stuart A. Newman, 7–8. Cambridge, MA: MIT Press, 2003.
Ohno, Susumu. "The Notion of the Cambrian Pananimalia Genome." *Proceedings of the National Academy of Sciences* 93 (1996): 8475–78.
Oliveri, Paolo, and Eric H. Davidson. "Built to Run, Not to Fail." *Science* 315 (2007): 1510–11.
Oliveri, P., Q. Tu, and E. H. Davidson. "Global Regulatory Logic for Specification of an Embryonic Cell Lineage." *Proceedings of the National Academy of Sciences* 105 (2008): 5955–62.
Paterson, John R., Diego C. García-Bellido, Michael S. Y. Lee, Glenn A. Brock, James B. Jago, and Gregory D. Edgecombe. "Acute Vision in the Giant Cambrian Predator *Anomalocaris* and the Origin of Compound Eyes." *Nature* 480 (2011): 237–40.
Rasmussen, Nicolas. "A New Model of Developmental Constraints as Applied to the *Drosophila* System." *Journal of Theoretical Biology* 127 (1987): 271–99.
Reidhaar-Olson, John, and Robert Sauer. "Functionally Acceptable Solutions in Two Alpha-Helical Regions of Lambda Repressor." *Proteins: Structure, Function, and Genetics* 7 (1990): 306–16.
Richter, Stefan, Martin Stein, Thomas Frace, and Nikolaus U. Szucsich. "The Arthropod Head." In *Arthropod Biology and Evolution: Molecules, Development, Morphology,* edited by A. Minelli et al., 223–40. Berlin: Springer-Verlag, 2013.
Rudel, David, and Ralf Sommer. "The Evolution of Developmental Mechanisms." *Developmental Biology* 264 (2003): 15–37.
Schützenberger, Marcel. "Algorithms and the Neo-Darwinian Theory of Evolution." In *Mathematical Challenges to the Neo-Darwinian Interpretation of Evolution,* edited by P. S. Moorhead and M. M. Kaplan, 9–11. Wistar Institute Symposium Monograph. New York: Liss, 1967.
Shannon, Claude. "A Mathematical Theory of Communication." *Bell System Technical Journal* 27 (1948): 370–423, 623–29.
Shaw, Robert. "Collateral Ancestor." *Encyclopedia of Genealogy,* www.eogen.com/collateralancestor.
Shubin, N. H., and C. R. Marshall. "Fossils, Genes, and the Origin of Novelty." *Paleobiology* 26, no. 4, supplement (2000): 335.

Smith, M. Paul, and David A. T. Harper. "Causes of the Cambrian Explosion." *Science* 341 (2013): 1355–56.

Sober, Elliott, and Michael Steele. "Testing the Hypothesis of Common Ancestry." *Journal of Theoretical Biology* 218 (2002): 395–408.

Swisher, Carl C. III, Yuan-qing Wang, Xiao-lin Wang, Xing Xu, and Yuan Wang. "Cretaceous Age for the Feathered Dinosaurs of Liaoning, China." *Nature* 400 (1999): 58–61.

Syvanen, Michael. "Evolutionary Implications of Horizontal Gene Transfer." *Annual Review of Genetics* 46 (2012): 339–56.

Waggoner, Benjamin M. "Phylogenetic Hypotheses of the Relationships of Arthropods to Precambrian and Cambrian Problematic Fossil Taxa." *Systematic Biology* 45, no. 2 (1996): 190–222.

Wheat, Christopher W., and Niklas Wahlberg. "Phylogenomic Insights into the Cambrian Explosion, the Colonization of Land and the Evolution of Flight in Arthropoda." *Systematic Biology* 62, no. 1 (2013): 93–109.

Whittington, Harry B., and Derek E. G. Briggs. "The Largest Cambrian Animal, Anomalocaris, Burgess Shale, British Columbia." *Philosophical Transactions of the Royal Society B* 309 (1985): 569–609.

Yong, Ed. "The Sharp Eyes of Anomalocaris, a Top Predator That Lived Half a Billion Years Ago," *Discover Magazine,* December 7, 2011, http://blogs.discovermagazine.com/notrocket science/2011/12/07/anomalocaris-sharp-eyes-predator/.

CREDITS AND PERMISSIONS

The inclusion of any figures, illustrations, photographs, diagrams, charts, or other types of images in this book should not be construed as an endorsement of the ideas and arguments contained in this book on the part of any copyright holders or creators of those images, other than the author of the book himself.

Figure 1.1: Tree of life image from Ernst Haeckel, *Volume II of Generelle Morphologie* (1866). Public Domain.

Figure 1.2a: Louis Agassiz photograph courtesy of Wikimedia Commons. Public Domain.

Figure 1.2b: Charles Darwin photograph courtesy of Wikimedia Commons. Original photograph taken by Henry Maull and John Fox. Public Domain.

Figure 1.3a: Drawing of brachiopod by Ray Braun based on information from Figure 18, Brusca, R. C., and Brusca, G. J. *Invertebrates,* 794. Sunderland, MA: Sinauer Associates, 1990; Figure 7.55, Mintz, L. W. *Historical Geology: The Science of a Dynamic Earth,* 130. 2nd ed. Columbus, Charles E. Merrill Publishing, 1977.

Figure 1.3b: Brachiopod fossil photograph showing internal remains courtesy of Paul Chien. Used with permission.

Figure 1.3c: Brachiopod fossil photograph © Colin Keates / DK Limited / Corbis. Used with permission.

Figure 1.4a: Drawing of trilobite by Ray Braun based on information from "Fossil Groups," University of Bristol, http://palaeo.gly.bris.ac.uk/palaeofiles/fossilgroups/trilobites/page2.htm.

Figure 1.4b: Trilobite fossil photograph courtesy of Illustra Media. Used with permission.

Figure 1.5: Drawing of tetracoral by Ray Braun based on information from "Convergence," http://www.znam.bg/com/action/showArticle;jsessionid=FA17A83CDA0EC25151F6B80869F07E49?encID=790&article=2059773460.

Figure 1.6: Drawing of geological timescale by Ray Braun based on information from Mintz, L. W. *Historical Geology: The Science of a Dynamic Earth,* back page. 2nd ed. Columbus, Charles E. Merrill Publishing, 1977; Gradstein, F. M., Ogg, J. G., Schmitz, M. D., Ogg, G. M. *The Geological Timescale 2012 Volumes 1 and 2.* Elsevier, 2012.

Figure 1.7: Drawing by Sir Thomas Dick-Lauder and published in Tyndall, J., "The Parallel Roads of Glen Roy." In *Fragments of Science: A Series of Detached Essays, Addresses, and Reviews, Volume I.* New York: D. Appleton and Company, 1915. Courtesy of Wikimedia Commons. Public Domain.

Figure 1.8: Drawing by Ray Braun based on information from a fossil exhibit that was on display at the California Academy of Sciences in the 1990s.

Figure 2.1: Photograph of the Burgess Shale © Thomas Kitchin & Victoria Hurst/All Canada Photos/Corbis. Used with permission.

Figure 2.2: Photograph of Charles Doolittle Walcott courtesy of Smithsonian Institution Archives. Image #84–16281. Used with permission.

Figure 2.3a: Drawing of *Marrella* by Ray Braun based on information from Figure 3.12, Gould, S. J. *Wonderful Life: The Burgess Shale and the Nature of History,* 114. New York: Norton, 1990.

Figure 2.3b: Photograph of *Marrella* fossil © User: Smith609 (Own Work) / Wikimedia Commons / CC-BY-SA-3.0. Used with permission; usage not intended to imply endorsement by the author/licensor of the work.

Figure 2.4a: Drawing of *Hallucigenia* by Ray Braun based on information from Figure 14.6,

Xian-guang, H., Aldridge, R. J., Bergström, J., Siveter, D. J., Siveter, D. J., and Xiang-hong, F. *The Cambrian Fossils of Chengjiang, China: The Flowering of Early Animal Life,* 88 Oxford: Blackwell Publishing, 2004; Figure 19(b), Conway Morris, S., *The Crucible of Creation: The Burgess Shale and the Rise of Animals,* 55. Oxford: Oxford University Press, 2000.

Figure 2.4b: Photograph of *Hallucigenia* fossil courtesy of Smithsonian Institution. Used with permission.

Figures 2.5a, 2.5b, and 2.5c: Chart drawn by Ray Braun based on data compiled from references in Chapter 2, Endnote 5.

Figure 2.6: Diagram drawn by Ray Braun based on information from Figure 1, Meyer, S. C., Ross, M., Nelson, P. and Chien, P. "The Cambrian Explosion: Biology's Big Bang." In *Darwinism, Design and Public Education,* edited by John Angus Campbell and Stephen C. Meyer, 325. East Lansing, MI: Michigan State University Press, 2003. Courtesy of Brian Gage.

Figure 2.7: Diagram drawn by Ray Braun based on original drawing by Art Battson. Courtesy of Art Battson.

Figure 2.8: Diagram drawn by Ray Braun based on Figure 1, Wiester, J., and Dehaan, R. F., "The Cambrian Explosion: The Fossil Record and Intelligent Design." In *Signs of Intelligence: Understanding Intelligent Design,* 149. William A. Dembski and James M. Kushiner, eds. Grand Rapids, MI: Brazos Press, 2001. Courtesy also of Art Battson.

Figure 2.9a: Drawing of *Opabinia* by Ray Braun based on information from Figure 3.21, Gould, S. J. *Wonderful Life: The Burgess Shale and the Nature of History,* 126. New York: Norton, 1990; Figure 173, Briggs, D., Erwin, D., and Collier, F. *The Fossils of the Burgess Shale,* 210. Washington: Smithsonian Institution Press, 1994.

Figure 2.9b: Photograph of *Opabinia* fossil from Walcott, C. D., "Middle Cambrian Branchiopoda, Malacostraca, Trilobita, and Merostomata." *Smithsonian Miscellaneous Collections,* Volume 57, Number 6 (Publication 2051), City of Washington, Published by the Smithsonian Institution, March 13, 1912. Public Domain.

Figure 2.10a: Drawing of *Anomalocaris* by Ray Braun based on information from Figure 264, Chen, J. Y., Zhou, G. Q., Zhu, M. Y., and Yeh, K. Y. *The Chengjiang Biota—A Unique Window of the Cambrian Explosion,* 197. Taichung, Taiwan: National Museum of Natural Science, 1996.

Figure 2.10b: Photograph of *Anomalocaris* fossil courtesy of J. Y. Chen. Source: Figure 265A, Chen, J. Y., Zhou, G. Q., Zhu, M. Y., and Yeh, K. Y. *The Chengjiang Biota—A Unique Window of the Cambrian Explosion,* 198. Taichung, Taiwan: National Museum of Natural Science, 1996. Used with permission.

Figure 2.11a: Tree of life diagram from Charles Darwin, *Origin of Species,* 1859. Courtesy of Wikimedia Commons. Public domain.

Figure 2.11b: Diagram drawn by Ray Braun.

Figure 2.12: Diagram drawn by Ray Braun.

Figure 3.1: Photograph of J. Y. Chen from *Icons of Evolution* Documentary, Coldwater Media, 2002. Copyright © Discovery Institute 2013. Courtesy of Discovery Institute. Used with permission.

Figure 3.2a: Photograph from Chengjiang fossil site courtesy of Illustra Media. Used With Permission.

Figure 3.2b: Photograph from Chengjiang fossil site courtesy of Paul Chien. Used With Permission.

Figure 3.2c: Photograph from Chengjiang fossil site courtesy of Paul Chien. Used With Permission.

Figure 3.3: Photograph of Harry Whittington from the Archives of the Museum of Comparative Zoology, Ernst Mayr Library, Harvard University. Used with permission.

Figure 3.4a: Drawing of *Nectocaris* by Ray Braun based on information from Smith, M. R., and Caron, J.-B., "Primitive soft-bodied cephalopods from the Cambrian," *Nature,* 465 (May 27, 2010): 469–472; *Nectocaris pteryx* at http://commons.wikimedia.org/wiki/File:Nectocaris_pteryx.JPG, courtesy of user: Stanton F. Fink at en.wikipedia / Wikimedia Commons / CC BY-SA 2.5.

Figure 3.4b: Photograph of *Nectocaris* fossil reprinted by permission from Macmillan Publishers Ltd: *Nature,* Figure 1, Smith, M. R., and Caron, J.-B., "Primitive soft-bodied cephalopods from the Cambrian," *Nature,* 465 (May 27, 2010): 469–472. Copyright 2010. Used with permission.

Figure 3.4c: Photograph of *Nectocaris* fossil reprinted by permission from Macmillan Publishers Ltd.: *Nature,* Figure 1, Smith, M. R., and Caron, J.-B., "Primitive soft-bodied cephalopods from the Cambrian," *Nature,* 465 (May 27, 2010): 469–472. Copyright 2010. Used with permission.

Figure 3.5a: Photograph of stromatolite fossil courtesy of user: Rygel, M. C., at en.wikipedia / Wikimedia Commons / CC-BY-SA-3.0. Used with permission; usage not intended to imply endorsement by the author/licensor of the work.

Figure 3.5b: Photograph of stromatolite fossil courtesy of American Association for the Advancement of Science, Figure 2B, Hoffman, P., "Algal Stromatolites: Use in Stratigraphic Correlation and Paleocurrent Determination," *Science,* 157 (September 1, 1967): 1043–45. Reprinted with permission from AAAS. Used with permission.

Figure 3.6a–1: Drawing of ctenophore by Ray Braun based on information from Figure 28, Chen, J. Y., and Zhou, G., "Biology of the Chengjiang fauna." In "The Cambrian Explosion and the Fossil Record," 33, *Bulletin of the National Museum of Natural Science,* 10:11–106. Edited by Chen, J. Y., Cheng, Y., and Iten, H. V., eds. Taiwan: National Museum of Natural Science, 1997.

Figure 3.6a–2: Photograph of ctenophorefossil courtesy of J. Y. Chen. Source: Figure 102, Chen, J. Y., Zhou, G. Q., Zhu, M. Y., and Yeh, K. Y. *The Chengjiang Biota—A Unique Window of the Cambrian Explosion,* 96. Taichung, Taiwan: National Museum of Natural Science, 1996. Used with permission.

Figure 3.6b–1: Drawing of phoronid by Ray Braun based on information from Figure 51, Chen, J. Y., and Zhou, G., "Biology of the Chengjiang fauna." In "The Cambrian Explosion and the Fossil Record,"45, *Bulletin of the National Museum of Natural Science,* 10:11–106. Edited by Chen, J. Y., Cheng, Y., and Iten, H. V., eds. Taiwan: National Museum of Natural Science, 1997.

Figure 3.6b–2: Photograph of phoronid fossil courtesy of J. Y. Chen. Source: Figure 49, Chen, J. Y., and Zhou, G., "Biology of the Chengjiang fauna." In "The Cambrian Explosion and the Fossil Record," 44, *Bulletin of the National Museum of Natural Science,* 10:11–106. Edited by Chen, J. Y., Cheng, Y., and Iten, H. V., eds. Taiwan: National Museum of Natural Science, 1997. Used with permission.

Figure 3.6c–1: Drawing of *Waptia* by Ray Braun based on information from Figure 110, Briggs, D., Erwin, D., and Collier, F. *The Fossils of the Burgess Shale,* 157. Washington: Smithsonian Institution Press, 1994.

Figure 3.6c–2: Photograph of *Waptia* fossil courtesy of Paul Chien. Used with permission.

Figure 3.6d–1: Drawing of priapulid worm by Ray Braun based on information from Figures 32A and 32B, Chen, J. Y., and Zhou, G. "Biology of the Chengjiang fauna." In "The Cambrian Explosion and the Fossil Record," 36, *Bulletin of the National Museum of Natural Science,* 10:11–106. Edited by Chen, J. Y., Cheng, Y., and Iten, H. V., eds. Taiwan: National Museum of Natural Science, 1997.

Figure 3.6d–2: Photograph of priapulid worm fossil courtesy of Paul Chien. Used with permission.

Figure 3.6e–1: Drawing of *Eldonia* by Ray Braun based on information from Figure 147, Chen, J. Y., Zhou, G. Q., Zhu, M. Y., and Yeh, K. Y. *The Chengjiang Biota—A Unique Window of the Cambrian Explosion,* 124. Taichung, Taiwan: National Museum of Natural Science, 1996; "Cambrian Café," http://cambrian-cafe.seesaa.net/archives/201003–1.html.

Figure 3.6e–2: Photograph of *Eldonia* fossil courtesy of J. Y. Chen. Source: Figure 148, Chen, J. Y., Zhou, G. Q., Zhu, M. Y., and Yeh, K. Y. *The Chengjiang Biota—A Unique Window of the Cambrian Explosion,* 125. Taichung, Taiwan: National Museum of Natural Science, 1996. Used with permission.

Figure 3.6f–1: Drawing of hyolith by Ray Braun based on information from Figure 172, Chen, J. Y., Zhou, G. Q., Zhu, M. Y., and Yeh, K. Y. *The Chengjiang Biota—A Unique Window of the Cambrian Explosion,* 132. Taichung, Taiwan: National Museum of Natural Science, 1996.

Figure 3.6f–2: Photograph of hyolith fossil courtesy of J. Y. Chen. Source: Figure 173A, Chen, J. Y., Zhou, G. Q., Zhu, M. Y., and Yeh, K. Y. *The Chengjiang Biota—A Unique Window of the Cambrian Explosion,* 139. Taichung, Taiwan: National Museum of Natural Science, 1996. Used with permission.

Figure 3.7a–1: Photograph of sponge embryo fossil courtesy of Paul Chien. Used with permission.

Figure 3.7a–2: Photograph of sponge embryo fossil courtesy of Paul Chien. Used with permission.

Figure 3.7b: Photograph of sponge embryo fossil courtesy of Paul Chien. Used with permission.

Figure 3.8: Diagram drawn by Ray Braun based on information in Figure 2, Meyer, S. C., Ross, M., Nelson, P., and Chien, P. "The Cambrian Explosion: Biology's Big Bang." In *Darwinism, Design and Public Education,* edited by John Angus Campbell and Stephen C. Meyer, 326. East Lansing, MI: Michigan State University Press, 2003. Courtesy of Brian Gage.

Figure 3.9a: Drawing of *Myllokunmingia* by Ray Braun based on information from Figure 2A, Shu, D. G., Lou, H. L., Conway Morris, S., Zhang, X. L., Hu, S. X., Chen, L., Han, J., Zhu, M., Li, Y., and Chen, L. Z., "Lower Cambrian Vertebrates from South China." *Nature,* 402 (1999): 42–46.

Figure 3.9b: Photograph of *Myllokunmingia* fossil reprinted by permission from Macmillan Publishers Ltd.: *Nature,* Figure 2A, Shu, D. G., Lou, H. L., Conway Morris, S., Zhang, X. L., Hu, S. X., Chen, L., Han, J., Zhu, M., Li, Y., and Chen L. Z., "Lower Cambrian Vertebrates from South China." *Nature,* 402 (1999): 42–46. Copyright 1999. Used with permission.

Figure 4.1a–1: Drawing of *Dickinsonia* by Ray Braun based on information from "Les premiers animaux de la Terre," L'Historie de la vie sur terre, April 23, 2009, http://titereine.centerblog.net /4-Les-premiers-animaux-de-la-Terre.

Figure 4.1a–2: Photograph of *Dickinsonia* fossil courtesy of Figure 2, Peterson, K. J., Cotton, J. A., Gehling, J. G., and Pisani, D., "The Ediacaran emergence of bilaterians: congruence between the genetic and the geological fossil records," *Philosophical Transactions of the Royal Society B,* 2008, 363 (1496): 1435–43, by permission of the Royal Society.

Figure 4.1b–1: Drawing of *Spriggina* by Ray Braun based on information from Spriggina flounensi C.jpg, Wikipedia, http://en.wikipedia.org/wiki/File:Spriggina_flounensi_C.jpg.

Figure 4.1b–2: Photograph of *Spriggina* fossil courtesy of Figure 2, Peterson, K. J., Cotton, J. A., Gehling, J. G., and Pisani, D., "The Ediacaran emergence of bilaterians: congruence between the genetic and the geological fossil records," *Philosophical Transactions of the Royal Society B,* 2008, 363 (1496): 1435–43, by permission of the Royal Society.

Figure 4.1c–1: Drawing of *Charnia* by Ray Braun based on information from Charnia_Species_ BW_by_avancna, http://avancna.deviantart.com/art/Charnia-Species-BW-101515874.

Figure 4.1c–2: Photograph of *Charnia masoni* fossil courtesy of user: Smith609 at en.wikipedia / Wikimedia Commons / CC-BY-2.5. Used with permission; usage not intended to imply endorsement by the author/licensor of the work.

Figure 4.2a: Photograph of *Arkarua* fossil courtesy of Figure 4B, Gehling, J. G., "Earliest known echinoderm—a new Ediacaran fossil from the Pound Subgroup of South Australia," *Alcheringa: An Australasian Journal of Palaeontology,* 11 (1987): 337–45. Used with permission; usage not intended to imply endorsement by the author / licensor of the work.

Figure 4.2b: Photograph of *Parvancorina* fossil courtesy of Figure 2, Peterson, K. J., Cotton, J. A., Gehling, J. G., and Pisani, D., "The Ediacaran emergence of bilaterians: congruence between the genetic and the geological fossil records," *Philosophical Transactions of the Royal Society B,* 2008, 363 (1496): 1435–43, by permission of the Royal Society.

Figure 4.3: Photograph of *Vernanimalcula* fossil courtesy of American Association for the Advancement of Science, from Figure 1b, Chen, J.-Y., Bottjer, D. J., Oliveri, P., Dornbos, S. Q., Gao, F., Ruffins, S., "Small Bilaterian Fossils from 40 to 55 Million Years Before the Cambrian," *Science,* 305 (July 9, 2004): 218–22. Reprinted with permission from AAAS.

Figure 5.1: Pentadactyl limb drawing courtesy of Jody F. Sjogren and Figure 4-1, Wells, J., *Icons of Evolution: Science or Myth?* Washington D.C.: Regnery, 2000. Copyright © Jody F. Sjogren 2000. Used with permission.

Figure 5.2: Drawing by Ray Braun based on information from Paul Nelson; Smith, A. B., and Peterson, K. J., "Dating the Time and origin of Major Clades," *Annual Review of Earth and Planetary Sciences,* 30 (2002): 65–88.

Figure 6.1: Drawing by Ray Braun based on information from Telford, M. J. *et al.* "The evolution of the Ecdysozoa," *Philosophical Transactions of the Royal Society B,* Vol. 363 (2008): 1529–37; Aguinaldo, A. M., Turbeville, J. M., Linford, L. S., Rivera, M. C., Garey, J. R., Raff, R. A., and Lake, J. A. "Evidence for a clade of nematodes, arthropods and other moulting animals," *Nature,* 387 (1997): 489–93; Mallatt, J. M., Garey, J. R., and Shultz, J. W., "Ecdysozoan phylogeny and Bayesian inference: first use of nearly complete 28S and 18S rRNA gene sequences to classify the arthropods and their kin,"*Molecular Phylogenetics and Evolution,* 31 (2004): 178–91; Halanych, K. M., "The New View of Animal Phylogeny," *Annual Review of Ecology and Systematics,* 35 (2004): 229–56; Roy, S. W., and Irimia, M., "Rare Genomic Characters Do Not Support Coelomata: Intron Loss/ Gain," *Molecular Biology and Evolution,* 25 (2008): 620–23; Hyman, L. H. *The Invertebrates. Vol. 1: Protozoa through Ctenophora.* New York: McGraw-Hill, 1940; Holton, T. A., and Pisani, D., "Deep Genomic-Scale Analyses of the Metazoa Reject Coelomata: Evidence from Single- and Multigene Families Analyzed Under a Supertree and Supermatrix Paradigm," *Genome Biology and Evolution,* 2 (2010): 310–24.

Figure 6.2: Drawing by Ray Braun and Paul Nelson based on information from Figure 1, Edgecombe, G. D., Giribet, G., Dunn, C. W., Hejnol, A., Kristensen, R. M., Neves, R. C., Rouse, G. W., Worsaae, K., and Sørensen, M. V., "Higher-level metazoan relationships: recent progress and remaining questions," *Organisms, Diversification, & Evolution,* 11 (June 2011): 151–72.

Figure 6.3: Drawing by Ray Braun and Paul Nelson based on information from Extavour, C. G., and Akam, M., "Mechanisms of germ cell specification across the metazoans: epigenesis and preformation," *Development,* 130 (2003): 5869–84.

Figure 6.4: Drawing by Ray Braun and Paul Nelson based on information from Figure 1, Extavour, C. G. M. "Evolution of the bilaterian germ line: lineage origin and modulation of specific mechanisms," *Integrative and Comparative Biology,* 47 (2007): 770–85.

Figure 6.5: Drawing by Ray Braun and Paul Nelson based on information from Willmer, P. G. *Invertebrate Relationships: Patterns in Animal Evolution,* 2–14. Cambridge: Cambridge University Press, 1990.

Figure 7.1a: Photograph of Stephen Jay Gould courtesy of Steve Liss / TIME& LIFE Images / Getty Images. Used with permission.

Figure 7.1b: Photograph of Niles Eldredge © Julian Dufort 2011. Used with permission.

Figure 7.2: Drawn by Ray Braun based on an original drawing by Brian Gage. Courtesy of Brian Gage.

Figure 7.3: Drawn by Ray Braun based on an original drawing by Brian Gage. Courtesy of Brian Gage.

Figure 8.1: Photograph of James Watson and Francis Crick courtesy of A. Barrington Brown/ Science Source. Used With Permission.

Figure 8.2: Drawn by Ray Braun based on an original drawing by Fred Hereen. Courtesy of Fred Hereen.

Figure 8.3: Drawn by Ray Braun based on Figure 10, designed by Brian Gage, in Meyer, S. C., Ross, M., Nelson, P., and Chien, P., "The Cambrian Explosion: Biology's Big Bang." In *Darwinism, Design and Public Education,* edited by John Angus Campbell and Stephen C. Meyer, 336. East Lansing, MI: Michigan State University Press, 2003.Courtesy of Brian Gage.

Figure 9.1: Photograph of Murray Eden courtesy of MIT Museum. Used with permission.

Figure 9.2: Drawn by Ray Braun.

Figure 9.3: Drawn by Ray Braun.

Figure 10.1: Photograph of Douglas Axe courtesy of Brittnay Landoe. Used with permission.

Figure 10.2: Drawing by Ray Braun based on information from Figures 3.39, 3.40, and 3.44 in Berg, J. M., Tymoczko, J. L., Stryer, L. *Biochemistry,* 60–61, 5th ed. New York, NY: W. H. Freeman and Co, 2002.

Figure 10.3: Drawing by Ray Braun based on information from Figure 21, designed by Brian Gage, in Meyer, S. C., Ross, M., Nelson, P., and Chien, P. "The Cambrian Explosion: Biology's Big Bang." In *Darwinism, Design and Public Education,* edited by John Angus Campbell and Stephen C. Meyer, 374. East Lansing, MI: Michigan State University Press, 2003. Courtesy of Brian Gage.

Figure 10.4: Drawn by Ray Braun.

Figure 11.1: Drawing by Ray Braun and Casey Luskin based on information from Kaessmann, H., "Origins, evolution, and phenotypic impact of new genes," *Genome Research,* 20 (2010): 1313–26; Long, M., Betrán, E., Thornton, K., and Wang W., "The Origin of New Genes: Glimpses from the Young and Old," *Nature Reviews Genetics,* 4 (November 2003): 865–75.

Figure 11.2: Drawn by Ray Braun based on information from Luskin, C., "The NCSE, Judge Jones, and Citation Bluffs about the Origin of New Functional Genetic Information," Discovery.org (March 2, 2010), http://www.discovery.org/a/14251.

Figure 12.1: Drawn by Ray Braun based on information from Figure 27, in Frazzetta, T. H. *Complex Adaptations in Evolving Populations,* 148. Sunderland, MA: Sinauer Associates, 1975.

Figure 12.2: Photograph of Michael Behe courtesy of Laszlo Bencze. Used with permission.

Figure 12.3: Drawing by Ray Braun based on information from "Powerball—Prizes and Odds," PowerBall, http://www.powerball.com/powerball/pb_prizes.asp.

Figure 12.4: Drawn by Ray Braun based on information from Figure 6, Behe, M. J., and Snoke, D. W., "Simulating evolution by gene duplication of protein features that require multiple amino acid residues," *Protein Science,* 13 (2004): 2651–64. Original image courtesy of John Wiley and Sons and *Protein Science*. Original image Copyright © 2004 The Protein Society. Used with permission.

Figure 12.5: Photograph of Ann Gauger courtesy of Laszlo Bencze. Used with permission.

Figure 12.6: Drawn by Ray Braun based on information from Figure 5A, Gauger, A. K., and Axe, D. D. "The Evolutionary Accessibility of New Enzyme Functions: A Case Study from the Biotin Pathway." *BIO-Complexity,* 2011 (1): 1–17. Original image courtesy of Ann Gauger and Douglas Axe.

Figure 13.1a: Photograph of Christiane Nüsslein-Volhard courtesy of User:Rama / Wikimedia Commons / CC BY-SA 2.0 FR. Used with permission; usage not intended to imply endorsement by the author/licensor of the work.

Figure 13.1b: Photograph of Eric Wieschaus courtesy of Matthias Kubisch / Wikimedia Commons / CC0 1.0. Public domain; usage not intended to imply endorsement by the author/licensor of the work.

Figure 13.2: Drawing by Ray Braun and Paul Nelson based on information from "Mutant Fruit Flies," Exploratorium, http://www.exploratorium.edu/exhibits/mutant_flies/mutant_flies.html.

Figure 13.3: Photograph of Paul Nelson courtesy of Paul Nelson. Used with permission.

Figure 13.4a: Copyright 2008 National Academy of Sciences U.S.A. Figure 1D, Oliveri, P., Tu, Q., and Davidson, E. H., "Global Regulatory Logic for Specification of an Embryonic Cell Lineage," *Proceedings of the National Academy of Sciences USA,* 105 (2008): 5955–62. Use of PNAS material does not imply any endorsement by PNAS or the National Academy of Sciences or the authors. Used with permission.

Figure 13.4b: Copyright 2008 National Academy of Sciences U.S.A. Figure 1E, Oliveri, P., Tu, Q., and Davidson, E. H., "Global Regulatory Logic for Specification of an Embryonic Cell Lineage," *Proceedings of the National Academy of Sciences USA,* 105 (2008): 5955–62. Use of PNAS material does not imply any endorsement by PNAS or the National Academy of Sciences or the authors. Used with permission.

Figure 13.4c: Copyright 2008 National Academy of Sciences U.S.A. Figure 7, Oliveri, P., Tu, Q., and Davidson, E. H., "Global Regulatory Logic for Specification of an Embryonic Cell Lineage," *Proceedings of the National Academy of Sciences USA,* 105 (2008): 5955–62. Use of PNAS material does not imply any endorsement by PNAS or the National Academy of Sciences or the authors. Used with permission.

Figure 14.1: Photograph of Jonathan Wells courtesy of Laszlo Bencze. Used with permission.

Figure 14.2: Diagram drawn by Ray Braun based on PowerPoint slides developed by Michael Keas.

Figure 14.3a: Drawn by Joseph Condeelis. Copyright © Discovery Institute 2013. Used with permission.

Figure 14.3b: Courtesy of The Company of Biologists. Source: Figure 1B from Smyth, J. T., DeHaven, W. I., Bird, G. S., and Putney, J. W., "Role of the microtubule cytoskeleton in the function of the store-operated Ca^{2+} channel activator STIM1," *Journal of Cell Science,* 120 (November 1, 2007): 3762–71. Used with permission.

Figure 15.1: Photograph of Stuart Kauffman courtesy of User: Teemu Rajala/ Wikimedia Commons / CC BY 3.0. Used with permission; usage not intended to imply endorsement by the author/licensor of the work.

Figure 15.2: Drawing of DPMs by Ray Braun based on information from Table 1, Stuart A. Newman and Ramray Bhat, "Dynamical patterning modules: physico-genetic determinants of morphological development and evolution," *Physical Biology,* 5 (1): 1–14 (April 9, 2008); Figure

11.1, Stuart A. Newman, "Dynamical Patterning Modules," in *Evolution: The Extended Synthesis,* 294, Massimo Pigliucci and Gerd B. Muller eds. (The MIT Press, 2010); Stuart A. Newman, "Physico-Genetic Determinants in the Evolution of Development," *Science,* 338 (October 12, 2012): 217–19.

Figure 16.1: Drawn by Ray Braun and Paul Nelson.

Figure 16.2: Copyright 2007 National Academy of Sciences, U.S.A. Figure 1, Prud'homme, B., Gompel, N., and Carroll, S. B., "Emerging principles of regulatory evolution," *Proceedings of the National Academy of Sciences (PNAS), USA,* 104 (May 15, 2007): 8605–12. Use of PNAS material does not imply any endorsement by PNAS or the National Academy of Sciences or the authors. Used with permission.

Figure 16.3: Reprinted from *Current Biology,* 11, Starling, E. B., and Cohen, S. M., "Limb development: Getting down to the ground state," R1025–R1027, Figure 1, Copyright 2001, with permission from Elsevier.

Figure 17.1: Photograph of Charles Thaxton courtesy of Charles Thaxton. Used with permission.

Figure 17.2: Drawing by Ray Braun based on information from Grotzinger, J., Jordan, T. H., Press, F., Siever, R. *Understanding Earth,* 32–33, 5th ed., New York: W. H. Freeman, 2007; Cox., A., and Hart, R. B. *Plate Tectonics: How It Works,* 19, Cambridge, MA: Blackwell Science, 1986; Lowrie, W. *Fundamentals of Geophysics,* 18–19, Cambridge, UK: Cambridge University Press, 1997.

Figure 17.3: Drawing by Ray Braun based on information from Sober, E., *Reconstructing the Past,* 4–5. Cambridge, MA: MIT Press, 1988.

Figure 18.1: Photograph of Douglas Erwin courtesy of Robyn Wishna / UPHOTO/Cornell University. Used with permission.

Figure 18.2: Drawn by Ray Braun based on information from Figure 2:2, Meyer, S. C., Minnich, S., Moneymaker, J., Nelson, P. A., and Seelke, R. *Explore Evolution: The Arguments for and Against Neo-Darwinism,* 44. Melbourne and London: Hill House, 2007.

Figure 18.3: Photograph of transistors courtesy of iStockphoto.com/ S230. © iStockphoto.com/ S230. Used with permission.

Figure 18.4: Diagram drawn by Ray Braun based on information from Figure 2, Nelson, P., and Wells, J., "Homology in Biology: Problem for Naturalistic Science and Prospect for Intelligent Design." In *Darwinism, Design and Public Education,* 317, edited by John Angus Campbell and Stephen C. Meyer, 303–322. East Lansing, MI: Michigan State University Press, 2003.

Figure 18.5: Diagram drawn by Ray Braun based on PowerPoint slides developed by Michael Keas.

Figure 18.6: Diagram drawn by Ray Braun based on PowerPoint slides developed by Michael Keas.

Figure 18.7: Diagram drawn by Ray Braun based on PowerPoint slides developed by Michael Keas.

Figure 19.1: Photograph of Richard Sternberg courtesy of Laszlo Bencze. Used with Permission.

Figure 19.2: Photograph of Moais courtesy of iStockphoto/Think-stock. Used with permission.

Figure 20.1: Photograph of trilobite fossil at Burgess Shale courtesy of Michael Melford / NATIONAL GEOGRAPHIC IMAGE COLLECTION / Getty Images. Used with permission.

Figure 20.2a: Photograph of Stephen C. Meyer and family copyright © 2013 Stephen C. Meyer.

Figure 20.2b: Photograph of mountainside near Burgess Shale copyright © 2013 Stephen C. Meyer.

Figure 21.1: Drawn by Ray Braun.

Figure 21.2: Drawn by Ray Braun.

Figure 21.3: Drawn by Ray Braun.

Figure 21.4: Diagram reprinted from Figure 1, *Arthropod Structure & Development*, 39, Gregory D. Edgecombe, "Arthropod phylogeny: An overview from the perspectives of morphology, molecular data and the fossil record," 74–87, Copyright 2010, with permission from Elsevier.

Figure 21.5: Diagram reprinted from Figure 3, *Studies in History and Philosophy of Science Part C: Studies in History and Philosophy of Biological and Biomedical Sciences*, 39, Keynyn Brysse, "From weird wonders to stem lineages: the second reclassification of the Burgess Shale fauna," 298–313, Copyright 2008, with permission from Elsevier.

INDEX

Page references followed by *fig* indicate an illustration.